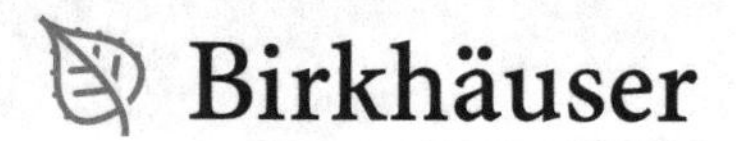
Birkhäuser

Tutorials, Schools, and Workshops in the Mathematical Sciences

This series will serve as a resource for the publication of results and developments presented at summer or winter schools, workshops, tutorials, and seminars. Written in an informal and accessible style, they present important and emerging topics in scientific research for PhD students and researchers. Filling a gap between traditional lecture notes, proceedings, and standard textbooks, the titles included in TSWMS present material from the forefront of research.

More information about this series at http://www.springer.com/series/15641

Ilker Inam · Engin Büyükaşık
Editors

Notes from the International Autumn School on Computational Number Theory

Editors
Ilker Inam
Department of Mathematics
Faculty of Arts and Science
Bilecik Şeyh Edebali University
Bilecik, Turkey

Engin Büyükaşık
Department of Mathematics
Izmir Institute of Technology
İzmir, Turkey

ISSN 2522-0969 ISSN 2522-0977 (electronic)
Tutorials, Schools, and Workshops in the Mathematical Sciences
ISBN 978-3-030-12557-8 ISBN 978-3-030-12558-5 (eBook)
https://doi.org/10.1007/978-3-030-12558-5

Library of Congress Control Number: 2019931507

Mathematics Subject Classification (2010): 11D61, 11J86, 11B39, 11B83, 11D59, 94B05, 94B15, 11G50, 13L05, 11F11, 11F33, 11R42, 11S40, 14G05, 11-04, 11Y99, 11M26, 11G25, 11G40, 14H52, 11G05, 11G07, 11L05, 11F30, 42B05, 11F25, 11F20, 11F37, 11-01, 11F67, 11Y16

This book is published under the imprint Birkhäuser, www.birkhauser-science.com by the registered company Springer Nature Switzerland AG
The registered company address is: Gewerbestrasse 11, 6330 Cham, Switzerland

For Cemre Tuna, I.I.

Preface

Computational number theory is a flourishing subject, touching a large variety of mathematical and application areas.

This volume contains lecture notes, complemented by some research and survey articles, related to the *International Autumn School on Computational Number Theory 2017*, a research school that was held at the Izmir Institute of Technology (IZTECH) in October and November 2017. Organized by Engin Büyükaşık, Ilker Inam, and Gabor Wiese, the school was supported by Bilecik Şeyh Edebali University, Izmir Institute of Technology, the Turkish Mathematical Society's MAD-Program, and the University of Luxembourg.

Very prominent objects in computational number theory are modular forms and L-functions: they are of fundamental theoretical importance (for example, for Wiles' proof of Fermat's Last Theorem) and at the same time amenable to explicit calculations.

Henri Cohen, a world leader in computational number theory, contributes two comprehensive sets of lecture notes to this volume: one on *Modular Forms* and one on *Computational Number Theory in Relation with L-functions*. This is complemented by Gabor Wiese's notes on *Computational Arithmetic of Modular Forms*, describing among other things how to compute modular forms via the modular symbols algorithm. Finally, Florian Luca's lecture notes on *Diophantine Equations* explain a core subject of number theory.

Bilecik, Turkey — Ilker Inam
İzmir, Turkey — Engin Büyükaşık

Acknowledgements

We wish to express our sincere gratitude to all speakers at the school for their excellent lectures and special thanks go to our anonymous referees for their careful reading. Moreover, we thank to the local organizing committee that consists of our students for all their efforts which made the school valuable.

Contents

Part I
Lecture Notes

An Introduction to Modular Forms

Henri Cohen

Abstract In this course, we introduce the main notions relative to the classical theory of modular forms. A complete treatise in a similar style can be found in the author's book joint with Strömberg (Cohen and Strömberg, Modular Forms: A Classical Approach, Graduate Studies in Math. 179, American Math. Soc. (2017) [1]).

1 Functional Equations

Let f be a complex function defined over some subset D of $\mathbb{C}$. A *functional equation* is some type of equation relating the value of f at any point $z \in D$ to some other point, for instance, $f(z+1) = f(z)$. If γ is some function from D to itself, one can ask more generally that $f(\gamma(z)) = f(z)$ for all $z \in D$ (or even $f(\gamma(z)) = v(\gamma, z)f(z)$ for some known function v). It is clear that $f(\gamma^m(z)) = f(z)$ for all $m \geq 0$, and even for all $m \in \mathbb{Z}$ if γ is invertible, and more generally the set of bijective functions u such that $f(u(z)) = f(z)$ forms a *group*.

Thus, the basic setting of functional equations (at least of the type that we consider) is that we have a group of transformations G of D, that we ask that $f(u(z)) = f(z)$ (or more generally $f(u(z)) = j(u, z)f(z)$ for some known j) for all $u \in G$ and $z \in D$, and we ask for some type of regularity condition on f such as continuity, meromorphy, or holomorphy.

Note that there is a trivial but essential way to construct from scratch functions f satisfying a functional equation of the above type: simply choose any function g and set $f(z) = \sum_{v \in G} g(v(z))$. Since G is a group, it is clear that *formally* $f(u(z)) = f(z)$ for $u \in G$. Of course there are convergence questions to be dealt with, but this is a fundamental construction, which we call *averaging* over the group.

We consider a few fundamental examples.

H. Cohen (✉)
Institut de Mathématiques de Bordeaux, Université de Bordeaux,
351 Cours de la Libération, 33405 Talence Cedex, France
e-mail: Henri.Cohen@math.u-bordeaux.fr

I. Inam and E. Büyükasik (eds.), *Notes from the International Autumn School on Computational Number Theory*, Tutorials, Schools, and Workshops in the Mathematical Sciences, https://doi.org/10.1007/978-3-030-12558-5_1

1.1 Fourier Series

We choose $D = \mathbb{R}$ and $G = \mathbb{Z}$ acting on $\mathbb{R}$ by translations. Thus, we ask that $f(x+1) = f(x)$ for all $x \in \mathbb{R}$. It is well known that this leads to the theory of *Fourier series*: if f satisfies suitable regularity conditions (we need not specify them here since in the context of modular forms they will be satisfied) then f has an expansion of the type

$$f(x) = \sum_{n\in\mathbb{Z}} a(n)e^{2\pi inx}\;,$$

absolutely convergent for all $x \in \mathbb{R}$, where the *Fourier coefficients* $a(n)$ are given by the formula

$$a(n) = \int_0^1 e^{-2\pi inx} f(x)\,dx\;,$$

which follows immediately from the orthonormality of the functions $e^{2\pi imx}$ (you may of course replace the integral from 0 to 1 by an integral from z to $z+1$ for any $z \in \mathbb{R}$).

An important consequence of this, easily proved, is the *Poisson summation formula*: define the *Fourier transform* of f by

$$\widehat{f}(x) = \int_{-\infty}^{\infty} e^{-2\pi ixt} f(t)\,dt\;.$$

We ignore all convergence questions, although of course they must be taken into account in any computation.

Consider the function $g(x) = \sum_{n\in\mathbb{Z}} f(x+n)$, which is exactly the averaging procedure mentioned above. Thus $g(x+1) = g(x)$, so g has a Fourier series, and an easy computation shows the following (again omitting any convergence or regularity assumptions):

Proposition 1.1 (Poisson summation) *We have*

$$\sum_{n\in\mathbb{Z}} f(x+n) = \sum_{m\in\mathbb{Z}} \widehat{f}(m)e^{2\pi imx}\;.$$

In particular,

$$\sum_{n\in\mathbb{Z}} f(n) = \sum_{m\in\mathbb{Z}} \widehat{f}(m)\;.$$

A typical application of this formula is to the ordinary Jacobi *theta function*: it is well known (prove it otherwise) that the function $e^{-\pi x^2}$ is invariant under Fourier transform. This implies the following:

Proposition 1.2 *If* $f(x) = e^{-a\pi x^2}$ *for some* $a > 0$, *then* $\widehat{f}(x) = a^{-1/2}e^{-\pi x^2/a}$.

Proof Simple change of variable in the integral. □

Corollary 1.3 *Define*

$$T(a) = \sum_{n \in \mathbb{Z}} e^{-a\pi n^2} .$$

We have the functional equation

$$T(1/a) = a^{1/2} T(a) .$$

Proof Immediate from the proposition and Poisson summation. □

This is historically the first example of modularity, which we will see in more detail below.

Exercise 1.4 Set $S = \sum_{n \geq 1} e^{-(n/10)^2}$.

1. Compute numerically S to 100 decimal digits, and show that it is apparently equal to $5\sqrt{\pi} - 1/2$.
2. Show that, in fact, S is not exactly equal to $5\sqrt{\pi} - 1/2$, and using the above corollary give a precise estimate for the difference.

Exercise 1.5 1. Show that the function $f(x) = 1/\cosh(\pi x)$ is also invariant under Fourier transform.

2. In a manner similar to the corollary, define

$$T_2(a) = \sum_{n \in \mathbb{Z}} 1/\cosh(\pi n a) .$$

Show that we have the functional equation

$$T_2(1/a) = a T_2(a) .$$

3. Show that, in fact, $T_2(a) = T(a)^2$ (this may be more difficult).
4. Do the same exercise as the previous one by noticing that $S = \sum_{n \geq 1} 1/\cosh(n/10)$ is very close to $5\pi - 1/2$.

Above we have mainly considered Fourier series of functions defined on $\mathbb{R}$. We now consider more generally functions f defined on $\mathbb{C}$ or a subset of $\mathbb{C}$. We again assume that $f(z+1) = f(z)$, i.e., that f is periodic of period 1. Thus (modulo regularity), f has a Fourier series, but the Fourier coefficients $a(n)$ now depend on $y = \Im(z)$:

$$f(x+iy) = \sum_{n \in \mathbb{Z}} a(n; y) e^{2\pi i n x} \quad \text{with} \quad a(n; y) = \int_0^1 f(x+iy) e^{-2\pi i n x}\, dx .$$

If we impose no extra condition on f, the *functions* $a(n; y)$ are quite arbitrary. But in almost all of our applications f will be *holomorphic*; this means that $\partial(f)(z)/\partial\overline{z} = 0$, or equivalently that $(\partial/\partial(x) + i\partial/\partial(y))(f) = 0$. Replacing in the Fourier expansion (recall that we do not worry about convergence issues) gives

$$\sum_{n\in\mathbb{Z}}(2\pi i n a(n;y) + i a'(n;y))e^{2\pi i n x} = 0\;,$$

and hence by uniqueness of the expansion, we obtain the differential equation $a'(n;y) = -2\pi n a(n;y)$, so that $a(n;y) = c(n)e^{-2\pi n y}$ for some constant $c(n)$. This allows us to write cleanly the Fourier expansion of a holomorphic function in the form

$$f(z) = \sum_{n\in\mathbb{Z}} c(n)e^{2\pi i n z}\;.$$

Note that if the function is only *meromorphic*, the region of convergence will be limited by the closest pole. Consider, for instance, the function $f(z) = 1/(e^{2\pi i z} - 1) = e^{-\pi i z}/(2i\sin(\pi z))$. If we set $y = \Im(z)$, we have $|e^{2\pi i z}| = e^{-2\pi y}$, so if $y > 0$, we have the Fourier expansion $f(z) = -\sum_{n\geq 0} e^{2\pi i n z}$, while if $y < 0$ we have the different Fourier expansion $f(z) = \sum_{n\leq -1} e^{2\pi i n z}$.

2 Elliptic Functions

The preceding section was devoted to periodic functions. We now assume that our functions are defined on some subset of $\mathbb{C}$ and assume that they are *doubly periodic*: this can be stated either by saying that there exist two $\mathbb{R}$-linearly independent complex numbers ω_1 and ω_2 such that $f(z + \omega_i) = f(z)$ for all z and $i = 1, 2$, or equivalently by saying that there exists a *rank* 2 *lattice* Λ in $\mathbb{C}$ (here $\mathbb{Z}\omega_1 + \mathbb{Z}\omega_2$) such that for any $\lambda \in \Lambda$, we have $f(z + \lambda) = f(z)$.

Note in passing that if $\omega_1/\omega_2 \in \mathbb{Q}$ this is equivalent to (single) periodicity, and if $\omega_1/\omega_2 \in \mathbb{R}\setminus\mathbb{Q}$ the set of periods would be dense so the only "doubly periodic" (at least continuous) functions would essentially reduce to functions of one variable. For a similar reason, there do not exist nonconstant continuous functions which are triply periodic.

In the case of simply periodic functions considered above there already existed some natural functions such as $e^{2\pi i n x}$. In the doubly periodic case no such function exists (at least on an elementary level), so we have to construct them, and for this we use the standard averaging procedure seen and used above. Here, the group is the lattice Λ, so we consider functions of the type $f(z) = \sum_{\omega\in\Lambda}\phi(z + \omega)$. For this to converge $\phi(z)$ must tend to 0 sufficiently fast as $|z|$ tends to infinity, and since this is a double sum (Λ is a two-dimensional lattice), it is easy to see by comparison with an integral (assuming $|\phi(z)|$ is regularly decreasing) that $|\phi(z)|$ should decrease at least like $1/|z|^\alpha$ for $\alpha > 2$. Thus, a first reasonable definition is to set

$$f(z) = \sum_{\omega \in \Lambda} \frac{1}{(z+\omega)^3} = \sum_{(m,n) \in \mathbb{Z}^2} \frac{1}{(z + m\omega_1 + n\omega_2)^3} .$$

This will indeed be a doubly periodic function, and by normal convergence it is immediate to see that it is a meromorphic function on $\mathbb{C}$ having only poles for $z \in \Lambda$, so this is our first example of an *elliptic function*, which is by definition a doubly periodic function which is meromorphic on $\mathbb{C}$. Note for future reference that since $-\Lambda = \Lambda$ this specific function f is odd: $f(-z) = -f(z)$.

However, this is not quite the basic elliptic function that we need. We can integrate term by term, as long as we choose constants of integration such that the integrated series continues to converge. To avoid stupid multiplicative constants, we integrate $-2f(z)$: all antiderivatives of $-2/(z+\omega)^3$ are of the form $1/(z+\omega)^2 + C(\omega)$ for some constant $C(\omega)$, and hence to preserve convergence we will choose $C(0) = 0$ and $C(\omega) = -1/\omega^2$ for $\omega \neq 0$: indeed, $|1/(z+\omega)^2 - 1/\omega^2|$ is asymptotic to $2|z|/|\omega^3|$ as $|\omega| \to \infty$, so we are again in the domain of normal convergence. We will thus define

$$\wp(z) = \frac{1}{z^2} + \sum_{\omega \in \Lambda \setminus \{0\}} \left(\frac{1}{(z+\omega)^2} - \frac{1}{\omega^2} \right) ,$$

the *Weierstrass $\wp$-function.*

By construction $\wp'(z) = -2f(z)$, where f is the function constructed above, so $\wp'(z+\omega) = \wp'(z)$ for any $\omega \in \Lambda$, and hence $\wp(z+\omega) = \wp(z) + D(\omega)$ for some constant $D(\omega)$ depending on ω but not on z. Note a slightly subtle point here: we use the fact that $\mathbb{C} \setminus \Lambda$ is *connected*. Do you see why?

Now as before it is clear that $\wp(z)$ is an even function: thus, setting $z = -\omega/2$ we have $\wp(\omega/2) = \wp(-\omega/2) + D(\omega) = \wp(\omega/2) + D(\omega)$, so $D(\omega) = 0$, and hence $\wp(z+\omega) = \wp(z)$ and $\wp$ is indeed an elliptic function. There is a mistake in this reasoning: do you see it?

Since $\wp$ has poles on Λ, we cannot reason as we do when $\omega/2 \in \Lambda$. Fortunately, this does not matter: since $\omega_i/2 \notin \Lambda$ for $i = 1, 2$, we have shown at least that $D(\omega_i) = 0$, and hence that $\wp(z+\omega_i) = \wp(z)$ for $i = 1, 2$, so $\wp$ is doubly periodic (so indeed $D(\omega) = 0$ for *all* $\omega \in \Lambda$).

The theory of elliptic functions is incredibly rich, and whole treatises have been written about them. Since this course is mainly about modular forms, we will simply summarize the main properties, and emphasize those that are relevant to us. All are proved using manipulation of power series and complex analysis, and all the proofs are quite straightforward. For instance,

Proposition 2.1 *Let f be a nonzero elliptic function with period lattice Λ as above, and denote by $P = P_a$ a "fundamental parallelogram" $P_a = \{z = a + x\omega_1 + y\omega_2,\ 0 \le x < 1,\ 0 \le y < 1\}$, where a is chosen so that the boundary of P_a does not contain any zeros or poles of f (see Fig. 1).*

1. *The number of zeros of f in P is equal to the number of poles (counted with multiplicity), and this number is called the* order *of f.*

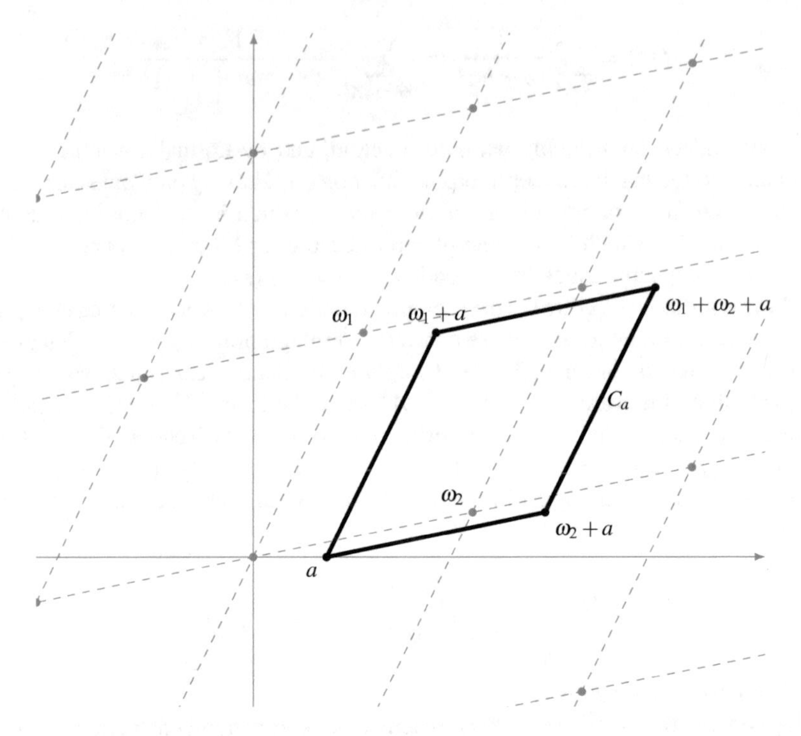

Fig. 1 Fundamental Parallelogram P_a

2. *The sum of the residues of f at the poles in P is equal to* 0.
3. *The sum of the zeros and poles of f in P belongs to Λ.*
4. *If f is nonconstant its order is at least* 2.

Proof For (1), (2), and (3), simply integrate $f(z)$, $f'(z)/f(z)$, and $zf'(z)/f(z)$ along the boundary of P and use the residue theorem. For (4), we first note that by (2) f cannot have order 1 since it would have a simple pole with residue 0. But it also cannot have order 0: this would mean that f has no pole, so it is an entire function, and since it is doubly periodic its values are those taken in the topological closure of P which is compact, so f is *bounded*. By a famous theorem of Liouville (of which this is the no less most famous application), it implies that f is constant, contradicting the assumption of (4). □

Note that clearly $\wp$ has order 2, and the last result shows that we cannot find an elliptic function of order 1. Note however the following:

Exercise 2.2 1. By integrating term by term the series defining $-\wp(z)$ show that if we define the *Weierstrass zeta function*

$$\zeta(z) = \frac{1}{z} + \sum_{\omega \in \Lambda \setminus \{0\}} \left(\frac{1}{z+\omega} - \frac{1}{\omega} + \frac{z}{\omega^2} \right),$$

this series converges normally on any compact subset of $\mathbb{C} \setminus \Lambda$ and satisfies $\zeta'(z) = -\wp(z)$.

2. Deduce that there exist constants η_1 and η_2 such that $\zeta(z+\omega_1) = \zeta(z) + \eta_1$ and $\zeta(z+\omega_2) = \zeta(z) + \eta_2$, so that if $\omega = m\omega_1 + n\omega_2$, we have $\zeta(z+\omega) = \zeta(z) + m\eta_1 + n\eta_2$. Thus, ζ (which would be of order 1) is not doubly periodic but only quasi-doubly periodic: this is called a *quasi-elliptic function*.
3. By integrating around the usual fundamental parallelogram, show the important relation due to Legendre:

$$\omega_1\eta_2 - \omega_2\eta_1 = \pm 2\pi i\ ,$$

the sign depending on the ordering of ω_1 and ω_2.

The main properties of $\wp$ that we want to mention are as follows: First, for z sufficiently small and $\omega \neq 0$, we can expand

$$\frac{1}{(z+\omega)^2} = \sum_{k\geq 0}(-1)^k(k+1)z^k\frac{1}{\omega^{k+2}}\ ,$$

so

$$\wp(z) = \frac{1}{z^2} + \sum_{k\geq 1}(-1)^k(k+1)z^k G_{k+2}(\Lambda)\ ,$$

where we have set

$$G_k(\Lambda) = \sum_{\omega\in\Lambda\setminus\{0\}}\frac{1}{\omega^k}\ ,$$

which are called *Eisenstein series* of weight k. Since Λ is symmetrical, it is clear that $G_k = 0$ if k is odd, so the expansion of $\wp(z)$ around $z = 0$ is given by

$$\wp(z) = \frac{1}{z^2} + \sum_{k\geq 1}(2k+1)z^{2k}G_{2k+2}(\Lambda)\ .$$

Second, one can show that *all* elliptic functions are simply rational functions in $\wp(z)$ and $\wp'(z)$, so we need not look any further in our construction.

Third, and this is probably one of the most important properties of $\wp(z)$, it satisfies a *differential equation* of order 1: the proof is as follows. Using the above Taylor expansion of $\wp(z)$, it is immediate to check that

$$F(z) = \wp'(z)^2 - (4\wp(z)^3 - g_2(\Lambda)\wp(z) - g_3(\Lambda))$$

has an expansion around $z = 0$ beginning with $F(z) = c_1 z + \cdots$, where we have set $g_2(\Lambda) = 60G_4(\Lambda)$ and $g_3(\Lambda) = 140G_6(\Lambda)$. In addition, F is evidently an elliptic function, and since it has no pole at $z = 0$ it has no poles on Λ, and hence no poles at all, so it has order 0. Thus, by Proposition 2.1 (4) f is constant, and since by construction it vanishes at 0 it is identically 0. Thus, $\wp$ satisfies the differential equation

$$\wp'(z)^2 = 4\wp(z)^3 - g_2(\Lambda)\wp(z) - g_3(\Lambda)\,.$$

A fourth and somewhat surprising property of the function $\wp(z)$ is connected to the theory of *elliptic curves*: the above differential equation shows that $(\wp(z), \wp'(z))$ parametrizes the cubic curve $y^2 = 4x^3 - g_2x - g_3$, which is the general equation of an elliptic curve (you do not need to know the theory of elliptic curves for what follows). Thus, if z_1 and z_2 are in $\mathbb{C} \setminus \Lambda$, the two points $P_i = (\wp(z_i), \wp'(z_i))$ for $i = 1, 2$ are on the curve, and hence if we draw the line through these two points (the tangent to the curve if they are equal), it is immediate to see from Proposition 2.1 (3) that the third point of intersection corresponds to the parameter $-(z_1 + z_2)$, and can of course be computed as a rational function of the coordinates of P_1 and P_2. It follows that $\wp(z)$ (and $\wp'(z)$) possess an *addition formula* expressing $\wp(z_1 + z_2)$ in terms of the $\wp(z_i)$ and $\wp'(z_i)$.

Exercise 2.3 Find this addition formula. You will have to distinguish the cases $z_1 = z_2$, $z_1 = -z_2$, and $z_1 \neq \pm z_2$.

An interesting corollary of the differential equation for $\wp(z)$, which we will prove in a different way below, is a *recursion* for the Eisenstein series $G_{2k}(\Lambda)$:

Proposition 2.4 *We have the recursion for $k \geq 4$:*

$$(k-3)(2k-1)(2k+1)G_{2k} = 3 \sum_{2 \leq j \leq k-2} (2j-1)(2(k-j)-1)G_{2j}G_{2(k-j)}\,.$$

Proof Taking the derivative of the differential equation and dividing by $2\wp'$, we obtain $\wp''(z) = 6\wp(z)^2 - g_2(\Lambda)/2$. If we set by convention $G_0(\Lambda) = -1$ and $G_2(\Lambda) = 0$, and for notational simplicity omit Λ which is fixed, we have $\wp(z) = \sum_{k \geq -1}(2k+1)z^{2k}G_{2k+2}$, so on the one hand

$$\wp''(z) = \sum_{k \geq -1} (2k+1)(2k)(2k-1)z^{2k-2}G_{2k+2}\,,$$

and on the other hand $\wp(z)^2 = \sum_{K \geq -2} a(K)z^{2K}$ with

$$a(K) = \sum_{k_1+k_2=K} (2k_1+1)(2k_2+1)G_{2k_1+2}G_{2k_2+2}\,.$$

Replacing in the differential equation, it is immediate to check that the coefficients agree up to z^2, and for $K \geq 2$, we have the identification

$$6 \sum_{\substack{k_1+k_2=K \\ k_i \geq -1}} (2k_1+1)(2k_2+1)G_{2k_1+2}G_{2k_2+2} = (2K+3)(2K+2)(2K+1)G_{2K+4}$$

which is easily seen to be equivalent to the recursion of the proposition using $G_0 = -1$ and $G_2 = 0$. □

For instance,

$$G_8 = \frac{3}{7}G_4^2 \quad G_{10} = \frac{5}{11}G_4G_6 \quad G_{12} = \frac{18G_4^3 + 25G_6^2}{143} ,$$

and more generally this implies that G_{2k} is a *polynomial* in G_4 and G_6 with rational coefficients which are *independent* of the lattice Λ.

As other corollary, we note that if we choose $\omega_2 = 1$ and $\omega_1 = iT$ with T tending to $+\infty$, then the definition $G_{2k}(\Lambda) = \sum_{(m,n)\in\mathbb{Z}^2\setminus\{(0,0)\}}(m\omega_1 + n\omega_2)^{-2k}$ implies that $G_{2k}(\Lambda)$ will tend to $\sum_{n\in\mathbb{Z}\setminus\{0\}} n^{-2k} = 2\zeta(2k)$, where ζ is the Riemann zeta function. If follows that for all $k \geq 2$, $\zeta(2k)$ is a polynomial in $\zeta(4)$ and $\zeta(6)$ with rational coefficients. Of course this is a weak but nontrivial result, since we know that $\zeta(2k)$ is a rational multiple of π^{2k}.

To finish this section on elliptic functions and make the transition to modular forms, we write explicitly $\Lambda = \Lambda(\omega_1, \omega_2)$ and by abuse of notation $G_{2k}(\omega_1, \omega_2) := G_{2k}(\Lambda(\omega_1, \omega_2))$, and we consider the dependence of G_{2k} on ω_1 and ω_2. We note two evident facts: first, $G_{2k}(\omega_1, \omega_2)$ is *homogeneous* of degree $-2k$: for any nonzero complex number λ, we have $G_{2k}(\lambda\omega_1, \lambda\omega_2) = \lambda^{-2k}G_{2k}(\omega_1, \omega_2)$. In particular, $G_{2k}(\omega_1, \omega_2) = \omega_2^{-2k}G_{2k}(\omega_1/\omega_2, 1)$. Second, a general $\mathbb{Z}$-basis of Λ is given by $(\omega_1', \omega_2') = (a\omega_1 + b\omega_2, c\omega_1 + d\omega_2)$ with a, b, c, and d integers such that $ad - bc = \pm 1$. If we choose an *oriented* basis such that $\Im(\omega_1/\omega_2) > 0$, we, in fact, have $ad - bc = 1$.

Thus, $G_{2k}(a\omega_1 + b\omega_2, c\omega_1 + d\omega_2) = G_{2k}(\omega_1, \omega_2)$, and using homogeneity this can be written as

$$(c\omega_1 + d\omega_2)^{-2k}G_{2k}\left(\frac{a\omega_1 + b\omega_2}{c\omega_1 + d\omega_2}, 1\right) = \omega_2^{-2k}G_{2k}\left(\frac{\omega_1}{\omega_2}, 1\right) .$$

Thus, if we set $\tau = \omega_1/\omega_2$ and by an additional abuse of notation abbreviate $G_{2k}(\tau, 1)$ to $G_{2k}(\tau)$, we have by definition

$$G_{2k}(\tau) = \sum_{(m,n)\in\mathbb{Z}^2\setminus\{(0,0)\}} (m\tau + n)^{-2k} ,$$

and we have shown the following *modularity* property:

Proposition 2.5 *For any* $\left(\begin{smallmatrix} a & b \\ c & d \end{smallmatrix}\right) \in \mathrm{SL}_2(\mathbb{Z})$, *the group of* 2×2 *integer matrices of determinant* 1, *and any* $\tau \in \mathbb{C}$ *with* $\Im(\tau) > 0$, *we have*

$$G_{2k}\left(\frac{a\tau + b}{c\tau + d}\right) = (c\tau + d)^{2k}G_{2k}(\tau) .$$

This will be our basic definition of (weak) modularity.

3 Modular Forms and Functions

3.1 Definitions

Let us introduce some notation:

• We denote by Γ the *modular group* $\mathrm{SL}_2(\mathbb{Z})$. Note that properly speaking the modular group should be the group of transformations $\tau \mapsto (a\tau + b)/(c\tau + d)$, which is isomorphic to the quotient of $\mathrm{SL}_2(\mathbb{Z})$ by the equivalence relation saying that M and $-M$ are equivalent, but for this course, we will stick to this definition. If $\gamma = \left(\begin{smallmatrix} a & b \\ c & d \end{smallmatrix}\right)$, we will, of course, write $\gamma(\tau)$ for $(a\tau + b)/(c\tau + d)$.

• The *Poincaré upper half-plane* $\mathcal{H}$ is the set of complex numbers τ such that $\Im(\tau) > 0$. Since for $\gamma = \left(\begin{smallmatrix} a & b \\ c & d \end{smallmatrix}\right) \in \Gamma$, we have $\Im(\gamma(\tau)) = \Im(\tau)/|c\tau + d|^2$, we see that Γ is a group of transformations of $\mathcal{H}$ (more generally so is $\mathrm{SL}_2(\mathbb{R})$, there is nothing special about $\mathbb{Z}$).

• The completed upper half-plane $\overline{\mathcal{H}}$ is by definition $\overline{\mathcal{H}} = \mathcal{H} \cup \mathbb{P}_1(\mathbb{Q}) = \mathcal{H} \cup \mathbb{Q} \cup \{i\infty\}$. Note that this is *not* the closure in the topological sense since we do not include any real irrational numbers.

Definition 3.1 Let $k \in \mathbb{Z}$ and let F be a function from $\mathcal{H}$ to $\mathbb{C}$.

1. We will say that F is *weakly modular* of weight k for Γ if for all $\gamma = \left(\begin{smallmatrix} a & b \\ c & d \end{smallmatrix}\right) \in \Gamma$ and all $\tau \in \mathcal{H}$, we have

$$F(\gamma(\tau)) = (c\tau + d)^k F(\tau) \; .$$

2. We will say that F is a modular *form* if, in addition, F is holomorphic on $\mathcal{H}$ and if $|F(\tau)|$ remains bounded as $\Im(\tau) \to \infty$.
3. We will say that F is a modular *cusp form* if it is a modular form such that $F(\tau)$ tends to 0 as $\Im(\tau) \to \infty$.

We make a number of immediate but important remarks.

Remarks 3.2 1. The Eisenstein series $G_{2k}(\tau)$ are basic examples of modular forms of weight $2k$, which are not cusp forms since $G_{2k}(\tau)$ tends to $2\zeta(2k) \neq 0$ when $\Im(\tau) \to \infty$.
2. With the present definition, it is clear that there are no nonzero modular forms of *odd weight* k, since if k is odd we have $(-c\tau - d)^k = -(c\tau + d)^k$ and $\gamma(\tau) = (-\gamma)(\tau)$. However, when considering modular forms defined on *subgroups* of Γ there may be modular forms of odd weight, so we keep the above definition.
3. Applying modularity to $\gamma = T = \left(\begin{smallmatrix} 1 & 1 \\ 0 & 1 \end{smallmatrix}\right)$, we see that $F(\tau + 1) = F(\tau)$, and hence F has a Fourier series expansion, and if F is holomorphic, by the remark made above in the section on Fourier series, we have an expansion $F(\tau) = \sum_{n\in\mathbb{Z}} a(n)e^{2\pi in\tau}$ with $a(n) = e^{2\pi ny} \int_0^1 F(x + iy)e^{-2\pi inx}\,dx$ for any $y > 0$. Thus, if $|F(x + iy)|$ remains bounded as $y \to \infty$ it follows that as $y \to \infty$, we have

$a(n) \leq Be^{2\pi ny}$ for a suitable constant B, so we deduce that $a(n) = 0$ whenever $n < 0$ since $e^{2\pi ny} \to 0$. Thus, if F is a modular *form*, we have $F(\tau) = \sum_{n\geq 0} a(n)e^{2\pi in\tau}$, and hence $\lim_{\Im(\tau)\to\infty} F(\tau) = a(0)$, so F is a cusp form if and only if $a(0) = 0$.

Definition 3.3 We will denote by $M_k(\Gamma)$, the vector space of modular forms of weight k on Γ (M for Modular of course), and by $S_k(\Gamma)$ the subspace of cusp forms (S for the German Spitzenform, meaning exactly cusp form).

Notation: for any matrix $\gamma = \left(\begin{smallmatrix} a & b \\ c & d \end{smallmatrix}\right)$ with $ad - bc > 0$, we will define the weight k *slash operator* $F|_k\gamma$ by

$$F|_k\gamma(\tau) = (ad - bc)^{k/2}(c\tau + d)^{-k}F(\gamma(\tau))\ .$$

The reason for the factor $(ad - bc)^{k/2}$ is that $\lambda\gamma$ has the same action on $\mathcal{H}$ as γ, so this makes the formula homogeneous. For instance, F is weakly modular of weight k if and only if $F|_k\gamma = F$ for all $\gamma \in \Gamma$.

We will also use the universal modular form convention of writing q for $e^{2\pi i\tau}$, so that a Fourier expansion is of the type $F(\tau) = \sum_{n\geq 0} a(n)q^n$. We use the additional convention that if α is any complex number, q^α will mean $e^{2\pi i\tau\alpha}$.

Exercise 3.4 Let $F(\tau) = \sum_{n\geq 0} a(n)q^n \in M_k(\Gamma)$, and let $\gamma = \left(\begin{smallmatrix} A & B \\ C & D \end{smallmatrix}\right)$ be a matrix in $M_2^+(\mathbb{Z})$, i.e., A, B, C, and D are integers and $\Delta = \det(\gamma) = AD - BC > 0$. Set $g = \gcd(A, C)$, let u and v be such that $uA + vC = g$, set $b = uB + vD$, and finally let $\zeta_\Delta = e^{2\pi i/\Delta}$. Prove the matrix identity

$$\begin{pmatrix} A & B \\ C & D \end{pmatrix} = \begin{pmatrix} A/g & -v \\ C/g & u \end{pmatrix} \begin{pmatrix} g & b \\ 0 & \Delta/g \end{pmatrix},$$

and deduce that we have the more general Fourier expansion

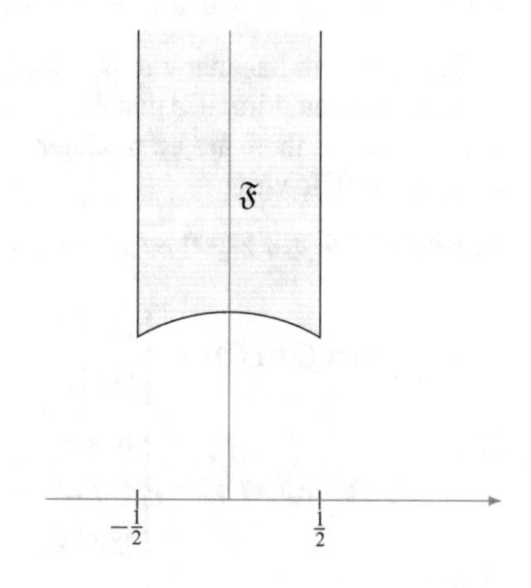

Fig. 2 The fundamental domain, $\mathfrak{F}$, of Γ

$$F|_k\gamma(\tau)=\frac{g^{k/2}}{\Delta^k}\sum_{n\ge0}\zeta_\Delta^{nbg}a(n)q^{g^2/\Delta}\;,$$

which is, of course, equal to F if $\Delta=1$, since then $g=1$.

3.2 Basic Results

The first fundamental result in the theory of modular forms is that these spaces are *finite dimensional.* The proof uses exactly the same method that we have used to prove the basic results on elliptic functions. We first note that there is a "fundamental domain" (which replaces the fundamental parallelogram, see Fig. 2) for the action of Γ on $\mathcal{H}$, given by

$$\mathfrak{F}=\{\tau\in\mathcal{H},\ -1/2\le\Re(\tau)<1/2,\ |\tau|\ge1\}\;.$$

The proof that this is a fundamental domain, in other words that any $\tau\in\mathcal{H}$ has a unique image by Γ belonging to $\mathfrak{F}$ is not very difficult and will be omitted. We then integrate $F'(z)/F(z)$ along the boundary of $\mathfrak{F}$, and using modularity, we obtain the following result:

Theorem 3.5 *Let $F\in M_k(\Gamma)$ be a nonzero modular form. For any $\tau_0\in\mathcal{H}$, denote by $v_{\tau_0}(F)$ the* valuation *of F at τ_0, i.e., the unique integer v such that $F(\tau)/(\tau-\tau_0)^v$ is holomorphic and nonzero at τ_0, and if $F(\tau)=G(e^{2\pi i\tau})$, define $v_{i\infty}(F)=v_0(G)$ (i.e., the number of first vanishing Fourier coefficients of F). We have the formula*

$$v_{i\infty}(F)+\sum_{\tau\in\mathfrak{F}}\frac{v_\tau(F)}{e_\tau}=\frac{k}{12}\;,$$

where $e_i=2$, $e_\rho=3$, and $e_\tau=1$ otherwise ($\rho=e^{2\pi i/3}$).

This theorem has many important consequences but, as already noted, the most important is that it implies that $M_k(\Gamma)$ is finite dimensional. First, it trivially implies that $k\ge0$, i.e., there are no modular *forms* of negative weight. In addition, it easily implies the following:

Corollary 3.6 *Let $k\ge0$ be an even integer. We have*

$$\dim(M_k(\Gamma))=\begin{cases}\lfloor k/12\rfloor & \text{if } k\equiv2 \pmod{12}\;,\\ \lfloor k/12\rfloor+1 & \text{if } k\not\equiv2 \pmod{12}\;,\end{cases}$$

$$\dim(S_k(\Gamma))=\begin{cases}0 & \text{if } k<12\;,\\ \lfloor k/12\rfloor-1 & \text{if } k\ge12,\ k\equiv2 \pmod{12}\;,\\ \lfloor k/12\rfloor & \text{if } k\ge12,\ k\not\equiv2 \pmod{12}\;.\end{cases}$$

Since the product of two modular forms is clearly a modular form (of weight the sum of the two weights), it is clear that $M_*(\Gamma) = \bigoplus_k M_k(\Gamma)$ (and similarly $S_*(\Gamma)$) is an algebra, whose structure is easily described as follows:

Corollary 3.7 *We have $M_*(\Gamma) = \mathbb{C}[G_4, G_6]$, and $S_*(\Gamma) = \Delta M_*(\Gamma)$, where Δ is the unique generator of the one-dimensional vector space $S_{12}(\Gamma)$ whose Fourier expansion begins with $\Delta = q + O(q^2)$.*

Thus, for instance, $M_0(\Gamma) = \mathbb{C}$, $M_2(\Gamma) = \{0\}$, $M_4(\Gamma) = \mathbb{C}G_4$, $M_6(\Gamma) = \mathbb{C}G_6$, $M_8(\Gamma) = \mathbb{C}G_8 = \mathbb{C}G_4^2$, $M_{10}(\Gamma) = \mathbb{C}G_{10} = \mathbb{C}G_4G_6$,

$$M_{12}(\Gamma) = \mathbb{C}G_{12} \oplus \mathbb{C}\Delta = \mathbb{C}G_4^3 \oplus \mathbb{C}G_6^2 .$$

In particular, we recover the fact proved differently that G_8 is a multiple of G_4^2 (the exact multiple being obtained by computing the Fourier expansions), G_{10} is a multiple of G_4G_6, G_{12} is a linear combination of G_4^3 and G_6^2. Also, we see that Δ is a linear combination of G_4^3 and G_6^2 (we will see this more precisely below).

A basic result on the structure of the modular group Γ is the following:

Proposition 3.8 *Set $T = \begin{pmatrix} 1 & 1 \\ 0 & 1 \end{pmatrix}$, which acts on $\mathcal{H}$ by the unit translation $\tau \mapsto \tau + 1$, and $S = \begin{pmatrix} 0 & -1 \\ 1 & 0 \end{pmatrix}$ which acts on $\mathcal{H}$ by the symmetry inversion $\tau \mapsto -1/\tau$. Then, Γ is generated by S and T, with relations generated by $S^2 = -I$ and $(ST)^3 = -I$ (I the identity matrix).*

There are several (easy) proofs of this fundamental result, which we do not give. Simply note that this proposition is essentially equivalent to the fact that the set $\mathfrak{F}$ described above is indeed a fundamental domain.

A consequence of this proposition is that to check whether some function F has the modularity property, it is sufficient to check that $F(\tau + 1) = F(\tau)$ and $F(-1/\tau) = \tau^k F(\tau)$.

Exercise 3.9 (*Bol's identity*). Let F be any continuous function defined on the upper half-plane $\mathcal{H}$, and define $I_0(F, a) = F$ and for any integer $m \geq 1$ and $a \in \overline{\mathcal{H}}$ set:

$$I_m(F, a)(\tau) = \int_a^\tau \frac{(\tau - z)^{m-1}}{(m-1)!} F(z)\, dz .$$

1. Show that $I_m(F, a)'(\tau) = I_{m-1}(F, a)(\tau)$, so that $I_m(F, a)$ is an mth antiderivative of F.
2. Let $\gamma \in \Gamma$, and assume that $k \geq 1$ is an integer. Show that

$$I_{k-1}(F, a)|_{2-k}\gamma = I_{k-1}(F|_k\gamma, \gamma^{-1}(a)) .$$

3. Deduce that if we set $F_a^* = I_{k-1}(F, a)$, then

$$D^{(k-1)}(F_a^*|_{2-k}\gamma) = F|_k\gamma ,$$

where $D = (1/2\pi i)d/d\tau = qd/dq$ is the basic differential operator that we will use (see Sect. 3.10).

4. Assume now that F is weakly modular of weight $k \geq 1$ and holomorphic on $\mathcal{H}$ (in particular, if $F \in M_k(\Gamma)$, but $|F|$ could be unbounded as $\Im(\tau) \to \infty$). Show that

$$(F_a^*|_{2-k}|\gamma)(\tau) = F_a^*(\tau) + P_{k-2}(\tau)\ ,$$

where P_{k-2} is the polynomial of degree less than or equal to $k-2$ given by

$$P_{k-2}(X) = \int_{\gamma^{-1}(a)}^{a} \frac{(X-z)^{k-2}}{(k-2)!} F(z)\, dz\ .$$

What this exercise shows is that the $(k-1)$st derivative of some function which behaves modularly in weight $2-k$ behaves modularly in weight k, and conversely that the $(k-1)$st antiderivative of some function which behaves modularly in weight k behaves modularly in weight k up to addition of a polynomial of degree at most $k-2$. This duality between weights k and $2-k$ is, in fact, a consequence of the *Riemann–Roch theorem*.

Note also that this exercise is the beginning of the fundamental theories of *periods* and of *modular symbols*.

Also, it is not difficult to generalize Bol's identity. For instance, applied to the Eisenstein series G_4 and using Proposition 3.13 below, we obtain:

Proposition 3.10 *1. Set*

$$F_4^*(\tau) = -\frac{\pi^3}{180}\left(\frac{\tau}{i}\right)^3 + \sum_{n\geq 1} \sigma_{-3}(n)q^n\ .$$

We have the functional equation

$$\tau^2 F_4^*(-1/\tau) = F_4^*(\tau) + \frac{\zeta(3)}{2}(1-\tau^2) - \frac{\pi^3}{36}\frac{\tau}{i}\ .$$

2. Equivalently, if we set

$$F_4^{**}(\tau) = -\frac{\pi^3}{180}\left(\frac{\tau}{i}\right)^3 - \frac{\pi^3}{72}\left(\frac{\tau}{i}\right) + \frac{\zeta(3)}{2} + \sum_{n\geq 1} \sigma_{-3}(n)q^n$$

we have the functional equation

$$F_4^{**}(-1/\tau) = \tau^{-2} F_4^{**}(\tau)\ .$$

Note that the appearance of $\zeta(3)$ comes from the fact that, up to a multiplicative constant, the L-function associated to G_4 is equal to $\zeta(s)\zeta(s-3)$, whose value at $s=3$ is equal to $-\zeta(3)/2$.

3.3 The Scalar Product

We begin by the following exercise:

Exercise 3.11 1. Denote by $d\mu = dxdy/y^2$ a measure on $\mathcal{H}$, where as usual x and y are the real and imaginary parts of $\tau \in \mathcal{H}$. Show that this measure is invariant under $\mathrm{SL}_2(\mathbb{R})$.

2. Let f and g be in $M_k(\Gamma)$. Show that the function $F(\tau) = f(\tau)\overline{g(\tau)}y^k$ is invariant under the modular group Γ.

It follows, in particular, from this exercise that if $F(\tau)$ is any integrable function which is invariant by the modular group Γ, the integral $\int_{\Gamma\backslash\mathcal{H}} F(\tau)d\mu$ makes sense if it converges. Since $\mathfrak{F}$ is a fundamental domain for the action of Γ on $\mathcal{H}$, this can also be written as $\int_{\mathfrak{F}} F(\tau)d\mu$. Thus, it follows from the second part that we can define

$$< f, g >= \int_{\Gamma\backslash\mathcal{H}} f(\tau)\overline{g(\tau)}y^k \frac{dxdy}{y^2} ,$$

whenever this converges.

It is immediate to show that a necessary and sufficient condition for convergence is that at least one of f and g be a cusp form, i.e., lies in $S_k(\Gamma)$. In particular, it is clear that this defines a *scalar product* on $S_k(\Gamma)$ called the Petersson scalar product. In addition, any cusp form in $S_k(\Gamma)$ is *orthogonal* to G_k with respect to this scalar product. It is instructive to give a sketch of the simple proof of this fact as given below:

Proposition 3.12 *If $f \in S_k(\Gamma)$, we have $< G_k, f >= 0$.*

Proof Recall that $G_k(\tau) = \sum_{(m,n)\in\mathbb{Z}^2\backslash\{(0,0)\}}(m\tau + n)^{-k}$. We split the sum according to the GCD of m and n: we let $d = \gcd(m, n)$, so that $m = dm_1$ and $n = dn_1$ with $\gcd(m_1, n_1) = 1$. It follows that

$$G_k(\tau) = 2\sum_{d\geq 1} d^{-k}E_k(\tau) = 2\zeta(k)E_k(\tau) ,$$

where $E_k(\tau) = (1/2)\sum_{\gcd(m,n)=1}(m\tau + n)^{-k}$. We thus need to prove that $< E_k, f >= 0$.

On the other hand, denote by Γ_∞ the group generated by T, i.e., translations $\left(\begin{smallmatrix}1 & b\\ 0 & 1\end{smallmatrix}\right)$ for $b \in \mathbb{Z}$. This acts by left multiplication on Γ, and it is immediate to check that a system of representatives for this action is given by matrices $\left(\begin{smallmatrix}u & v\\ m & n\end{smallmatrix}\right)$, where $\gcd(m, n) = 1$ and u and v are chosen arbitrarily (but only once for each pair (m, n)) such that $un - vm = 1$. It follows that, we can write

$$E_k(\tau) = \sum_{\gamma\in\Gamma_\infty\backslash\Gamma} (m\tau + n)^{-k} ,$$

where it is understood that $\gamma = \begin{pmatrix} u & v \\ m & n \end{pmatrix}$ (the factor 1/2 has disappeared since γ and $-\gamma$ have the same action on $\mathscr{H}$).

Thus

$$< E_k, f > = \int_{\Gamma\backslash\mathscr{H}} \sum_{\gamma\in\Gamma_\infty\backslash\Gamma} (m\tau+n)^{-k}\overline{f(\tau)}y^k \frac{dxdy}{y^2}$$
$$= \sum_{\gamma\in\Gamma_\infty\backslash\Gamma} \int_{\Gamma\backslash\mathscr{H}} (m\tau+n)^{-k}\overline{f(\tau)}y^k \frac{dxdy}{y^2} .$$

Now note that by modularity $f(\tau) = (m\tau+n)^{-k} f(\gamma(\tau))$, and since $\Im(\gamma(\tau)) = \Im(\tau)/|m\tau+n|^2$, it follows that

$$(m\tau+n)^{-k}\overline{f(\tau)}y^k = \overline{f(\gamma(\tau))}\Im(\gamma(\tau))^k .$$

Thus, since $d\mu = dxdy/y^2$ is an invariant measure, we have

$$< E_k, f > = \sum_{\gamma\in\Gamma_\infty\backslash\Gamma} \int_{\Gamma\backslash\mathscr{H}} \overline{f(\gamma(\tau))}\Im(\gamma(\tau))^k d\mu = \int_{\Gamma_\infty\backslash\mathscr{H}} \overline{f(\tau)}y^k \frac{dxdy}{y^2} .$$

Since Γ_∞ is simply the group of integer translations, a fundamental domain for $\Gamma_\infty\backslash\mathscr{H}$ is simply the vertical strip $[0, 1] \times [0, \infty[$, so that

$$< E_k, f >= \int_0^\infty y^{k-2}dy \int_0^1 \overline{f(x+iy)}dx ,$$

which trivially vanishes since the inner integral is simply the conjugate of the constant term in the Fourier expansion of f, which is 0 since $f \in S_k(\Gamma)$.

The above procedure (replacing the complicated fundamental domain of $\Gamma\backslash\mathscr{H}$ by the trivial one of $\Gamma_\infty\backslash\mathscr{H}$) is very common in the theory of modular forms and is called *unfolding*.

3.4 Fourier Expansions

The Fourier expansions of the Eisenstein series $G_{2k}(\tau)$ are easy to compute. The result is the following:

Proposition 3.13 *For $k \geq 4$ even, we have the Fourier expansion*

$$G_k(\tau) = 2\zeta(k) + 2\frac{(2\pi i)^k}{(k-1)!}\sum_{n\geq 1}\sigma_{k-1}(n)q^n ,$$

where $\sigma_{k-1}(n) = \sum_{d|n,\ d>0} d^{k-1}$.

Since we know that when k is even $2\zeta(k) = -(2\pi i)^k B_k/k!$, where B_k is the k-th Bernoulli number defined by

$$\frac{t}{e^t - 1} = \sum_{k\geq 0} \frac{B_k}{k!} t^k ,$$

it follows that $G_k = 2\zeta(k)E_k$, with

$$E_k(\tau) = 1 - \frac{2k}{B_k} \sum_{n\geq 1} \sigma_{k-1}(n) q^n .$$

This is the normalization of Eisenstein series that we will use. For instance,

$$E_4(\tau) = 1 + 240 \sum_{n\geq 1} \sigma_3(n) q^n ,$$
$$E_6(\tau) = 1 - 504 \sum_{n\geq 1} \sigma_5(n) q^n ,$$
$$E_8(\tau) = 1 + 480 \sum_{n\geq 1} \sigma_7(n) q^n .$$

In particular, the relations given above which follow from the dimension formula become much simpler and are obtained simply by looking at the first terms in the Fourier expansion:

$$E_8 = E_4^2 , \quad E_{10} = E_4 E_6 , \quad E_{12} = \frac{441 E_4^3 + 250 E_6^2}{691} , \quad \Delta = \frac{E_4^3 - E_6^2}{1728} .$$

Note that the relation $E_4^2 = E_8$ (and the others) implies a highly nontrivial relation between the sum of divisors function: if we set by convention $\sigma_3(0) = 1/240$, so that $E_4(\tau) = \sum_{n\geq 0} \sigma_3(n) q^n$, we have

$$E_8(\tau) = E_4^2(\tau) = 240^2 \sum_{n\geq 0} q^n \sum_{0\leq m\leq n} \sigma_3(m)\sigma_3(n-m) ,$$

so that by identification $\sigma_7(n) = 120 \sum_{0\leq m\leq n} \sigma_3(m)\sigma_3(n-m)$, so

$$\sigma_7(n) = \sigma_3(n) + 120 \sum_{1\leq m\leq n-1} \sigma_3(m)\sigma_3(n-m) .$$

It is quite difficult (but not impossible) to prove this directly, i.e., without using at least indirectly the theory of modular forms.

Exercise 3.14 Find a similar relation for $\sigma_9(n)$ using $E_{10} = E_4 E_6$.

This type of reasoning is one of the reasons for which the theory of modular forms is so important (and lots of fun!): if you have a modular form F, you can usually express it in terms of a completely explicit basis of the space to which it belongs since spaces of modular forms are *finite dimensional* (in the present example, the space is one dimensional), and deduce highly nontrivial relations for the Fourier coefficients. We will see a further example of this below for the number $r_k(n)$ of representations of an integer n as a sum of k squares.

Exercise 3.15 1. Prove that for any $k \in \mathbb{C}$, we have the identity

$$\sum_{n\geq 1} \sigma_k(n)q^n = \sum_{n\geq 1} \frac{n^k q^n}{1-q^n}\,,$$

the right-hand side being called a *Lambert series*.

2. Set $F(k) = \sum_{n\geq 1} n^k/(e^{2\pi n} - 1)$. Using the Fourier expansions given above, compute explicitly $F(5)$ and $F(9)$.
3. Using Proposition 3.10, compute explicitly $F(-3)$.
4. Using Proposition 3.23 below, compute explicitly $F(1)$.

Note that in this exercise, we only compute $F(k)$ for $k \equiv 1 \pmod 4$. It is also possible but more difficult to compute $F(k)$ for $k \equiv 3 \pmod 4$. For instance, we have

$$F(3) = \frac{\Gamma(1/4)^8}{80(2\pi)^6} - \frac{1}{240}\,.$$

3.5 Obtaining Modular Forms by Averaging

We have mentioned at the beginning of this course that one of the ways to obtain functions satisfying functional equations is to use *averaging* over a suitable group or set: we have seen this for periodic functions in the form of the Poisson summation formula, and for doubly periodic functions in the construction of the Weierstrass $\wp$-function. We can do the same for modular forms, but we must be careful in two different ways. First, we do not want *invariance* by Γ, but we want an automorphy factor $(c\tau + d)^k$. This is easily dealt with by noting that $(d/d\tau)(\gamma(\tau)) = (c\tau + d)^{-2}$: indeed, if ϕ is some function on $\mathcal{H}$, we can define

$$F(\tau) = \sum_{\gamma\in\Gamma} \phi(\gamma(\tau))((d/d\tau)(\gamma(\tau)))^{k/2}\,.$$

Exercise 3.16 Ignoring all convergence questions, by using the chain rule $(f \circ g)' = (f' \circ g)g'$ show that for all $\delta = \left(\begin{smallmatrix} A & B \\ C & D \end{smallmatrix}\right) \in \Gamma$, we have

$$F(\delta(\tau)) = (C\tau + D)^k F(\tau)\,.$$

But the second important way in which we must be careful is that the above construction rarely converges. There are, however, examples where it does converge:

Exercise 3.17 Let $\phi(\tau) = \tau^{-m}$, so that

$$F(\tau) = \sum_{\gamma=\left(\begin{smallmatrix} a & b \\ c & d \end{smallmatrix}\right)\in\Gamma} \frac{1}{(a\tau+b)^m(c\tau+d)^{k-m}} .$$

Show that if $2 \leq m \leq k-2$ and $m \neq k/2$, this series converges normally on any compact subset of $\mathcal{H}$ (i.e., it is majorized by a convergent series with positive terms), so defines a modular form in $M_k(\Gamma)$.

Note that the series converges also for $m = k/2$, but this is more difficult.

One of the essential reasons for non-convergence of the function F is the trivial observation that for a given pair of coprime integers (c, d) there are infinitely many elements $\gamma \in \Gamma$ having (c, d) as their second row. Thus, in general, it seems more reasonable to define

$$F(\tau) = \sum_{\gcd(c,d)=1} \phi(\gamma_{c,d}(\tau))(c\tau+d)^{-k} ,$$

where $\gamma_{c,d}$ is *any fixed* matrix in Γ with second row equal to (c, d). However, we need this to make sense: if $\gamma_{c,d} = \left(\begin{smallmatrix} a & b \\ c & d \end{smallmatrix}\right) \in \Gamma$ is one such matrix, it is clear that the general matrix having second row equal to (c, d) is $T^n \left(\begin{smallmatrix} a & b \\ c & d \end{smallmatrix}\right) = \left(\begin{smallmatrix} a+nc & b+nd \\ c & d \end{smallmatrix}\right)$, and as usual $T = \left(\begin{smallmatrix} 1 & 1 \\ 0 & 1 \end{smallmatrix}\right)$ is translation by 1: $\tau \mapsto \tau + 1$. Thus, an essential necessary condition for our series to make any kind of sense is that the function ϕ be *periodic* of period 1.

The simplest such function is, of course, the constant function 1.

Exercise 3.18 (*See the proof of Proposition* 3.12.) Show that

$$F(\tau) = \sum_{\gcd(c,d)=1} (c\tau+d)^{-k} = 2E_k(\tau) ,$$

where E_k is the normalized Eisenstein series defined above.

But by the theory of Fourier series, we know that periodic functions of period 1 are (infinite) linear combinations of the functions $e^{2\pi i n\tau}$. This leads to the definition of *Poincaré series*:

$$P_k(n;\tau) = \frac{1}{2} \sum_{\gcd(c,d)=1} \frac{e^{2\pi i n \gamma_{c,d}(\tau)}}{(c\tau+d)^k} ,$$

where we note that we can choose any matrix $\gamma_{c,d}$ with bottom row (c, d) since the function $e^{2\pi i n\tau}$ is 1-periodic, so that $P_k(n;\tau) \in M_k(\Gamma)$.

Exercise 3.19 Assume that $k \geq 4$ is even.

1. Show that if $n < 0$, the series defining P_k diverges (wildly in fact).
2. Note that $P_k(0; \tau) = E_k(\tau)$, so that $\lim_{\tau \to i\infty} P_k(0; \tau) = 1$. Show that if $n > 0$, the series converges normally and that we have $\lim_{\tau \to i\infty} P_k(n; \tau) = 0$. Thus, in fact, $P_k(n; \tau) \in S_k(\Gamma)$ if $n > 0$.
3. By using the same *unfolding method* as in Proposition 3.12, show that if $f = \sum_{n \geq 0} a(n)q^n \in M_k(\Gamma)$ and $n > 0$, we have

$$< P_k(n), f >= \frac{(k-2)!}{(4\pi n)^{k-1}} a(n) .$$

It is easy to show that, in fact, the $P_k(n)$ *generate* $S_k(\Gamma)$. We can also compute their *Fourier expansions* as we have done for E_k, but they involve Bessel functions and Kloosterman sums.

3.6 The Ramanujan Delta Function

Recall that by definition Δ is the generator of the one-dimensional space $S_{12}(\Gamma)$ whose Fourier coefficient of q^1 is normalized to be equal to 1. By simple computation, we find the first terms in the Fourier expansion of Δ:

$$\Delta(\tau) = q - 24q^2 + 252q^3 - 1472q^4 + \cdots ,$$

with no apparent formula for the coefficients. The nth coefficient is denoted $\tau(n)$ (no confusion with $\tau \in \mathcal{H}$), and called Ramanujan's tau function, and Δ itself is called Ramanujan's Delta function.

Of course, using $\Delta = (E_4^3 - E_6^2)/1728$ and expanding the powers, one can give a complicated but explicit formula for $\tau(n)$ in terms of the functions σ_3 and σ_5, but this is far from being the best way to compute them. In fact, the following exercise already gives a much better method.

Exercise 3.20 Let D be the differential operator $(1/(2\pi i))d/d\tau = qd/dq$.

1. Show that the function $F = 4E_4D(E_6) - 6E_6D(E_4)$ is a modular form of weight 12, then by looking at its constant term show that it is a cusp form, and finally compute the constant c such that $F = c \cdot \Delta$.
2. Deduce the formula

$$\tau(n) = \frac{n}{12}(5\sigma_3(n) + 7\sigma_5(n)) + 70 \sum_{1 \leq m \leq n-1} (2n - 5m)\sigma_3(m)\sigma_5(n-m) .$$

3. Deduce, in particular, the congruences $\tau(n) \equiv n\sigma_5(n) \equiv n\sigma_1(n) \pmod 5$ and $\tau(n) \equiv n\sigma_3(n) \pmod 7$.

Although there are much faster methods, this is already a very reasonable way to compute $\tau(n)$.

The cusp form Δ is one of the most important functions in the theory of modular forms. Its first main property, which is not at all apparent from its definition, is that it has a *product expansion*:

Theorem 3.21 *We have*

$$\Delta(\tau) = q\prod_{n\ge1}(1-q^n)^{24}\,.$$

Proof We are not going to give a complete proof, but sketch a method which is one of the most natural to obtain the result.

We start backward, from the product $R(\tau)$ on the right-hand side. The logarithm transforms products into sums, but in the case of *functions* f, the *logarithmic derivative* f'/f (more precisely $D(f)/f$, where $D = qd/dq$) also does this, and it is also more convenient. We have

$$D(R)/R = 1 - 24\sum_{n\ge1}\frac{nq^n}{1-q^n} = 1 - 24\sum_{n\ge1}\sigma_1(n)q^n$$

as is easily seen by expanding $1/(1-q^n)$ as a geometric series. This is exactly the case $k = 2$ of the Eisenstein series E_k, which we have excluded from our discussion for convergence reasons, so we come back to our series G_{2k} (we will divide by the normalizing factor $2\zeta(2) = \pi^2/3$ at the end), and introduce a convergence factor due to Hecke, setting

$$G_{2,s}(\tau) = \sum_{(m,n)\in\mathbb{Z}^2\setminus\{(0,0)\}}(m\tau+n)^{-2}|m\tau+n|^{-2s}\,.$$

As above this converges for $\Re(s) > 0$, it satisfies

$$G_{2,s}(\gamma(\tau)) = (c\tau+d)^2|c\tau+d|^{2s}G_{2,s}(\tau)$$

and hence, in particular, is periodic of period 1. It is straightforward to compute its Fourier expansion, which we will not do here, and the Fourier expansion shows that $G_{2,s}$ has an *analytic continuation* to the whole complex plane. In particular, the limit as $s \to 0$ makes sense; if we denote it by $G_2^*(\tau)$, by continuity it will of course satisfy $G_2^*(\gamma(\tau)) = (c\tau+d)^2G_2^*(\tau)$, and the analytic continuation of the Fourier expansion that has been computed gives

$$G_2^*(\tau) = \frac{\pi^2}{3}\left(1 - \frac{3}{\pi\Im(\tau)} - 24\sum_{n\ge1}\sigma_1(n)q^n\right)\,.$$

Note the essential fact that there is now a *nonanalytic term* $3/(\pi\Im(\tau))$. We will, of course, set the following definition:

Definition 3.22 We define

$$E_2(\tau) = 1 - 24\sum_{n\ge1}\sigma_1(n)q^n \quad\text{and}\quad E_2^*(\tau) = E_2(\tau) - \frac{3}{\pi\Im(\tau)}\ .$$

Thus, $E_2(\tau) = D(R)/R$, $G_2^*(\tau) = (\pi^2/3)E_2^*(\tau)$, and we have the following:

Proposition 3.23 *For any* $\gamma = \left(\begin{smallmatrix} a & b \\ c & d \end{smallmatrix}\right) \in \Gamma$, *we have* $E_2^*(\gamma(\tau)) = (c\tau + d)^2 E_2^*(\tau)$. *Equivalently,*

$$E_2(\gamma(\tau)) = (c\tau + d)^2 E_2(\tau) + \frac{12}{2\pi i}c(c\tau + d)\ .$$

Proof The first result has been seen above, and the second follows from the formula $\Im(\gamma(\tau)) = \Im(\tau)/|c\tau + d|^2$. □

Exercise 3.24 Show that

$$E_2(\tau) = -24\left(-\frac{1}{24} + \sum_{m\ge1}\frac{m}{q^{-m} - 1}\right)\ .$$

Proof of the theorem. We can now prove the theorem on the product expansion of Δ: noting that $(d/d\tau)\gamma(\tau) = 1/(c\tau + d)^2$, the above formulas imply that if we set $S = R(\gamma(\tau))$, we have

$$\begin{aligned}\frac{D(S)}{S} &= \frac{D(R)}{R}(\gamma(\tau))(d/d\tau)(\gamma(\tau)) \\ &= (c\tau + d)^{-2}E_2(\gamma(\tau)) = E_2(\tau) + \frac{12}{2\pi i}\frac{c}{c\tau + d} \\ &= \frac{D(R)}{R}(\tau) + 12\frac{D(c\tau + d)}{c\tau + d}\ .\end{aligned}$$

By integrating and exponentiating, it follows that

$$R(\gamma(\tau)) = (c\tau + d)^{12}R(\tau)\ ,$$

and since clearly R is holomorphic on $\mathcal{H}$ and tends to 0 as $\Im(\tau) \to \infty$ (i.e., as $q \to 0$), it follows that R is a cusp form of weight 12 on Γ, and since $S_{12}(\Gamma)$ is one-dimensional and the coefficient of q^1 in R is 1, we have $R = \Delta$, proving the theorem. □

Exercise 3.25 We have shown in passing that $D(\Delta) = E_2\Delta$. Expanding the Fourier expansion of both sides, show that we have the recursion

$$(n - 1)\tau(n) = -24\sum_{1\le m\le n-1}\sigma_1(m)\tau(n - m)\ .$$

Exercise 3.26 1. Let $F \in M_k(\Gamma)$, and for some *squarefree* integer N set

$$G(\tau) = \sum_{d|N} \mu(d) d^{k/2} F(d\tau) ,$$

where μ is the Möbius function. Show that $G|_k W_N = \mu(N) G$, where $W_N = \begin{pmatrix} 0 & -1 \\ N & 0 \end{pmatrix}$ is the so-called *Fricke involution*.

2. Show that if $N > 1$, the same result is true for $F = E_2$, although E_2 is only quasi-modular.
3. Deduce that if $\mu(N) = (-1)^{k/2-1}$, we have $G(i/\sqrt{N}) = 0$.
4. Applying this to E_2 and using Exercise 3.24, deduce that if $\mu(N) = 1$ and $N > 1$, we have

$$\sum_{\gcd(m,N)=1} \frac{m}{e^{2\pi m/\sqrt{N}} - 1} = \frac{\phi(N)}{24} ,$$

where $\phi(N)$ is Euler's totient function.

5. Using directly the functional equation of E_2^*, show that for $N = 1$ there is an additional term $-1/(8\pi)$, i.e., that

$$\sum_{m \geq 1} \frac{m}{e^{2\pi m} - 1} = \frac{1}{24} - \frac{1}{8\pi} .$$

3.7 *Product Expansions and the Dedekind Eta Function*

We continue our study of product expansions. We first mention an important identity due to Jacobi, the triple product identity, as well as some consequences:

Theorem 3.27 (Triple product identity) *If $|q| < 1$ and $u \neq 0$, we have*

$$\prod_{n \geq 1} (1 - q^n)(1 - q^n u) \prod_{n \geq 0} (1 - q^n/u) = \sum_{k \geq 0} (-1)^k (u^k - u^{-(k+1)}) q^{k(k+1)/2} .$$

Proof (sketch): Denote by $L(q, u)$ the left-hand side. We have clearly $L(q, u/q) = -uL(q, u)$, and since one can write $L(q, u) = \sum_{k \in \mathbb{Z}} a_k(q) u^k$ this implies the recursion $a_k(q) = -q^k a_{k-1}(q)$, so $a_k(q) = (-1)^k q^{k(k+1)/2} a_0(q)$, and separating $k \geq 0$ and $k < 0$ this shows that

$$L(q, u) = a_0(q) \sum_{k \geq 0} (-1)^k (u^k - u^{-(k+1)}) q^{k(k+1)/2} .$$

The slightly longer part is to show that $a_0(q) = 1$: this is done by setting $u = i/q^{1/2}$ and $u = 1/q^{1/2}$, which after a little computation implies that $a(q^4) = a(q)$, and from there it is immediate to deduce that $a(q)$ is a constant, and equal to 1. □

To give the next corollaries, we need to define the *Dedekind eta function* $\eta(\tau)$, by

$$\eta(\tau)=q^{1/24}\prod_{n\ge1}(1-q^n)\;,$$

(recall that $q^\alpha=e^{2\pi i\alpha\tau}$). Thus, by definition $\eta(\tau)^{24}=\Delta(\tau)$. Since $\Delta(-1/\tau)=\tau^{12}\Delta(\tau)$, it follows that $\eta(-1/\tau)=c\cdot(\tau/i)^{1/2}\eta(\tau)$ for some 24th root of unity c (where we always use the principal determination of the square root), and since we see from the infinite product that $\eta(i)\neq0$, replacing τ by i shows that, in fact, $c=1$. Thus, η satisfies the two basic modular equations

$$\eta(\tau+1)=e^{2\pi i/24}\eta(\tau)\quad\text{and}\quad\eta(-1/\tau)=(\tau/i)^{1/2}\eta(\tau)\;.$$

Of course, we have more generally

$$\eta(\gamma(\tau))=v_\eta(\gamma)(c\tau+d)^{1/2}\eta(\tau)$$

for any $\gamma\in\Gamma$, with a complicated 24th root of unity $v_\eta(\gamma)$, so η is in some (reasonable) sense a modular form of weight $1/2$, similar to the function θ that we introduced at the very beginning.

The triple product identity immediately implies the following two identities:

Corollary 3.28 *We have*

$$\eta(\tau)=q^{1/24}\left(1+\sum_{k\ge1}(-1)^k(q^{k(3k-1)/2}+q^{k(3k+1)/2})\right)\quad\textit{and}$$
$$\eta(\tau)^3=q^{1/8}\sum_{k\ge0}(-1)^k(2k+1)q^{k(k+1)/2}\;.$$

Proof In the triple product identity, replace (u,q) by $(1/q,q^3)$, we obtain

$$\prod_{n\ge1}(1-q^{3n})(1-q^{3n-1})\prod_{n\ge0}(1-q^{3n+1})=\sum_{k\ge0}(-1)^k(q^{-k}-q^{k+1})q^{3k(k+1)/2}\;.$$

The left-hand side is clearly equal to $\eta(\tau)$, and the right-hand side to

$$\begin{aligned}&1-q+\sum_{k\ge1}(-1)^k(q^{k(3k+1)/2}-q^{(k+1)(3k+2)/2})\\&=1+\sum_{k\ge1}(-1)^kq^{k(3k+1)/2}-q+\sum_{k\ge2}(-1)^kq^{k(3k-1)/2}\;,\end{aligned}$$

giving the formula for $\eta(\tau)$. For the second formula, divide the triple product identity by $1-1/u$ and make $u\to1$. □

Thus, the first few terms are

$$\prod_{n\geq 1}(1-q^n) = 1 - q - q^2 + q^5 + q^7 - q^{12} - q^{15} + \cdots$$

$$\prod_{n\geq 1}(1-q^n)^3 = 1 - 3q + 5q^3 - 7q^6 + 9q^{10} - 11q^{15} + \cdots .$$

The first identity was proved by L. Euler.

Exercise 3.29 1. Show that $24\Delta D(\eta) = \eta D(\Delta)$, and using the explicit Fourier expansion of η, deduce the recursion

$$\sum_{k\in\mathbb{Z}}(-1)^k(75k^2 + 25k + 2 - 2n)\tau\left(n - \frac{k(3k+1)}{2}\right) = 0 .$$

2. Similarly, from $8\Delta D(\eta^3) = \eta^3 D(\Delta)$ deduce the recursion

$$\sum_{k\in\mathbb{Z}}(-1)^k(2k+1)(9k^2 + 9k + 2 - 2n)\tau\left(n - \frac{k(k+1)}{2}\right) = 0 .$$

Exercise 3.30 Define the *q-Pochhammer symbol* $(q)_n$ by $(q)_n = (1-q)(1-q^2)\cdots(1-q^n)$.

1. Set $f(a,q) = \prod_{n\geq 1}(1-aq^n)$, and define coefficients $c_n(q)$ by setting $f(a,q) = \sum_{n\geq 0} c_n(q)a^n$. Show that $f(a,q) = (1-aq)f(aq,q)$, deduce that $c_n(q)(1-q^n) = -q^n c_{n-1}(q)$ and finally the identity

$$\prod_{n\geq 1}(1-aq^n) = \sum_{n\geq 0}(-1)^n a^n q^{n(n+1)/2}/(q)_n .$$

2. Write in terms of the Dedekind eta function, the identities obtained by specializing to $a = 1$, $a = -1$, $a = -1/q$, $a = q^{1/2}$, and $a = -q^{1/2}$.
3. Similarly, prove the identity

$$1/\prod_{n\geq 1}(1-aq^n) = \sum_{n\geq 0} a^n q^n/(q)_n ,$$

and once again write in terms of the Dedekind eta function, the identities obtained by specializing to the same five values of a.
4. By multiplying two of the above identities and using the triple product identity, prove the identity

$$\frac{1}{\prod_{n\geq 1}(1-q^n)} = \sum_{n\geq 0}\frac{q^{n^2}}{(q)_n^2} .$$

Note that this last series is the generating function of the *partition function* $p(n)$, so if one wants to make a table of $p(n)$ up to $n = 10{,}000$, say, using the left-hand side would require 10,000 terms, while using the right-hand side only requires 100.

3.8 Computational Aspects of the Ramanujan τ Function

Since its introduction, the Ramanujan tau function $\tau(n)$ has fascinated number theorists. For instance, there is a conjecture due to D. H. Lehmer that $\tau(n) \neq 0$, and an even stronger conjecture (which would imply the former) that for every prime p we have $p \nmid \tau(p)$ (on probabilistic grounds, the latter conjecture is probably false).

To test these conjectures as well as others, it is an interesting computational challenge to *compute* $\tau(n)$ for large n (because of Ramanujan's first two conjectures, i.e., Mordell's theorem that we will prove in Sect. 4 below, it is sufficient to compute $\tau(p)$ for p *prime*).

We can have two distinct goals. The first is to compute a *table* of $\tau(n)$ for $n \leq B$, where B is some (large) bound. The second is to compute *individual values* of $\tau(n)$, equivalently of $\tau(p)$ for p prime.

Consider first the construction of a *table*. The use of the first recursion given in the above exercise needs $O(n^{1/2})$ operations per value of $\tau(n)$, and hence $O(B^{3/2})$ operations in all to have a table for $n \leq B$.

However, it is well known that the *Fast Fourier Transform* (FFT) allows one to compute products of power series in essentially linear time. Thus, using Corollary 3.28, we can directly write the power series expansion of η^3, and use the FFT to compute its eighth power $\eta^{24} = \Delta$. This will require $O(B\log(B))$ operations, so it is much faster than the preceding method; it is essentially optimal since one needs $O(B)$ time simply to write the result.

Using large computer resources, especially in memory, it is reasonable to construct a table up to $B = 10^{12}$, but not much more. Thus, the problem of computing *individual* values of $\tau(p)$ is important. We have already seen one such method in Exercise 3.20 above, which gives a method for computing $\tau(n)$ in time $O(n^{1+\varepsilon})$ for any $\varepsilon > 0$.

A deep and important theorem of B. Edixhoven, J.-M. Couveignes, et al., says that it is possible to compute $\tau(p)$ in time *polynomial* in $\log(p)$, and, in particular, in time $O(p^{\varepsilon})$ for any $\varepsilon > 0$. Unfortunately, this algorithm is not at all practical, and at least for now, completely useless for us. The only practical and important application is for the computation of $\tau(p)$ modulo some small prime numbers ℓ (typically $\ell < 50$, so far from being sufficient to apply the Chinese Remainder Theorem).

However, there exists an algorithm which takes time $O(n^{1/2+\varepsilon})$ for any $\varepsilon > 0$, so much better than the one of Exercise 3.20, and which is very practical. It is based on the use of the Eichler–Selberg *trace formula*, together with the computation of *Hurwitz class numbers* $H(N)$ (essentially the class numbers of imaginary quadratic orders counted with suitable multiplicity): if we set $H_3(N) = H(4N) + 2H(N)$ (note that $H(4N)$ can be computed in terms of $H(N)$), then for p prime

$$\tau(p) = 28p^6 - 28p^5 - 90p^4 - 35p^3 - 1$$
$$- 128 \sum_{1\le t<p^{1/2}} t^6(4t^4 - 9pt^2 + 7p^2)H_3(p - t^2) .$$

See [1] Exercise 12.13 of Chap. 12 for details. Using this formula and a cluster, it should be reasonable to compute $\tau(p)$ for p of the order of 10^{16}.

3.9 Modular Functions and Complex Multiplication

Although the terminology is quite unfortunate, we cannot change it. By definition, a modular *function* is a function F from $\mathcal{H}$ to $\mathbb{C}$ which is weakly modular of weight 0 (so that $F(\gamma(\tau)) = F(\tau)$, in other words is *invariant* under Γ, or equivalently defines a function from $\Gamma\backslash\mathcal{H}$ to $\mathbb{C}$), meromorphic, including at ∞. This last statement requires some additional explanation, but in simple terms, this means that the Fourier expansion of F has only finitely many Fourier coefficients for negative powers of q: $F(\tau) = \sum_{n\ge n_0} a(n)q^n$, for some (possibly negative) n_0.

A trivial way to obtain modular functions is simply to take the quotient of two modular forms having the same weight. The most important is the j-function defined by

$$j(\tau) = \frac{E_4^3(\tau)}{\Delta(\tau)} ,$$

whose Fourier expansion begins by

$$j(\tau) = \frac{1}{q} + 744 + 196884q + 21493760q^2 + \cdots$$

Indeed, one can easily prove the following theorem:

Theorem 3.31 *Let F be a meromorphic function on $\mathcal{H}$. The following are equivalent:*

1. *F is a modular function.*
2. *F is the quotient of two modular forms of equal weight.*
3. *F is a rational function of j.*

Exercise 3.32 1. Noting that Theorem 3.5 is valid more generally for modular functions (with $v_\tau(f) = -r < 0$ if f has a pole of order r at τ) and using the specific properties of $j(\tau)$, compute $v_\tau(f)$ for the functions $j(\tau)$, $j(\tau) - 1728$, and $D(j)(\tau)$, at the points $\rho = e^{2\pi i/3}$, i, $i\infty$, and τ_0 for τ_0 distinct from these three special points.

2. Set $f = f(a, b, c) = D(j)^a/(j^b(j - 1728)^c)$. Show that f is a modular *form* if and only if $2c \le a$, $3b \le 2a$, and $b + c \ge a$, and give similar conditions for f to be a *cusp form*.

3. Show that $E_4 = f(2, 1, 1)$, $E_6 = f(3, 2, 1)$, and $\Delta = f(6, 4, 3)$, so that, for instance, $D(j) = -E_{14} = -E_4^2 E_6/\Delta$.

An important theory linked to modular functions is the theory of *complex multiplication*, which deserves a course in itself. We simply mention one of the basic results.

We will say that a complex number $\tau \in \mathcal{H}$ is a CM point (CM for Complex Multiplication), if it belongs to an imaginary quadratic field, or equivalently if there exist integers a, b, and c with $a \neq 0$ such that $a\tau^2 + b\tau + c = 0$. The first basic theorem is the following:

Theorem 3.33 *If τ is a CM point, then $j(\tau)$ is an algebraic integer.*

Note that this theorem has two parts: the first and most important part is that $j(\tau)$ is algebraic. This is, in fact, easy to prove. The second part is that it is an algebraic *integer*, and this is more difficult. Since any modular function f is a rational function of j, it follows that if this rational function has algebraic coefficients then $f(\tau)$ will be algebraic (but not necessarily integral). Another immediate consequence is the following:

Corollary 3.34 *Let τ be a CM point and define $\Omega_\tau = \eta(\tau)^2$, where η is as usual the Dedekind eta function. For any modular form f of weight k (in fact, f can also be meromorphic), the number $f(\tau)/\Omega_\tau^k$ is algebraic. In fact, $E_4(\tau)/\Omega_\tau^4$ and $E_6(\tau)/\Omega_\tau^6$ are always algebraic* integers.

But the importance of this theorem lies in algebraic number theory. We give the following theorem without explaining the necessary notions:

Theorem 3.35 *Let τ be a CM point, $D = b^2 - 4ac$ its* discriminant, *where we choose* $\gcd(a, b, c) = 1$ *and $K = \mathbb{Q}(\sqrt{D})$. Then, $K(j(\tau))$ is the maximal abelian unramified extension of K, the* Hilbert class field *of K, and $\mathbb{Q}(j(\tau))$ is the* ring class field *of discriminant D. In particular, the degree of the minimal polynomial of the algebraic integer $j(\tau)$ is equal to the* class number $h(D)$ *of the order of discriminant D, and its algebraic conjugates are given by an explicit formula called the* Shimura reciprocity law.

Examples:

$$j((1+i\sqrt{3})/2) = 0 = 1728 - 3(24)^2$$
$$j(i) = 1728 = 12^3 = 1728 - 4(0)^2$$
$$j((1+i\sqrt{7})/2) = -3375 = (-15)^3 = 1728 - 7(27)^2$$
$$j(i\sqrt{2}) = 8000 = 20^3 = 1728 + 8(28)^2$$
$$j((1+i\sqrt{11})/2) = -32768 = (-32)^3 = 1728 - 11(56)^2$$
$$j((1+i\sqrt{163})/2) = -262537412640768000 = (-640320)^3$$

$$
\begin{aligned}
&= 1728 - 163(40133016)^2 \\
j(i\sqrt{3}) &= 54000 = 2(30)^3 = 1728 + 12(66)^2 \\
j(2i) &= 287496 = (66)^3 = 1728 + 8(189)^2 \\
j((1+3i\sqrt{3})/2) &= -12288000 = -3(160)^3 = 1728 - 3(2024)^2 \\
j((1+i\sqrt{15})/2) &= \frac{-191025 - 85995\sqrt{5}}{2} \\
&= \frac{1-\sqrt{5}}{2}\left(\frac{75+27\sqrt{5}}{2}\right)^3 = 1728 - 3\left(\frac{273+105\sqrt{5}}{2}\right)^2
\end{aligned}
$$

Note that we give the results in the above form since show that the functions $j^{1/3}$ and $(j-1728)^{1/2}$ also have interesting arithmetic properties.

The example with $D = -163$ is particularly spectacular:

Exercise 3.36 Using the above table, show that

$$(e^{\pi\sqrt{163}} - 744)^{1/3} = 640320 - \varepsilon\ ,$$

with $0 < \varepsilon < 10^{-24}$, and more precisely that ε is approximately equal to $65628e^{-(5/3)\pi\sqrt{163}}$ (note that $65628 = 196884/3$).

Exercise 3.37 1. Using once again the example of 163, compute heuristically a few terms of the Fourier expansion of j assuming that it is of the form $1/q + \sum_{n\geq 0} c(n)q^n$ with $c(n)$ reasonably small integers using the following method. Set $q = -e^{-\pi\sqrt{163}}$, and let $J = (-640320)^3$ be the exact value of $j((-1+i\sqrt{163})/2)$. By computing $J - 1/q$, one notices that the result is very close to 744, so we guess that $c(0) = 744$. We then compute $(J - 1/q - c(0))/q$ and note that once again the result is close to an integer, giving $c(1)$, and so on. Go as far as you can with this method.

2. Do the same for 67 instead of 163. You will find the same Fourier coefficients (but you can go less far).
3. On the other hand, do the same for 58, starting with J equal to the integer close to $e^{\pi\sqrt{58}}$. You will find a *different* Fourier expansion: it corresponds, in fact, to another modular function, this time defined on a subgroup of Γ, called a *Hauptmodul*.
4. Try to find other rational numbers D such that $e^{\pi\sqrt{D}}$ is close to an integer, and do the same exercise for them (an example where D is not integral is $89/3$).

3.10 Derivatives of Modular Forms

If we differentiate the modular equation $f((a\tau+b)/(c\tau+d)) = (c\tau+d)^k f(\tau)$ with $\left(\begin{smallmatrix} a & b \\ c & d \end{smallmatrix}\right) \in \Gamma$ using the operator $D = (1/(2\pi i))d/d\tau$ (which gives simpler formulas than $d/d\tau$ since $D(q^n) = nq^n$), we easily obtain

$$D(f)\left(\frac{a\tau+b}{c\tau+d}\right) = (c\tau+d)^{k+2}\left(D(f)(\tau) + \frac{k}{2\pi i}\frac{c}{c\tau+d}f(\tau)\right) .$$

Thus, the derivative of a weakly modular form of weight k looks like one of weight $k+2$, except that there is an extra term. This term vanishes if $k=0$, so the derivative of a modular function of weight 0 is indeed modular of weight 2 (we have seen above the example of $j(\tau)$ which satisfies $D(j) = -E_{14}/\Delta$).

If $k > 0$ and we really want a true weakly modular form of weight $k+2$ there are two ways to do this. The first one is called the *Serre derivative*:

Exercise 3.38 Using Proposition 3.23, show that if f is weakly modular of weight k, then $D(f) - (k/12)E_2 f$ is weakly modular of weight $k+2$. In particular, if $f \in M_k(\Gamma)$, then $SD_k(f) := D(f) - (k/12)E_2 f \in M_{k+2}(\Gamma)$.

The second method is to set $D^*(f) := D(f) - (k/(4\pi\Im(\tau)))f$ since by Proposition 3.23, we have $D^*(f) = SD_k(f) - (k/12)E_2^* f$. This loses holomorphy, but is very useful in certain contexts.

Note that if more than one modular form is involved, there are more ways to make new modular forms using derivatives.

Exercise 3.39 1. For $i = 1, 2$ let $f_i \in M_{k_i}(\Gamma)$. By considering the modular function $f_1^{k_2}/f_2^{k_1}$ of weight 0, show that

$$k_2 f_2 D(f_1) - k_1 f_1 D(f_2) \in S_{k_1+k_2+2}(\Gamma) .$$

Note that this generalizes Exercise 3.20.

2. Compute constants a, b, and c (depending on k_1 and k_2 and not all 0) such that

$$[f_1, f_2]_2 = aD^2(f_1) + bD(f_1)D(f_2) + cD^2(f_2) \in S_{k_1+k_2+4}(\Gamma) .$$

This gives the first two of the so-called *Rankin–Cohen* brackets.

As an application of derivatives of modular forms, we give a proof of a theorem of Siegel. We begin by the following:

Lemma 3.40 *Let a and b be nonnegative integers such that $4a + 6b = 12r + 2$. The constant term of the Fourier expansion of $F_r(a, b) = E_4^a E_6^b/\Delta^r$ vanishes.*

Proof By assumption $F_r(a, b)$ is a meromorphic modular form of weight 2. Since $D(\sum_{n\ge n_0} a(n)q^n) = \sum_{n\ge n_0} na(n)q^n$, it is sufficient to find a modular function $G_r(a, b)$ of weight 0 such that $F_r(a, b) = D(G_r(a, b))$ (recall that the derivative of a modular function of weight 0 is still modular). We prove this by an induction first on r, then on b. Recall that by Exercise 3.32, we have $D(j) = -E_{14}/\Delta = -E_4^2 E_6/\Delta$, and since $4a + 6b = 14$ has only the solution $(a, b) = (2, 1)$ the result is true for $r = 1$. Assume it is true for $r - 1$. We now do a recursion on b, noting that since $2a + 3b = 6r + 1$, b is odd. Note that $D(j^r) = rj^{r-1}D(j) = -rE_4^{3r-1}E_6/\Delta^r$, so the constant term of $F_r(a, 1)$ indeed vanishes. However, since $E_4^3 - E_6^2 = 1728\Delta$, if $a \geq 3$, we have

$$F_r(a-3, b+2) = E_4^{a-3}E_6^b(E_4^3 - 1728\Delta)/\Delta^r = F_r(a, b) - 1728F_{r-1}(a-3, b) ,$$

proving that the result is true for r by induction on b since we assumed it true for $r - 1$. □

We can now prove (part of) Siegel's theorem:

Theorem 3.41 *For* $r = \dim(M_k(\Gamma))$ *define coefficients* c_i^k *by*

$$\frac{E_{12r-k+2}}{\Delta^r} = \sum_{i \geq -r} c_i^k q^i ,$$

where by convention we set $E_0 = 1$*. Then, for any* $f = \sum_{n \geq 0} a(n) \in M_k(\Gamma)$*, we have the relation*

$$\sum_{0 \leq n \leq r} c_{-n}^k a(n) = 0 .$$

In addition, we have $c_0^k \neq 0$*, so that* $a(0) = \sum_{1 \leq n \leq r}(c_{-n}^k/c_0^k)a(n)$ *is a linear combination with* rational coefficients *of the* $a(n)$ *for* $1 \leq n \leq r$*.*

Proof First note that by Corollary 3.6, we have $r \geq (k-2)/12$ (with equality only if $k \equiv 2 \pmod{12}$), so the definition of the coefficients c_i^k makes sense. Note also that since the Fourier expansion of $E_{12r-k+2}$ begins with $1 + O(q)$ and that of Δ^r by $q^r + O(q^{r+1})$, that of the quotient begins with $q^{-r} + O(q^{1-r})$ (in particular, $c_{-r}^k = 1$). The proof of the first part is now immediate: the modular form $fE_{12r-k+2}$ belongs to $M_{12r+2}(\Gamma)$, so by Corollary 3.7 is a linear combination of $E_4^a E_6^b$ with $4a + 6b = 12r + 2$. It follows from the lemma that the constant term of $fE_{12r-k+2}/\Delta^r$ vanishes, and this constant term is equal to $\sum_{0 \leq n \leq r} c_{-n}^k a(n)$, proving the first part of the theorem. The fact that $c_0^k \neq 0$ (which is of course essential) is a little more difficult and will be omitted, see [1] Theorem 9.5.1. □

This theorem has (at least) two consequences. First, a theoretical one: if one can construct a modular form whose constant term is some interesting quantity and whose Fourier coefficients $a(n)$ are rational, this shows that the interesting quantity is also rational. This is what allowed Siegel to show that the value at negative integers of Dedekind zeta functions of totally real number fields are rational, see Sect. 7.2. Second, a practical one: it allows to compute explicitly the constant coefficient $a(0)$ in terms of the $a(n)$, giving interesting formulas, see again Sect. 7.2.

4 Hecke Operators: Ramanujan's Discoveries

We now come to one of the most amazing and important discoveries on modular forms due to S. Ramanujan, which has led to the modern development of the subject. Recall that we set

$$\Delta(\tau) = q \prod_{m \geq 1} (1 - q^m)^{24} = \sum_{n \geq 1} \tau(n) q^n \ .$$

We have $\tau(2) = -24$, $\tau(3) = 252$, and $\tau(6) = -6048 = -24 \cdot 252$, so that $\tau(6) = \tau(2)\tau(3)$. After some more experiments, Ramanujan conjectured that if m and n are coprime, we have $\tau(mn) = \tau(m)\tau(n)$. Thus, by decomposing an integer into products of prime powers, assuming this conjecture, we are reduced to the study of $\tau(p^k)$ for p prime.

Ramanujan then noticed that $\tau(4) = -1472 = (-24)^2 - 2^{11} = \tau(2)^2 - 2^{11}$, and again after some experiments he conjectured that $\tau(p^2) = \tau(p)^2 - p^{11}$, and more generally that $\tau(p^{k+1}) = \tau(p)\tau(p^k) - p^{11}\tau(p^{k-1})$. Thus, $u_k = \tau(p^k)$ satisfies a linear recurrence relation

$$u_{k+1} - \tau(p)u_k + p^{11}u_{k-1} = 0 \ ,$$

and since $u_0 = 1$ the sequence is entirely determined by the value of $u_1 = \tau(p)$. It is well known that the behavior of a linear recurrent sequence is determined by its *characteristic polynomial*. Here, it is equal to $X^2 - \tau(p)X + p^{11}$, and the third of Ramanujan's conjectures is that the discriminant of this equation is always negative, or equivalently that $|\tau(p)| < p^{11/2}$.

Note that if α_p and β_p are the roots of the characteristic polynomial (necessarily distinct since we cannot have $|\tau(p)| = p^{11/2}$), then $\tau(p^k) = (\alpha_p^{k+1} - \beta_p^{k+1})/(\alpha_p - \beta_p)$, and the last conjecture says that α_p and β_p are *complex conjugate*, and, in particular, of modulus *equal* to $p^{11/2}$.

These conjectures are all true. The first two (multiplicativity and recursion) were proved by L. Mordell only 1 year after Ramanujan formulated them, and indeed the proof is quite easy (in fact, we will prove them below). The third conjecture $|\tau(p)| < p^{11/2}$ is extremely hard, and was only proved by P. Deligne in 1970 using the whole machinery developed by the school of A. Grothendieck to solve the Weil conjectures.

The main idea of Mordell, which was generalized later by E. Hecke, is to introduce certain linear operators (now called Hecke operators) on spaces of modular forms, to prove that they satisfy the multiplicativity and recursion properties (this is, in general, much easier than to prove this on numbers), and finally to use the fact that $S_{12}(\Gamma) = \mathbb{C}\Delta$ is of dimension 1, so that necessarily Δ is an *eigenform* of the Hecke operators whose eigenvalues are exactly its Fourier coefficients.

Although there are more natural ways of introducing them, we will define the Hecke operator $T(n)$ on $M_k(\Gamma)$ directly by its action on Fourier expansions $T(n)(\sum_{m\geq 0} a(m)q^m) = \sum_{m\geq 0} b(m)q^m$, where

$$b(m) = \sum_{d\mid\gcd(m,n)} d^{k-1}a(mn/d^2)\ .$$

Note that we can consider this definition as purely formal, apart from the presence of the integer k this is totally unrelated to the possible fact that $\sum_{m\geq 0} a(m)q^m \in M_k(\Gamma)$.

A simple but slightly tedious combinatorial argument shows that these operators satisfy

$$T(n)T(m) = \sum_{d\mid\gcd(n,m)} d^{k-1}T(nm/d^2)\ .$$

In particular, if m and n are coprime, we have $T(n)T(m) = T(nm)$ (multiplicativity), and if p is a prime and $k \geq 1$, we have $T(p^k)T(p) = T(p^{k+1}) + p^{k-1}T(p^{k-1})$ (recursion). This shows that these operators are indeed good candidates for proving the first two of Ramanujan's conjectures.

We need to show the essential fact that they preserve $M_k(\Gamma)$ and $S_k(\Gamma)$ (the latter will follow from the former since by the above definition $b(0) = \sum_{d\mid n} d^{k-1}a(0) = a(0)\sigma_{k-1}(n) = 0$ if $a(0) = 0$). By recursion and multiplicativity, it is sufficient to show this for $T(p)$ with p prime. Now, if $F(\tau) = \sum_{m\geq 0} a(m)q^m$, $T(p)(F)(\tau) = \sum_{m\geq 0} b(m)q^m$ with $b(m) = a(mp)$ if $p \nmid m$, and $b(m) = a(mp) + p^{k-1}a(m/p)$ if $p \mid m$.

On the other hand, let us compute $G(\tau) = \sum_{0\leq j<p} F((\tau + j)/p)$. Replacing directly in the Fourier expansion, we have

$$G(\tau) = \sum_{m\geq 0} a(m)q^{m/p} \sum_{0\leq j<p} e^{2\pi imj/p}\ .$$

The inner sum is a complete geometric sum which vanishes unless $p \mid m$, in which case it is equal to p. Thus, changing m into pm, we have $G(\tau) = p\sum_{m\geq 0} a(pm)q^m$. On the other hand, we have trivially $\sum_{p\mid m} a(m/p)q^m = \sum_{m\geq 0} a(m)q^{pm} = F(p\tau)$. Replacing both of these formulas in the formula for $T(p)(F)$, we see that

$$T(p)(F)(\tau) = p^{k-1}F(p\tau) + \frac{1}{p}\sum_{0\leq j<p} F\left(\frac{\tau + j}{p}\right)\ .$$

Exercise 4.1 Show more generally that

$$T(n)(F)(\tau) = \sum_{ad=n} a^{k-1}\frac{1}{d}\sum_{0\leq b<d} F\left(\frac{a\tau + b}{d}\right)\ .$$

It is now easy to show that $T(p)F$ is modular: replace τ by $\gamma(\tau)$ in the above formula and make a number of elementary manipulations to prove modularity. In fact, since Γ is generated by $\tau \mapsto \tau+1$ and $\tau \mapsto -1/\tau$, it is immediate to check modularity for these two maps on the above formula.

As mentioned above, the proof of the first two Ramanujan conjectures is now immediate: since $T(n)$ acts on the one-dimensional space $S_{12}(\Gamma)$, we must have $T(n)(\Delta) = c \cdot \Delta$ for some constant c. Replacing in the definition of $T(n)$, we thus have for all m $c\tau(m) = \sum_{d \mid \gcd(n,m)} d^{11}\tau(nm/d^2)$. Choosing $m = 1$ and using $\tau(1) = 1$ shows that $c = \tau(n)$, so that

$$\tau(n)\tau(m) = \sum_{d \mid \gcd(n,m)} d^{11}\tau(nm/d^2)$$

which implies (and is equivalent to) the first two conjectures of Ramanujan.

Denote by $P_k(n)$ the *characteristic polynomial* of the linear map $T(n)$ on $S_k(\Gamma)$. A strong form of the so-called Maeda's conjecture states that for $n > 1$ the polynomial $P_k(n)$ is *irreducible*. This has been tested up to very large weights.

Exercise 4.2 The above proof shows that the Hecke operators also preserve the space of modular *functions*, so by Theorem 3.31, the image of $j(\tau)$ will be a rational function in j:

1. Show, for instance, that

$$T(2)(j) = j^2/2 - 744j + 81000 \quad \text{and}$$
$$T(3)(j) = j^3/3 - 744j^2 + 356652j - 12288000\ .$$

2. Set $J = j - 744$, i.e., j with no term in q^0 in its Fourier expansion. Deduce that

$$T(2)(J) = J^2/2 - 196884 \quad \text{and}$$
$$T(3)(J) = J^3/3 - 196884J - 21493760\ ,$$

and observe that the coefficients that we obtain are exactly the Fourier coefficients of J.

3. Prove that $T(n)(j)$ is a *polynomial* in j. Does the last observation generalize?

5 Euler Products, Functional Equations

5.1 Euler Products

The case of Δ is quite special, in that the modular form space to which it naturally belongs, $S_{12}(\Gamma)$, is only one dimensional. As can easily be seen from the dimension formula, this occurs (for cusp forms) only for $k = 12, 16, 18, 20, 22$, and 26 (there are no nonzero cusp forms in weight 14 and the space is of dimension 2 in weight 24), and thus the evident cusp forms ΔE_{k-12} for these values of k (setting $E_0 = 1$) are generators of the space $S_k(\Gamma)$, so are eigenforms of the Hecke operators and share exactly the same properties as Δ, with p^{11} replaced by p^{k-1}.

When the dimension is greater than 1, we must work slightly more. From the formulas given above, it is clear that the $T(n)$ forms a *commutative algebra* of operators on the finite-dimensional vector space $S_k(\Gamma)$. In addition, we have seen above that there is a natural *scalar product* on $S_k(\Gamma)$. One can show the not completely trivial fact that $T(n)$ is Hermitian for this scalar product, and hence, in particular, is diagonalizable. It follows by an easy and classical result of linear algebra that these operators are *simultaneously diagonalizable*, i.e., there exists a basis F_i of forms in $S_k(\Gamma)$ such that $T(n)F_i = \lambda_i(n)F_i$ for all n and i. Identifying Fourier coefficients as we have done above for Δ shows that if $F_i = \sum_{n\geq 1} a_i(n)q^n$, we have $a_i(n) = \lambda_i(n)a_i(1)$. This implies first that $a_i(1) \neq 0$, otherwise F_i would be identically zero, so that by dividing by $a_i(1)$ we can always *normalize* the eigenforms so that $a_i(1) = 1$, and second, as for Δ, that $a_i(n) = \lambda_i(n)$, i.e., the eigenvalues are exactly the Fourier coefficients. In addition, since the $T(n)$ are Hermitian, these eigenvalues are real for any embedding into $\mathbb{C}$, and hence are *totally real*, in other words their minimal polynomial has only real roots. Finally, using Theorem 3.5, it is immediate to show that the field generated by the $a_i(n)$ is finite dimensional over $\mathbb{Q}$, i.e., is a number field.

Exercise 5.1 Consider the space $S = S_{24}(\Gamma)$, which is the smallest weight where the dimension is greater than 1, here 2. By the structure theorem given above, it is generated, for instance, by Δ^2 and ΔE_4^3. Compute the matrix of the operator $T(2)$ on this basis of S, diagonalize this matrix, so find the *eigenfunctions* of $T(2)$ on S (the prime number 144169 should occur). Check that these eigenfunctions are also eigenfunctions of $T(3)$.

Thus, let $F = \sum_{n\geq 1} a(n)q^n$ be a *normalized* eigenfunction for all the Hecke operators in $S_k(\Gamma)$ (for instance, $F = \Delta$ with $k = 12$), and consider the *Dirichlet series*

$$L(F, s) = \sum_{n\geq 1} \frac{a(n)}{n^s} ,$$

for the moment formally, although we will show below that it converges for $\Re(s)$ sufficiently large. The multiplicativity property of the coefficients ($a(nm) = a(n)a(m)$ if $\gcd(n, m) = 1$, coming from that of the $T(n)$) is *equivalent* to the fact that we have an *Euler product* (a product over primes)

$$L(F, s) = \prod_{p\in P} L_p(F, s) \quad \text{with} \quad L_p(F, s) = \sum_{j\geq 0} \frac{a(p^j)}{p^{js}} ,$$

where we will always denote by P the set of prime numbers.

The additional recursion property $a(p^{j+1}) = a(p)a(p^j) - p^{k-1}a(p^{j-1})$ is equivalent to the identity

$$L_p(F, s) = \frac{1}{1 - a(p)p^{-s} + p^{k-1}p^{-2s}}$$

(multiply both sides by the denominator to check this). We have thus proved the following theorem:

Theorem 5.2 *Let $F = \sum_{n\geq 1} a(n)q^n \in S_k(\Gamma)$ be an eigenfunction of all Hecke operators. We have an Euler product*

$$L(F, s) = \sum_{n\geq 1} \frac{a(n)}{n^s} = \prod_{p\in P} \frac{1}{1 - a(p)p^{-s} + p^{k-1}p^{-2s}} .$$

Note that we have not really used the fact that F is a cusp form: the above theorem is still valid if $F = F_k$ is the normalized Eisenstein series

$$F_k(\tau) = -\frac{B_k}{2k} E_k(\tau) = -\frac{B_k}{2k} + \sum_{n\geq 1} \sigma_{k-1}(n)q^n ,$$

which is easily seen to be a normalized eigenfunction for all Hecke operators. In fact,

Exercise 5.3 Let $a \in \mathbb{C}$ be any complex number and let as usual $\sigma_a(n) = \sum_{d\mid n} d^a$.

1. Show that

$$\sum_{n\geq 1} \frac{\sigma_a(n)}{n^s} = \zeta(s-a)\zeta(s) = \prod_{p\in P} \frac{1}{1 - \sigma_a(p)p^{-s} + p^a p^{-2s}} ,$$

with $\sigma_a(p) = p^a + 1$.

2. Show that

$$\sigma_a(m)\sigma_a(n) = \sum_{d\mid \gcd(m,n)} d^a \sigma_a\left(\frac{mn}{d^2}\right) ,$$

so that, in particular, F_k is indeed a normalized eigenfunction for all Hecke operators.

5.2 Analytic Properties of L-Functions

Everything that we have done up to now is purely formal, i.e., we do not need to assume convergence. However, in the sequel, we will need to prove some analytic results, and for this, we need to prove convergence for certain values of s. We begin with the following easy bound, due to Hecke:

Proposition 5.4 *Let $F = \sum_{n\ge1} a(n)q^n \in S_k(\Gamma)$ be a cusp form (not necessarily an eigenform). There exists a constant $c > 0$ (depending on F) such that for all n, we have $|a(n)| \le cn^{k/2}$.*

Proof The trick is to consider the function $g(\tau) = |F(\tau)\Im(\tau)^{k/2}|$: since we have seen that $\Im(\gamma(\tau)) = \Im(\tau)/|c\tau + d|^2$, it follows that $g(\tau)$ is *invariant* under Γ. It follows that $\sup_{\tau\in\mathscr{H}} g(\tau) = \sup_{\tau\in\mathfrak{F}} g(\tau)$, where $\mathfrak{F}$ is the fundamental domain used above. Now because of the Fourier expansion and the fact that F is a cusp form, $|F(\tau)| = O(e^{-2\pi\Im(\tau)})$ as $\Im(\tau) \to \infty$, so $g(\tau)$ tends to 0 also. It immediately follows that g is *bounded* on $\mathfrak{F}$, and hence on $\mathscr{H}$, so that there exists a constant $c_1 > 0$ such that $|F(\tau)| \le c_1\Im(\tau)^{-k/2}$ for all τ.

We can now easily prove Hecke's bound: from the Fourier series section, we know that for any $y > 0$

$$a(n) = e^{2\pi ny} \int_0^1 F(x + iy)e^{-2\pi inx}\, dx\ ,$$

so that $|a(n)| \le c_1e^{2\pi ny}y^{-k/2}$, and choosing $y = 1/n$ proves the proposition with $c = e^{2\pi}c_1$. □

The following corollary is now clear:

Corollary 5.5 *The L-function of a cusp form of weight k converges absolutely (and uniformly on compact subsets) for $\Re(s) > k/2 + 1$.*

Remark 5.6 Deligne's deep result mentioned above on the third Ramanujan conjecture implies that we have the following optimal bound: there exists $c > 0$ such that $|a(n)| \le c\sigma_0(n)n^{(k-1)/2}$, and, in particular, $|a(n)| = O(n^{(k-1)/2+\varepsilon})$ for all $\varepsilon > 0$. This implies that the L-function of a cusp form converges absolutely and uniformly on compact subsets, in fact, also for $\Re(s) > (k + 1)/2$.

Exercise 5.7 Define for all $s \in \mathbb{C}$ the function $\sigma_s(n)$ by $\sigma_s(n) = \sum_{d|n} d^s$ if $n \in \mathbb{Z}_{>0}$, $\sigma_s(0) = \zeta(-s)/2$ (and $\sigma_s(n) = 0$ otherwise). Set

$$S(s_1, s_2; n) = \sum_{0\le m\le n} \sigma_{s_1}(m)\sigma_{s_2}(n - m)\ .$$

1. Compute $S(s_1, s_2; n)$ exactly in terms of $\sigma_{s_1+s_2+1}(n)$ for $(s_1, s_2) = (3, 3)$ and $(3, 5)$, and also for $(s_1, s_2) = (1, 1), (1, 3), (1, 5)$, and $(1, 7)$ by using properties of the function E_2.
2. Using Hecke's bound for cusp forms, show that if s_1 and s_2 are odd positive integers the ratio $S(s_1, s_2; n)/\sigma_{s_1+s_2+1}(n)$ tends to a limit $L(s_1, s_2)$ as $n \to \infty$, and compute this limit in terms of Bernoulli numbers. In addition, give an estimate for the *error term* $|S(s_1, s_2; n)/\sigma_{s_1+s_2+1}(n) - L(s_1, s_2)|$.
3. Using the values of the Riemann zeta function at even positive integers in terms of Bernoulli numbers, show that if s_1 and s_2 are odd positive integers, we have

$$L(s_1, s_2) = \frac{\zeta(s_1+1)\zeta(s_2+1)}{(s_1+s_2+1)\binom{s_1+s_2}{s_1}\zeta(s_1+s_2+2)} .$$

4. (A little project.) *Define* $L(s_1, s_2)$ by the above formula for all s_1, s_2 in $\mathbb{C}$ for which it makes sense, interpreting $\binom{s_1+s_2}{s_1}$ as $\Gamma(s_1+s_2+1)/(\Gamma(s_1+1)\Gamma(s_2+1))$. Check on a computer whether it still seems to be true that

$$S(s_1, s_2; n)/\sigma_{s_1+s_2+1}(n) \to L(s_1, s_2) .$$

Try to *prove* it for $s_1 = s_2 = 2$, and then for general s_1, s_2. If you succeed, give also an estimate for the error term analogous to the one obtained above.

We now do some (elementary) analysis.

Proposition 5.8 *Let $F \in S_k(\Gamma)$. For $\Re(s) > k/2 + 1$, we have*

$$(2\pi)^{-s}\Gamma(s)L(F, s) = \int_0^\infty F(it)t^{s-1}\,dt .$$

Proof Using $\Gamma(s) = \int_0^\infty e^{-t}t^{s-1}\,dt$, this is trivial by uniform convergence which insures that we can integrate term by term. □

Corollary 5.9 *The function $L(F, s)$ is a holomorphic function which can be analytically continued to the whole of $\mathbb{C}$. In addition, if we set $\Lambda(F, s) = (2\pi)^{-s}\Gamma(s) L(F, s)$, we have the functional equation $\Lambda(F, k - s) = i^{-k}\Lambda(F, s)$.*

Note that in our case k is even, so that $i^{-k} = (-1)^{k/2}$, but we prefer writing the constant as above so as to be able to use a similar result in odd weight, which occurs in more general situations.

Proof Indeed, splitting the integral at 1, changing t into $1/t$ in one of the integrals, and using modularity shows immediately that

$$(2\pi)^{-s}\Gamma(s)L(F, s) = \int_1^\infty F(it)(t^{s-1} + i^k t^{k-1-s})\,dt .$$

Since the integral converges absolutely and uniformly for all s (recall that $F(it)$ tends exponentially fast to 0 when $t \to \infty$), this immediately implies the corollary. □

As an aside, note that the integral formula used in the above proof is a very efficient numerical method to compute $L(F, s)$, since the series obtained on the right by term-by-term integration is exponentially convergent. For instance,

Exercise 5.10 Let $F(\tau) = \sum_{n\geq 1} a(n)q^n$ be the Fourier expansion of a cusp form of weight k on Γ. Using the above formula, show that the value of $L(F, k/2)$ at the center of the "critical strip" $0 \leq \Re(s) \leq k$ is given by the following exponentially convergent series:

$$L(F,k/2)=(1+(-1)^{k/2})\sum_{n\geq 1}\frac{a(n)}{n^{k/2}}e^{-2\pi n}P_{k/2}(2\pi n)\;,$$

where $P_{k/2}(X)$ is the polynomial

$$P_{k/2}(X)=\sum_{0\leq j<k/2}X^j/j!=1+X/1!+X^2/2!+\cdots+X^{k/2-1}/(k/2-1)!\;.$$

Note, in particular, that if $k\equiv 2\pmod 4$, we have $L(F,k/2)=0$. Prove this directly.

Exercise 5.11 1. Prove that if F is not necessarily a cusp form, we have $|a(n)|\leq cn^{k-1}$ for some $c>0$.
2. Generalize the proposition and the integral formulas so that they are also valid for non-cusp forms; you will have to add polar parts of the type $1/s$ and $1/(s-k)$.
3. Show that $L(F,s)$ still extends to the whole of $\mathbb{C}$ with functional equation, but that it has a pole, simple, at $s=k$, and compute its residue. In passing, show that $L(F,0)=-a(0)$.

5.3 Special Values of L-Functions

A general "paradigm" on L-functions, essentially due to P. Deligne, is that if some "natural" L-function has both an Euler product and functional equations similar to the above, then for suitable integral "special points" the value of the L-function should be a certain (a priori transcendental) number ω times an algebraic number.

In the case of modular forms, this is a theorem of Yu. Manin.

Theorem 5.12 *Let F be a normalized eigenform in $S_k(\Gamma)$, and denote by K the number field generated by its Fourier coefficients. There exist two nonzero complex numbers ω_+ and ω_- such that for $1\leq j\leq k-1$ integral, we have*

$$\Lambda(F,j)/\omega_{(-1)^j}\in K\;,$$

where we recall that $\Lambda(F,s)=(2\pi)^{-s}\Gamma(s)L(F,s)$.

In addition, $\omega_\pm$ can be chosen such that $\omega_+\omega_-=<F,F>$.

In other words, for j odd, we have $L(F,j)/\omega_-\in K$ while for j even, we have $L(F,j)/\omega_+\in K$.

For instance, in the case $F=\Delta$, if we choose $\omega_-=\Lambda(F,3)$ and $\omega_+=\Lambda(F,2)$, we have

$$(\Lambda(F,j))_{1\leq j\leq 11\ odd}=(1620/691,1,9/14,9/14,1,1620/691)\omega_-$$
$$(\Lambda(F,j))_{1\leq j\leq 11\ even}=(1,25/48,5/12,25/48,1)\omega_+\;,$$

and $\omega_+\omega_-=(8192/225)<F,F>$.

Exercise 5.13 (*see also Exercise* 3.9). For $F \in S_k(\Gamma)$ define the *period polynomial* $P(F,X)$ by

$$P(F;X) = \int_0^{i\infty} (X-\tau)^{k-2} F(\tau)\,d\tau\ .$$

1. For $\gamma \in \Gamma$ show that

$$P(F;X)|_{2-k} = \int_{\gamma^{-1}(0)}^{\gamma^{-1}(i\infty)} (X-\tau)^{k-2} F(\tau)\,d\tau\ .$$

2. Show that $P(F;X)$ satisfies

$$P(F;X)|_{2-k}S + P(F;X) = 0 \quad \text{and}$$
$$P(F;X)|_{2-k}(ST)^2 + P(F;X)|_{2-k}(ST) + P(F;X) = 0\ .$$

3. Show that

$$P(F;X) = -\sum_{j=0}^{k-2}(-i)^{k-1-j}\binom{k-2}{j}\Lambda(F,k-1-j)X^j\ .$$

4. If $F = \Delta$, using Manin's theorem above show that up to the multiplicative constant ω_+, $\Re(P(F;X))$ factors completely in $\mathbb{Q}[X]$ as a product of linear polynomials, and show a similar result for $\Im(P(F;X))$ after omitting the extreme terms involving 691.

5.4 Nonanalytic Eisenstein Series and Rankin–Selberg

If we replace the expression $(c\tau+d)^k$ by $|c\tau+d|^{2s}$ for some complex number s, we can also obtain functions which are invariant by Γ, although they are nonanalytic. More precisely:

Definition 5.14 Write as usual $y = \Im(\tau)$. For $\Re(s) > 1$, we define

$$G(s)(\tau) = \sum_{(c,d)\in\mathbb{Z}^2\setminus\{(0,0)\}} \frac{y^s}{|c\tau+d|^{2s}} \quad \text{and}$$

$$E(s)(\tau) = \sum_{\gamma\in\Gamma_\infty\backslash\Gamma} \Im(\gamma(\tau))^s = \frac{1}{2}\sum_{\gcd(c,d)=1} \frac{y^s}{|c\tau+d|^{2s}}\ .$$

This is again an *averaging* procedure, and it follows that $G(s)$ and $E(s)$ are *invariant* under Γ. In addition, as in the case of the holomorphic Eisenstein series

G_k and E_k, it is clear that $G(s) = 2\zeta(2s)E(s)$. One can also easily compute their Fourier expansion, and the result is as follows:

Proposition 5.15 *Set $\Lambda(s) = \pi^{-s/2}\Gamma(s/2)\zeta(s)$. We have the Fourier expansion*

$$\Lambda(2s)E(s) = \Lambda(2s)y^s + \Lambda(2-2s)y^{1-s} + 4y^{1/2}\sum_{n\geq 1}\frac{\sigma_{2s-1}(n)}{n^{s-1/2}}K_{s-1/2}(2\pi ny)\cos(2\pi nx)\;.$$

In the above, $K_\nu(x)$ is a K-Bessel function which we do not define here. The main properties that we need is that it tends to 0 exponentially (more precisely $K_\nu(x) \sim (\pi/(2x))^{1/2}e^{-x}$ as $x \to \infty$) and that $K_{-\nu} = K_\nu$. It follows from the above Fourier expansion that $E(s)$ has an *analytic continuation* to the whole complex plane, that it satisfies the functional equation $\mathscr{E}(1-s) = \mathscr{E}(s)$, where we set $\mathscr{E}(s) = \Lambda(2s)E(s)$, and that $E(s)$ has a unique pole, at $s = 1$, which is simple with residue $3/\pi$, independent of τ.

Exercise 5.16 Using the properties of the Riemann zeta function $\zeta(s)$, show this last property, i.e., that $E(s)$ has a unique pole, at $s = 1$, which is simple with residue $3/\pi$, independent of τ.

There are many reasons for introducing these non-holomorphic Eisenstein series, but for us the main reason is that they are fundamental in *unfolding* methods. Recall that using unfolding, in Proposition 3.12 we showed that E_k (or G_k) was orthogonal to any cusp form. In the present case, we obtain a different kind of result called a *Rankin–Selberg convolution.* Let f and g be in $M_k(\Gamma)$, one of them being a cusp form. Since $E(s)$ is invariant by Γ, the scalar product $< E(s)f, g >$ makes sense, and the following proposition gives its value:

Proposition 5.17 *Let $f(\tau) = \sum_{n\geq 0} a(n)q^n$ and $g(\tau) = \sum_{n\geq 0} b(n)q^n$ be in $M_k(\Gamma)$, with at least one being a cusp form. For $\Re(s) > 1$, we have*

$$< E(s)f, g > = \frac{\Gamma(s+k-1)}{(4\pi)^{s+k-1}}\sum_{n\geq 1}\frac{a(n)\overline{b(n)}}{n^{s+k-1}}\;.$$

Proof We essentially copy the proof of Proposition 3.12 so we skip the details: setting temporarily $F(\tau) = f(\tau)\overline{g(\tau)}y^k$ which is invariant by Γ, we have

$$\begin{aligned}< E(s)f, g > &= \int_{\Gamma\backslash\mathscr{H}}\sum_{\gamma\in\Gamma_\infty\backslash\Gamma}\Im(\gamma(\tau))^s F(\gamma(\tau))\,d\mu\\ &= \sum_{\Gamma_\infty\backslash\mathscr{H}}\Im(\tau)^s F(\tau)\,d\mu\\ &= \int_0^\infty y^{s+k-2}\int_0^1 F(x+iy)\,dx\,dy\;.\end{aligned}$$

The inner integral is equal to the constant term in the Fourier expansion of F, and hence is equal to $\sum_{n\ge1} a(n)\overline{b(n)}e^{-4\pi ny}$ (note that by assumption one of f and g is a cusp form, so the term $n = 0$ vanishes), and the proposition follows. □

Corollary 5.18 *For $\Re(s) > k$ set*

$$R(f,g)(s) = \sum_{n\ge1} \frac{a(n)\overline{b(n)}}{n^s}\;.$$

1. *$R(f,g)(s)$ has an analytic continuation to the whole complex plane and satisfies the functional equation $\mathcal{R}(2k-1-s) = \mathcal{R}(s)$ with*

$$\mathcal{R}(s) = \Lambda(2s-2k+1)(4\pi)^{-s}\Gamma(s)R(f,g)(s)\;.$$

2. *$R(f,g)(s)$ has a single pole, which is simple, at $s = k$ with residue*

$$\frac{3}{\pi}\frac{(4\pi)^k}{(k-1)!} < f, g > \;.$$

Proof This immediately follows from the corresponding properties of $E(s)$: we have

$$\Lambda(2s-2k+2)(4\pi)^{-s}\Gamma(s)R(f,g)(s) = < \mathcal{E}(s-k+1)f, g > \;,$$

and the right-hand side has an analytic continuation to $\mathbb{C}$, and is invariant when changing s into $2k-1-s$. In addition, by the proposition $E(s-k+1) = \mathcal{E}(s-k+1)/\Lambda(2s-2k+2)$ has a single pole, which is simple, at $s = k$, with residue $3/\pi$, so $R(f,g)(s)$ also has a single pole, which is simple, at $s = k$ with residue $\dfrac{3}{\pi}\dfrac{(4\pi)^k}{(k-1)!} < f, g >$. □

It is an important fact (see Theorem 7.9 of my notes on L-functions in the present volume) that L-functions having analytic continuation and standard functional equations can be very efficiently computed at any point in the complex plane (see the note after the proof of Corollary 5.9 for the special case of $L(F,s)$). Thus, the above corollary gives a very efficient method for computing Petersson scalar products.

Note that the *holomorphic* Eisenstein series $E_k(\tau)$ can also be used to give Rankin–Selberg convolutions, but now between forms of different weights.

Exercise 5.19 Let $f = \sum_{n\ge0} a(n)q^n \in M_\ell(\Gamma)$ and $g = \sum_{n\ge0} b(n)q^n \in M_{k+\ell}(\Gamma)$, at least one being a cusp form. Using exactly the same unfolding method as in the above proposition or as in Proposition 3.12, show that

$$< E_k f, g > = \frac{(k+\ell-2)!}{(4\pi)^{k+\ell-1}} \sum_{n\ge1} \frac{a(n)\overline{b(n)}}{n^{k+\ell-1}}\;.$$

6 Modular Forms on Subgroups of Γ

6.1 Types of Subgroups

We have used as basic definition of (weak) modularity $F|_k\gamma = F$ for all $\gamma \in \Gamma$. But there is no reason to restrict to Γ: we could very well ask the same modularity condition for some group G of transformations of $\mathcal{H}$ different from Γ.

There are many types of such groups, and they have been classified: for us, we will simply distinguish three types, with no justification. For any such group G, we can talk about a fundamental domain, similar to $\mathfrak{F}$ that we have drawn above (I do not want to give a rigorous definition here). We can distinguish essentially three types of such domains, corresponding to three types of groups.

The first type is when the domain (more precisely its closure) is *compact*: we say in that case that G is *cocompact*. It is equivalent to saying that it does not have any "cusp" such as $i\infty$ in the case of G. These groups are very important, but we will not consider them here.

The second type is when the domain is not compact (i.e., it has cusps), but it has *finite volume* for the measure $d\mu = dxdy/y^2$ on $\mathcal{H}$ defined in Exercise 3.11. Such a group is said to have finite *covolume*, and the main example is $G = \Gamma$ that we have just considered, and hence also evidently all the subgroups of Γ of *finite index*.

Exercise 6.1 Show that the covolume of the modular group Γ is finite and equal to $\pi/3$.

The third type is when the volume is infinite: a typical example is the group Γ_∞ generated by integer translations, i.e., the set of matrices $\left(\begin{smallmatrix} 1 & n \\ 0 & 1 \end{smallmatrix}\right)$. A fundamental domain is then any vertical strip in $\mathcal{H}$ of width 1, which can trivially be shown to have infinite volume. These groups are not important (at least for us) for the following reason: they would have "too many" modular forms. For instance, in the case of Γ_∞ a "modular form" would simply be a holomorphic periodic function of period 1, and we come back to the theory of Fourier series, much less interesting.

We will, therefore, restrict to groups of the second type, which are called *Fuchsian groups of the first kind*. In fact, for this course, we will even restrict to subgroups G of Γ of *finite index*.

However, even with this restriction, it is still necessary to distinguish two types of subgroups: the so-called *congruence subgroups*, and the others, of course, called non-congruence subgroups. The theory of modular forms on non-congruence subgroups is quite a difficult subject and active research is being done on them. One annoying aspect is that they apparently do not have a theory of Hecke operators.

Thus, will restrict even more to congruence subgroups. We give the following definitions:

Definition 6.2 Let $N \geq 1$ be an integer.

1. We define

$$\Gamma(N) = \{\gamma = \begin{pmatrix} a & b \\ c & d \end{pmatrix} \in \Gamma,\ \gamma \equiv \begin{pmatrix} 1 & 0 \\ 0 & 1 \end{pmatrix} \pmod{N}\}\,,$$

$$\Gamma_1(N) = \{\gamma = \begin{pmatrix} a & b \\ c & d \end{pmatrix} \in \Gamma,\ \gamma \equiv \begin{pmatrix} 1 & * \\ 0 & 1 \end{pmatrix} \pmod{N}\}\,,$$

$$\Gamma_0(N) = \{\gamma = \begin{pmatrix} a & b \\ c & d \end{pmatrix} \in \Gamma,\ \gamma \equiv \begin{pmatrix} * & * \\ 0 & * \end{pmatrix} \pmod{N}\}\,,$$

where the congruences are component-wise and $*$ indicates that no congruence is imposed.

2. A subgroup of Γ is said to be a *congruence subgroup* if it contains $\Gamma(N)$ for some N, and the smallest such N is called the *level* of the subgroup.

It is clear that $\Gamma(N) \subset \Gamma_1(N) \subset \Gamma_0(N)$, and it is trivial to prove that $\Gamma(N)$ is normal in Γ (hence, in any subgroup of Γ containing it), that $\Gamma_1(N)/\Gamma(N) \simeq \mathbb{Z}/N\mathbb{Z}$ (with the map $\left(\begin{smallmatrix} a & b \\ c & d \end{smallmatrix}\right) \mapsto b \bmod N$), and that $\Gamma_1(N)$ is normal in $\Gamma_0(N)$ with $\Gamma_0(N)/\Gamma_1(N) \simeq (\mathbb{Z}/N\mathbb{Z})^*$ (with the map $\left(\begin{smallmatrix} a & b \\ c & d \end{smallmatrix}\right) \mapsto d \bmod N$).

If G is a congruence subgroup of level N, we have $\Gamma(N) \subset G$, so (whatever the definition) a modular form on G will, in particular, be on $\Gamma(N)$. Because of the above isomorphisms, it is not difficult to reduce the study of forms on $\Gamma(N)$ to those on $\Gamma_1(N)$, and the latter to forms on $\Gamma_0(N)$, except that we have to add a slight "twist" to the modularity property. Thus, for simplicity, we will restrict to modular forms on $\Gamma_0(N)$.

6.2 Modular Forms on Subgroups

In view of the definition given for Γ, it is natural to say that F is weakly modular of weight k on $\Gamma_0(N)$ if for all $\gamma \in \Gamma_0(N)$, we have $F|_k\gamma = F$, where we recall that if $\gamma = \left(\begin{smallmatrix} a & b \\ c & d \end{smallmatrix}\right)$, then $F|_k\gamma(\tau) = (c\tau + d)^{-k}F(\tau)$. To obtain a modular *form*, we need also to require that F is holomorphic on $\mathcal{H}$, plus some additional technical condition "at infinity". In the case of the full modular group Γ, this condition was that $F(\tau)$ remains bounded as $\Im(\tau) \to \infty$. In the case of a subgroup, this condition is not sufficient (it is easy to show that if we do not require an additional condition the corresponding space will, in general, be infinite dimensional). There are several equivalent ways of giving the additional condition. One is the following: writing as usual $\tau = x + iy$, we require that there exists N such that in the strip $-1/2 \le x \le 1/2$, we have $|F(\tau)| \le y^N$ as $y \to \infty$ and $|F(\tau)| \le y^{-N}$ as $y \to 0$ (since F is 1-periodic, there is no loss of generality in restricting to the strip).

It is easily shown that if F is weakly modular and holomorphic, then the above inequalities imply that $|F(\tau)|$ is, in fact, *bounded* as $y \to \infty$ (but, in general, *not* as $y \to 0$), so the first condition is exactly the one that we gave in the case of the full modular group.

Similarly, we can define a *cusp form* by asking that in the above strip $|F(\tau)|$ tends to 0 as $y \to \infty$ and as $y \to 0$.

Exercise 6.3 If $F \in M_k(\Gamma)$ show that the second condition $|F(\tau)| \le y^{-N}$ as $y \to 0$ is satisfied.

Now that we have a solid definition of modular form, we can try to proceed as in the case of the full modular group. A number of things can easily be generalized. It is always convenient to choose a system of representatives (γ_j) of right cosets for $\Gamma_0(N)$ in Γ, so that

$$\Gamma = \bigsqcup_j \Gamma_0(N)\gamma_j \ .$$

For instance, if $\mathfrak{F}$ is the fundamental domain of Γ seen above, one can choose $\mathscr{D} = \bigsqcup \gamma_j(\mathfrak{F})$ as fundamental domain for $\Gamma_0(N)$. The theorem that we gave on valuations generalizes immediately:

$$\sum_{\tau \in \overline{\mathscr{D}}} \frac{v_\tau(F)}{e_\tau} = \frac{k}{12}[\Gamma : \Gamma_0(N)] \ ,$$

where $\overline{\mathscr{D}}$ is $\mathscr{D}$ to which is added a finite number of "cusps" (we do not explain this; it is *not* the topological closure), $e_\tau = 2$ (resp., 3) if τ is Γ-equivalent to i (resp., to ρ), and $e_\tau = 1$ otherwise, and we can then deduce the dimension of $M_k(\Gamma_0(N))$ and $S_k(\Gamma_0(N))$ as we did for Γ:

Theorem 6.4 *We have $M_0(\Gamma_0(N)) = \mathbb{C}$ (i.e., the only modular forms of weight 0 are the constants) and $S_0(\Gamma_0(N)) = \{0\}$. For $k \ge 2$ even, we have*

$$\begin{aligned} \dim(M_k(\Gamma_0(N))) &= A_1 - A_{2,3} - A_{2,4} + A_3 \quad \textit{and} \\ \dim(S_k(\Gamma_0(N))) &= A_1 - A_{2,3} - A_{2,4} - A_3 + \delta_{k,2} \ , \end{aligned}$$

where $\delta_{k,2}$ is the Kronecker symbol (1 if $k = 2$, 0 otherwise) and the A_i are given as follows:

$$\begin{aligned} A_1 &= \frac{k-1}{12} N \prod_{p \mid N} \left(1 + \frac{1}{p}\right) \ , \\ A_{2,3} &= \left(\frac{k-1}{3} - \left\lfloor \frac{k}{3} \right\rfloor\right) \prod_{p \mid N} \left(1 + \left(\frac{-3}{p}\right)\right) \quad \textit{if } 9 \nmid N, \ \ 0 \ \textit{otherwise,} \\ A_{2,4} &= \left(\frac{k-1}{4} - \left\lfloor \frac{k}{4} \right\rfloor\right) \prod_{p \mid N} \left(1 + \left(\frac{-4}{p}\right)\right) \quad \textit{if } 4 \nmid N, \ \ 0 \ \textit{otherwise,} \\ A_3 &= \frac{1}{2} \sum_{d \mid N} \phi(\gcd(d, N/d)) \ . \end{aligned}$$

6.3 Examples of Modular Forms on Subgroups

We give a few examples of modular forms on subgroups. First, note the following easy lemma:

Lemma 6.5 *If* $F \in M_k(\Gamma_0(N))$*, then for any* $m \in \mathbb{Z}_{\geq 1}$*, we have* $F(m\tau) \in M_k(\Gamma_0(mN))$.

Proof Trivial since when $\left(\begin{smallmatrix} a & b \\ c & d \end{smallmatrix}\right) \in \Gamma_0(mN)$ one can write $(m(a\tau+b)/(c\tau+d)) = (a(m\tau)+mb)/((c/m)\tau+d)$. □

Thus, we can already construct many forms on subgroups, but in a sense they are not very interesting, since they are "old" in a precise sense that we will define below.

A second more interesting example is Eisenstein series: there are more general Eisenstein series than those that we have seen for Γ, but we simply give the following important example: using a similar proof to the above lemma we can construct Eisenstein series of *weight* 2 as follows. Recall that $E_2(\tau) = 1 - 24\sum_{n\geq 1}\sigma_1(n)q^n$ is not quite modular, and that $E_2^*(\tau) = E_2(\tau) - 3/(\pi\Im(\tau))$ is weakly modular (but of course non-holomorphic). Consider the function $F(\tau) = NE_2(N\tau) - E_2(\tau)$, analogous to the construction of the lemma with a correction term.

We have the evident but crucial fact that we also have $F(\tau) = NE_2^*(N\tau) - E_2^*(\tau)$ (since $\Im(\tau)$ is multiplied by N), so F is also weakly modular on $\Gamma_0(N)$, but since it is holomorphic, we have thus constructed a (nonzero) modular form of weight 2 on $\Gamma_0(N)$.

A third important example is provided by theta series. This would require a book in itself, so we restrict to the simplest case. We have seen in Corollary 1.3 that the function $T(a) = \sum_{n\in\mathbb{Z}} e^{-a\pi n^2}$ satisfies $T(1/a) = a^{1/2}T(a)$, which looks like (and is) a modularity condition. This was for $a > 0$ real. Let us generalize and for $\tau \in \mathcal{H}$ set

$$\theta(\tau) = \sum_{n\in\mathbb{Z}} q^{n^2} = \sum_{n\in\mathbb{Z}} e^{2\pi i n^2\tau} ,$$

so that, for instance, we simply have $T(a) = \theta(ia/2)$. The proof of the functional equation for T that we gave using Poisson summation is still valid in this more general case and shows that

$$\theta(-1/(4\tau)) = (2\tau/i)^{1/2}\theta(\tau) .$$

On the other hand, the definition trivially shows that $\theta(\tau+1) = \theta(\tau)$. If we denote by W_4 the matrix $\left(\begin{smallmatrix} 0 & -1 \\ 4 & 0 \end{smallmatrix}\right)$ corresponding to the map $\tau \mapsto -1/(4\tau)$ and as usual $T = \left(\begin{smallmatrix} 1 & 1 \\ 0 & 1 \end{smallmatrix}\right)$, we thus have $\theta|_{1/2}W_4 = c\theta$ and $\theta_{1/2}T = \theta$ for some eighth root of unity c. (Note: we always use the principal determination of the square roots; if you are uncomfortable with this, simply square everything, this is what we will do below anyway.) This implies that if we let Γ_θ be the intersection of Γ with the group generated by W_4 and T (as transformations of $\mathcal{H}$), then for all $\gamma \in \Gamma_\theta$ we will have

$\theta|_{1/2}\gamma = c(\gamma)\theta$ for some eighth root of unity $c(\gamma)$, but in fact $c(\gamma)$ is a fourth root of unity which we will give explicitly below.

One can easily describe this group Γ_θ, and in particular show that it contains $\Gamma_0(4)$ as a subgroup of index 2. This implies that $\theta^4 \in M_2(\Gamma_0(4))$, and more generally of course $\theta^{4m} \in M_{2m}(\Gamma_0(4))$.

As one of the most famous application of the finite dimensionality of modular form spaces, solve the following exercise:

Exercise 6.6 1. Using the dimension formulas, show that $2E_2(2\tau) - E_2(\tau)$ together with $4E_2(4\tau) - E_2(\tau)$ form a basis of $M_2(\Gamma_0(4))$.
2. Using the Fourier expansion of E_2, deduce an explicit formula for the Fourier expansion of θ^4, and hence that $r_4(n)$, the number of representations of n as a sum of four squares (in $\mathbb{Z}$, all permutations counted) is given for $n \geq 1$ by the formula
$$r_4(n) = 8(\sigma_1(n) - 4\sigma_1(n/4))\ ,$$
where it is understood that $\sigma_1(x) = 0$ if $x \notin \mathbb{Z}$. In particular, show that this trivially implies Lagrange's theorem that every integer is a sum of four squares.
3. Similarly, show that $r_8(n)$, the nth Fourier coefficient of θ^8, is given for $n \geq 1$ by
$$r_8(n) = 16(\sigma_3(n) - 2\sigma_3(n/2) + 16\sigma_3(n/4))\ .$$

Remark 6.7 Using more general methods one can give "closed" formulas for $r_k(n)$ for $k = 1, 2, 3, 4, 5, 6, 7, 8$, and 10, see, e.g., [1].

6.4 Hecke Operators and L-Functions

We can introduce the same Hecke operators as before, but to have a reasonable definition we must add a coprimality condition: we define $T(n)(\sum_{m\geq 0} a(m)q^m) = \sum_{m\geq 0} b(m)q^m$, with
$$b(m) = \sum_{\substack{d \mid \gcd(m,n) \\ \gcd(d,N)=1}} d^{k-1}a(mn/d^2)\ .$$

This additional condition $\gcd(d, N) = 1$ is, of course, automatically satisfied if n is coprime to N, but not otherwise.

One then shows exactly like in the case of the full modular group that
$$T(n)T(m) = \sum_{\substack{d \mid \gcd(n,m) \\ \gcd(d,N)=1}} d^{k-1}T(nm/d^2)\ ,$$
that they preserve modularity, so, in particular, the $T(n)$ form a commutative algebra of operators on $S_k(\Gamma_0(N))$. And this is where the difficulties specific to subgroups of

Γ begin: in the case of Γ, we stated (without proof nor definition) that the $T(n)$ were *Hermitian* with respect to the Petersson scalar product, and deduced the existence of eigenforms for *all* Hecke operators. Unfortunately, here the same proof shows that the $T(n)$ are Hermitian when n is coprime to N, but *not* otherwise.

It follows that there exist common eigenforms for the $T(n)$, but *only* for n coprime to N, which creates difficulties.

An analogous problem occurs for *Dirichlet characters*: if χ is a Dirichlet character modulo N, it may, in fact, come by natural extension from a character modulo M for some divisor $M \mid N$, $M < N$. The characters which have nice properties, in particular, with respect to the functional equation of their L-functions, are the *primitive* characters, for which such an M does not exist.

A similar but slightly more complicated thing can be done for modular forms. It is clear that if $M \mid N$ and $F \in M_k(\Gamma_0(M))$, then, of course, $F \in M_k(\Gamma_0(N))$. More generally, by Lemma 6.5, for any $d \mid N/M$, we have $F(d\tau) \in M_k(\Gamma_0(N))$. Thus, we want to exclude such "oldforms". However, it is not sufficient to say that a newform is not an oldform. The correct definition is to define a newform as a form which is *orthogonal* to the space of oldforms with respect to the scalar product, and, of course, the new space is the space of newforms. Note that in the case of Dirichlet characters, this orthogonality condition (for the standard scalar product of two characters) is automatically satisfied so need not be added.

This theory was developed by Atkin–Lehner–Li, and the new space $S_k^{\text{new}}(\Gamma_0(N))$ can be shown to have all the nice properties that we require. Although not trivial, one can prove that it has a basis of common eigenforms for *all* Hecke operators, not only those with n coprime to N. More precisely, one shows that in the new space an eigenform for the $T(n)$ for all n coprime to N is automatically an eigenform for *any* operator which commutes with all the $T(n)$, such as, of course, the $T(m)$ for $\gcd(m, N) > 1$.

In addition, we have not really lost anything by restricting to the new space, since it is easy to show that

$$S_k(\Gamma_0(N)) = \bigoplus_{M \mid N} \bigoplus_{d \mid N/M} B(d) S_k^{\text{new}}(\Gamma_0(M)) ,$$

where $B(d)$ is the operator sending $F(\tau)$ to $F(d\tau)$. Note that the sums in the above formula are *direct* sums.

Exercise 6.8 The above formula shows that

$$\dim(S_k(\Gamma_0(N))) = \sum_{M \mid N} \sigma_0(N/M) \dim(S_k^{\text{new}}(\Gamma_0(M))) ,$$

where $\sigma_0(n)$ is the number of divisors of n.

1. Using the Möbius inversion formula, show that if we define an arithmetic function β by $\beta(p) = -2$, $\beta(p^2) = 1$, and $\beta(p^k) = 0$ for $k \geq 3$, and extend by multiplicativity ($\beta(\prod p_i^{v_i}) = \prod \beta(p_i^{v_i})$), we have the following dimension formula for the

new space:

$$\dim(S_k^{\text{new}}(\Gamma_0(N))) = \sum_{M \mid N} \beta(N/M) \dim(S_k(\Gamma_0(M))) .$$

2. Using Theorem 6.4, deduce a direct formula for the dimension of the new space.

Proposition 6.9 *Let $F \in S_k(\Gamma_0(N))$ and $W_N = \left(\begin{smallmatrix} 0 & -1 \\ N & 0 \end{smallmatrix}\right)$.*

1. We have $F|_k W_N \in S_k(\Gamma_0(N))$, where

$$F|_k W_N(\tau) = N^{-k/2} \tau^{-k} F(-1/(N\tau)) .$$

2. If F is an eigenform (in the new space), then $F|_k W_N = \pm F$ for a suitable sign $\pm$.

Proof (1): This simply follows from the fact that W_N *normalizes* $\Gamma_0(N)$: $W_N^{-1}\Gamma_0(N) W_N = \Gamma_0(N)$ as can easily be checked, and the same result would be true for any other normalizing operator such as the *Atkin–Lehner* operators which we will not define. The operator W_N is called the *Fricke involution*.

(2): It is easy to show that W_N commutes with all Hecke operators $T(n)$ when $\gcd(n, N) = 1$, so by what we have mentioned above, if F is an eigenform in the new space, it is automatically an eigenform for W_N, and since W_N acts as an involution, its eigenvalues are ± 1. □

The eigenforms can again be normalized with $a(1) = 1$, and their L-function has an Euler product, of a slightly more general shape:

$$L(F, s) = \prod_{p \nmid N} \frac{1}{1 - a(p)p^{-s} + p^{k-1}p^{-2s}} \prod_{p \mid N} \frac{1}{1 - a(p)p^{-s}} .$$

Proposition 5.8 is, of course, still valid, but is not the correct normalization to obtain a functional equation. We replace it by

$$N^{s/2}(2\pi)^{-s}\Gamma(s)L(F, s) = \int_0^\infty F(it/N^{1/2})t^{s-1}\, dt ,$$

which, of course, is trivial from the proposition by replacing t by $t/N^{1/2}$. Indeed, thanks to the above proposition, we split the integral at $t = 1$, and using the action of W_N, we deduce the following proposition:

Proposition 6.10 *Let $F \in S_k^{new}(\Gamma_0(N))$ be an eigenform for all Hecke operators, and write $F|_k W_N = \varepsilon F$ for some $\varepsilon = \pm 1$. The L-function $L(F, s)$ extends to a holomorphic function in $\mathbb{C}$, and if we set $\Lambda(F, s) = N^{s/2}(2\pi)^{-s}\Gamma(s)L(F, s)$, we have the functional equation*

$$\Lambda(F, k - s) = \varepsilon i^{-k} \Lambda(F, s) .$$

Proof Indeed, the trivial change of variable t into $1/t$ proves the formula

$$N^{s/2}(2\pi)^{-s}\Gamma(s)L(F,s)=\int_1^{\infty}F(it/N^{1/2})(t^{s-1}+\varepsilon i^k t^{k-1-s})\,dt\ ,$$

from which the result follows. □

Once again, we leave to the reader to check that if $F(\tau)=\sum_{n\ge 1}a(n)q^n$, we have

$$L(F,k/2)=(1+\varepsilon(-1)^{k/2})\sum_{n\ge 1}\frac{a(n)}{n^{k/2}}e^{-2\pi n/N^{1/2}}P_{k/2}(2\pi n/N^{1/2})\ .$$

6.5 Modular Forms with Characters

Consider again the problem of sums of squares, in other words of the powers of $\theta(\tau)$. We needed to raise it to a power which is a multiple of 4 so as to have a pure modularity property as we defined it above. But consider the function $\theta^2(\tau)$. The same proof that we mentioned for θ^4 shows that for any $\gamma=\left(\begin{smallmatrix}a&b\\c&d\end{smallmatrix}\right)\in\Gamma_0(4)$, we have

$$\theta^2(\gamma(\tau))=\left(\frac{-4}{d}\right)(c\tau+d)\theta^2(\tau)\ ,$$

where $\left(\frac{-4}{d}\right)$ is the Legendre–Kronecker character (in this specific case equal to $(-1)^{(d-1)/2}$ since d is odd, being coprime to c). Thus, it satisfies a modularity property, except that it is "twisted" by $\left(\frac{-4}{d}\right)$. Note that the equation makes sense since if we change γ into $-\gamma$ (which does not change $\gamma(\tau)$), then $(c\tau+d)$ is changed into $-(c\tau+d)$, and $\left(\frac{-4}{d}\right)$ is changed into $\left(\frac{-4}{-d}\right)=-\left(\frac{-4}{d}\right)$. It is thus essential that the multiplier that we put in front of $(c\tau+d)^k$, here $\left(\frac{-4}{d}\right)$, has the same parity as k.

We mentioned above that the study of modular forms on $\Gamma_1(N)$ could be reduced to those on $\Gamma_0(N)$ "with a twist". Indeed, more precisely it is trivial to show that

$$M_k(\Gamma_1(N))=\bigoplus_{\chi(-1)=(-1)^k}M_k(\Gamma_0(N),\chi)\ ,$$

where χ ranges through all Dirichlet characters modulo N of the specified parity, and where $M_k(\Gamma_0(N),\chi)$ is defined as the space of functions F satisfying

$$F(\gamma(\tau))=\chi(d)(c\tau+d)^kF(\tau)$$

for all $\gamma=\left(\begin{smallmatrix}a&b\\c&d\end{smallmatrix}\right)\in\Gamma_0(N)$, plus the usual holomorphy and conditions at the cusps (note that $\gamma\mapsto\chi(d)$ are the group homomorphism from $\Gamma_0(N)$ to $\mathbb{C}^*$ which induces the abovementioned isomorphism from $\Gamma_0(N)/\Gamma_1(N)$ to $(\mathbb{Z}/N\mathbb{Z})^*$).

Exercise 6.11 1. Show that a system of coset representatives of $\Gamma_1(N)\backslash\Gamma_0(N)$ is given by matrices $M_d = \left(\begin{smallmatrix} u & -v \\ N & d \end{smallmatrix}\right)$, where $0 \le d < N$ such that $\gcd(d, N) = 1$ and u and v are such that $ud + vN = 1$.

2. Let $f \in M_k(\Gamma_1(N))$. Show that in the above decomposition of $M_k(\Gamma_1(N))$, we have $f = \sum_{\chi(-1)=(-1)^k} f_\chi$ with

$$f_\chi = \sum_{0\le d<N,\ \gcd(d,N)=1} \overline{\chi(d)} f|_k M_d \,.$$

These spaces are just as nice as the spaces $M_k(\Gamma_0(N))$ and share exactly the same properties. They have finite dimension (which we do not give), there are Eisenstein series, Hecke operators, newforms, Euler products, L-functions, etc. An excellent rule of thumb is simply to replace any formula containing d^{k-1} (or p^{k-1}) by $\chi(d)d^{k-1}$ (or $\chi(p)p^{k-1}$). In fact, in the Euler product of the L-function of an eigenform, we do not need to distinguish $p \nmid N$ and $p \mid N$ since we have

$$L(F, s) = \prod_{p\in P} \frac{1}{1 - a(p)p^{-s} + \chi(p)p^{k-1-2s}} \,,$$

and $\chi(p) = 0$ if $p \mid N$ since χ is a character modulo N.

Thus, for instance, $\theta^2 \in M_1(\Gamma_0(4), \chi_{-4})$, more generally $\theta^{4m+2} \in M_{2m+1}(\Gamma_0(4), \chi_{-4})$, where we use the notation χ_D for the Legendre–Kronecker symbol $\left(\frac{D}{d}\right)$.

The space $M_1(\Gamma_0(4), \chi_{-4})$ has dimension 1, generated by the single Eisenstein series

$$1 + 4\sum_{n\ge1} \sigma_0^{(-4)}(n)q^n \,, \quad \text{where} \quad \sigma_{k-1}^{(D)}(n) = \sum_{d\mid n} \left(\frac{D}{d}\right) d^{k-1}$$

according to our rule of thumb (which does not tell us the constant 4). Comparing constant coefficients, we deduce that $r_2(n) = 4\sigma_0^{(-4)}(n)$, where as usual $r_2(n)$ is the number of representations of n as a sum of two squares. This formula was in essence discovered by Fermat.

For $r_6(n)$, we must work slightly more: $\theta^6 \in M_3(\Gamma_0(4), \chi_{-4})$, and this space has dimension 2, generated by two Eisenstein series. The first is the natural "rule of thumb" one (which again does not give us the constant)

$$F_1 = 1 - 4\sum_{n\ge1} \sigma_2^{(-4)}(n)q^n \,,$$

and the second is

$$F_2 = \sum_{n\ge1} \sigma_2^{(-4,*)}(n)q^n \,,$$

where

$$\sigma_{k-1}^{(D,*)} = \sum_{d \mid n} \left(\frac{D}{n/d} \right) d^{k-1} ,$$

a sort of dual to $\sigma_{k-1}^{(D)}$ (these are my notation). Since $\theta^6 = 1 + 12q + \cdots$, comparing the Fourier coefficients of 1 and q shows that $\theta^6 = F_1 + 16F_2$, so we deduce that

$$r_6(n) = -4\sigma_2^{(-4)}(n) + 16\sigma_2^{(-4,*)}(n) = \sum_{d \mid n} \left(16 \left(\frac{-4}{n/d} \right) - 4 \left(\frac{-4}{d} \right) \right) d^2 .$$

6.6 Remarks on Dimension Formulas and Galois Representations

The explicit dimension formulas alluded to above are valid for $k \in \mathbb{Z}$ *except* for $k = 1$; in addition, thanks to the theorems mentioned below, we also have explicit dimension formulas for $k \in 1/2 + \mathbb{Z}$. Thus, the theory of modular forms of weight 1 is very special, and their general construction is more difficult.

This is also reflected in the construction of *Galois representations* attached to modular eigenforms, which is an important and deep subject that we will not mention in this course, except to say the following: in weight $k \geq 2$ these representations are ℓ-adic (or modulo ℓ), i.e., with values in $\mathrm{GL}_2(\mathbb{Q}_\ell)$ (or $\mathrm{GL}_2(\mathbb{F}_\ell)$), while in weight 1 they are *complex* representations, i.e., with values in $\mathrm{GL}_2(\mathbb{C})$. The construction in weight 2 is quite old, and comes directly from the construction of the so-called *Tate module* $T(\ell)$ attached to an Abelian variety (more precisely the Jacobian of a modular curve), while the construction in higher weight, due to Deligne, is much deeper since it implies the third Ramanujan conjecture $|\tau(p)| < p^{11/2}$. Finally, the case of weight 1 is due to Deligne–Serre, in fact, using the construction for $k \geq 2$ and congruences.

6.7 Origins of Modular Forms

Modular forms are all pervasive in mathematics, physics, and combinatorics. We just want to mention the most important constructions:

- Historically, the first modular forms were probably *theta functions* (this dates back to J. Fourier at the end of the eighteenth century in his treatment of the *heat equation*) such as $\theta(\tau)$ seen above, and more generally theta functions associated with *lattices*. These functions can have integral or half-integral weight (see below) depending on whether the number of variables which occurs (equivalently, the dimension of the lattice) is even or odd. Later, these theta functions were generalized by introducing *spherical polynomials* associated with the lattice.

For example, the theta function associated to the lattice $\mathbb{Z}^2$ is simply $f(\tau) = \sum_{(x,y)\in\mathbb{Z}^2} q^{x^2+y^2}$, which is clearly equal to θ^2, so belongs to $M_1(\Gamma_0(4), \chi_{-4})$. But, we can also consider, for instance

$$f_5(\tau) = \sum_{(x,y)\in\mathbb{Z}^2} (x^4 - 6x^2y^2 + y^4)q^{x^2+y^2} ,$$

and show that $f_5 \in S_5(\Gamma_0(4), \chi_{-4})$.

Exercise 6.12 1. Using the notation and results of Exercise 3.39, show that $[\theta, \theta]_2 = cf_5$ for a suitable constant c, so that, in particular, $f_5 \in S_5(\Gamma_0(4), \chi_{-4})$.
2. Show that the polynomial $P(x, y) = x^4 - 6x^2y^2 + y^4$ is a *spherical polynomial*, in other words that $D(P) = 0$, where D is the Laplace differential operator $D = \partial^2/\partial^2 x + \partial^2/\partial^2 y$.

- The second occurrence of modular forms is probably *Eisenstein series*, which, in fact, are the first that we encountered in this course. We have only seen the most basic Eisenstein series G_k (or normalized versions) on the full modular group and a few on $\Gamma_0(4)$, but there are very general constructions over any space such as $M_k(\Gamma_0(N), \chi)$. Their Fourier expansions can easily be explicitly computed and are similar to what we have given above. More difficult is the case when k is only half-integral, but this can also be done.
  As we have seen, an important generalization of Eisenstein series is *Poincaré series*, which can also be defined over any space as above.
- A third important construction of modular forms comes from the Dedekind eta function $\eta(\tau)$ defined above. In itself it has a complicated *multiplier system*, but if we define an *eta quotient* as $F(\tau) = \prod_{m\in I} \eta(m\tau)^{r_m}$ for a certain set I of positive integers and exponents $r_m \in \mathbb{Z}$, then it is not difficult to write necessary and sufficient conditions for F to belong to some $M_k(\Gamma_0(N), \chi)$. The first example that we have met is, of course, the Ramanujan delta function $\Delta(\tau) = \eta(\tau)^{24}$. Other examples are, for instance, $\eta(\tau)\eta(23\tau) \in S_1(\Gamma_0(23), \chi_{-23})$, $\eta(\tau)^2\eta(11\tau)^2 \in S_2(\Gamma_0(11))$, and $\eta(2\tau)^{30}/\eta(\tau)^{12} \in S_9(\Gamma_0(8), \chi_{-4})$.
- Closely related to eta-quotients are q-identities involving the q-Pochhammer symbol $(q)_n$ and generalizing those seen in Exercise 3.30, many of which give modular forms not related to the eta function.
- A much deeper construction comes from algebraic geometry: by the modularity theorem of Wiles et al., to any elliptic curve defined over $\mathbb{Q}$ is associated a modular form in $S_2(\Gamma_0(N))$ which is a normalized Hecke eigenform, where N is the so-called *conductor* of the curve. For instance, the eta quotient of level 11 just seen above is the modular form associated to the isogeny class of the elliptic curve of conductor 11 with equation $y^2 + y = x^3 - x^2 - 10x - 20$.

7 More General Modular Forms

In this brief section, we will describe modular forms of a more general kind than those seen up to now.

7.1 Modular Forms of Half-Integral Weight

Coming back again to the function θ, the formulas seen above suggest that θ itself must be considered a modular form, of weight $1/2$. We have already mentioned that

$$\theta^2(\gamma(\tau)) = \left(\frac{-4}{d}\right)(c\tau+d)\theta^2(\tau)\;.$$

But what about θ itself? For this, we must be very careful about the determination of the square root:

Notation: $z^{1/2}$ will *always* denote the principal determination of the square root, i.e., such that $-\pi/2 < \mathrm{Arg}(z^{1/2}) \le \pi/2$. For instance, $(2i)^{1/2} = 1+i$, $(-1)^{1/2} = i$. Warning: we do not, in general, have $(z_1z_2)^{1/2} = z_1^{1/2}z_2^{1/2}$, but only up to sign. As a second notation, when k is odd, $z^{k/2}$ will *always* denote $(z^{1/2})^k$ and *not* $(z^k)^{1/2}$ (for instance, $(2i)^{3/2} = (1+i)^3 = -2+2i$, while $((2i)^3)^{1/2} = 2-2i$).

Thus, let us try and take the square root of the modularity equation for θ^2:

$$\theta(\gamma(\tau)) = v(\gamma,\tau)\left(\frac{-4}{d}\right)^{1/2}(c\tau+d)^{1/2}\;,$$

where $v(\gamma,\tau) = \pm 1$ and may depend on γ and τ. A detailed study of Gauss sums shows that $v(\gamma,\tau) = \left(\frac{-4c}{d}\right)$, the general Kronecker symbol, so that the modularity equation for θ is, for any $\gamma \in \Gamma_0(4)$:

$$\theta(\gamma(\tau)) = v_\theta(\gamma)(c\tau+d)^{1/2}\theta(\tau) \quad\text{with}\quad v_\theta(\gamma) = \left(\frac{c}{d}\right)\left(\frac{-4}{d}\right)^{-1/2}\;.$$

Note that there is something very subtle going on here: this complicated *theta multiplier system* $v_\theta(\gamma)$ must satisfy a complicated *cocycle relation* coming from the trivial identity $\theta((\gamma_1\gamma_2)(\tau)) = \theta(\gamma_1(\gamma_2(\tau)))$ which can be shown to be equivalent to the general *quadratic reciprocity law*.

The following definition is due to G. Shimura:

Definition 7.1 Let $k \in 1/2+\mathbb{Z}$. A function F from $\mathcal{H}$ to $\mathbb{C}$ will be said to be a modular form of (half-integral) weight k on $\Gamma_0(N)$ with character χ if for all $\gamma = \left(\begin{smallmatrix} a & b \\ c & d\end{smallmatrix}\right) \in \Gamma_0(N)$, we have

$$F(\gamma(\tau)) = v_\theta(\gamma)^{2k}\chi(d)(c\tau+d)^kF(\tau)\;,$$

and if the usual holomorphy and conditions at the cusps are satisfied (equivalently if $F^2 \in M_{2k}(\Gamma_0(N), \chi^2\chi_{-4})$).

Note that if $k \in 1/2 + \mathbb{Z}$, we have $v_\theta(\gamma)^{4k} = \chi_{-4}$, which explains the extra factor χ_{-4} in the above definition.

Since $v_\theta(\gamma)$ is defined only for $\gamma \in \Gamma_0(4)$, we need $\Gamma_0(N) \subset \Gamma_0(4)$, in other words $4 \mid N$. In addition, by definition $v_\theta(\gamma)(c\tau + d)^{1/2} = \theta(\gamma(\tau))/\theta(\tau)$ is invariant if we change γ into $-\gamma$, so if $k \in 1/2 + \mathbb{Z}$, the same is true of $v_\theta(\gamma)^{2k}(c\tau + d)^k$, and hence it follows that in the above definition, we must have $\chi(-d) = \chi(d)$, i.e., χ must be an *even* character ($\chi(-1) = 1$).

As usual, we denote by $M_k(\Gamma_0(N), \chi)$ and $S_k(\Gamma_0(N), \chi)$, the spaces of modular and cusp forms. The theory is more difficult than the theory in integral weight, but is now well developed. We mention a few items as follows:

1. There is an explicit but more complicated *dimension formula* due to J. Oesterlé and the author.
2. By a theorem of Serre–Stark, modular forms of weight $1/2$ are simply linear combinations of *unary theta functions* generalizing the function θ above.
3. One can easily construct Eisenstein series, but the computation of their Fourier expansion, due to Shimura and the author, is more complicated.
4. As usual, if we can express θ^m solely in terms of Eisenstein series, this leads to explicit formulas for $r_m(n)$, the number of representation of n as a sum of m squares. Thus, we obtain explicit formulas for $r_3(n)$ (due to Gauss), $r_5(n)$ (due to Smith and Minkowski), and $r_7(n)$, so if we complement the formulas in integral weight, we have explicit formulas for $r_m(n)$ for $1 \le m \le 8$ and $m = 10$.
5. The deeper part of the theory, which is specific to the half-integral weight case, is the existence of *Shimura lifts* from $M_k(\Gamma_0(N), \chi)$ to $M_{2k-1}(\Gamma_0(N/2), \chi^2)$, the description of the *Kohnen subspace* $S_k^+(\Gamma_0(N), \chi)$ which allows both the Shimura lift to go down to level $N/4$, and also to define a suitable Atkin–Lehner type new space, and the deep results of Waldspurger, which nicely complement the work of Shimura on lifts.

We could try to find other types of interesting modularity properties than those coming from θ. For instance, we have seen that the Dedekind eta function is a modular form of weight $1/2$ (not in Shimura's sense), and more precisely it satisfies the following modularity equation, now for any $\gamma \in \Gamma$:

$$\eta(\gamma(\tau)) = v_\eta(\gamma)(c\tau + d)^{1/2}\eta(\tau) ,$$

where $v_\eta(\gamma)$ is a very complicated 24-th root of unity. We could, of course, define η-modular forms of half-integral weight $k \in 1/2 + \mathbb{Z}$ by requiring $F(\gamma(\tau)) = v_\eta(\gamma)^{2k}(c\tau + d)^k F(\tau)$, but it can be shown that this would not lead to any interesting theory (more precisely the only interesting functions would be *eta-quotients* $F(\tau) = \prod_m \eta(m\tau)^{r_m}$, which can be studied directly without any new theory.

Note that there are functional relations between η and θ.

Proposition 7.2 *We have*

$$\theta(\tau) = \frac{\eta^2(\tau+1/2)}{\eta(2\tau+1)} = \frac{\eta^5(2\tau)}{\eta^2(\tau)\eta^2(4\tau)}\;.$$

Exercise 7.3 1. Prove these relations in the following way: first show that the right-hand sides satisfy the same modularity equations as θ for $T = \left(\begin{smallmatrix}1 & 1\\ 0 & 1\end{smallmatrix}\right)$ and $W_4 = \left(\begin{smallmatrix}0 & -1\\ 4 & 0\end{smallmatrix}\right)$, so, in particular, that they are weakly modular on $\Gamma_0(4)$, and second show that they are really modular forms, in other words, that they are holomorphic on $\mathscr{H}$ and at the cusps.
2. Using the definition of η, deduce two *product expansions* for $\theta(\tau)$.

We could also try to study modular forms of fractional or even real weight k not integral or half-integral, but this would lead to functions with no interesting *arithmetical* properties.

In a different direction, we can relax the condition of holomorphy (or meromorphy) and ask that the functions be eigenfunctions of the *hyperbolic Laplace operator*

$$\Delta = -y^2\left(\frac{\partial^2}{\partial^2 x} + \frac{\partial^2}{\partial^2 y}\right) = -4y^2\frac{\partial^2}{\partial\tau\partial\overline{\tau}}$$

which can be shown to be invariant under Γ (more generally, under $\mathrm{SL}_2(\mathbb{R})$) together with suitable boundedness conditions. This leads to the important theory of *Maass forms*. The case of the eigenvalue 0 reduces to ordinary modular forms since $\Delta(F) = 0$ is equivalent to F being a linear combination of a holomorphic and antiholomorphic (i.e., conjugate to a holomorphic) function, each of which will be modular or conjugate of modular.

The case of the eigenvalue $1/4$ also leads to functions having nice arithmetical properties, but all other eigenvalues give functions with (conjecturally) transcendental coefficients, but these functions are useful in number theory for other reasons which we cannot explain here. Note that a famous conjecture of Selberg asserts that for *congruence subgroups*, there are no eigenvalues λ with $0 < \lambda < 1/4$. For instance, for the full modular group, the smallest nonzero eigenvalue is $\lambda = 91.1412\cdots$, which is quite large.

Exercise 7.4 Using the fact that Δ is invariant under Γ, show that $\Delta(\Im(\gamma(\tau))) = s(1-s)\Im(\gamma(\tau))$ and deduce that the non-holomorphic Eisenstein series $E(s)$ introduced in Definition 5.14 is an eigenfunction of the hyperbolic Laplace operator with eigenvalue $s(1-s)$ (note that it does not satisfy the necessary boundedness conditions, so it is not a Maass form: the functions $E(s)$ with $\Re(s) = 1/2$ constitute what is called the *continuous spectrum*, and the Maass forms the *discrete spectrum* of Δ acting on $\Gamma\backslash\mathscr{H}$).

7.2 Modular Forms in Several Variables

The last generalization that we want to mention (there are much more!) is to several variables. The natural idea is to consider holomorphic functions from $\mathcal{H}^r$ to $\mathbb{C}$, now for some $r > 1$, satisfying suitable modularity properties. If we simply ask that $\gamma \in \Gamma$ (or some subgroup) acts component-wise, we will not obtain anything interesting. The right way to do it, introduced by Hilbert–Blumenthal, is to consider a *totally real* number field K of degree r, and denote by Γ_K the group of matrices $\gamma = \left(\begin{smallmatrix} a & b \\ c & d \end{smallmatrix}\right) \in \mathrm{SL}_2(\mathbb{Z}_K)$, where $\mathbb{Z}_K$ is the ring of algebraic integers of K (we could also consider the larger group $\mathrm{GL}_2(\mathbb{Z}_K)$, which leads to a very similar theory). Such a γ has r *embeddings* γ_i into $\mathrm{SL}_2(\mathbb{R})$, which we will denote by $\gamma_i = \left(\begin{smallmatrix} a_i & b_i \\ c_i & d_i \end{smallmatrix}\right)$, and the correct definition is to ask that

$$F(\gamma_1(\tau_1), \cdots, \gamma_r(\tau_r)) = (c_1\tau_1 + d_1)^k \cdots (c_r\tau_r + d_r)^k F(\tau_1, \ldots, \tau_r) .$$

Note that the restriction to totally real number fields is due to the fact that for γ_i to preserve the upper half-plane, it is necessary that $\gamma_i \in \mathrm{SL}_2(\mathbb{R})$. Note also that the γ_i are *not* independent, they are conjugates of a single $\gamma \in \mathrm{SL}_2(\mathbb{Z}_K)$.

A holomorphic function satisfying the above is called a *Hilbert–Blumenthal* modular form (of *parallel weight k*, one can also consider forms where the exponents for the different embeddings are not equal), or more simply a Hilbert modular form (note that there are no "conditions at infinity", since one can prove that they are automatically satisfied unless $K = \mathbb{Q}$).

Since $T = \left(\begin{smallmatrix} 1 & 1 \\ 0 & 1 \end{smallmatrix}\right) \in \mathrm{SL}_2(\mathbb{Z}_K)$ is equal to all its conjugates, such modular forms have Fourier expansions, but using the action of $\left(\begin{smallmatrix} 1 & \alpha \\ 0 & 1 \end{smallmatrix}\right)$ with $\alpha \in \mathbb{Z}_K$, it is easy to show that these expansions are of a special type, involving the *codifferent* $\mathfrak{d}^{-1}$ of K, which is the fractional ideal of $x \in K$ such that $\mathrm{Tr}(x\mathbb{Z}_K) \subset \mathbb{Z}$, where Tr denotes the trace.

One can construct Eisenstein series, here called Hecke–Eisenstein series, and compute their Fourier expansion. One of the important consequences of this computation is that it gives an explicit formula for the value $\zeta_K(1-k)$ of the *Dedekind zeta function* of K at negative integers (hence, by the functional equation of ζ_K, also at positive even integers), and, in particular, it proves that these values are *rational numbers*, a theorem due to C.-L. Siegel as an immediate consequence of Theorem 3.41. An example is as follows:

Proposition 7.5 *Let $K = \mathbb{Q}(\sqrt{D})$ be a real quadratic field with D a fundamental discriminant. Then*

1. *We have*

$$\zeta_K(-1) = \frac{1}{60} \sum_{|s|<\sqrt{D}} \sigma_1\left(\frac{D-s^2}{4}\right) ,$$

$$\zeta_K(-3) = \frac{1}{120} \sum_{|s|<\sqrt{D}} \sigma_3\left(\frac{D-s^2}{4}\right) .$$

2. *We also have formulas such as*

$$\sum_{|s|<\sqrt{D}} \sigma_1(D-s^2) = 60\left(9-2\left(\frac{D}{2}\right)\right)\zeta_K(-1)\,,$$

$$\sum_{|s|<\sqrt{D}} \sigma_3(D-s^2) = 120\left(129-8\left(\frac{D}{2}\right)\right)\zeta_K(-3)\,.$$

We can, of course, reformulate these results in terms of L-functions by using $L(\chi_D, -1) = -12\zeta_K(-1)$ and $L(\chi_D, -3) = 120\zeta_K(-3)$, where as usual χ_D is the quadratic character modulo D.

Exercise 7.6 Using Exercise 6.6 and the above formulas, show that the number $r_5(D)$ of representations of D as a sum of five squares is given by

$$r_5(D) = 480\left(5-2\left(\frac{D}{2}\right)\right)\zeta_K(-1) = -40\left(5-2\left(\frac{D}{2}\right)\right)L(\chi_D, -1)\,.$$

Note that this formula can be generalized to arbitrary D, and is due to Smith and (much later) to Minkowski. There also exists a similar formula for $r_7(D)$: when $-D$ (*not* D) is a fundamental discriminant

$$r_7(D) = -28\left(41-4\left(\frac{D}{2}\right)\right)L(\chi_{-D}, -2)\,.$$

Note also that if we restrict to the *diagonal* $\tau_1 = \cdots = \tau_r$, a Hilbert modular form of (parallel) weight k gives rise to an ordinary modular form of weight kr.

We finish this section with some terminology with no explanation: if K is *not* a totally real number field, one can also define modular forms, but they will not be defined on products of the upper half-plane $\mathcal{H}$ alone, but will also involve the *hyperbolic* 3-*space* $\mathcal{H}_3$. Such forms are called *Bianchi* modular forms.

A different generalization, close to the Weierstrass $\wp$-function seen above, is the theory of *Jacobi forms*, due to M. Eichler and D. Zagier. One of the many interesting aspects of this theory is that it mixes in nontrivial way properties of forms of integral weight with forms of half-integral weight.

Finally, we mention *Siegel modular forms*, introduced by C.-L. Siegel, which are defined on higher dimensional *symmetric spaces*, on which the *symplectic groups* $\mathrm{Sp}_{2n}(\mathbb{R})$ act. The case $n = 1$ gives ordinary modular forms, and the next simplest, $n = 2$, is closely related to Jacobi forms since the Fourier coefficients of Siegel modular forms of degree 2 can be expressed in terms of Jacobi forms.

8 Some Pari/GP Commands

There exist three software packages which are able to compute with modular forms: `magma`, `Sage`, and `Pari/GP` since the spring of 2018. We give here some basic `Pari/GP` commands with little or no explanation (which is available by typing `?` or `??`): we encourage the reader to read the tutorial `tutorial-mf` available with the distribution and to practice with the package since it is an excellent way to learn about modular forms. All commands begin with the prefix `mf`, with the exception of `lfunmf` which more properly belongs to the L-function package.

Creation of modular forms: `mfDelta` (Ramanujan Delta), `mfTheta` (ordinary theta function), `mfEk` (normalized Eisenstein series E_k), more generally `mfeisenstein`, `mffrometaquo` (eta-quotients), `mffromqf` (theta function of lattices with or without spherical polynomial), `mffromell` (from elliptic curves over $\mathbb{Q}$), etc...

Arithmetic operations: `mfcoefs` (Fourier coefficients at infinity), `mflinear` (linear combination, so including addition/subtraction and scalar multiplication), `mfmul`, `mfdiv`, `mfpow` (clear), etc...

Modular operations: `mfbd`, `mftwist`, `mfhecke`, `mfatkin`, `mfderivE2`, `mfbracket`, etc...

Creation of modular form *spaces*: `mfinit`, `mfdim` (dimension of the space), `mfbasis` (random basis of the space), `mftobasis` (decomposition of a form on the `mfbasis`), `mfeigenbasis` (basis of normalized eigenforms).

Searching for modular forms with given Fourier coefficients:
`mfeigensearch`, `mfsearch`.

Expansion of $F|_k\gamma$: `mfslashexpansion`.

Numerical functions: `mfeval` (evaluation at a point in $\mathcal{H}$ or at a cusp), `mfcuspval` (valuation at a cusp), `mfsymboleval` (computation of integrals over paths in the completed upper half-plane), `mfpetersson` (Petersson scalar product), `lfunmf` (L-function associated to a modular form), etc...

Note that for now `Pari/GP` is the only package for which these last functions (beginning with `mfslashexpansion`) are implemented.

9 Suggestions for Further Reading

The literature on modular forms is vast, so I will only mention the books which I am familiar with and that in my opinion will be very useful to the reader. Note that the classic book [4] is absolutely remarkable, but may be difficult for a beginning course.

In addition to the recent book [1] by Strömberg and the author (which, of course, I strongly recommend!!!), I also highly recommend the paper [5], which is essentially a small book. Perhaps the most classical reference is [3]. The more recent book [2] is more advanced since its ultimate goal is to explain the modularity theorem of Wiles et al.

References

1. H. Cohen and F. Strömberg, *Modular Forms: A Classical Approach*, Graduate Studies in Math. **179**, American Math. Soc., (2017).
2. F. Diamond and J. Shurman, *A first course in modular forms*, Graduate Texts in Math. **228**, Springer (2005),
3. T. Miyake, *Modular Forms*, Springer (1989).
4. G. Shimura, *Introduction to the arithmetic theory of automorphic functions*, Publ. Math. Soc. Japan **11**, Princeton University Press (1994) (reprinted from the 1971 original).
5. D. Zagier, *Elliptic modular forms and their applications*, in "The 1-2-3 of modular forms", Universitext, Springer (2008), pp. 1–103.

Computational Arithmetic of Modular Forms

Gabor Wiese

Abstract These course notes are about computing modular forms and some of their arithmetic properties. Their aim is to explain and prove the modular symbols algorithm in as elementary and as explicit terms as possible, and to enable the devoted student to implement it over any ring (such that a sufficient linear algebra theory is available in the chosen computer algebra system). The chosen approach is based on group cohomology and along the way the needed tools from homological algebra are provided.

1 Motivation and Survey

This section serves as an introduction to the topics of the course. We will briefly review the theory of modular forms and Hecke operators. Then we will define the modular symbols formalism and state a theorem by Eichler and Shimura establishing a link between modular forms and modular symbols. This link is the central ingredient since the modular symbols algorithm for the computation of modular forms is entirely based on it. In this introduction, we shall also be able to give an outline of this algorithm.

1.1 Theory: Brief Review of Modular Forms and Hecke Operators

Congruence Subgroups

We first recall the standard congruence subgroups of $\mathrm{SL}_2(\mathbb{Z})$. By N we shall always denote a positive integer.

G. Wiese (✉)
Mathematics Research Unit, University of Luxembourg, Maison du nombre,
6 avenue de la Fonte, 4364 Esch-sur-Alzette, Luxembourg
e-mail: gabor.wiese@uni.lu

I. Inam and E. Büyükasik (eds.), *Notes from the International Autumn School on Computational Number Theory*, Tutorials, Schools, and Workshops in the Mathematical Sciences, https://doi.org/10.1007/978-3-030-12558-5_2

Consider the group homomorphism

$$\mathrm{SL}_2(\mathbb{Z}) \to \mathrm{SL}_2(\mathbb{Z}/N\mathbb{Z}).$$

By Exercise 1.22 it is surjective. Its kernel is called the principal congruence subgroup of level N and denoted $\Gamma(N)$.

The group $\mathrm{SL}_2(\mathbb{Z}/N\mathbb{Z})$ acts naturally on $(\mathbb{Z}/N\mathbb{Z})^2$ (by multiplying the matrix with a vector). We look at the orbit and the stabiliser of $\binom{1}{0}$. The orbit is

$$\mathrm{SL}_2(\mathbb{Z}/N\mathbb{Z})\binom{1}{0} = \{\binom{a}{c} \mid a, c \text{ generate } \mathbb{Z}/N\mathbb{Z}\}$$

because the determinant is 1. We also point out that the orbit of $\binom{1}{0}$ can and should be viewed as the set of elements in $(\mathbb{Z}/N\mathbb{Z})^2$ which are of precise (additive) order N. We now consider the stabiliser of $\binom{1}{0}$ and define the group $\Gamma_1(N)$ as the preimage of that stabiliser group in $\mathrm{SL}_2(\mathbb{Z})$. Explicitly, this means that $\Gamma_1(N)$ consists of those matrices in $\mathrm{SL}_2(\mathbb{Z})$ whose reduction modulo N is of the form $\left(\begin{smallmatrix} 1 & * \\ 0 & 1 \end{smallmatrix}\right)$.

The group $\mathrm{SL}_2(\mathbb{Z}/N\mathbb{Z})$ also acts on $\mathbb{P}^1(\mathbb{Z}/N\mathbb{Z})$, the projective line over $\mathbb{Z}/N\mathbb{Z}$, which one can define as the tuples $(a : c)$ with $a, c \in \mathbb{Z}/N\mathbb{Z}$ such that $\langle a, c\rangle = \mathbb{Z}/N\mathbb{Z}$ modulo the equivalence relation given by multiplication by an element of $(\mathbb{Z}/N\mathbb{Z})^\times$. The action is the natural one (we should actually view $(a : c)$ as a column vector, as above). The orbit of $(1 : 0)$ for this action is $\mathbb{P}^1(\mathbb{Z}/N\mathbb{Z})$. The preimage in $\mathrm{SL}_2(\mathbb{Z})$ of the stabiliser group of $(1 : 0)$ is called $\Gamma_0(N)$. Explicitly, it consists of those matrices in $\mathrm{SL}_2(\mathbb{Z})$ whose reduction is of the form $\left(\begin{smallmatrix} * & * \\ 0 & * \end{smallmatrix}\right)$. We also point out that the quotient of $\mathrm{SL}_2(\mathbb{Z}/N\mathbb{Z})$ modulo the stabiliser of $(1 : 0)$ corresponds to the set of cyclic subgroups of precise order N in $(\mathbb{Z}/N\mathbb{Z})^2$. These observations are at the base of defining level structures for elliptic curves.

It is clear that $\Gamma_1(N)$ is a normal subgroup of $\Gamma_0(N)$ and that the map

$$\Gamma_0(N)/\Gamma_1(N) \xrightarrow{\left(\begin{smallmatrix} a & b \\ c & d \end{smallmatrix}\right)\mapsto a \mod N} (\mathbb{Z}/N\mathbb{Z})^\times$$

is a group isomorphism.

The quotient $\Gamma_0(N)/\Gamma_1(N)$ will be important in the sequel because it will act on modular forms and modular symbols for $\Gamma_1(N)$. For that purpose, we shall often consider characters (i.e. group homomorphisms) of the form

$$\chi : (\mathbb{Z}/N\mathbb{Z})^\times \to \mathbb{C}^\times.$$

We shall also often extend χ to a map $(\mathbb{Z}/N\mathbb{Z}) \to \mathbb{C}$ by imposing $\chi(r) = 0$ if $(r, N) \neq 1$.

On the number theory side, the group $(\mathbb{Z}/N\mathbb{Z})^\times$ enters as the Galois group of a cyclotomic extension. More precisely, by class field theory or Exercise 1.23 we have the isomorphism

$$\mathrm{Gal}(\mathbb{Q}(\zeta_N)/\mathbb{Q}) \xrightarrow{\mathrm{Frob}_\ell \mapsto \ell} (\mathbb{Z}/N\mathbb{Z})^\times$$

for all primes $\ell \nmid N$. By Frob_ℓ we denote (a lift of) the Frobenius endomorphism $x \mapsto x^\ell$, and by ζ_N we denote any primitive N-th root of unity. We shall, thus, later on also consider χ as a character of $\mathrm{Gal}(\mathbb{Q}(\zeta_N)/\mathbb{Q})$. The name *Dirichlet character* (here of *modulus* N) is common usage for both.

Modular Forms

We now recall the definitions of modular forms. Standard references are [5, 10], but I still vividly recommend [9], which gives a concise and yet rather complete introduction. We denote by

$$\mathbb{H} = \{z \in \mathbb{C} | \mathrm{im}(z) > 0\}$$

the upper half plane. The set of cusps is by definition $\mathbb{P}^1(\mathbb{Q}) = \mathbb{Q} \cup \{\infty\}$. The group $\mathrm{PSL}_2(\mathbb{Z})$ acts on $\mathbb{H}$ by Möbius transforms; more explicitly, for $M = \left(\begin{smallmatrix} a & b \\ c & d \end{smallmatrix}\right) \in \mathrm{SL}_2(\mathbb{Z})$ and $z \in \mathbb{H} \cup \mathbb{P}^1(\mathbb{Q})$ one sets

$$M.z = \frac{az+b}{cz+d}. \tag{1}$$

For $M = \left(\begin{smallmatrix} a & b \\ c & d \end{smallmatrix}\right)$ an integer matrix with non-zero determinant, an integer k and a function $f : \mathbb{H} \to \mathbb{C}$, we put

$$(f|_k M)(z) = (f|M)(z) := f\big(M.z\big)\frac{\det(M)^{k-1}}{(cz+d)^k}.$$

Fix integers $k \geq 1$ and $N \geq 1$. A function

$$f : \mathbb{H} \to \mathbb{C}$$

given by a convergent power series (the $a_n(f)$ are complex numbers)

$$f(z) = \sum_{n=0}^{\infty} a_n(f)(e^{2\pi i z})^n = \sum_{n=0}^{\infty} a_n(f) q^n \quad \text{with } q(z) = e^{2\pi i z}$$

is called a *modular form of weight k for $\Gamma_1(N)$* if

(i) $(f|_k \left(\begin{smallmatrix} a & b \\ c & d \end{smallmatrix}\right))(z) = f(\frac{az+b}{cz+d})(cz+d)^{-k} = f(z)$ for all $\left(\begin{smallmatrix} a & b \\ c & d \end{smallmatrix}\right) \in \Gamma_1(N)$, and
(ii) the function $(f|_k \left(\begin{smallmatrix} a & b \\ c & d \end{smallmatrix}\right))(z) = f(\frac{az+b}{cz+d})(cz+d)^{-k}$ admits a limit when z tends to $i\infty$ (we often just write ∞) for all $\left(\begin{smallmatrix} a & b \\ c & d \end{smallmatrix}\right) \in \mathrm{SL}_2(\mathbb{Z})$ (this condition is called f *is holomorphic at the cusp* a/c).

We use the notation $\mathrm{M}_k(\Gamma_1(N)\,;\,\mathbb{C})$. If we replace (ii) by

(ii)' the function $(f|_k \left(\begin{smallmatrix} a & b \\ c & d \end{smallmatrix}\right))(z) = f(\frac{az+b}{cz+d})(cz+d)^{-k}$ is a holomorphic function and the limit $f(\frac{az+b}{cz+d})(cz+d)^{-k}$ is 0 when z tends to $i\infty$,

then f is called a *cusp form*. For these, we introduce the notation $\mathrm{S}_k(\Gamma_1(N)\,;\,\mathbb{C})$.

Let us now suppose that we are given a Dirichlet character χ of modulus N as above. Then we replace (i) as follows:

(i)' $f(\frac{az+b}{cz+d})(cz+d)^{-k} = \chi(d)f(z)$ for all $\left(\begin{smallmatrix} a & b \\ c & d \end{smallmatrix}\right) \in \Gamma_0(N)$.

Functions satisfying this condition are called *modular forms* (respectively, *cusp forms* if they satisfy (ii)') *of weight k, character χ and level N*. The notation $\mathrm{M}_k(N, \chi\,;\,\mathbb{C})$ (respectively, $\mathrm{S}_k(N, \chi\,;\,\mathbb{C})$) will be used.

All these are finite dimensional $\mathbb{C}$-vector spaces. For $k \geq 2$, there are dimension formulae, which one can look up in [20]. We, however, point the reader to the fact that for $k = 1$ nearly nothing about the dimension is known (except that it is smaller than the respective dimension for $k = 2$; it is believed to be much smaller, but only very weak results are known to date).

Hecke Operators

At the base of everything that we will do with modular forms are the Hecke operators and the diamond operators. One should really define them more conceptually (e.g. geometrically), but this takes some time. Here is a definition by formulae.

If a is an integer coprime to N, by Exercise 1.24 we may let σ_a be a matrix in $\Gamma_0(N)$ such that

$$\sigma_a \equiv \left(\begin{smallmatrix} a^{-1} & 0 \\ 0 & a \end{smallmatrix}\right) \mod N. \tag{2}$$

We define the *diamond operator* $\langle a \rangle$ (you see the diamond in the notation, with some phantasy) by the formula

$$\langle a \rangle f = f|_k \sigma_a.$$

If $f \in \mathrm{M}_k(N, \chi\,;\,\mathbb{C})$, then we have by definition $\langle a \rangle f = \chi(a) f$. The diamond operators give a group action of $(\mathbb{Z}/N\mathbb{Z})^\times$ on $\mathrm{M}_k(\Gamma_1(N)\,;\,\mathbb{C})$ and on $\mathrm{S}_k(\Gamma_1(N)\,;\,\mathbb{C})$, and the $\mathrm{M}_k(N, \chi\,;\,\mathbb{C})$ and $\mathrm{S}_k(N, \chi\,;\,\mathbb{C})$ are the χ-eigenspaces for this action. We thus have the isomorphism

$$\mathrm{M}_k(\Gamma_1(N)\,;\,\mathbb{C}) \cong \bigoplus_{\chi} \mathrm{M}_k(N, \chi\,;\,\mathbb{C})$$

for χ running through the characters of $(\mathbb{Z}/N\mathbb{Z})^\times$ (and similarly for the cuspidal spaces).

Let ℓ be a prime. We let

$$\mathcal{R}_\ell := \{\left(\begin{smallmatrix} 1 & r \\ 0 & \ell \end{smallmatrix}\right) | 0 \leq r \leq \ell - 1\} \cup \{\sigma_\ell \left(\begin{smallmatrix} \ell & 0 \\ 0 & 1 \end{smallmatrix}\right)\}, \qquad \text{if } \ell \nmid N \tag{3}$$

$$\mathcal{R}_\ell := \{\left(\begin{smallmatrix} 1 & r \\ 0 & \ell \end{smallmatrix}\right) | 0 \leq r \leq \ell - 1\}, \qquad \text{if } \ell \mid N \tag{4}$$

We use these sets to define the *Hecke operator* T_ℓ acting on f as above as follows:

$$f|_k T_\ell := T_\ell f := \sum_{\delta \in \mathcal{R}_\ell} f|_k \delta.$$

Lemma 1.1 *Suppose $f \in \mathrm{M}_k(N, \chi\,;\,\mathbb{C})$. Recall that we have extended χ so that $\chi(\ell) = 0$ if ℓ divides N. We have the formula*

$$a_n(T_\ell f) = a_{\ell n}(f) + \ell^{k-1}\chi(\ell)a_{n/\ell}(f).$$

In the formula, $a_{n/\ell}(f)$ is to be read as 0 if ℓ does not divide n.

Proof Exercise 1.25. □

The Hecke operators for composite n can be defined as follows (we put T_1 to be the identity):

$$\begin{aligned} T_{\ell^{r+1}} &= T_\ell \circ T_{\ell^r} - \ell^{k-1}\langle \ell \rangle T_{\ell^{r-1}} && \text{for all primes } \ell \text{ and } r \geq 1, \\ T_{uv} &= T_u \circ T_v && \text{for coprime positive integers } u, v. \end{aligned} \tag{5}$$

We derive the very important formula (valid for every n)

$$a_1(T_n f) = a_n(f). \tag{6}$$

It is the only formula that we will really need.

From Lemma 1.1 and the above formulae, it is also evident that the Hecke operators commute among one another. By Exercise 1.26 eigenspaces for a collection of operators (i.e. each element of a given set of Hecke operators acts by scalar multiplication) are respected by all Hecke operators. Hence, it makes sense to consider modular forms which are eigenvectors for every Hecke operator. These are called *Hecke eigenforms*, or often just *eigenforms*. Such an eigenform f is called *normalised* if $a_1(f) = 1$. We shall consider eigenforms in more detail in the following section.

Finally, let us point out the formula (for ℓ prime and $\ell \equiv d \mod N$)

$$\ell^{k-1}\langle d \rangle = T_\ell^2 - T_{\ell^2}. \tag{7}$$

Hence, the diamond operators can be expressed as $\mathbb{Z}$-linear combinations of Hecke operators. Note that divisibility is no trouble since we may choose ℓ_1, ℓ_2, both congruent to d modulo N satisfying an equation $1 = \ell_1^{k-1}r + \ell_2^{k-1}s$ for appropriate $r, s \in \mathbb{Z}$.

Hecke Algebras and the q-Pairing

We now quickly introduce the concept of Hecke algebras. It will be treated in more detail in later sections. In fact, *when we claim to compute modular forms with the modular symbols algorithm, we are really computing Hecke algebras.* In the couple of lines to follow, we show that the Hecke algebra is the dual of modular forms, and hence all knowledge about modular forms can—in principle—be derived from the Hecke algebra.

For the moment, we define the *Hecke algebra* of $\mathrm{M}_k(\Gamma_1(N)\,;\,\mathbb{C})$ as the sub-$\mathbb{C}$-algebra inside the endomorphism ring of the $\mathbb{C}$-vector space $\mathrm{M}_k(\Gamma_1(N)\,;\,\mathbb{C})$ generated by all Hecke operators and all diamond operators. We make similar definitions for $\mathrm{S}_k(\Gamma_1(N)\,;\,\mathbb{C})$, $\mathrm{M}_k(N,\chi\,;\,\mathbb{C})$ and $\mathrm{S}_k(N,\chi\,;\,\mathbb{C})$. Let us introduce the pieces of notation

$$\mathbb{T}_{\mathbb{C}}(\mathrm{M}_k(\Gamma_1(N)\,;\,\mathbb{C})),\ \mathbb{T}_{\mathbb{C}}(\mathrm{S}_k(\Gamma_1(N)\,;\,\mathbb{C})),\ \mathbb{T}_{\mathbb{C}}(\mathrm{M}_k(N,\chi\,;\,\mathbb{C})) \text{ and } \mathbb{T}_{\mathbb{C}}(\mathrm{S}_k(N,\chi\,;\,\mathbb{C})),$$

respectively. We now define a bilinear pairing, which we call the *(complex) q-pairing*, as

$$\mathrm{M}_k(N,\chi\,;\,\mathbb{C}) \times \mathbb{T}_{\mathbb{C}}(\mathrm{M}_k(N,\chi\,;\,\mathbb{C})) \to \mathbb{C},\quad (f,T) \mapsto a_1(Tf)$$

(compare with Eq. 6).

Lemma 1.2 *Suppose $k \geq 1$. The complex q-pairing is perfect, as is the analogous pairing for $\mathrm{S}_k(N,\chi\,;\,\mathbb{C})$. In particular,*

$$\mathrm{M}_k(N,\chi\,;\,\mathbb{C}) \cong \mathrm{Hom}_{\mathbb{C}}(\mathbb{T}_{\mathbb{C}}(\mathrm{M}_k(N,\chi\,;\,\mathbb{C})),\mathbb{C}),\quad f \mapsto (T \mapsto a_1(Tf))$$

and similarly for $\mathrm{S}_k(N,\chi\,;\,\mathbb{C})$. For $\mathrm{S}_k(N,\chi\,;\,\mathbb{C})$, the inverse is given by sending ϕ to $\sum_{n=1}^{\infty}\phi(T_n)q^n$.

Proof Let us first recall that a pairing over a field is perfect if and only if it is non-degenerate. That is what we are going to check. It follows from Eq. 6 like this. If for all n we have $0 = a_1(T_nf) = a_n(f)$, then $f = 0$ (this is immediately clear for cusp forms; for general modular forms at the first place we can only conclude that f is a constant, but since $k \geq 1$, non-zero constants are not modular forms). Conversely, if $a_1(Tf) = 0$ for all f, then $a_1(T(T_nf)) = a_1(T_nTf) = a_n(Tf) = 0$ for all f and all n, whence $Tf = 0$ for all f. As the Hecke algebra is defined as a subring in the endomorphism of $\mathrm{M}_k(N,\chi\,;\,\mathbb{C})$ (resp. the cusp forms), we find $T = 0$, proving the non-degeneracy. □

The perfectness of the q-pairing is also called the *existence of a q-expansion principle*. Due to its central role for this course, we repeat and emphasise that the Hecke algebra is the linear dual of the space of modular forms.

Lemma 1.3 *Let f in $\mathrm{M}_k(\Gamma_1(N)\,;\,\mathbb{C})$ be a normalised eigenform. Then*

$$T_nf = a_n(f)f \quad \text{for all } n \in \mathbb{N}.$$

Moreover, the natural map from the above duality gives a bijection

$$\{\text{Normalised eigenforms in } \mathrm{M}_k(\Gamma_1(N)\,;\,\mathbb{C})\} \leftrightarrow \mathrm{Hom}_{\mathbb{C}-algebra}(\mathbb{T}_{\mathbb{C}}(\mathrm{M}_k(\Gamma_1(N)\,;\,\mathbb{C})),\mathbb{C}).$$

Similar results hold, of course, also in the presence of χ.

Proof Exercise 1.27. □

1.2 Theory: The Modular Symbols Formalism

In this section, we give a definition of formal modular symbols, as implemented in MAGMA and like the one in [6, 16, 20], except that we do not factor out torsion, but intend a common treatment for all rings.

Contrary to the texts just mentioned, we prefer to work with the group

$$\mathrm{PSL}_2(\mathbb{Z}) = \mathrm{SL}_2(\mathbb{Z})/\langle -1\rangle,$$

since it will make some of the algebra much simpler and since it has a very simple description as a free product (see later). The definitions of modular forms could have been formulated using $\mathrm{PSL}_2(\mathbb{Z})$ instead of $\mathrm{SL}_2(\mathbb{Z})$, too (Exercise 1.28).

We introduce some definitions and pieces of notation to be used in all the course.

Definition 1.4 Let R be a ring, Γ a group and V a left $R[\Gamma]$-module. The Γ-invariants of V are by definition

$$V^{\Gamma} = \{v \in V | g.v = v \;\forall\; g \in \Gamma\} \subseteq V.$$

The Γ-coinvariants of V are by definition

$$V_{\Gamma} = V/\langle v - g.v | g \in \Gamma, v \in V\rangle.$$

If $H \leq \Gamma$ is a finite subgroup, we define the norm of H as

$$N_H = \sum_{h \in H} h \in R[\Gamma].$$

Similarly, if $g \in \Gamma$ is an element of finite order n, we define the norm of g as

$$N_g = N_{\langle g\rangle} = \sum_{i=0}^{n-1} g^i \in R[\Gamma].$$

Please look at the important Exercise 1.29 for some properties of these definitions. We shall make use of the results of this exercise in the section on group cohomology.

For the rest of this section, we let R be a commutative ring with unit and Γ be a subgroup of finite index in $\mathrm{PSL}_2(\mathbb{Z})$. For the time being, we allow general modules; so we let V be a left $R[\Gamma]$-module. Recall that $\mathrm{PSL}_2(\mathbb{Z})$ acts on $\mathbb{H} \cup \mathbb{P}^1(\mathbb{Q})$ by Möbius transformations, as defined earlier. A generalised version of the definition below appeared in [23].

Definition 1.5 We define the R-modules

$$\mathcal{M}_R := R[\{\alpha, \beta\} | \alpha, \beta \in \mathbb{P}^1(\mathbb{Q})]/\langle \{\alpha, \alpha\}, \{\alpha, \beta\} + \{\beta, \gamma\} + \{\gamma, \alpha\} | \alpha, \beta, \gamma \in \mathbb{P}^1(\mathbb{Q}) \rangle$$

and

$$\mathcal{B}_R := R[\mathbb{P}^1(\mathbb{Q})].$$

We equip both with the natural left Γ-action. Furthermore, we let

$$\mathcal{M}_R(V) := \mathcal{M}_R \otimes_R V \quad \text{and} \quad \mathcal{B}_R(V) := \mathcal{B}_R \otimes_R V$$

for the left diagonal Γ-action.

(a) We call the Γ-coinvariants

$$\mathcal{M}_R(\Gamma, V) := \mathcal{M}_R(V)_\Gamma = \mathcal{M}_R(V)/\langle (x - gx) | g \in \Gamma, x \in \mathcal{M}_R(V) \rangle$$

the space of (Γ, V)-modular symbols.

(b) We call the Γ-coinvariants

$$\mathcal{B}_R(\Gamma, V) := \mathcal{B}_R(V)_\Gamma = \mathcal{B}_R(V)/\langle (x - gx) | g \in \Gamma, x \in \mathcal{B}_R(V) \rangle$$

the space of (Γ, V)-boundary symbols.

(c) We define the *boundary map* as the map

$$\mathcal{M}_R(\Gamma, V) \to \mathcal{B}_R(\Gamma, V)$$

which is induced from the map $\mathcal{M}_R \to \mathcal{B}_R$ sending $\{\alpha, \beta\}$ to $\{\beta\} - \{\alpha\}$.

(d) The kernel of the boundary map is denoted by $\mathcal{CM}_R(\Gamma, V)$ and is called *the space of cuspidal (Γ, V)-modular symbols.*

(e) The image of the boundary map inside $\mathcal{B}_R(\Gamma, V)$ is denoted by $\mathcal{E}_R(\Gamma, V)$ and is called *the space of (Γ, V)-Eisenstein symbols.*

The reader is now invited to prove that the definition of $\mathcal{M}_R(\Gamma, V)$ behaves well with respect to base change (Exercise 1.30).

The Modules $V_n(R)$ and $V_n^\chi(R)$

Let R be a ring. We put $V_n(R) = R[X, Y]_n \cong \mathrm{Sym}^n(R^2)$ (see Exercise 1.31). By $R[X, Y]_n$ we mean the homogeneous polynomials of degree n in two variables with coefficients in the ring R. By $\mathrm{Mat}_2(\mathbb{Z})_{\det \neq 0}$ we denote the monoid of integral 2×2-matrices with non-zero determinant (for matrix multiplication), i.e., $\mathrm{Mat}_2(\mathbb{Z})_{\det \neq 0} = \mathrm{GL}_2(\mathbb{Q}) \cap \mathbb{Z}^{2 \times 2}$. Then $V_n(R)$ is a $\mathrm{Mat}_2(\mathbb{Z})_{\det \neq 0}$-module in several natural ways.

One can give it the structure of a left $\mathrm{Mat}_2(\mathbb{Z})_{\det \neq 0}$-module via the polynomials by putting

$$(\left(\begin{smallmatrix} a & b \\ c & d \end{smallmatrix}\right).f)(X,Y) = f\big((X,Y)\left(\begin{smallmatrix} a & b \\ c & d \end{smallmatrix}\right)\big) = f\big((aX+cY, bX+dY)\big).$$

Merel and Stein, however, consider a different one, and that is the one implemented in MAGMA, namely

$$(\left(\begin{smallmatrix} a & b \\ c & d \end{smallmatrix}\right).f)(X,Y) = f\big((\left(\begin{smallmatrix} a & b \\ c & d \end{smallmatrix}\right))^{\iota}\left(\begin{smallmatrix} X \\ Y \end{smallmatrix}\right)\big) = f\big(\left(\begin{smallmatrix} d & -b \\ -c & a \end{smallmatrix}\right)\left(\begin{smallmatrix} X \\ Y \end{smallmatrix}\right)\big) = f\big(\left(\begin{smallmatrix} dX-bY \\ -cX+aY \end{smallmatrix}\right)\big).$$

Here, ι denotes Shimura's main involution whose definition can be read off from the line above (note that M^{ι} is the inverse of M if M has determinant 1). Fortunately, both actions are isomorphic due to the fact that the transpose of $((\begin{smallmatrix} a & b \\ c & d \end{smallmatrix}))^{\iota}\left(\begin{smallmatrix} X \\ Y \end{smallmatrix}\right)$ is equal to $(X,Y)\sigma^{-1}\left(\begin{smallmatrix} a & b \\ c & d \end{smallmatrix}\right)\sigma$, where $\sigma = \left(\begin{smallmatrix} 0 & 1 \\ -1 & 0 \end{smallmatrix}\right)$. More precisely, we have the isomorphism $V_n(R) \xrightarrow{f \mapsto \sigma^{-1}.f} V_n(R)$, where the left-hand side module carries 'our' action and the right-hand side module carries the other one. By $\sigma^{-1}.f$ we mean 'our' $\sigma^{-1}.f$.

Of course, there is also a natural right action by $\mathrm{Mat}_2(\mathbb{Z})_{\det\neq 0}$, namely

$$(f.\left(\begin{smallmatrix} a & b \\ c & d \end{smallmatrix}\right))(\left(\begin{smallmatrix} X \\ Y \end{smallmatrix}\right)) = f(\left(\begin{smallmatrix} a & b \\ c & d \end{smallmatrix}\right)\left(\begin{smallmatrix} X \\ Y \end{smallmatrix}\right)) = f(\left(\begin{smallmatrix} aX+bY \\ cX+dY \end{smallmatrix}\right)).$$

By the standard inversion trick, also both left actions described above can be turned into right ones.

Let now $(\mathbb{Z}/N\mathbb{Z})^{\times} \to R^{\times}$ be a Dirichlet character, which we shall also consider as a character $\chi : \Gamma_0(N) \xrightarrow{\left(\begin{smallmatrix} a & b \\ c & d \end{smallmatrix}\right)\mapsto a} (\mathbb{Z}/N\mathbb{Z})^{\times} \xrightarrow{\chi} R^{\times}$. By R^{χ} we denote the $R[\Gamma_0(N)]$-module which is defined to be R with the $\Gamma_0(N)$-action through χ, i.e. $\left(\begin{smallmatrix} a & b \\ c & d \end{smallmatrix}\right).r = \chi(a)r = \chi^{-1}(d)r$ for $\left(\begin{smallmatrix} a & b \\ c & d \end{smallmatrix}\right) \in \Gamma_0(N)$ and $r \in R$.

For use with Hecke operators, we extend this action to matrices $\left(\begin{smallmatrix} a & b \\ c & d \end{smallmatrix}\right) \in \mathbb{Z}^{2\times 2}$ which are congruent to an upper triangular matrix modulo N (but not necessarily of determinant 1). Concretely, we also put $\left(\begin{smallmatrix} a & b \\ c & d \end{smallmatrix}\right).r = \chi(a)r$ for $r \in R$ in this situation. Sometimes, however, we want to use the coefficient d in the action. In order to do so, we let $R^{\iota,\chi}$ be R with the action $\left(\begin{smallmatrix} a & b \\ c & d \end{smallmatrix}\right).r = \chi(d)r$ for matrices as above. In particular, the $\Gamma_0(N)$-actions on $R^{\iota,\chi}$ and $R^{\chi^{-1}}$ coincide.

Note that due to $(\mathbb{Z}/N\mathbb{Z})^{\times}$ being an abelian group, the same formulae as above make R^{χ} also into a right $R[\Gamma_0(N)]$-module. Hence, putting

$$(f \otimes r).\left(\begin{smallmatrix} a & b \\ c & d \end{smallmatrix}\right) = (f|_k\left(\begin{smallmatrix} a & b \\ c & d \end{smallmatrix}\right)) \otimes \left(\begin{smallmatrix} a & b \\ c & d \end{smallmatrix}\right)r$$

makes $\mathrm{M}_k(\Gamma_1(N)\,;\,\mathbb{C}) \otimes_{\mathbb{C}} \mathbb{C}^{\chi}$ into a right $\Gamma_0(N)$-module and we have the description (Exercise 1.32)

$$\mathrm{M}_k(N,\chi\,;\,\mathbb{C}) = (\mathrm{M}_k(\Gamma_1(N)\,;\,\mathbb{C}) \otimes_{\mathbb{C}} \mathbb{C}^{\chi})^{(\mathbb{Z}/N\mathbb{Z})^{\times}} \tag{8}$$

and similarly for $\mathrm{S}_k(N,\chi\,;\,\mathbb{C})$.

We let

$$V_n^{\chi}(R) := V_n(R) \otimes_R R^{\chi} \text{ and } V_n^{\iota,\chi}(R) := V_n(R) \otimes_R R^{\iota,\chi}$$

equipped with the natural diagonal left $\Gamma_0(N)$-actions. Note that unfortunately these modules are in general not $\mathrm{SL}_2(\mathbb{Z})$-modules, but we will not need that. Note, moreover, that if $\chi(-1) = (-1)^n$, then minus the identity acts trivially on $V_n^\chi(R)$ and $V_n^{\iota,\chi}(R)$, whence we consider these modules also as $\Gamma_0(N)/\{\pm 1\}$-modules.

The Modular Symbols Formalism for Standard Congruence Subgroups

We now specialise the general set-up on modular symbols that we have used so far to the precise situation needed for establishing relations with modular forms.

So we let $N \geq 1$, $k \geq 2$ be integers and fix a character $\chi : (\mathbb{Z}/N\mathbb{Z})^\times \to R^\times$, which we also sometimes view as a group homomorphism $\Gamma_0(N) \to R^\times$ as above. We impose that $\chi(-1) = (-1)^k$.

We define

$$\mathcal{M}_k(N, \chi\,;\, R) := \mathcal{M}_R(\Gamma_0(N)/\{\pm 1\}, V_{k-2}^\chi(R)),$$

$$\mathcal{C}\mathcal{M}_k(N, \chi\,;\, R) := \mathcal{C}\mathcal{M}_R(\Gamma_0(N)/\{\pm 1\}, V_{k-2}^\chi(R)),$$

$$\mathcal{B}_k(N, \chi\,;\, R) := \mathcal{B}_R(\Gamma_0(N)/\{\pm 1\}, V_{k-2}^\chi(R))$$

and

$$\mathcal{E}_k(N, \chi\,;\, R) := \mathcal{E}_R(\Gamma_0(N)/\{\pm 1\}, V_{k-2}^\chi(R)).$$

We make the obvious analogous definitions for $\mathcal{M}_k(\Gamma_1(N)\,;\, R)$, etc.

Let

$$\eta := \begin{pmatrix} -1 & 0 \\ 0 & 1 \end{pmatrix}. \tag{9}$$

Because of

$$\eta \begin{pmatrix} a & b \\ c & d \end{pmatrix} \eta = \begin{pmatrix} a & -b \\ -c & d \end{pmatrix}$$

we have

$$\eta\Gamma_1(N)\eta = \Gamma_1(N) \quad \text{and} \quad \eta\Gamma_0(N)\eta = \Gamma_0(N).$$

We can use the matrix η to define an involution (also denoted by η) on the various modular symbols spaces. We just use the diagonal action on $\mathcal{M}_R(V) := \mathcal{M}_R \otimes_R V$, provided, of course, that η acts on V. On $V_{k-2}(R)$ we use the usual $\mathrm{Mat}_2(\mathbb{Z})_{\det \neq 0}$-action, and on $V_{k-2}^\chi(R) = V_{k-2}(R) \otimes R^\chi$ we let η only act on the first factor. We will denote by the superscript $^+$ the subspace invariant under this involution, and by the superscript $^-$ the anti-invariant one. We point out that there are other very good definitions of $+$-spaces and $-$-spaces. For instance, in many applications it can be of advantage to define the $+$-space as the η-coinvariants, rather than the η-invariants. In particular, for modular symbols, where we are using quotients and coinvariants all the time, this alternative definition is more suitable. The reader should just think about the differences between these two definitions. Note that here we are not following the conventions of [20], p. 141. Our action just seems more natural than adding an extra minus sign.

Hecke Operators

The aim of this part is to state the definition of Hecke operators and diamond operators on formal modular symbols $\mathcal{M}_k(N, \chi\,;\, R)$ and $\mathcal{C}\mathcal{M}_k(N, \chi\,;\, R)$. One immediately sees that it is very similar to the one on modular forms. One can get a different insight in the defining formulae by seeing how they are derived from a double coset formulation in Sect. 7.

The definition given here is also explained in detail in [20]. We should also mention the very important fact that one can transfer Hecke operators in an explicit way to Manin symbols using Heilbronn matrices. We shall not do this explicitly in this course. This point is discussed in detail in [16, 20].

We now give the definition only for T_ℓ for a prime ℓ and the diamond operators. The T_n for composite n can be computed from those by the formulae already stated in (5). Notice that the $R[\Gamma_0(N)]$-action on $V_{k-2}^{\chi}(R)$ (for the usual conventions, in particular, $\chi(-1) = (-1)^k$) extends naturally to an action of the semi-group generated by $\Gamma_0(N)$ and $\mathcal{R}_\ell$ (see Eq. 3). Thus, this semi-group acts on $\mathcal{M}_k(N, \chi\,;\, R)$ (and the cusp space) by the diagonal action on the tensor product. Let $x \in \mathcal{M}_k(\Gamma_1(N)\,;\, R)$ or $x \in \mathcal{M}_k(N, \chi\,;\, R)$. We put

$$T_\ell x = \sum_{\delta \in \mathcal{R}_\ell} \delta.x.$$

If a is an integer coprime to N, we define the diamond operator as

$$\langle a \rangle x = \sigma_a x$$

with σ_a as in Eq. (2). When $x = (m \otimes v \otimes 1)_{\Gamma_0(N)/\{\pm 1\}} \in \mathcal{M}_k(N, \chi\,;\, R)$ for $m \in \mathcal{M}_R$ and $v \in V_{k-2}$, we have $\langle a \rangle x = (\sigma_a m \otimes \sigma_a v) \otimes \chi(a^{-1}))_{\Gamma_0(N)/\{\pm 1\}} = x$, thus $(\sigma_a(m \otimes v) \otimes 1)_{\Gamma_0(N)/\{\pm 1\}} = \chi(a)(m \otimes v \otimes 1)_{\Gamma_0(N)/\{\pm 1\}}$.

As in the section on Hecke operators on modular forms, we define Hecke algebras on modular symbols in a very similar way. We will take the freedom of taking arbitrary base rings (we will do that for modular forms in the next section, too).

Thus for any ring R we let $\mathbb{T}_R(\mathcal{M}_k(\Gamma_1(n)\,;\, R))$ be the R-subalgebra of the R-endomorphism algebra of the R-module $\mathcal{M}_k(\Gamma_1(n)\,;\, R)$ generated by the Hecke operators T_n. For a character $\chi : \mathbb{Z}/N\mathbb{Z} \to R^\times$, we make a similar definition. We also make a similar definition for the cuspidal subspace and the $+$- and $-$-spaces.

The following fact will be obvious from the description of modular symbols as Manin symbols (see Theorem 5.7), which will be derived in a later chapter. Here, we already want to use it.

Proposition 1.6 *The R-modules $\mathcal{M}_k(\Gamma_1(N)\,;\, R)$, $\mathcal{C}\mathcal{M}_k(\Gamma_1(N)\,;\, R)$, $\mathcal{M}_k(N, \chi\,;\, R)$, $\mathcal{C}\mathcal{M}_k(N, \chi\,;\, R)$ are finitely presented.*

Corollary 1.7 *Let R be a Noetherian ring. The Hecke algebras $\mathbb{T}_R(\mathcal{M}_k(\Gamma_1(N)\,;\, R))$, $\mathbb{T}_R(\mathcal{C}\mathcal{M}_k(\Gamma_1(N)\,;\, R))$, $\mathbb{T}_R(\mathcal{M}_k(N, \chi\,;\, R))$ and $\mathbb{T}_R(\mathcal{C}\mathcal{M}_k(N, \chi\,;\, R))$ are finitely presented R-modules.*

Proof This follows from Proposition 1.6 since the endomorphism ring of a finitely generated module is finitely generated and submodules of finitely generated modules over Noetherian rings are finitely generated. Furthermore, over a Noetherian ring, finitely generated implies finitely presented. □

This very innocent looking corollary will give—together with the Eichler–Shimura isomorphism—that coefficient fields of normalised eigenforms are number fields. We next prove that the formation of Hecke algebras for modular symbols behaves well with respect to flat base change. We should have in mind the example $R = \mathbb{Z}$ or $R = \mathbb{Z}[\chi] := \mathbb{Z}[\chi(n) : n \in \mathbb{N}]$ (i.e. the ring of integers of the cyclotomic extension of $\mathbb{Q}$ generated by the values of χ or, equivalently, $\mathbb{Z}[e^{2\pi i/r}]$ where r is the order of χ) and $S = \mathbb{C}$.

Proposition 1.8 *Let R be a Noetherian ring and $R \to S$ a flat ring homomorphism.*

(a) The natural map

$$\mathbb{T}_R(\mathcal{M}_k(\Gamma_1(N)\,;\,R)) \otimes_R S \cong \mathbb{T}_S(\mathcal{M}_k(\Gamma_1(N)\,;\,S))$$

is an isomorphism of S-algebras.

(b) The natural map

$$\mathrm{Hom}_R(\mathbb{T}_R(\mathcal{M}_k(\Gamma_1(N)\,;\,R)), R) \otimes_R S \cong \mathrm{Hom}_S(\mathbb{T}_S(\mathcal{M}_k(\Gamma_1(N)\,;\,S)), S)$$

is an isomorphism of S-modules.

(c) The map

$$\mathrm{Hom}_R(\mathbb{T}_R(\mathcal{M}_k(\Gamma_1(N)\,;\,R)), S) \xrightarrow{\phi \mapsto (T \otimes s \mapsto \phi(T)s)} \mathrm{Hom}_S(\mathbb{T}_R(\mathcal{M}_k(\Gamma_1(N)\,;\,R)) \otimes_R S, S)$$

is also an isomorphism of S-modules.

(d) Suppose in addition that R is an integral domain and S a field extension of the field of fractions of R. Then the natural map

$$\mathbb{T}_R(\mathcal{M}_k(\Gamma_1(N)\,;\,R)) \otimes_R S \to \mathbb{T}_R(\mathcal{M}_k(\Gamma_1(N)\,;\,S)) \otimes_R S$$

is an isomorphism of S-algebras.

For a character $\chi : (\mathbb{Z}/N\mathbb{Z})^\times \to R^\times$, similar results hold. Similar statements also hold for the cuspidal subspace.

Proof We only prove the proposition for $M := \mathcal{M}_k(\Gamma_1(N)\,;\,R)$. The arguments are exactly the same in the other cases.

(a) By Exercise 1.30 it suffices to prove

$$\mathbb{T}_R(M) \otimes_R S \cong \mathbb{T}_S(M \otimes_R S).$$

Due to flatness and the finite presentation of M the natural homomorphism

$$\mathrm{End}_R(M) \otimes_R S \to \mathrm{End}_S(M \otimes_R S)$$

is an isomorphism (see [13], Prop. 2.10). By definition, the Hecke algebra $\mathbb{T}_R(M)$ is an R-submodule of $\mathrm{End}_R(M)$. As injections are preserved by flat morphisms, we obtain the injection

$$\mathbb{T}_R(M) \otimes_R S \hookrightarrow \mathrm{End}_R(M) \otimes_R S \cong \mathrm{End}_S(M \otimes_R S).$$

The image is equal to $\mathbb{T}_S(M \otimes_R S)$, since all Hecke operators are hit, establishing (a).

(b) follows from the same citation from [13] as above.

(c) Suppose that under the map from Statement (c) $\phi \in \mathrm{Hom}_R(\mathbb{T}_R(M), S)$ is mapped to the zero map. Then $\phi(T)s = 0$ for all T and all $s \in S$. In particular with $s = 1$ we get $\phi(T) = 0$ for all T, whence ϕ is the zero map, showing injectivity. Suppose now that $\psi \in \mathrm{Hom}_S(\mathbb{T}_R(M) \otimes_R S, S)$ is given. Call ϕ the composite $\mathbb{T}_R(M) \to \mathbb{T}_R(M) \otimes_R S \xrightarrow{\psi} S$. Then ψ is the image of ϕ, showing surjectivity.

(d) We first define

$$N := \ker\left(M \xrightarrow{\pi : m \mapsto m \otimes 1} M \otimes_R S\right).$$

We claim that N consists only of R-torsion elements. Let $x \in N$. Then $x \otimes 1 = 0$. If $rx \neq 0$ for all $r \in R - \{0\}$, then the map $R \xrightarrow{r \mapsto rx} N$ is injective. We call F the image to indicate that it is a free R-module. Consider the exact sequence of R-modules:

$$0 \to F \to M \to M/F \to 0.$$

From flatness we get the exact sequence

$$0 \to F \otimes_R S \to M \otimes_R S \to M/F \otimes_R S \to 0.$$

But, $F \otimes_R S$ is 0, since it is generated by $x \otimes 1 \in M \otimes_R S$. However, F is free, whence $F \otimes_R S$ is also S. This contradiction shows that there is some $r \in R - \{0\}$ with $rx = 0$.

As N is finitely generated, there is some $r \in R - \{0\}$ such that $rN = 0$. Moreover, N is characterised as the set of elements $x \in M$ such that $rx = 0$. For, we already know that $x \in N$ satisfies $rx = 0$. If, conversely, $rx = 0$ with $x \in M$, then $0 = rx \otimes 1/r = x \otimes 1 \in M \otimes_R S$.

Every R-linear (Hecke) operator T on M clearly restricts to N, since $rTn = Trn = T0 = 0$. Suppose now that T acts as 0 on $M \otimes_R S$. We claim that then $rT = 0$ on all of M. Let $m \in M$. We have $0 = T\pi m = \pi Tm$. Thus $Tm \in N$ and, so, $rTm = 0$, as claimed. In other words, the kernel of the homomorphism $\mathbb{T}_R(M) \to \mathbb{T}_R(M \otimes_R S)$ is killed by r. This homomorphism is surjective, since by definition

$\mathbb{T}_R(M \otimes_R S)$ is generated by all Hecke operators acting on $M \otimes_R S$. Tensoring with S kills the torsion and the statement follows. □

Some words of warning are necessary. It is essential that $R \to S$ is a flat homomorphism. A similar result for $\mathbb{Z} \to \mathbb{F}_p$ is not true in general. I call this a 'faithfulness problem', since then $\mathcal{M}_k(\Gamma_1(N)\,;\,\mathbb{F}_p)$ is not a faithful module for $\mathbb{T}_\mathbb{Z}(\mathcal{M}_k(\Gamma_1(N)\,;\,\mathbb{C})) \otimes_\mathbb{Z} \mathbb{F}_p$. Some effort goes into finding k and N, where this module is faithful. See, for instance, [22]. Moreover, $\mathcal{M}_k(\Gamma_1(N)\,;\,R)$ need not be a free R-module and can contain torsion. Please have a look at Exercise 1.33 now to find out whether one can use the $+$- and the $-$-space in the proposition.

1.3 Theory: The Modular Symbols Algorithm

The Eichler–Shimura Theorem

At the basis of the modular symbols algorithm is the following theorem by Eichler, which was extended by Shimura. One of our aims in this lecture is to provide a proof for it. In this introduction, however, we only state it and indicate how the modular symbols algorithm can be derived from it.

Theorem 1.9 (Eichler–Shimura) *There are isomorphisms respecting the Hecke operators*

(a) $\mathrm{M}_k(N, \chi\,;\,\mathbb{C})) \oplus \mathrm{S}_k(N, \chi\,;\,\mathbb{C})^\vee \cong \mathcal{M}_k(N, \chi\,;\,\mathbb{C})$,
(b) $\mathrm{S}_k(N, \chi\,;\,\mathbb{C})) \oplus \mathrm{S}_k(N, \chi\,;\,\mathbb{C})^\vee \cong \mathcal{CM}_k(N, \chi\,;\,\mathbb{C})$,
(c) $\mathrm{S}_k(N, \chi\,;\,\mathbb{C}) \cong \mathcal{CM}_k(N, \chi\,;\,\mathbb{C})^+$.

Similar isomorphisms hold for modular forms and modular symbols on $\Gamma_1(N)$ and $\Gamma_0(N)$.

Proof Later in this lecture (Theorems 5.9 and 6.15, Corollary 7.30). □

Corollary 1.10 *Let R be a subring of $\mathbb{C}$ and $\chi : (\mathbb{Z}/N\mathbb{Z})^\times \to R^\times$ a character. Then there is the natural isomorphism*

$$\mathbb{T}_R(\mathrm{M}_k(N, \chi\,;\,\mathbb{C})) \cong \mathbb{T}_R(\mathcal{M}_k(N, \chi\,;\,\mathbb{C})).$$

A similar result holds cusp forms, and also for $\Gamma_1(N)$ without a character as well as for $\Gamma_0(N)$.

Proof We only prove this for the full space of modular forms. The arguments in the other cases are very similar. Theorem 1.9 tells us that the R-algebra generated by the Hecke operators inside the endomorphism ring of $\mathrm{M}_k(N, \chi\,;\,\mathbb{C})$ equals the R-algebra generated by the Hecke operators inside the endomorphism ring of $\mathcal{M}_k(N, \chi\,;\,\mathbb{C})$, i.e. the assertion to be proved. To see this, one just needs to see that the algebra generated by all Hecke operators on $\mathrm{M}_k(N, \chi\,;\,\mathbb{C}) \oplus \mathrm{S}_k(N, \chi\,;\,\mathbb{C})^\vee$ is the same as the one generated by all Hecke operators on $\mathrm{M}_k(N, \chi\,;\, \mathbb{C})$, which follows from the fact that if some Hecke operator T annihilates the full space of modular forms, then it also annihilates the dual of the cusp space. □

The following corollary of the Eichler–Shimura theorem is of utmost importance for the theory of modular forms. It says that Hecke algebras of modular forms have an integral structure (take $R = \mathbb{Z}$ or $R = \mathbb{Z}[\chi]$). We will say more on this topic in the next section.

Corollary 1.11 *Let R be a subring of $\mathbb{C}$ and $\chi : (\mathbb{Z}/N\mathbb{Z})^\times \to R^\times$ a character. Then the natural map*

$$\mathbb{T}_R(\mathrm{M}_k(N, \chi\,;\, \mathbb{C})) \otimes_R \mathbb{C} \cong \mathbb{T}_\mathbb{C}(\mathrm{M}_k(N, \chi\,;\, \mathbb{C}))$$

is an isomorphism. A similar result holds cusp forms, and also for $\Gamma_1(N)$ without a character as well as for $\Gamma_0(N)$.

Proof We again stick to the full space of modular forms. Tensoring the isomorphism from Corollary 1.10 with $\mathbb{C}$ we get

$$\begin{aligned}\mathbb{T}_R(\mathrm{M}_k(N, \chi\,;\, \mathbb{C})) \otimes_R \mathbb{C} &\cong \mathbb{T}_R(\mathcal{M}_k(N, \chi\,;\, \mathbb{C})) \otimes_R \mathbb{C} \\ &\cong \mathbb{T}_\mathbb{C}(\mathcal{M}_k(N, \chi\,;\, \mathbb{C})) \cong \mathbb{T}_\mathbb{C}(\mathrm{M}_k(N, \chi\,;\, \mathbb{C})),\end{aligned}$$

using Proposition 1.8(d) and again Theorem 1.9. □

The next corollary is at the base of the modular symbols algorithm, since it describes modular forms in linear algebra terms involving only modular symbols.

Corollary 1.12 *Let R be a subring of $\mathbb{C}$ and $\chi : (\mathbb{Z}/N\mathbb{Z})^\times \to R^\times$ a character. Then we have the isomorphisms*

$$\begin{aligned}\mathrm{M}_k(N, \chi\,;\, \mathbb{C}) &\cong \mathrm{Hom}_R(\mathbb{T}_R(\mathcal{M}_k(N, \chi\,;\, R)), R) \otimes_R \mathbb{C} \\ &\cong \mathrm{Hom}_R(\mathbb{T}_R(\mathcal{M}_k(N, \chi\,;\, R)), \mathbb{C}) \qquad \text{and} \\ \mathrm{S}_k(N, \chi\,;\, \mathbb{C}) &\cong \mathrm{Hom}_R(\mathbb{T}_R(\mathcal{CM}_k(N, \chi\,;\, R)), R) \otimes_R \mathbb{C} \\ &\cong \mathrm{Hom}_R(\mathbb{T}_R(\mathcal{CM}_k(N, \chi\,;\, R)), \mathbb{C}).\end{aligned}$$

Similar results hold for $\Gamma_1(N)$ without a character and also for $\Gamma_0(N)$.

Proof This follows from Corollaries 1.10, 1.11, Proposition 1.8 and Lemma 1.2. □

Please look at Exercise 1.34 to find out which statement should be included in this corollary concerning the $+$-spaces. Here is another important consequence of the Eichler–Shimura theorem.

Corollary 1.13 *Let $f = \sum_{n=1}^\infty a_n(f)q^n \in \mathrm{S}_k(\Gamma_1(N)\,;\, \mathbb{C})$ be a normalised Hecke eigenform. Then $\mathbb{Q}_f := \mathbb{Q}(a_n(f) | n \in \mathbb{N})$ is a number field of degree less than or equal to* $\dim_\mathbb{C} \mathrm{S}_k(\Gamma_1(N)\,;\, \mathbb{C})$.

If f has Dirichlet character χ, then $\mathbb{Q}_f$ is a finite field extension of $\mathbb{Q}(\chi)$ of degree less than or equal to $\dim_{\mathbb{C}} \mathrm{S}_k(N, \chi\,;\,\mathbb{C})$. *Here $\mathbb{Q}(\chi)$ is the extension of $\mathbb{Q}$ generated by all the values of χ.*

Proof It suffices to apply the previous corollaries with $R = \mathbb{Q}$ or $R = \mathbb{Q}(\chi)$ and to remember that normalised Hecke eigenforms correspond to algebra homomorphisms from the Hecke algebra into $\mathbb{C}$. □

Sketch of the Modular Symbols Algorithm

It may now already be quite clear how the modular symbols algorithm for computing cusp forms proceeds. We give a very short sketch.

Algorithm 1.14 Input: A field $K \subset \mathbb{C}$, integers $N \geq 1$, $k \geq 2$, P, a character $\chi : (\mathbb{Z}/N\mathbb{Z})^\times \to K^\times$.
Output: A basis of the space of cusp forms $\mathrm{S}_k(N, \chi\,;\,\mathbb{C})$; each form is given by its standard q-expansion with precision P.

(1) create $M := \mathcal{CM}_k(N, \chi\,;\,K)$.
(2) $L \leftarrow []$ (empty list), $n \leftarrow 1$.
(3) repeat
(4)   compute T_n on M.
(5)   join T_n to the list L.
(6) $\mathbb{T} \leftarrow$ the K-algebra generated by all $T \in L$.
(7) $n \leftarrow n + 1$
(8) until $\dim_K(\mathbb{T}) = \dim_{\mathbb{C}} \mathrm{S}_k(N, \chi\,;\,\mathbb{C})$
(9) compute a K-basis B of $\mathbb{T}$.
(10) compute the basis $B^\vee$ of $\mathbb{T}^\vee$ dual to B.
(11) for ϕ in $B^\vee$ do
(12)   output $\sum_{n=1}^{P} \phi(T_n) q^n \in K[q]$.
(13) end for.

We should make a couple of remarks concerning this algorithm. Please remember that there are dimension formulae for $\mathrm{S}_k(N, \chi\,;\,\mathbb{C})$, which can be looked up in [20]. It is clear that the repeat-until loop will stop, due to Corollary 1.12. We can even give an upper bound as to when it stops at the latest. That is the so-called Sturm bound, which is the content of the following proposition.

Proposition 1.15 (Sturm) *Let $f \in \mathrm{M}_k(N, \chi\,;\,\mathbb{C})$ such that $a_n(f) = 0$ for all $n \leq \frac{k\mu}{12}$, where $\mu = N \prod_{l \mid N \text{ prime}} (1 + \frac{1}{l})$.*
Then $f = 0$.

Proof Apply Corollary 9.20 of [20] with $\mathfrak{m} = (0)$. □

Corollary 1.16 *Let K, N, χ etc. as in the algorithm. Then $\mathbb{T}_K(\mathcal{CM}_k(N, \chi\,;\,K))$ can be generated as a K-vector space by the operators T_n for $1 \leq n \leq \frac{k\mu}{12}$.*

Proof Exercise 1.35. □

We shall see later how to compute eigenforms and how to decompose the space of modular forms in a 'sensible' way.

1.4 Theory: Number Theoretic Applications

We close this survey and motivation section by sketching some number theoretic applications.

Galois Representations Attached to Eigenforms

We mention the sad fact that until 2006 only the one-dimensional representations of $\mathrm{Gal}(\overline{\mathbb{Q}}/\mathbb{Q})$ were well understood. In the case of finite image one can use the Kronecker–Weber theorem, which asserts that any cyclic extension of $\mathbb{Q}$ is contained in a cyclotomic field. This is generalised by global class field theory to one-dimensional representations of $\mathrm{Gal}(\overline{\mathbb{Q}}/K)$ for each number field K. Since we now have a proof of Serre's modularity conjecture [17] (a theorem by Khare, Wintenberger [15]), we also know a little bit about two-dimensional representations of $\mathrm{Gal}(\overline{\mathbb{Q}}/\mathbb{Q})$, but, replacing $\mathbb{Q}$ by any other number field, all one has is conjectures.

The great importance of modular forms for modern number theory is due to the fact that one may attach a two-dimensional representation of the Galois group of the rationals to each normalised cuspidal eigenform. The following theorem is due to Shimura for $k = 2$ and due to Deligne for $k \geq 2$.

Until the end of this section, we shall use the language of Galois representations (e.g. irreducible, unramified, Frobenius element, cyclotomic character) without introducing it. It will not be used elsewhere. The meanwhile quite old lectures by Darmon, Diamond and Taylor are still an excellent introduction to the subject [7].

Theorem 1.17 *Let $k \geq 2$, $N \geq 1$, p a prime, and $\chi : (\mathbb{Z}/N\mathbb{Z})^\times \to \mathbb{C}^\times$ a character. Then to any normalised eigenform $f \in \mathrm{S}_k(N, \chi\,; \mathbb{C})$ with $f = \sum_{n\geq 1} a_n(f)q^n$ one can attach a Galois representation, i.e. a continuous group homomorphism,*

$$\rho_f : \mathrm{Gal}(\overline{\mathbb{Q}}/\mathbb{Q}) \to \mathrm{GL}_2(\overline{\mathbb{Q}}_p)$$

such that

(i) ρ_f is irreducible,
(ii) $\det(\rho_f(c)) = -1$ for any complex conjugation $c \in \mathrm{Gal}(\overline{\mathbb{Q}}/\mathbb{Q})$ (one says that ρ_f is odd*),*
(iii) for all primes $\ell \nmid Np$ the representation ρ_f is unramified at ℓ,

$$\mathrm{tr}(\rho_f(\mathrm{Frob}_\ell)) = a_\ell(f) \quad \textit{and} \quad \det(\rho_f(\mathrm{Frob}_\ell)) = \ell^{k-1}\chi(\ell).$$

In the statement, Frob_ℓ *denotes a Frobenius element at ℓ.*

By choosing a $\rho(\mathrm{Gal}(\overline{\mathbb{Q}}/\mathbb{Q}))$-stable lattice in $\overline{\mathbb{Q}}_p^2$ and applying reduction and semi-simplification one obtains the following consequence.

Theorem 1.18 *Let $k \geq 2$, $N \geq 1$, p a prime, and $\chi : (\mathbb{Z}/N\mathbb{Z})^\times \to \mathbb{C}^\times$ a character. Then to any normalised eigenform $f \in \mathrm{S}_k(N, \chi\,; \mathbb{C})$ with $f = \sum_{n\geq 1} a_n(f)q^n$ and to any prime ideal $\mathfrak{P}$ of the ring of integers $\mathcal{O}_f$ of $\mathbb{Q}_f = \mathbb{Q}(a_n(f) : n \in \mathbb{N})$*

with residue characteristic p (and silently a fixed embedding $\mathcal{O}_f/\mathfrak{P} \hookrightarrow \overline{\mathbb{F}}_p$), one can attach a Galois representation, i.e. a continuous group homomorphism (for the discrete topology on $\mathrm{GL}_2(\overline{\mathbb{F}}_p)$*),*

$$\rho_f : \mathrm{Gal}(\overline{\mathbb{Q}}/\mathbb{Q}) \to \mathrm{GL}_2(\overline{\mathbb{F}}_p)$$

such that

(i) ρ_f *is semi-simple,*
(ii) $\det(\rho_f(c)) = -1$ *for any complex conjugation* $c \in \mathrm{Gal}(\overline{\mathbb{Q}}/\mathbb{Q})$ *(one says that* ρ_f *is* odd*),*
(iii) for all primes $\ell \nmid Np$ *the representation* ρ_f *is unramified at* ℓ*,*

$$\mathrm{tr}(\rho_f(\mathrm{Frob}_\ell)) \equiv a_\ell(f) \mod \mathfrak{P} \quad \textit{and} \quad \det(\rho_f(\mathrm{Frob}_\ell)) \equiv \ell^{k-1}\chi(\ell) \mod \mathfrak{P}.$$

Translation to Number Fields

Proposition 1.19 *Let* f, $\mathbb{Q}_f$, $\mathfrak{P}$ *and* ρ_f *be as in Theorem 1.18. Then the following hold:*

(a) The image of ρ_f *is finite and its image is contained in* $\mathrm{GL}_2(\mathbb{F}_{p^r})$ *for some* r*.*
(b) The kernel of ρ_f *is an open subgroup of* $\mathrm{Gal}(\overline{\mathbb{Q}}/\mathbb{Q})$ *and is hence of the form* $\mathrm{Gal}(\overline{\mathbb{Q}}/K)$ *for some Galois number field* K*. Thus, we can and do consider* $\mathrm{Gal}(K/\mathbb{Q})$ *as a subgroup of* $\mathrm{GL}_2(\mathbb{F}_{p^r})$*.*
(c) The characteristic polynomial of Frob_ℓ *(more precisely, of* $\mathrm{Frob}_{\Lambda/\ell}$ *for any prime* Λ *of* K *dividing* ℓ*) is equal to* $X^2 - a_\ell(f)X + \chi(\ell)\ell^{k-1} \mod \mathfrak{P}$ *for all primes* $\ell \nmid Np$*.*

Proof Exercise 1.36. □

To appreciate the information obtained from the $a_\ell(f) \mod \mathfrak{P}$, the reader is invited to do Exercise 1.37 now.

Images of Galois Representations

One can also often tell what the Galois group $\mathrm{Gal}(K/\mathbb{Q})$ is as an abstract group. There are not so many possibilities, as we see from the following theorem.

Theorem 1.20 (Dickson) *Let* p *be a prime and* H *a finite subgroup of* $\mathrm{PGL}_2(\overline{\mathbb{F}}_p)$*. Then a conjugate of* H *is isomorphic to one of the following groups:*

- *finite subgroups of the upper triangular matrices,*
- $\mathrm{PSL}_2(\mathbb{F}_{p^r})$ *or* $\mathrm{PGL}_2(\mathbb{F}_{p^r})$ *for* $r \in \mathbb{N}$*,*
- *dihedral groups* D_r *for* $r \in \mathbb{N}$ *not divisible by* p*,*
- A_4*,* A_5 *or* S_4*.*

For modular forms, there are several results mostly by Ribet concerning the groups that occur as images [18]. Roughly speaking, they say that the image is 'as big as possible' for almost all $\mathfrak{P}$ (for a given f). For modular forms without CM and inner twists (we do not define these notions in this course) this means that if G is the image, then G modulo scalars is equal to $\mathrm{PSL}_2(\mathbb{F}_{p^r})$ or $\mathrm{PGL}_2(\mathbb{F}_{p^r})$, where $\mathbb{F}_{p^r}$ is the extension of $\mathbb{F}_p$ generated by the $a_n(f) \mod \mathfrak{P}$.

An interesting question is to study which groups (i.e. which $\mathrm{PSL}_2(\mathbb{F}_{p^r})$) actually occur. It would be nice to prove that all of them do, since—surprisingly—the simple groups $\mathrm{PSL}_2(\mathbb{F}_{p^r})$ are still resisting a lot to all efforts to realise them as Galois groups over $\mathbb{Q}$ in the context of inverse Galois theory.

Serre's Modularity Conjecture

Serre's modularity conjecture is the following. Let p be a prime and $\rho : \mathrm{Gal}(\overline{\mathbb{Q}}/\mathbb{Q}) \to \mathrm{GL}_2(\overline{\mathbb{F}}_p)$ be a continuous, odd, irreducible representation.

- Let N_ρ be the (outside of p) conductor of ρ (defined by a formula analogous to the formula for the Artin conductor, except that the local factor for p is dropped).
- Let k_ρ be the integer defined by [17].
- Let χ_ρ be the prime-to-p part of $\det \circ \rho$ considered as a character $(\mathbb{Z}/N_\rho\mathbb{Z})^\times \times (\mathbb{Z}/p\mathbb{Z})^\times \to \overline{\mathbb{F}}_p^\times$.

Theorem 1.21 (Khare, Wintenberger, Kisin: Serre's Modularity Conjecture) *Let p be a prime and $\rho : \mathrm{Gal}(\overline{\mathbb{Q}}/\mathbb{Q}) \to \mathrm{GL}_2(\overline{\mathbb{F}}_p)$ be a continuous, odd, irreducible representation.*

Then there exists a normalised eigenform

$$f \in \mathrm{S}_{k_\rho}(N_\rho, \chi_\rho\,; \mathbb{C})$$

such that ρ is isomorphic to the Galois representation

$$\rho_f : \mathrm{Gal}(\overline{\mathbb{Q}}/\mathbb{Q}) \to \mathrm{GL}_2(\overline{\mathbb{F}}_p)$$

attached to f by Theorem 1.18.

Serre's modularity conjecture implies that we can compute (in principle, at least) arithmetic properties of all Galois representations of the type in Serre's conjecture by computing the mod p Hecke eigenforms they come from. Conceptually, Serre's modularity conjecture gives an explicit description of all irreducible, odd and continuous 'mod p' representations of $\mathrm{Gal}(\overline{\mathbb{Q}}/\mathbb{Q})$ and, thus, in a sense generalises class field theory.

Edixhoven et al. [12] have succeeded in giving an algorithm which computes the actual Galois representation attached to a mod p modular form. Hence, with Serre's conjecture we have a way of—in principle—obtaining all information on 2-dimensional irreducible, odd continuous representations of $\mathrm{Gal}(\overline{\mathbb{Q}}/\mathbb{Q})$.

1.5 Theory: Exercises

Exercise 1.22 (a) The group homomorphism

$$\mathrm{SL}_2(\mathbb{Z}) \to \mathrm{SL}_2(\mathbb{Z}/N\mathbb{Z})$$

given by reducing the matrices modulo N is surjective.
(b) Check the bijections

$$\mathrm{SL}_2(\mathbb{Z})/\Gamma_1(N) = \{(\begin{smallmatrix} a \\ c \end{smallmatrix}) \,|\, \langle a, c\rangle = \mathbb{Z}/N\mathbb{Z}\}$$

and

$$\mathrm{SL}_2(\mathbb{Z})/\Gamma_0(N) = \mathbb{P}^1(\mathbb{Z}/N\mathbb{Z}),$$

which were given in the beginning.

Exercise 1.23 Let N be an integer and $\zeta_N \in \mathbb{C}$ any primitive N-th root of unity. Prove that the map

$$\mathrm{Gal}(\mathbb{Q}(\zeta_N)/\mathbb{Q}) \xrightarrow{\mathrm{Frob}_\ell \mapsto \ell} (\mathbb{Z}/N\mathbb{Z})^\times$$

(for all primes $\ell \nmid N$) is an isomorphism.

Exercise 1.24 Prove that a matrix σ_a as in Eq. 2 exists.

Exercise 1.25 Prove Lemma 1.1. See also [10, Proposition 5.2.2].

Exercise 1.26 (a) Let K be a field, V a vector space and T_1, T_2 two commuting endomorphisms of V, i.e. $T_1 T_2 = T_2 T_1$. Let $\lambda_1 \in K$ and consider the λ_1-eigenspace of T_1, i.e. $V_1 = \{v | T_1 v = \lambda_1 v\}$. Prove that $T_2 V_1 \subseteq V_1$.
(b) Suppose that $\mathrm{M}_N(\Gamma_1(k)\,;\,\mathbb{C})$ is non-zero. Prove that it contains a Hecke eigenform.

Exercise 1.27 Prove Lemma 1.3.
Hint: use the action of Hecke operators explicitly described on q-expansions.

Exercise 1.28 Check that it makes sense to replace $\mathrm{SL}_2(\mathbb{Z})$ by $\mathrm{PSL}_2(\mathbb{Z})$ in the definition of modular forms.

Hint: for the transformation rule: if -1 is not in the congruence subgroup in question, there is nothing to show; if -1 is in it, one has to verify that it acts trivially. Moreover, convince yourself that the holomorphy at the cusps does not depend on replacing a matrix by its negative.

Exercise 1.29 Let R be a ring, Γ a group and V a left $R[\Gamma]$-module.

(a) Define the augmentation ideal I_Γ by the exact sequence

$$0 \to I_\Gamma \to R[\Gamma] \xrightarrow{\gamma \mapsto 1} R \to 1.$$

Prove that I_Γ is the ideal in $R[\Gamma]$ generated by the elements $1 - g$ for $g \in \Gamma$.

(b) Conclude that $V_\Gamma = V/I_\Gamma V$.
(c) Conclude that $V_\Gamma \cong R \otimes_{R[\Gamma]} V$.
(d) Suppose that $\Gamma = \langle T \rangle$ is a cyclic group (either finite or infinite (isomorphic to $(\mathbb{Z}, +)$)). Prove that I_Γ is the ideal generated by $(1 - T)$.
(e) Prove that $V^\Gamma \cong \mathrm{Hom}_{R[\Gamma]}(R, V)$.

Exercise 1.30 Let R, Γ and V as in Definition 1.5 and let $R \to S$ be a ring homomorphism.

(a) Prove that

$$\mathcal{M}_R(\Gamma, V) \otimes_R S \cong \mathcal{M}_S(\Gamma, V \otimes_R S).$$

(b) Suppose $R \to S$ is flat. Prove a similar statement for the cuspidal subspace.
(c) Are similar statements true for the boundary or the Eisenstein space? What about the +- and the −-spaces?

Exercise 1.31 Prove that the map

$$\mathrm{Sym}^n(R^2) \to R[X, Y]_n, \quad \binom{a_1}{b_1} \otimes \cdots \otimes \binom{a_n}{b_n} \mapsto (a_1X + b_1Y) \cdots (a_nX + b_nY)$$

is an isomorphism, where $\mathrm{Sym}^n(R^2)$ is the n-th symmetric power of R^2, which is defined as the quotient of $\underbrace{R^2 \otimes_R \cdots \otimes_R R^2}_{n\text{-times}}$ by the span of all elements $v_1 \otimes \cdots \otimes v_n - v_{\sigma(1)} \otimes \cdots \otimes v_{\sigma(n)}$ for all σ in the symmetric group on the letters $\{1, 2, \ldots, n\}$.

Exercise 1.32 Prove Eq. 8.

Exercise 1.33 Can one use +- or −-spaces in Proposition 1.8? What could we say if we defined the +-space as $M/(1 - \eta)M$ with M standing for some space of modular symbols?

Exercise 1.34 Which statements in the spirit of Corollary 1.12(b) are true for the +-spaces?

Exercise 1.35 Prove Corollary 1.16.

Exercise 1.36 Prove Proposition 1.19.

Exercise 1.37 In how far is a conjugacy class in $\mathrm{GL}_2(\mathbb{F}_{p^r})$ determined by its characteristic polynomial? Same question as above for a subgroup $G \subset \mathrm{GL}_2(\mathbb{F}_{p^r})$.

1.6 Computer Exercises

Computer Exercise 1.38 (a) Create a list L of all primes in between 234325 and 3479854? How many are there?

(b) For $n = 2, 3, 4, 5, 6, 7, 997$ compute for each $a \in \mathbb{Z}/n\mathbb{Z}$ how often it appears as a residue in the list L.

Computer Exercise 1.39 In this exercise, you verify the validity of the prime number theorem.

(a) Write a function `NumberOfPrimes` with the following specifications. Input: Positive integers a, b with $a \leq b$. Output: The number of primes in $[a, b]$.
(b) Write a function `TotalNumberOfPrimes` with the following specifications. Input: Positive integers x, s. Output: A list $[n_1, n_2, n_3, \ldots, n_m]$ such that n_i is the number of primes between 1 and $i \cdot s$ and m is the largest integer smaller than or equal to x/s.
(c) Compare the output of `TotalNumberOfPrimes` with the predictions of the prime number theorem: Make a function that returns the list $[r_1, r_2, \ldots, r_m]$ with $r_i = \frac{si}{\log si}$. Make a function that computes the quotient of two lists of 'numbers'.
(d) Play with these functions. What do you observe?

Computer Exercise 1.40 Write a function `ValuesInField` with: Input: a unitary polynomial f with integer coefficients and K a finite field. Output: the set of values of f in K.

Computer Exercise 1.41 (a) Write a function `BinaryExpansion` that computes the binary expansion of a positive integer. Input: positive integer n. Output: list of 0's and 1's representing the binary expansion.
(b) Write a function `Expo` with: Input: two positive integers a, b. Output a^b. You must not use the in-built function a^b, but write a sensible algorithm making use of the binary expansion of b. The only arithmetic operations allowed are multiplications.
(c) Write similar functions using the expansion with respect to a general base d.

Computer Exercise 1.42 In order to contemplate recursive algorithms, the monks in Hanoi used to play the following game. First they choose a degree of contemplation, i.e. a positive integer n. Then they create three lists:

$$L_1 := [n, n-1, \ldots, 2, 1]; \; L_2 := []; \; L_3 := [];$$

The aim is to exchange L_1 and L_2. However, the monks may only perform the following step: Remove the last element from one of the lists and append it to one of the other lists, subject to the important condition that in all steps all three lists must be descending.

Contemplate how the monks can achieve their goal. Write a procedure with input n that plays the game. After each step, print the number of the step, the three lists and test whether all lists are still descending.

[Hint: For recursive procedures, i.e. procedures calling themselves, in MAGMA one must put the command `forward my_procedure` in front of the definition of `my_procedure`.]

Computer Exercise 1.43 This exercise concerns the normalised cuspidal eigenforms in weight 2 and level 23.

(a) What is the number field K generated by the coefficients of each of the two forms?
(b) Compute the characteristic polynomials of the first 100 Fourier coefficients of each of the two forms.
(c) Write a function that for a given prime p computes the reduction modulo p of the characteristic polynomials from the previous point and their factorisation.
(d) Now use modular symbols over $\mathbb{F}_p$ for a given p. Compare the results.
(e) Now do the same for weight 2 and level 37. In particular, try $p = 2$. What do you observe? What could be the reason for this behaviour?

Computer Exercise 1.44 Implement Algorithm 1.14.

2 Hecke Algebras

An important point made in the previous section is that for computing modular forms, one computes Hecke algebras. This perspective puts Hecke algebras in its centre. The present section is written from that point of view. Starting from Hecke algebras, we define modular forms with coefficients in arbitrary rings, we study integrality properties and also present results on the structure of Hecke algebras, which are very useful for studying the arithmetic of modular forms.

It is essential for studying arithmetic properties of modular forms to have some flexibility for the coefficient rings. For instance, when studying mod p Galois representations attached to modular forms, it is often easier and sometimes necessary to work with modular forms whose q-expansions already lie in a finite field. Moreover, the concept of congruences of modular forms only gets its seemingly correct framework when working over rings such as extensions of finite fields or rings like $\mathbb{Z}/p^n\mathbb{Z}$.

There is a very strong theory of modular forms over a general ring R that uses algebraic geometry over R. One can, however, already get very far if one just defines modular forms over R as the R-linear dual of the $\mathbb{Z}$-Hecke algebra of the holomorphic modular forms, i.e. by taking q-expansions with coefficients in R. In this course we shall only use this. Precise definitions will be given in a moment. A priori it is maybe not clear whether non-trivial modular forms with q-expansions in the integers exist at all. The situation is as good as it could possibly be: the modular forms with q-expansion in the integers form a lattice in the space of all modular forms (at least for $\Gamma_1(N)$ and $\Gamma_0(N)$; if we are working with a Dirichlet character, the situation is slightly more involved). This is an extremely useful and important fact, which we shall derive from the corollaries of the Eichler–Shimura isomorphism given in the previous section.

Hecke algebras of modular forms over R are finitely generated R-modules. This leads us to a study, belonging to the theory of Commutative Algebra, of finite R-

algebras, that is, R-algebras that are finitely generated as R-modules. We shall prove structure theorems when R is a discrete valuation ring or a finite field. Establishing back the connection with modular forms, we will, for example, see that the maximal ideals of Hecke algebras correspond to Galois conjugacy classes of normalised eigenforms, and, for instance, the notion of a congruence can be expressed as a maximal prime containing two minimal ones.

2.1 Theory: Hecke Algebras and Modular Forms over Rings

We start by recalling and slightly extending the concept of Hecke algebras of modular forms. It is of utmost importance for our treatment of modular forms over general rings and their computation. In fact, as pointed out a couple of times, we will compute Hecke algebras and not modular forms. We shall assume that $k \geq 1$ and $N \geq 1$.

As in the introduction, we define the *Hecke algebra* of $\mathrm{M}_k(\Gamma_1(N)\,;\,\mathbb{C})$ as the subring (i.e. the $\mathbb{Z}$-algebra) inside the endomorphism ring of the $\mathbb{C}$-vector space $\mathrm{M}_k(\Gamma_1(N)\,;\,\mathbb{C})$ generated by all Hecke operators. Remember that due to Formula 7 all diamond operators are contained in the Hecke algebra. Of course, we make similar definitions for $\mathrm{S}_k(\Gamma_1(N)\,;\,\mathbb{C})$ and use the notations $\mathbb{T}_{\mathbb{Z}}(\mathrm{M}_k(\Gamma_1(N)\,;\,\mathbb{C}))$ and $\mathbb{T}_{\mathbb{Z}}(\mathrm{S}_k(\Gamma_1(N)\,;\,\mathbb{C}))$.

If we are working with modular forms with a character, we essentially have two possibilities for defining the Hecke algebra, namely, first as above as the $\mathbb{Z}$-algebra generated by all Hecke operators inside the endomorphism ring of the $\mathbb{C}$-vector space $\mathrm{M}_k(N, \chi\,;\,\mathbb{C})$ (notation $\mathbb{T}_{\mathbb{Z}}(\mathrm{M}_k(N, \chi\,;\,\mathbb{C}))$) or, second, as the $\mathbb{Z}[\chi]$-algebra generated by the Hecke operators inside $\mathrm{End}_{\mathbb{C}}(\mathrm{M}_k(N, \chi\,;\,\mathbb{C}))$ (notation $\mathbb{T}_{\mathbb{Z}[\chi]}(\mathrm{M}_k(N, \chi\,;\,\mathbb{C}))$); similarly for the cusp forms. Here $\mathbb{Z}[\chi]$ is the ring extension of $\mathbb{Z}$ generated by all values of χ, it is the integer ring of $\mathbb{Q}(\chi)$. For two reasons we prefer the second variant. The first reason is that we needed to work over $\mathbb{Z}[\chi]$ (or its extensions) for modular symbols. The second reason is that on the natural $\mathbb{Z}$-structure inside $\mathrm{M}_k(\Gamma_1(N)\,;\,\mathbb{C})$ the decomposition into $(\mathbb{Z}/N\mathbb{Z})^\times$-eigenspaces can only be made after a base change to $\mathbb{Z}[\chi]$. So, the $\mathbb{C}$-dimension of $\mathrm{M}_k(N, \chi\,;\,\mathbb{C})$ equals the $\mathbb{Q}[\chi]$-dimension of $\mathbb{T}_{\mathbb{Q}[\chi]}(\mathrm{M}_k(N, \chi\,;\,\mathbb{C}))$ and not the $\mathbb{Q}$-dimension of $\mathbb{T}_{\mathbb{Q}}(\mathrm{M}_k(N, \chi\,;\,\mathbb{C}))$.

Lemma 2.1 *(a) The $\mathbb{Z}$-algebras $\mathbb{T}_{\mathbb{Z}}(\mathrm{M}_k(\Gamma_1(N)\,;\,\mathbb{C}))$ and $\mathbb{T}_{\mathbb{Z}}(\mathrm{M}_k(N, \chi\,;\,\mathbb{C}))$ are free $\mathbb{Z}$-modules of finite rank; the same holds for the cuspidal Hecke algebras.*
(b) The $\mathbb{Z}[\chi]$-algebra $\mathbb{T}_{\mathbb{Z}[\chi]}(\mathrm{M}_k(N, \chi\,;\,\mathbb{C}))$ is a torsion-free finitely generated $\mathbb{Z}[\chi]$-module; the same holds for the cuspidal Hecke algebra.

Proof (a) Due to the corollaries of the Eichler–Shimura theorem (Corollary 1.11) we know that these algebras are finitely generated as $\mathbb{Z}$-modules. As they lie inside a vector space, they are free (using the structure theory of finitely generated modules over principal ideal domains).

(b) This is like (a), except that $\mathbb{Z}[\chi]$ need not be a principal ideal domain, so that we can only conclude torsion-freeness, but not freeness. □

Modular Forms over Rings

Let $k \geq 1$ and $N \geq 1$. Let R be any $\mathbb{Z}$-algebra (ring). We now use the q-pairing to define modular (cusp) forms over R. We let

$$\begin{aligned} \mathrm{M}_k(\Gamma_1(N)\,;\, R) &:= \mathrm{Hom}_{\mathbb{Z}}(\mathbb{T}_{\mathbb{Z}}(\mathrm{M}_k(\Gamma_1(N)\,;\, \mathbb{C})), R) \\ &\cong \mathrm{Hom}_R(\mathbb{T}_{\mathbb{Z}}(\mathrm{M}_k(\Gamma_1(N)\,;\, \mathbb{C})) \otimes_{\mathbb{Z}} R, R). \end{aligned}$$

We stress the fact that Hom_R denotes the homomorphisms as R-modules (and not as R-algebras; those will appear later). The isomorphism is proved precisely as in Proposition 1.8(c), where we did not use the flatness assumption. Every element f of $\mathrm{M}_k(\Gamma_1(N)\,;\, R)$ thus corresponds to a $\mathbb{Z}$-linear function $\Phi : \mathbb{T}_{\mathbb{Z}}(\mathrm{M}_k(\Gamma_1(N)\,;\, \mathbb{C})) \to R$ and is uniquely identified by its *formal q-expansion*

$$f = \sum_n \Phi(T_n) q^n = \sum_n a_n(f) q^n \in R[[q]].$$

We note that $\mathbb{T}_{\mathbb{Z}}(\mathrm{M}_k(\Gamma_1(N)\,;\, \mathbb{C}))$ acts naturally on $\mathrm{Hom}_{\mathbb{Z}}(\mathbb{T}_{\mathbb{Z}}(\mathrm{M}_k(\Gamma_1(N)\,;\, \mathbb{C})), R)$, namely by

$$(T.\Phi)(S) = \Phi(TS) = \Phi(ST). \tag{10}$$

This means that the action of $\mathbb{T}_{\mathbb{Z}}(\mathrm{M}_k(\Gamma_1(N)\,;\, \mathbb{C}))$ on $\mathrm{M}_k(\Gamma_1(N)\,;\, R)$ gives the same formulae as usual on formal q-expansions. For cusp forms we make the obvious analogous definition, i.e.

$$\begin{aligned} \mathrm{S}_k(\Gamma_1(N)\,;\, R) &:= \mathrm{Hom}_{\mathbb{Z}}(\mathbb{T}_{\mathbb{Z}}(\mathrm{S}_k(\Gamma_1(N)\,;\, \mathbb{C})), R) \\ &\cong \mathrm{Hom}_R(\mathbb{T}_{\mathbb{Z}}(\mathrm{S}_k(\Gamma_1(N)\,;\, \mathbb{C})) \otimes_{\mathbb{Z}} R, R). \end{aligned}$$

We caution the reader that for modular forms which are not cusp forms there also ought to be some 0th coefficient in the formal q-expansion, for example, for recovering the classical holomorphic q-expansion. Of course, for cusp forms, we do not need to worry.

Now we turn our attention to modular forms with a character. Let $\chi : (\mathbb{Z}/N\mathbb{Z})^\times \to \mathbb{C}^\times$ be a Dirichlet character and $\mathbb{Z}[\chi] \to R$ a ring homomorphism. We now proceed analogously to the treatment of modular symbols for a Dirichlet character. We work with $\mathbb{Z}[\chi]$ as the base ring (and not $\mathbb{Z}$). We let

$$\begin{aligned} \mathrm{M}_k(N, \chi\,;\, R) &:= \mathrm{Hom}_{\mathbb{Z}[\chi]}(\mathbb{T}_{\mathbb{Z}[\chi]}(\mathrm{M}_k(N, \chi\,;\, \mathbb{C})), R) \\ &\cong \mathrm{Hom}_R(\mathbb{T}_{\mathbb{Z}[\chi]}(\mathrm{M}_k(N, \chi\,;\, \mathbb{C})) \otimes_{\mathbb{Z}[\chi]} R, R) \end{aligned}$$

and similarly for the cusp forms.

We remark that these definitions of $\mathrm{M}_k(\Gamma_1(N)\,;\, \mathbb{C})$, $\mathrm{M}_k(N, \chi\,;\, \mathbb{C})$ etc. agree with those from Sect. 1; thus, it is justified to use the same pieces of notation. As a special case, we get that $\mathrm{M}_k(\Gamma_1(N)\,;\, \mathbb{Z})$ precisely consists of those holomorphic modular forms in $\mathrm{M}_k(\Gamma_1(N)\,;\, \mathbb{C})$ whose q-expansions take values in $\mathbb{Z}$.

If $\mathbb{Z}[\chi] \xrightarrow{\pi} R = \mathbb{F}$ with $\mathbb{F}$ a finite field of characteristic p or $\overline{\mathbb{F}}_p$, we call $\mathrm{M}_k(N, \chi\,;\, \mathbb{F})$ the space of *mod p modular forms of weight k, level N and character* χ. Of course, for the cuspidal space similar statements are made and we use similar notation.

We furthermore extend the notation for Hecke algebras introduced in Sect. 1 as follows. If S is an R-algebra and M is an S-module admitting the action of Hecke operators T_n for $n \in \mathbb{N}$, then we let $\mathbb{T}_R(M)$ be the R-subalgebra of $\mathrm{End}_S(M)$ generated by all T_n for $n \in \mathbb{N}$.

We now study base change properties of modular forms over R.

Proposition 2.2 *(a) Let $\mathbb{Z} \to R \to S$ be ring homomorphisms. Then the following statements hold.*

(i) The natural map

$$\mathrm{M}_k(\Gamma_1(N)\,;\, R) \otimes_R S \to \mathrm{M}_k(\Gamma_1(N)\,;\, S)$$

is an isomorphism.

(ii) The evaluation pairing

$$\mathrm{M}_k(\Gamma_1(N)\,;\, R) \times \mathbb{T}_\mathbb{Z}(\mathrm{M}_k(\Gamma_1(N)\,;\, \mathbb{C})) \otimes_\mathbb{Z} R \to R$$

is the q-pairing and it is perfect.

(iii) The Hecke algebra $\mathbb{T}_R(\mathrm{M}_k(\Gamma_1(N)\,;\, R))$ is naturally isomorphic to $\mathbb{T}_\mathbb{Z}(\mathrm{M}_k(\Gamma_1(N)\,;\, \mathbb{C})) \otimes_\mathbb{Z} R$.

(b) If $\mathbb{Z}[\chi] \to R \to S$ are flat, then Statement (i) holds for $\mathrm{M}_k(N, \chi\,;\, R)$.

(c) If $\mathbb{T}_{\mathbb{Z}[\chi]}(\mathrm{M}_k(N, \chi\,;\, \mathbb{C}))$ is a free $\mathbb{Z}[\chi]$-module and $\mathbb{Z}[\chi] \to R \to S$ are ring homomorphisms, statements (i)-(iii) hold for $\mathrm{M}_k(N, \chi\,;\, R)$.

Proof (a) We use the following general statement, in which M is assumed to be a free finitely generated R-module and N, T are R-modules:

$$\mathrm{Hom}_R(M, N) \otimes_R T \cong \mathrm{Hom}_R(M, N \otimes_R T).$$

To see this, just see M as $\bigoplus R$ and pull the direct sum out of the Hom, do the tensor product, and put the direct sum back into the Hom.

(i) Write $\mathbb{T}_\mathbb{Z}$ for $\mathbb{T}_\mathbb{Z}(\mathrm{M}_k(\Gamma_1(N)\,;\, \mathbb{C}))$. It is a free $\mathbb{Z}$-module by Lemma 2.1. We have

$$\mathrm{M}_k(\Gamma_1(N)\,;\, R) \otimes_R S = \mathrm{Hom}_\mathbb{Z}(\mathbb{T}_\mathbb{Z}, R) \otimes_R S,$$

which by the above is isomorphic to $\mathrm{Hom}_\mathbb{Z}(\mathbb{T}_\mathbb{Z}, R \otimes_R S)$ and hence to $\mathrm{M}_k(\Gamma_1(N)\,;\, S)$.

(ii) The evaluation pairing $\mathrm{Hom}_\mathbb{Z}(\mathbb{T}_\mathbb{Z}, \mathbb{Z}) \times \mathbb{T}_\mathbb{Z} \to \mathbb{Z}$ is perfect, since $\mathbb{T}_\mathbb{Z}$ is free as a $\mathbb{Z}$-module. The result follows from (i) by tensoring with R.

(iii) We consider the natural map

$$\mathbb{T}_\mathbb{Z} \otimes_\mathbb{Z} R \to \mathrm{End}_R(\mathrm{Hom}_R(\mathbb{T}_\mathbb{Z} \otimes_\mathbb{Z} R, R))$$

and show that it is injective. Its image is by definition $\mathbb{T}_R(\mathrm{M}_k(\Gamma_1(N)\,;\,R))$. Let T be in the kernel. Then $\phi(T) = 0$ for all $\phi \in \mathrm{Hom}_R(\mathbb{T}_\mathbb{Z} \otimes_\mathbb{Z} R, R)$. As the pairing in (ii) is perfect and, in particular, non-degenerate, $T = 0$ follows.

(b) Due to flatness we have

$$\mathrm{Hom}_R(\mathbb{T}_{\mathbb{Z}[\chi]} \otimes_{\mathbb{Z}[\chi]} R, R) \otimes_R S \cong \mathrm{Hom}_S(\mathbb{T}_{\mathbb{Z}[\chi]} \otimes_{\mathbb{Z}[\chi]} S, S),$$

as desired.

(c) The same arguments as in (a) work. □

Galois Conjugacy Classes

By the definition of the Hecke action in Eq. (10), the normalised Hecke eigenforms in $\mathrm{M}_k(\Gamma_1(N)\,;\,R)$ are precisely the set of $\mathbb{Z}$-algebra homomorphisms inside $\mathrm{Hom}_\mathbb{Z}(\mathbb{T}_\mathbb{Z}(\mathrm{M}_k(\Gamma_1(N)\,;\,\mathbb{C})), R)$, where the normalisation means that the identity operator T_1 is sent to 1. Such an algebra homomorphism Φ is often referred to as a *system of eigenvalues*, since the image of each T_n corresponds to an eigenvalue of T_n, namely to $\Phi(T_n) = a_n(f)$ (if f corresponds to Φ).

Let us now consider a perfect field K (if we are working with a Dirichlet character, we also want that K admits a ring homomorphism $\mathbb{Z}[\chi] \to K$). Denote by $\overline{K}$ an algebraic closure, so that we have

$$\begin{aligned}\mathrm{M}_k(\Gamma_1(N)\,;\,\overline{K}) &= \mathrm{Hom}_\mathbb{Z}(\mathbb{T}_\mathbb{Z}(\mathrm{M}_k(\Gamma_1(N)\,;\,\mathbb{C})), \overline{K}) \\ &\cong \mathrm{Hom}_K(\mathbb{T}_\mathbb{Z}(\mathrm{M}_k(\Gamma_1(N)\,;\,\mathbb{C})) \otimes_\mathbb{Z} K, \overline{K}).\end{aligned}$$

We can compose any $\Phi \in \mathrm{Hom}_\mathbb{Z}(\mathbb{T}_\mathbb{Z}(\mathrm{M}_k(\Gamma_1(N)\,;\,\mathbb{C})), \overline{K})$ by any field automorphism $\sigma : \overline{K} \to \overline{K}$ fixing K. Thus, we obtain an action of the absolute Galois group $\mathrm{Gal}(\overline{K}/K)$ on $\mathrm{M}_k(\Gamma_1(N)\,;\,\overline{K})$ (on formal q-expansions, we only need to apply σ to the coefficients). All this works similarly for the cuspidal subspace, too.

Like this, we also obtain a $\mathrm{Gal}(\overline{K}/K)$-action on the normalised eigenforms, and can hence speak about *Galois conjugacy classes of eigenforms*.

Proposition 2.3 *We have the following bijective correspondences:*

$$\begin{aligned}\mathrm{Spec}(\mathbb{T}_K(\cdot)) &\overset{1-1}{\leftrightarrow} \mathrm{Hom}_{K\text{-}alg}(\mathbb{T}_K(\cdot), \overline{K})/\,\mathrm{Gal}(\overline{K}/K) \\ &\overset{1-1}{\leftrightarrow} \{\textit{normalised eigenf. in}\cdot\}/\,\mathrm{Gal}(\overline{K}/K)\end{aligned}$$

and with $K = \overline{K}$

$$\mathrm{Spec}(\mathbb{T}_{\overline{K}}(\cdot)) \overset{1-1}{\leftrightarrow} \mathrm{Hom}_{\overline{K}\text{-}alg}(\mathbb{T}_{\overline{K}}(\cdot), \overline{K}) \overset{1-1}{\leftrightarrow} \{\textit{normalised eigenforms in}\cdot\}.$$

Here, $\cdot$ stands for either $\mathrm{M}_k(\Gamma_1(N)\,;\,\overline{K})$, $\mathrm{S}_k(\Gamma_1(N)\,;\,\overline{K})$ or the respective spaces with a Dirichlet character.

We recall that Spec of a ring is the set of prime ideals. In the next section we will see that in $\mathbb{T}_K(\cdot)$ and $\mathbb{T}_{\overline{K}}(\cdot)$ all prime ideals are already maximal (it is an easy consequence of the finite dimensionality).

Proof Exercise 2.19. □

We repeat that the coefficients of any eigenform f in $\mathrm{M}_k(N, \chi\,;\, \overline{K})$ lie in a finite extension of K, namely in $\mathbb{T}_K(\mathrm{M}_k(N, \chi\,;\, K))/\mathfrak{m}$, when $\mathfrak{m}$ is the maximal ideal corresponding to the conjugacy class of f.

Let us note that the above discussion applies to $\overline{K} = \mathbb{C}$, $\overline{K} = \overline{\mathbb{Q}}$, $\overline{K} = \overline{\mathbb{Q}}_p$, as well as to $\overline{K} = \overline{\mathbb{F}}_p$. In the next sections, we will also take into account the finer structure of Hecke algebras over $\mathcal{O}$, or rather over the completion of $\mathcal{O}$ at one prime.

2.1.1 Some Commutative Algebra

In this section, we leave the special context of modular forms for a moment and provide quite useful results from commutative algebra that will be applied to Hecke algebras in the sequel.

We start with a simple case which we will prove directly. Let $\mathbb{T}$ be an *Artinian* algebra, i.e. an algebra in which every descending chain of ideals becomes stationary. Our main example will be finite dimensional algebras over a field. That those are Artinian is obvious, since in every proper inclusion of ideals the dimension diminishes.

For any ideal $\mathfrak{a}$ of $\mathbb{T}$ the sequence $\mathfrak{a}^n$ becomes stationary, i.e. $\mathfrak{a}^n = \mathfrak{a}^{n+1}$ for all n 'big enough'. Then we will use the notation $\mathfrak{a}^\infty$ for $\mathfrak{a}^n$.

Proposition 2.4 *Let $\mathbb{T}$ be an Artinian ring.*

(a) Every prime ideal of $\mathbb{T}$ is maximal.
(b) There are only finitely many maximal ideals in $\mathbb{T}$.
(c) Let $\mathfrak{m}$ be a maximal ideal of $\mathbb{T}$. It is the only maximal ideal containing $\mathfrak{m}^\infty$.
(d) Let $\mathfrak{m} \neq \mathfrak{n}$ be two maximal ideals. For any $k \in \mathbb{N}$ and $k = \infty$ the ideals $\mathfrak{m}^k$ and $\mathfrak{n}^k$ are coprime.
(e) The Jacobson radical $\bigcap_{\mathfrak{m}\in\mathrm{Spec}(\mathbb{T})} \mathfrak{m}$ is equal to the nilradical and consists of the nilpotent elements.
(f) We have $\bigcap_{\mathfrak{m}\in\mathrm{Spec}(\mathbb{T})} \mathfrak{m}^\infty = (0)$.
(g) (Chinese Remainder Theorem) The natural map

$$\mathbb{T} \xrightarrow{a\mapsto(\ldots,a+\mathfrak{m}^\infty,\ldots)} \prod_{\mathfrak{m}\in\mathrm{Spec}(\mathbb{T})} \mathbb{T}/\mathfrak{m}^\infty$$

is an isomorphism.
(h) For every maximal ideal $\mathfrak{m}$, the ring $\mathbb{T}/\mathfrak{m}^\infty$ is local with maximal ideal $\mathfrak{m}$ and is hence isomorphic to $\mathbb{T}_\mathfrak{m}$, the localisation of $\mathbb{T}$ at $\mathfrak{m}$.

Proof (a) Let $\mathfrak{p}$ be a prime ideal of $\mathbb{T}$. The quotient $\mathbb{T} \twoheadrightarrow \mathbb{T}/\mathfrak{p}$ is an Artinian integral domain, since ideal chains in $\mathbb{T}/\mathfrak{p}$ lift to ideal chains in $\mathbb{T}$. Let $0 \neq x \in \mathbb{T}/\mathfrak{p}$. We have $(x)^n = (x)^{n+1} = (x)^\infty$ for some n big enough. Hence, $x^n = yx^{n+1}$ with some $y \in \mathbb{T}/\mathfrak{p}$ and so $xy = 1$, as $\mathbb{T}/\mathfrak{p}$ is an integral domain.

(b) Assume there are infinitely many maximal ideals, number a countable subset of them by $\mathfrak{m}_1, \mathfrak{m}_2, \ldots$. Form the descending ideal chain

$$\mathfrak{m}_1 \supset \mathfrak{m}_1 \cap \mathfrak{m}_2 \supset \mathfrak{m}_1 \cap \mathfrak{m}_2 \cap \mathfrak{m}_3 \supset \ldots.$$

This chain becomes stationary, so that for some n we have

$$\mathfrak{m}_1 \cap \cdots \cap \mathfrak{m}_n = \mathfrak{m}_1 \cap \cdots \cap \mathfrak{m}_n \cap \mathfrak{m}_{n+1}.$$

Consequently, $\mathfrak{m}_1 \cap \cdots \cap \mathfrak{m}_n \subset \mathfrak{m}_{n+1}$. We claim that there is $i \in \{1, 2, \ldots, n\}$ with $\mathfrak{m}_i \subset \mathfrak{m}_{n+1}$. Due to the maximality of $\mathfrak{m}_i$ we obtain the desired contradiction. To prove the claim we assume that $\mathfrak{m}_i \not\subseteq \mathfrak{m}_{n+1}$ for all i. Let $x_i \in \mathfrak{m}_i - \mathfrak{m}_{n+1}$ and $y = x_1 \cdot x_2 \cdots x_n$. Then $y \in \mathfrak{m}_1 \cap \cdots \cap \mathfrak{m}_n$, but $y \notin \mathfrak{m}_{n+1}$ due to the primality of $\mathfrak{m}_{n+1}$, giving a contradiction.

(c) Let $\mathfrak{m} \in \operatorname{Spec}(\mathbb{T})$ be a maximal ideal. Assume that $\mathfrak{n}$ is a different maximal ideal with $\mathfrak{m}^\infty \subset \mathfrak{n}$. Choose $x \in \mathfrak{m}$. Some power $x^r \in \mathfrak{m}^\infty$ and, thus, $x^r \in \mathfrak{n}$. As $\mathfrak{n}$ is prime, $x \in \mathfrak{n}$ follows, implying $\mathfrak{m} \subseteq \mathfrak{n}$, contradicting the maximality of $\mathfrak{m}$.

(d) Assume that $I := \mathfrak{m}^k + \mathfrak{n}^k \neq \mathbb{T}$. Then I is contained in some maximal ideal $\mathfrak{p}$. Hence, $\mathfrak{m}^\infty$ and $\mathfrak{n}^\infty$ are contained in $\mathfrak{p}$, whence by (c), $\mathfrak{m} = \mathfrak{n} = \mathfrak{p}$; contradiction.

(e) It is a standard fact from Commutative Algebra that the nilradical (the ideal of nilpotent elements) is the intersection of the minimal prime ideals.

(f) For $k \in \mathbb{N}$ and $k = \infty$, (d) implies

$$\bigcap_{\mathfrak{m} \in \operatorname{Spec}(\mathbb{T})} \mathfrak{m}^k = \prod_{\mathfrak{m} \in \operatorname{Spec}(\mathbb{T})} \mathfrak{m}^k = (\prod_{\mathfrak{m} \in \operatorname{Spec}(\mathbb{T})} \mathfrak{m})^k = (\bigcap_{\mathfrak{m} \in \operatorname{Spec}(\mathbb{T})} \mathfrak{m})^k.$$

By (e) we know that $\bigcap_{\mathfrak{m} \in \operatorname{Spec}(\mathbb{T})} \mathfrak{m}$ is the nilradical. It can be generated by finitely many elements $a_1, \ldots, a_n$ all of which are nilpotent. So a high enough power of $\bigcap_{\mathfrak{m} \in \operatorname{Spec}(\mathbb{T})} \mathfrak{m}$ is zero.

(g) The injectivity follows from (f). It suffices to show that the elements of the form $(0, \ldots, 0, 1, 0, \ldots, 0)$ are in the image of the map. Suppose the 1 is at the place belonging to $\mathfrak{m}$. Due to coprimeness (d) for any maximal ideal $\mathfrak{n} \neq \mathfrak{m}$ we can find $a_\mathfrak{n} \in \mathfrak{n}^\infty$ and $a_\mathfrak{m} \in \mathfrak{m}^\infty$ such that $1 = a_\mathfrak{m} + a_\mathfrak{n}$. Let $x := \prod_{\mathfrak{n} \in \operatorname{Spec}(\mathbb{T}), \mathfrak{n} \neq \mathfrak{m}} a_\mathfrak{n}$. We have $x \in \prod_{\mathfrak{n} \in \operatorname{Spec}(\mathbb{T}), \mathfrak{n} \neq \mathfrak{m}} \mathfrak{n}^\infty$ and $x = \prod_{\mathfrak{n} \in \operatorname{Spec}(\mathbb{T}), \mathfrak{n} \neq \mathfrak{m}} (1 - a_\mathfrak{m}) \equiv 1 \mod \mathfrak{m}$. Hence, the map sends x to $(0, \ldots, 0, 1, 0, \ldots, 0)$, proving the surjectivity.

(h) By (c), the only maximal ideal of $\mathbb{T}$ containing $\mathfrak{m}^\infty$ is $\mathfrak{m}$. Consequently, $\mathbb{T}/\mathfrak{m}^\infty$ is a local ring with maximal ideal the image of $\mathfrak{m}$. Let $s \in \mathbb{T} - \mathfrak{m}$. As $s + \mathfrak{m}^\infty \notin \mathfrak{m}/\mathfrak{m}^\infty$, the element $s + \mathfrak{m}^\infty$ is a unit in $\mathbb{T}/\mathfrak{m}^\infty$. Thus, the map

$$\mathbb{T}_\mathfrak{m} \xrightarrow{\frac{y}{s} \mapsto ys^{-1} + \mathfrak{m}^\infty} \mathbb{T}/\mathfrak{m}^\infty$$

is well-defined. It is clearly surjective. Suppose $\frac{y}{s}$ maps to 0. Since the image of s is a unit, $y \in \mathfrak{m}^\infty$ follows. The element x constructed in (g) is in $\prod_{\mathfrak{n} \in \mathrm{Spec}(\mathbb{T}), \mathfrak{n} \neq \mathfrak{m}} \mathfrak{n}^\infty$, but not in $\mathfrak{m}$. By (f) and (d), $(0) = \prod_{\mathfrak{m} \in \mathrm{Spec}(\mathbb{T})} \mathfrak{m}^\infty$. Thus, $y \cdot x = 0$ and also $\frac{y}{s} = \frac{yx}{sx} = 0$, proving the injectivity. □

A useful and simple way to rephrase a product decomposition as in (g) is to use idempotents. In concrete terms, the idempotents of $\mathbb{T}$ (as in the proposition) are precisely the elements of the form $(\ldots, x_\mathfrak{m}, \ldots)$ with $x_\mathfrak{m} \in \{0, 1\} \subseteq \mathbb{T}/\mathfrak{m}^\infty$.

Definition 2.5 Let $\mathbb{T}$ be a ring. An *idempotent of* $\mathbb{T}$ is an element e that satisfies $e^2 = e$. Two idempotents e, f are *orthogonal* if $ef = 0$. An idempotent e is *primitive* if $e\mathbb{T}$ is a local ring. A set of idempotents $\{e_1, \ldots, e_n\}$ is said to be *complete* if $1 = \sum_{i=1}^n e_i$.

In concrete terms for $\mathbb{T} = \prod_{\mathfrak{m} \in \mathrm{Spec}(\mathbb{T})} \mathbb{T}/\mathfrak{m}^\infty$, a complete set of primitive pairwise orthogonal idempotents is given by

$$(1, 0, \ldots, 0), (0, 1, 0, \ldots, 0), \ldots, (0, \ldots, 0, 1, 0), (0, \ldots, 0, 1).$$

In Exercise 2.20, you are asked (among other things) to prove that in the above case $\mathfrak{m}^\infty$ is a principal ideal generated by an idempotent.

Below we will present an algorithm for computing a complete set of primitive pairwise orthogonal idempotents for an Artinian ring.

We now come to a more general setting, namely working with a finite algebra $\mathbb{T}$ over a complete local ring instead of a field. We will lift the idempotents of the reduction of $\mathbb{T}$ (for the maximal ideal of the complete local ring) to idempotents of $\mathbb{T}$ by Hensel's lemma. This gives us a proposition very similar to Proposition 2.4.

Proposition 2.6 (Hensel's lemma) *Let R be a ring that is complete with respect to the ideal $\mathfrak{m}$ and let $f \in R[X]$ be a polynomial. If*

$$f(a) \equiv 0 \mod (f'(a))^2\mathfrak{m}$$

for some $a \in R$, then there is $b \in R$ such that

$$f(b) = 0 \text{ and } b \equiv a \mod f'(a)\mathfrak{m}.$$

If $f'(a)$ is not a zero divisor, then b is unique with these properties.

Proof [13], Theorem 7.3. □

Recall that the *height* of a prime ideal $\mathfrak{p}$ in a ring R is the supremum among all $n \in \mathbb{N}$ such that there are inclusions of prime ideals $\mathfrak{p}_0 \subsetneq \mathfrak{p}_1 \subsetneq \cdots \subsetneq \mathfrak{p}_{n-1} \subsetneq \mathfrak{p}$. The *Krull dimension* of R is the supremum of the heights of the prime ideals of R.

Proposition 2.7 *Let $\mathcal{O}$ be an integral domain of characteristic zero which is a finitely generated $\mathbb{Z}$-module. Write $\widehat{\mathcal{O}}$ for the completion of $\mathcal{O}$ at a maximal prime of $\mathcal{O}$ and* *denote by $\mathbb{F}$ the residue field and by $\widehat{K}$ the fraction field of $\widehat{\mathcal{O}}$. Let furthermore $\mathbb{T}$ be a commutative $\mathcal{O}$-algebra which is finitely generated as an $\mathcal{O}$-module. For any ring homomorphism $\mathcal{O} \to S$ write $\mathbb{T}_S$ for $\mathbb{T} \otimes_{\mathcal{O}} S$. Then the following statements hold.*

(a) *The Krull dimension of $\mathbb{T}_{\widehat{\mathcal{O}}}$ is less than or equal to 1, i.e. between any prime ideal and any maximal ideal $\mathfrak{p} \subset \mathfrak{m}$ there is no other prime ideal. The maximal ideals of $\mathbb{T}_{\widehat{\mathcal{O}}}$ correspond bijectively under taking pre-images to the maximal ideals of $\mathbb{T}_{\mathbb{F}}$. Primes $\mathfrak{p}$ of height 0 (i.e. those that do not contain any other prime ideal) which are properly contained in a prime of height 1 (i.e. a maximal prime) of $\mathbb{T}_{\widehat{\mathcal{O}}}$ are in bijection with primes of $\mathbb{T}_{\widehat{K}}$ under extension (i.e. $\mathfrak{p}\mathbb{T}_{\widehat{K}}$), for which the notation $\mathfrak{p}^e$ will be used.*
Under the correspondences, one has

$$\mathbb{T}_{\mathbb{F},\mathfrak{m}} \cong \mathbb{T}_{\widehat{\mathcal{O}},\mathfrak{m}} \otimes_{\widehat{\mathcal{O}}} \mathbb{F}$$

and

$$\mathbb{T}_{\widehat{\mathcal{O}},\mathfrak{p}} \cong \mathbb{T}_{\widehat{K},\mathfrak{p}^e}.$$

(b) *The algebra $\mathbb{T}_{\widehat{\mathcal{O}}}$ decomposes as*

$$\mathbb{T}_{\widehat{\mathcal{O}}} \cong \prod_{\mathfrak{m}} \mathbb{T}_{\widehat{\mathcal{O}},\mathfrak{m}},$$

where the product runs over the maximal ideals $\mathfrak{m}$ of $\mathbb{T}_{\widehat{\mathcal{O}}}$.

(c) *The algebra $\mathbb{T}_{\mathbb{F}}$ decomposes as*

$$\mathbb{T}_{\mathbb{F}} \cong \prod_{\mathfrak{m}} \mathbb{T}_{\mathbb{F},\mathfrak{m}},$$

where the product runs over the maximal ideals $\mathfrak{m}$ of $\mathbb{T}_{\mathbb{F}}$.

(d) *The algebra $\mathbb{T}_{\widehat{K}}$ decomposes as*

$$\mathbb{T}_{\widehat{K}} \cong \prod_{\mathfrak{p}} \mathbb{T}_{\widehat{K},\mathfrak{p}^e} \cong \prod_{\mathfrak{p}} \mathbb{T}_{\widehat{\mathcal{O}},\mathfrak{p}},$$

where the products run over the minimal prime ideals $\mathfrak{p}$ of $\mathbb{T}_{\widehat{\mathcal{O}}}$ which are contained in a prime ideal of height 1.

Proof We first need that $\widehat{\mathcal{O}}$ has Krull dimension 1. This, however, follows from the fact that $\mathcal{O}$ has Krull dimension 1, as it is an integral extension of $\mathbb{Z}$, and the correspondence between the prime ideals of a ring and its completion. As $\mathbb{T}_{\widehat{\mathcal{O}}}$ is a finitely generated $\widehat{\mathcal{O}}$-module, $\mathbb{T}_{\widehat{\mathcal{O}}}/\mathfrak{p}$ with a prime $\mathfrak{p}$ is an integral domain which is a finitely generated $\widehat{\mathcal{O}}/(\mathfrak{p} \cap \widehat{\mathcal{O}})$-module. Hence, it is either a finite field (when the prime ideal $\mathfrak{p} \cap \widehat{\mathcal{O}}$ is the unique maximal ideal of $\widehat{\mathcal{O}}$) or a finite extension of $\widehat{\mathcal{O}}$ (when $\mathfrak{p} \cap \widehat{\mathcal{O}} = 0$ so that the structure map $\widehat{\mathcal{O}} \to \mathbb{T}_{\widehat{\mathcal{O}}}/\mathfrak{p}$ is injective). This proves that the height of $\mathfrak{p}$ is less than or equal to 1. The correspondences and the isomorphisms of Part (a) are the subject of Exercise 2.21.

We have already seen Parts (c) and (d) in Lemma 2.4. Part (b) follows from (c) by applying Hensel's lemma (Proposition 2.6) to the idempotents of the decomposition of (c). We follow [13], Corollary 7.5, for the details. Since $\widehat{\mathcal{O}}$ is complete with respect to some ideal $\mathfrak{p}$, so is $\mathbb{T}_{\widehat{\mathcal{O}}}$ as it is a finitely generated $\widehat{\mathcal{O}}$-module. Hence, we may use Hensel's lemma in $\mathbb{T}_{\widehat{\mathcal{O}}}$. Given an idempotent $\overline{e}$ of $\mathbb{T}_{\mathbb{F}}$, we will first show that it lifts to a unique idempotent of $\mathbb{T}_{\widehat{\mathcal{O}}}$. Let e be any lift of $\overline{e}$ and let $f(X) = X^2 - X$ be a polynomial annihilating $\overline{e}$. We have that $f'(e) = 2e - 1$ is a unit since $(2e-1)^2 = 4e^2 - 4e + 1 \equiv 1 \mod \mathfrak{p}$. Hensel's lemma now gives us a unique root $e_1 \in \mathbb{T}_{\widehat{\mathcal{O}}}$ of f, i.e. an idempotent, lifting $\overline{e}$.

We now lift every element of a set of pairwise orthogonal idempotents of $\mathbb{T}_{\mathbb{F}}$. It now suffices to show that the lifted idempotents are also pairwise orthogonal (their sum is 1; otherwise we would get a contradiction in the correspondences in (a): there cannot be more idempotents in $\mathbb{T}_{\widehat{\mathcal{O}}}$ than in $\mathbb{T}_{\mathbb{F}}$). As their reductions are orthogonal, a product $e_i e_j$ of lifted idempotents is in $\mathfrak{p}$. Hence, $e_i e_j = e_i^d e_j^d \in \mathfrak{p}^d$ for all d, whence $e_i e_j = 0$, as desired. □

2.1.2 Commutative Algebra of Hecke Algebras

Let $k \geq 1$, $N \geq 1$ and $\chi : (\mathbb{Z}/N\mathbb{Z})^\times \to \mathbb{C}^\times$. Moreover, let p be a prime, $\mathcal{O} := \mathbb{Z}[\chi]$, $\mathfrak{P}$ a maximal prime of $\mathcal{O}$ above p, and let $\mathbb{F}$ be the residue field of $\mathcal{O}$ modulo $\mathfrak{P}$. We let $\widehat{\mathcal{O}}$ denote the completion of $\mathcal{O}$ at $\mathfrak{P}$. Moreover, the field of fractions of $\widehat{\mathcal{O}}$ will be denoted by $\widehat{K}$ and an algebraic closure by $\overline{\widehat{K}}$. For $\mathbb{T}_{\mathcal{O}}(\mathrm{M}_k(N, \chi\,;\,\mathbb{C}))$ we only write $\mathbb{T}_{\mathcal{O}}$ for short, and similarly over other rings. We keep using the fact that $\mathbb{T}_{\mathcal{O}}$ is finitely generated as an $\mathcal{O}$-module. We shall now apply Proposition 2.7 to $\mathbb{T}_{\widehat{\mathcal{O}}}$.

Proposition 2.8 *The Hecke algebras $\mathbb{T}_{\mathcal{O}}$ and $\mathbb{T}_{\widehat{\mathcal{O}}}$ are pure of Krull dimension* 1, *i.e. every maximal prime contains some minimal prime ideal.*

Proof It suffices to prove that $\mathbb{T}_{\widehat{\mathcal{O}}}$ is pure of Krull dimension 1 because completion of $\mathbb{T}_{\mathcal{O}}$ at a maximal ideal of $\mathcal{O}$ does not change the Krull dimension. First note that $\widehat{\mathcal{O}}$ is pure of Krull dimension 1 as it is an integral extension of $\mathbb{Z}_p$ (and the Krull dimension is an invariant in integral extensions). With the same reasoning, $\mathbb{T}_{\widehat{\mathcal{O}}}$ is of Krull dimension 1; we have to see that it is pure. According to Proposition 2.7, $\mathbb{T}_{\widehat{\mathcal{O}}}$ is the direct product of finite local $\widehat{\mathcal{O}}$-algebras $\mathbb{T}_i$. As each $\mathbb{T}_i$ embeds into a finite dimensional matrix algebra over $\widehat{K}$, it admits a simultaneous eigenvector (after possibly a finite extension of $\widehat{K}$) for the standard action of the matrix algebra on the corresponding $\widehat{K}$-vector space and the map φ sending an element of $\mathbb{T}_i$ to its eigenvalue is non-trivial and its kernel is a prime ideal strictly contained in the maximal ideal of $\mathbb{T}_i$. To see this, notice that the eigenvalues are integral, i.e. lie in the valuation ring of a finite extension of $\widehat{K}$, and can hence be reduced modulo the maximal ideal. The kernel of φ followed by the reduction map is the required maximal ideal. This proves that the height of the maximal ideal is 1. □

By Proposition 2.7, minimal primes of $\mathbb{T}_{\widehat{\mathcal{O}}}$ correspond to the maximal primes of $\mathbb{T}_{\widehat{K}}$ and hence to $\mathrm{Gal}(\overline{\widehat{K}}/\widehat{K})$-conjugacy classes of eigenforms in $\mathrm{M}_k(N, \chi\,;\, \overline{\widehat{K}})$. By a brute force identification of $\overline{\widehat{K}} = \overline{\mathbb{Q}}_p$ with $\mathbb{C}$ we may still think about these eigenforms as the usual holomorphic ones (the Galois conjugacy can then still be seen as conjugacy by a decomposition group above p inside the absolute Galois group of the field of fractions of $\mathcal{O}$).

Again by Proposition 2.7, maximal prime ideals of $\mathbb{T}_{\widehat{\mathcal{O}}}$ correspond to the maximal prime ideals of $\mathbb{T}_{\mathbb{F}}$ and hence to $\mathrm{Gal}(\overline{\mathbb{F}}/\mathbb{F})$-conjugacy classes of eigenforms in $\mathrm{M}_k(N, \chi\,;\, \overline{\mathbb{F}})$.

The spectrum of $\mathbb{T}_{\widehat{\mathcal{O}}}$ allows one to phrase very elegantly when conjugacy classes of eigenforms are congruent modulo a prime above p. Let us first explain what that means. Normalised eigenforms f take their coefficients $a_n(f)$ in rings of integers of number fields ($\mathbb{T}_{\mathcal{O}}/\mathfrak{m}$, when $\mathfrak{m}$ is the kernel of the $\mathcal{O}$-algebra homomorphism $\mathbb{T}_{\mathcal{O}} \to \mathbb{C}$, given by $T_n \mapsto a_n(f)$), so they can be reduced modulo primes above p (for which we will often just say 'reduced modulo p'). The reduction modulo a prime above p of the q-expansion of a modular form f in $\mathrm{M}_k(N, \chi\,;\, \mathbb{C})$ is the formal q-expansion of an eigenform in $\mathrm{M}_k(N, \chi\,;\, \overline{\mathbb{F}})$.

If two normalised eigenforms f, g in $\mathrm{M}_k(N, \chi\,;\, \mathbb{C})$ or $\mathrm{M}_k(N, \chi\,;\, \overline{\widehat{K}})$ reduce to the same element in $\mathrm{M}_k(N, \chi\,;\, \overline{\mathbb{F}})$, we say that they are *congruent modulo p*.

Due to Exercise 2.22, we may speak about *reductions modulo p* of $\mathrm{Gal}(\overline{\widehat{K}}/\widehat{K})$-conjugacy classes of normalised eigenforms to $\mathrm{Gal}(\overline{\mathbb{F}}/\mathbb{F})$-conjugacy classes. We hence say that two $\mathrm{Gal}(\overline{\widehat{K}}/\widehat{K})$-conjugacy classes, say corresponding to normalised eigenforms f, g, respectively, minimal ideals $\mathfrak{p}_1$ and $\mathfrak{p}_2$ of $\mathbb{T}_{\widehat{\mathcal{O}}}$, are *congruent modulo p*, if they reduce to the same $\mathrm{Gal}(\overline{\mathbb{F}}/\mathbb{F})$-conjugacy class.

Proposition 2.9 *The* $\mathrm{Gal}(\overline{\widehat{K}}/\widehat{K})$*-conjugacy classes belonging to minimal primes* $\mathfrak{p}_1$ *and* $\mathfrak{p}_2$ *of* $\mathbb{T}_{\widehat{\mathcal{O}}}$ *are congruent modulo p if and only if they are contained in a common maximal prime* $\mathfrak{m}$ *of* $\mathbb{T}_{\widehat{\mathcal{O}}}$.

Proof Exercise 2.23. □

We mention the fact that if f is a newform belonging to the maximal ideal $\mathfrak{m}$ of the Hecke algebra $\mathbb{T} := \mathbb{T}_{\mathbb{Q}}(S_k(\Gamma_1(N), \mathbb{C}))$, then $\mathbb{T}_{\mathfrak{m}}$ is isomorphic to $\mathbb{Q}_f = \mathbb{Q}(a_n | n \in \mathbb{N})$. This follows from newform (Atkin–Lehner) theory (see [10, §5.6–5.8]), which implies that the Hecke algebra on the newspace is diagonalisable, so that it is the direct product of the coefficient fields.

We include here the famous Deligne–Serre lifting lemma [8, Lemme 6.11], which we can easily prove with the tools developed so far.

Proposition 2.10 (Deligne–Serre lifting lemma) *Any normalised eigenform* $\overline{f} \in \mathrm{S}_k(\Gamma_1(N)\,;\, \overline{\mathbb{F}}_p)$ *is the reduction of a normalised eigenform* $f \in \mathrm{S}_k(\Gamma_1(N)\,;\, \mathbb{C})$.

Proof Let $\mathbb{T}_{\mathbb{Z}} = \mathbb{T}_{\mathbb{Z}}(\mathrm{S}_k(\Gamma_1(N)\,;\, \mathbb{C}))$. By definition, $\overline{f}$ is a ring homomorphism $\mathbb{T}_{\mathbb{Z}} \to \overline{\mathbb{F}}_p$ and its kernel is a maximal ideal $\mathfrak{m}$ of $\mathbb{T}_{\mathbb{Z}}$. According to Proposition 2.8, the Hecke algebra is pure of Krull dimension one, hence $\mathfrak{m}$ is of height 1, meaning that it strictly contains a minimal prime ideal $\mathfrak{p} \subset \mathbb{T}_{\mathbb{Z}}$. Let f be the composition of the maps in the first line of the diagram:

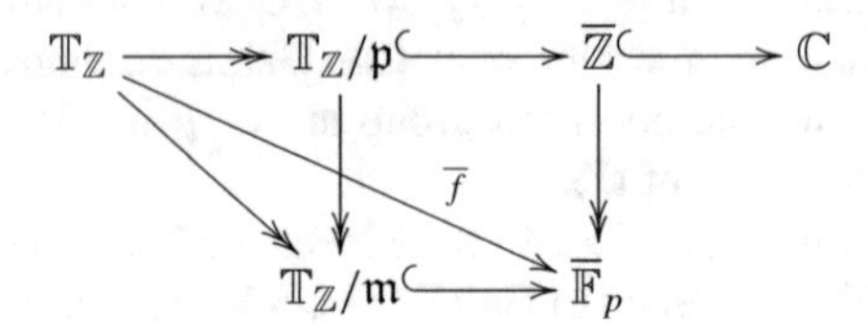

where all surjections and all injections are the natural ones, and the map $\overline{\mathbb{Z}} \twoheadrightarrow \overline{\mathbb{F}}_p$ is chosen in order to make the diagram commutative. Note that $\overline{f}$ is a ring homomorphism and thus a normalised eigenform in $\mathrm{S}_k(\Gamma_1(N)\,;\,\mathbb{C})$. By the diagram, its reduction is $\overline{f}$. □

2.2 Algorithms and Implementations: Localisation Algorithms

Let K be a perfect field, $\overline{K}$ an algebraic closure and A a finite dimensional commutative K-algebra. In the context of Hecke algebras we would like to compute a local decomposition of A as in Proposition 2.7.

2.2.1 Primary Spaces

Definition 2.11 An A-module V which is finite dimensional as K-vector space is called a *primary space* for A if the minimal polynomial for all $a \in A$ is a prime power in $K[X]$.

Lemma 2.12 *(a) A is local if and only if the minimal polynomial of a (in $K[X]$) is a prime power for all $a \in A$.*

(b) Let V be an A-module which is finite dimensional as K-vector space and which is a primary space for A. Then the image of A in $\mathrm{End}_K(V)$ *is a local algebra.*

(c) Let V be an A-module which is finite dimensional as K-vector space and let $a_1, \ldots, a_n$ be generators of the algebra A. Suppose that for $i \in \{1, \ldots, n\}$ the minimal polynomial a_i on V is a power of $(X - \lambda_i)$ in $K[X]$ for some $\lambda_i \in K$ (e.g. if $K = \overline{K}$). Then the image of A in $\mathrm{End}_K(V)$ *is a local algebra.*

Proof (a) Suppose first that A is local and take $a \in A$. Let $\phi_a : K[X] \to A$ be the homomorphism of K-algebras defined by sending X to a. Let (f) be the kernel with f monic, so that by definition f is the minimal polynomial of a. Hence, $K[X]/(f) \hookrightarrow A$, whence $K[X]/(f)$ is local, as it does not contain any non-trivial idempotent. Thus, f cannot have two different prime factors.

Conversely, if A were not local, we would have an idempotent $e \notin \{0, 1\}$. The minimal polynomial of e is $X(X-1)$, which is not a prime power.

(b) follows directly. For (c) one can use the following. Suppose that $(a-\lambda)^r V = 0$ and $(b-\mu)^s V = 0$. Then $((a+b)-(\lambda+\mu))^{r+s} V = 0$, as one sees by rewriting $((a+b)-(\lambda+\mu)) = (a-\lambda)+(b-\mu)$ and expanding out. From this it also follows that $(ab-\lambda\mu)^{2(r+s)} V = 0$ by rewriting $ab-\lambda\mu = (a-\lambda)(b-\mu)+\lambda(b-\mu)+\mu(a-\lambda)$. □

We warn the reader that algebras such that a set of generators acts primarily need not be local, unless they are defined over an algebraically closed field, as we have seen in Part (c) above. In Exercise 2.24 you are asked to find an example.

The next proposition, however, tells us that an algebra over a field having a basis consisting of primary elements is local. I found the idea for that proof in [11].

Proposition 2.13 *Let K be a field of characteristic 0 or a finite field. Let A be a finite dimensional commutative algebra over K and let $a_1, \ldots, a_n$ be a K-basis of A with the property that the minimal polynomial of each a_i is a power of a prime polynomial $p_i \in K[X]$.*

Then A is local.

Proof We assume that A is not local and take a decomposition $\alpha : A \xrightarrow{\sim} \prod_{j=1}^{r} A_j$ with $r \geq 2$. Let K_j be the residue field of A_j and consider the finite dimensional K-algebra $\overline{A} := \prod_{j=1}^{r} K_j$. Write $\overline{a_1}, \ldots, \overline{a_n}$ for the images of the a_i in $\overline{A}$. They form a K-basis. In order to have access to the components, also write $\overline{a_i} = (\overline{a_{i,1}}, \ldots, \overline{a_{i,r}})$. Since the minimal polynomial of an element in a product is the lowest common multiple of the minimal polynomials of the components, the assumption implies that, for each $i = 1, \ldots, r$, the minimal polynomial of $a_{i,j}$ is independent of j; call it $p_i \in K[X]$. Let N/K be the splitting field of the polynomials $p_1, \ldots, p_r$. This means that N is the normal closure of K_j over K for any j. As a particular case, note that $N = K_j$ for all j if K is a finite field since finite extensions of finite fields are automatically normal. Now consider the trace $\mathrm{Tr}_{N/K}$ and note that $\mathrm{Tr}_{N/K}(\overline{a_i})$ is a diagonal element in $\overline{A}$ for all $i = 1, \ldots, r$ since the components $\overline{a_{i,j}}$ are roots of the same minimal polynomial. Consequently, $\mathrm{Tr}_{N/K}(\overline{a})$ is a diagonal element for all $\overline{a} \in \overline{A}$ since the $\overline{a_i}$ form a K-basis of $\overline{A}$.

In order to come to a contradiction, it now suffices to produce an element the trace of which is not diagonal. By Exercise 2.25 there is $x \in K_1$ such that $\mathrm{Tr}_{N/K}(x) \neq 0$. Then the element $(x, 0, \ldots, 0) \in \overline{A}$ clearly provides an example of an element with non-diagonal trace. □

Lemma 2.14 *Let A be a local finite dimensional commutative algebra over a perfect field K. Let $a_1, \ldots, a_n$ be a set of K-algebra generators of A such that the minimal polynomial of each a_i is a prime polynomial. Then A is a field.*

Proof As the a_i are simultaneously diagonalisable over a separable closure (considering the algebra as a matrix algebra) due to their minimal polynomials being squarefree (using here the perfectness of K), so are sums and products of the a_i. Hence, 0 is the only nilpotent element in A. As the maximal ideal in an Artinian local algebra is the set of nilpotent elements, the lemma follows. □

Proposition 2.15 *Let A be a local finite dimensional commutative algebra over a perfect field K. Let $a_1, \ldots, a_n$ be a set of K-algebra generators of A. Let $p_i^{e_i}$ be the minimal polynomial of a_i (see Lemma 2.12).*

Then the maximal ideal $\mathfrak{m}$ of A is generated by $\{p_1(a_1), \ldots, p_n(a_n)\}$.

Proof Let $\mathfrak{a}$ be the ideal generated by $\{p_1(a_1), \ldots, p_n(a_n)\}$. The quotient $A/\mathfrak{a}$ is generated by the images of the a_i, call them $\overline{a_i}$. As $p_i(a_i) \in \mathfrak{a}$, it follows $p_i(\overline{a_i}) = 0$, whence the minimal polynomial of $\overline{a_i}$ equals the prime polynomial p_i. By Lemma 2.14, we know that $A/\mathfrak{a}$ is a field, whence $\mathfrak{a}$ is the maximal ideal. □

2.2.2 Algorithm for Computing Common Primary Spaces

It may help to think about finite dimensional commutative algebras over a field as algebras of matrices. Then the localisation statements of this section just mean choosing a basis such that one obtains block matrices.

By a *common primary space* for commuting matrices we mean a direct summand of the underlying vector space on which the minimal polynomials of the given matrices are prime powers. By Proposition 2.13, a common primary space of a basis of a matrix algebra is a local factor of the algebra.

By a *generalised eigenspace* for commuting matrices we mean a vector subspace of the underlying vector space on which the minimal polynomial of the given matrices are irreducible. Allowing base changes to extension fields, the matrices restricted to the generalised eigenspace are diagonalisable.

In this section, we present a straight forward algorithm for computing common primary spaces and common generalised eigenspaces.

Algorithm 2.16 <u>Input</u>: list `ops` of commuting operators acting on the K-vector space V.

<u>Output</u>: list of the common primary spaces inside V for all operators in `ops`.

(1) List := [V];
(2) for T in `ops` do
(3) newList := [];
(4) for W in List do
(5) compute the minimal polynomial $f \in K[X]$ of T restricted to W.
(6) factor f over K into its prime powers $f(X) = \prod_{i=1}^{n} p_i(X)^{e_i}$.
(7) if n equals 1, then
(8) append W to newList,
(9) else for i := 1 to n do
(10) compute $\widetilde{W}$ as the kernel of $p_i(T|_W)^\alpha$ with $\alpha = e_i$ for common primary spaces or $\alpha = 1$ for common generalised eigenspaces.

(11) append $\widetilde{W}$ to newList.
(12) end for; end if;
(13) end for;
(14) List := newList;
(15) end for;
(16) return List and stop.

2.2.3 Algorithm for Computing Idempotents

Using Algorithm 2.16 it is possible to compute a complete set of orthogonal primitive idempotents for A. We now sketch a direct algorithm.

Algorithm 2.17 Input: matrix M.
Output: complete set of orthogonal primitive idempotents for the matrix algebra generated by M and 1.

(1) compute the minimal polynomial f of M.
(2) factor it $f = (\prod_{i=1}^{n} p_i^{e_i})X^e$ over K with p_i distinct irreducible polynomials different from X.
(3) List := [];
(4) for $i = 1$ to n do
(5) $g := f/p_i^{e_i}$;
(6) $M_1 := g(M)$. If we think about M_1 in block form, then there is only one non-empty block on the diagonal, the rest is zero. In the next steps this block is replaced by the identity.
(7) compute the minimal polynomial h of M_1.
(8) strip possible factors X from h and normalise h so that $h(0) = 1$.
(9) append $1 - h(M_1)$ to List. Note that $h(M_1)$ is the identity matrix except at the block corresponding to p_i, which is zero. Thus $1 - h(M_1)$ is the idempotent being zero everywhere and being the identity in the block corresponding to p_i.
(10) end for;
(11) if $e > 0$ then
(12) append $1 - \sum_{e \in \text{List}} e$ to List.
(13) end if;
(14) return List and stop.

The algorithm for computing a complete set of orthogonal primitive idempotents for a commutative matrix algebra consists of multiplying together the idempotents of every matrix in a basis. See Computer Exercise 2.31.

2.3 Theoretical Exercises

Exercise 2.18 Use your knowledge on modular forms to prove that a modular form $f = \sum_{n=0}^{\infty} a_n(f)q^n$ of weight $k \geq 1$ and level N (and Dirichlet character χ) is uniquely determined by $\sum_{n=1}^{\infty} a_n(f)q^n$.

Exercise 2.19 Prove Proposition 2.3.

Hint: use that the kernel of a ring homomorphism into an integral domain is a prime ideal; moreover, use that all prime ideals in the Hecke algebra in the exercise are maximal; finally, use that field homomorphisms can be extended to separable extensions (using here that K is perfect).

Exercise 2.20 Let $\mathbb{T}$ be an Artinian ring.

(a) Let $\mathfrak{m}$ be a maximal ideal of $\mathbb{T}$. Prove that $\mathfrak{m}^\infty$ is a principal ideal generated by an idempotent. Call it $e_\mathfrak{m}$.
(b) Prove that the idempotents $1 - e_\mathfrak{m}$ and $1 - e_\mathfrak{n}$ for different maximal ideals $\mathfrak{m}$ and $\mathfrak{n}$ are orthogonal.
(c) Prove that the set $\{1 - e_\mathfrak{m} | \mathfrak{m} \in \mathrm{Spec}(\mathbb{T})\}$ forms a complete set of pairwise orthogonal idempotents.

Hint: see [2, §8].

Exercise 2.21 Prove the correspondences and the isomorphisms from Part (a) of Proposition 2.7.

Hint: you only need basic reasonings from commutative algebra.

Exercise 2.22 Let $f, g \in \mathrm{M}_k(N, \chi\,;\, \overline{\widehat{K}})$ be normalised eigenforms that we assume to be $\mathrm{Gal}(\overline{\widehat{K}}/\widehat{K})$-conjugate. Prove that their reductions modulo p are $\mathrm{Gal}(\overline{\mathbb{F}}/\mathbb{F})$-conjugate.

Exercise 2.23 Prove Proposition 2.9.

Hint: it suffices to write out the definitions.

Exercise 2.24 Find a non-local algebra A over a field K (of your choice) such that A is generated as a K-algebra by $a_1, \ldots, a_n$ having the property that the minimal polynomial of each a_i is a power of an irreducible polynomial in $K[X]$.

Exercise 2.25 Let K be a field of characteristic 0 or a finite field. Let L be a finite extension of K with Galois closure N over K. Show that there is an element $x \in L$ with $\mathrm{Tr}_{N/K}(x) \neq 0$.

Exercise 2.26 Let A be a commutative matrix algebra over a perfect field K. Suppose that the minimal polynomial of each element of a generating set is the power of a prime polynomial (i.e. it is primary).

Show that there exist base change matrices such that the base changed algebra consists only of lower triangular matrices. You may and you may have to extend scalars to a finite extension of K. In Computer Exercise 2.33 you are asked to find and implement an algorithm computing such base change matrices.

2.4 Computer Exercises

Computer Exercise 2.27 Change Algorithm 1.14 (see Computer Exercise 1.44) so that it works for modular forms over a given ring R.

Computer Exercise 2.28 Let A be a commutative matrix algebra over a perfect field K.

(a) Write an algorithm to test whether A is local.
(b) Suppose A is local. Write an algorithm to compute its maximal ideal.

Computer Exercise 2.29 Let A be a commutative algebra over a field K. The regular representation is defined as the image of the injection

$$A \to \mathrm{End}_K(A), \quad a \mapsto (b \mapsto a \cdot b).$$

Write a function computing the regular representation.

Computer Exercise 2.30 Implement Algorithm 2.16. Also write a function that returns the local factors as matrix algebras (possibly using regular representations).

Computer Exercise 2.31 (a) Implement Algorithm 2.17.
(b) Let S be a set of idempotents. Write a function selecting a subset of S consisting of pairwise orthogonal idempotents such that the subset spans S (all idempotents in S can be obtained as sums of elements in the subset).
(c) Write a function computing a complete set of pairwise orthogonal idempotents for a commutative matrix algebra A over a field by multiplying together the idempotents of the matrices in a basis and selecting a subset as in (b).
(d) Use Computer Exercise 2.28 to compute the maximal ideals of A.

Computer Exercise 2.32 Let A be a commutative matrix algebra over a perfect field K. Suppose that A is a field (for instance obtained as the quotient of a local A by its maximal ideal computed in Computer Exercise 2.28). Write a function returning an irreducible polynomial p such that A is $K[X]/(p)$.

If possible, the algorithm should not use factorisations of polynomials. It is a practical realisation of Kronecker's primitive element theorem.

Computer Exercise 2.33 Let A be a commutative matrix algebra over a perfect field K. Suppose that the minimal polynomial of each element of a generating set is the power of a prime polynomial (i.e. it is primary).

Write a function computing base change matrices such that the base changed algebra consists only of lower triangular matrices (cf. Exercise 2.26).

3 Homological Algebra

In this section, we provide the tools from homological algebra that will be necessary for the modular symbols algorithm (in its group cohomological version). A good reference is [21].

We will be sloppy about categories. When we write category below, we really mean abelian category, since we obviously need the existence of kernels, images, quotients etc. For what we have in mind, we should really understand the word category not in its precise mathematical sense but as a placeholder for $R-$ modules, or (co-)chain complexes of $R-$ modules and other categories from everyday life.

3.1 *Theory: Categories and Functors*

Definition 3.1 A *category* $\mathscr{C}$ consists of the following data:

- a class $\mathrm{obj}(\mathscr{C})$ of *objects*,
- a set $\mathrm{Hom}_{\mathscr{C}}(A, B)$ of *morphisms* for every ordered pair (A, B) of objects,
- an *identity morphism* $\mathrm{id}_A \in \mathrm{Hom}_{\mathscr{C}}(A, A)$ for every object A, and
- a *composition function*

  $$\mathrm{Hom}_{\mathscr{C}}(A, B) \times \mathrm{Hom}_{\mathscr{C}}(B, C) \to \mathrm{Hom}_{\mathscr{C}}(A, C), \quad (f, g) \mapsto g \circ f$$

  for every ordered triple (A, B, C) of objects

such that

- (Associativity) $(h \circ g) \circ f = h \circ (g \circ f)$ for all $f \in \mathrm{Hom}_{\mathscr{C}}(A, B)$, $g \in \mathrm{Hom}_{\mathscr{C}}(B, C)$, $h \in \mathrm{Hom}_{\mathscr{C}}(C, D)$ and
- (Unit Axiom) $\mathrm{id}_B \circ f = f = f \circ \mathrm{id}_A$ for $f \in \mathrm{Hom}_{\mathscr{C}}(A, B)$.

Example 3.2 Examples of categories are

- Sets: objects are sets, morphisms are maps.
- Let R be a not necessarily commutative ring. Left R-modules ($R-$ modules): objects are R-modules, morphisms are R-module homomorphisms. This is the category we are going to work with most of the time. Note that the category of $\mathbb{Z}$-modules is the category of abelian groups.
- Right R-modules (modules $-\ R$): as above.

Definition 3.3 Let $\mathscr{C}$ and $\mathscr{D}$ be categories. A *covariant/contravariant functor* $F: \mathscr{C} \to \mathscr{D}$ is

- a rule $\mathrm{obj}(\mathscr{C}) \to \mathrm{obj}(\mathscr{D})$, $C \mapsto F(C)$ and
- a rule $\begin{cases} \text{covariant:} & \mathrm{Hom}_{\mathscr{C}}(C_1, C_2) \to \mathrm{Hom}_{\mathscr{D}}(F(C_1), F(C_2)),\ f \mapsto F(f) \\ \text{contravariant:} & \mathrm{Hom}_{\mathscr{C}}(C_1, C_2) \to \mathrm{Hom}_{\mathscr{D}}(F(C_2), F(C_1)),\ f \mapsto F(f) \end{cases}$

such that

- $F(\mathrm{id}_C) = \mathrm{id}_{F(C)}$ and
- $\begin{cases} \text{covariant:} & F(g \circ f) = F(g) \circ F(f) \\ \text{contravariant:} & F(g \circ f) = F(f) \circ F(g) \end{cases}$

Example 3.4 • Let $M \in \mathrm{obj}(R - \text{modules})$. Define

$$\mathrm{Hom}_R(M, \cdot) : R - \text{modules} \to \mathbb{Z} - \text{modules}, \quad A \mapsto \mathrm{Hom}_R(M, A).$$

This is a covariant functor.
• Let $M \in \mathrm{obj}(R - \text{modules})$. Define

$$\mathrm{Hom}_R(\cdot, M) : R - \text{modules} \to \mathbb{Z} - \text{modules}, \quad A \mapsto \mathrm{Hom}_R(A, M).$$

This is a contravariant functor.
• Let $M \in \mathrm{obj}(R - \text{modules})$. Define

$$\cdot \otimes_R M : \text{modules} - R \to \mathbb{Z} - \text{modules}, \quad A \mapsto A \otimes_R M.$$

This is a covariant functor.
• Let $M \in \mathrm{obj}(\text{modules} - R)$. Define

$$M \otimes_R \cdot : R - \text{modules} \to \mathbb{Z} - \text{modules}, \quad A \mapsto M \otimes_R A.$$

This is a covariant functor.

Definition 3.5 Let $\mathscr{C}$ and $\mathscr{D}$ be categories and $F, G : \mathscr{C} \to \mathscr{D}$ be both covariant or both contravariant functors. A *natural transformation* $\alpha : F \Rightarrow G$ is a collection of morphisms $\alpha = (\alpha_C)_{C \in \mathscr{C}} : F(C) \to G(C)$ in $\mathscr{D}$ for $C \in \mathscr{C}$ such that for all morphisms $f : C_1 \to C_2$ in $\mathscr{C}$ the following diagram commutes:
covariant: contravariant:

$$\begin{array}{ccc} F(C_1) & \xrightarrow{F(f)} & F(C_2) \\ {\scriptstyle \alpha_{C_1}}\downarrow & & \downarrow{\scriptstyle \alpha_{C_2}} \\ G(C_1) & \xrightarrow{G(f)} & G(C_2) \end{array} \qquad \begin{array}{ccc} F(C_1) & \xleftarrow{F(f)} & F(C_2) \\ {\scriptstyle \alpha_{C_1}}\downarrow & & \downarrow{\scriptstyle \alpha_{C_2}} \\ G(C_1) & \xleftarrow{G(f)} & G(C_2). \end{array}$$

Example 3.6 Let R be a not necessarily commutative ring and let $A, B \in \mathrm{obj}(R - \text{modules})$ together with a morphism $A \to B$. Then there are natural transformations $\mathrm{Hom}_R(B, \cdot) \Rightarrow \mathrm{Hom}_R(A, \cdot)$ and $\mathrm{Hom}_R(\cdot, A) \Rightarrow \mathrm{Hom}_R(\cdot, B)$ as well as $\cdot \otimes_R A \Rightarrow \cdot \otimes_R B$ and $A \otimes_R \cdot \Rightarrow B \otimes_R \cdot$.

Proof Exercise 3.31. □

Definition 3.7 • A covariant functor $F : \mathscr{C} \to \mathscr{D}$ is called *left-exact*, if for every exact sequence

$$0 \to A \to B \to C$$

the sequence

$$0 \to F(A) \to F(B) \to F(C)$$

is also exact.

- A contravariant functor $F : \mathscr{C} \to \mathscr{D}$ is called *left-exact*, if for every exact sequence

$$A \to B \to C \to 0$$

the sequence

$$0 \to F(C) \to F(B) \to F(A)$$

is also exact.

- A covariant functor $F : \mathscr{C} \to \mathscr{D}$ is called *right-exact*, if for every exact sequence

$$A \to B \to C \to 0$$

the sequence

$$F(A) \to F(B) \to F(C) \to 0$$

is also exact.

- A contravariant functor $F : \mathscr{C} \to \mathscr{D}$ is called *right-exact*, if for every exact sequence

$$0 \to A \to B \to C$$

the sequence

$$F(C) \to F(B) \to F(A) \to 0$$

is also exact.

- A covariant or contravariant functor is *exact* if it is both left-exact and right-exact.

Example 3.8 Both functors $\mathrm{Hom}_R(\cdot, M)$ and $\mathrm{Hom}_R(M, \cdot)$ for $M \in \mathrm{obj}(R - \mathrm{modules})$ are left-exact. Both functors $\cdot \otimes_R M$ for $M \in \mathrm{obj}(R - \mathrm{modules})$ and $M \otimes_R \cdot$ for $M \in \mathrm{obj}(\mathrm{modules} - R)$ are right-exact.

Proof Exercise 3.32. □

Definition 3.9 Let R be a not necessarily commutative ring. A left R-module P is called *projective* if the functor $\mathrm{Hom}_R(P, \cdot)$ is exact. A left R-module I is called *injective* if the functor $\mathrm{Hom}_R(\cdot, I)$ is exact.

Lemma 3.10 *Let R be a not necessarily commutative ring and let P be a left R-module. Then P is projective if and only if P is a direct summand of some free R-module. In particular, free modules are projective.*

Proof Exercise 3.33. □

3.2 Theory: Complexes and Cohomology

Definition 3.11 A *(right) chain complex* $C_\bullet$ in the category R – modules is a collection of objects $C_n \in \text{obj}(R - \text{modules})$ for $n \geq m$ for some $m \in \mathbb{Z}$ together with homomorphisms $C_{n+1} \xrightarrow{\partial_{n+1}} C_n$, i.e.

$$\cdots \to C_{n+1} \xrightarrow{\partial_{n+1}} C_n \xrightarrow{\partial_n} C_{n-1} \to \cdots \to C_{m+2} \xrightarrow{\partial_{m+2}} C_{m+1} \xrightarrow{\partial_{m+1}} C_m \xrightarrow{\partial_m} 0,$$

such that

$$\partial_n \circ \partial_{n+1} = 0$$

for all $n \geq m$. The group of *n-cycles* of this chain complex is defined as

$$\mathrm{Z}_n(C_\bullet) = \ker(\partial_n).$$

The group of *n-boundaries* of this chain complex is defined as

$$\mathrm{B}_n(C_\bullet) = \mathrm{im}(\partial_{n+1}).$$

The *n-th homology group* of this chain complex is defined as

$$\mathrm{H}_n(C_\bullet) = \ker(\partial_n)/\,\mathrm{im}(\partial_{n+1}).$$

The chain complex $C_\bullet$ is *exact* if $\mathrm{H}_n(C_\bullet) = 0$ for all n. If $C_\bullet$ is exact and $m = -1$, one often says that $C_\bullet$ is a *resolution* of C_{-1}.

A *morphism of right chain complexes* $\phi_\bullet : C_\bullet \to D_\bullet$ is a collection of homomorphisms $\phi_n : C_n \to D_n$ for $n \in \mathbb{N}_0$ such that all the diagrams

$$\begin{array}{ccc} C_{n+1} & \xrightarrow{\partial_{n+1}} & C_n \\ \downarrow{\scriptstyle \phi_{n+1}} & & \downarrow{\scriptstyle \phi_n} \\ D_{n+1} & \xrightarrow{\partial_{n+1}} & D_n \end{array}$$

are commutative.

If all ϕ_n are injective, we regard $C_\bullet$ as a sub-chain complex of $D_\bullet$. If all ϕ_n are surjective, we regard $D_\bullet$ as a quotient complex of $C_\bullet$.

Definition 3.12 A *(right) cochain complex* $C^\bullet$ in the category R – modules is a collection of objects $C^n \in \text{obj}(R - \text{modules})$ for $n \geq m$ for some $m \in \mathbb{Z}$ together with homomorphisms $C^n \xrightarrow{\partial^{n+1}} C^{n+1}$, i.e.

$$0 \xrightarrow{\partial^m} C^m \xrightarrow{\partial^{m+1}} C^{m+1} \xrightarrow{\partial^{m+2}} C^{m+2} \to \cdots \to C^{n-1} \xrightarrow{\partial^n} C^n \xrightarrow{\partial^{n+1}} C^{n+1} \to \cdots,$$

such that

$$\partial^{n+1} \circ \partial^n = 0$$

for all $n \geq m$. The group of *n-cocycles* of this cochain complex is defined as

$$\mathrm{Z}^n(C_\bullet) = \ker(\partial^{n+1}).$$

The group of *n-coboundaries* of this cochain complex is defined as

$$\mathrm{B}^n(C_\bullet) = \mathrm{im}(\partial_n).$$

The *n-th cohomology group* of this cochain complex is defined as

$$\mathrm{H}^n(C^\bullet) = \ker(\partial^{n+1})/\,\mathrm{im}(\partial^n).$$

The cochain complex $C^\bullet$ is *exact* if $\mathrm{H}^n(C_\bullet) = 0$ for all n. If $C^\bullet$ is exact and $m = -1$, one often says that $C^\bullet$ is a *resolution* of C^{-1}.

A *morphism of right cochain complexes* $\phi^\bullet : C^\bullet \to D^\bullet$ is a collection of homomorphisms $\phi^n : C^n \to D^n$ for $n \in \mathbb{N}_0$ such that all the diagrams

$$\begin{array}{ccc} C^n & \xrightarrow{\partial^{n+1}} & C^{n+1} \\ {\scriptstyle\phi^n}\downarrow & & \downarrow{\scriptstyle\phi^{n+1}} \\ D^n & \xrightarrow{\partial^{n+1}} & D^{n+1} \end{array}$$

are commutative.

If all ϕ^n are injective, we regard $C^\bullet$ as a sub-chain complex of $D^\bullet$. If all ϕ^n are surjective, we regard $D^\bullet$ as a quotient complex of $C^\bullet$.

In Exercise 3.34 you are asked to define kernels, cokernels and images of morphisms of cochain complexes and to show that morphisms of cochain complexes induce natural maps on the cohomology groups. In fact, cochain complexes of R-modules form an abelian category.

Example: Standard Resolution of a Group

Let G be a group and R a commutative ring. Write G^n for the n-fold direct product $G \times \cdots \times G$ and equip $R[G^n]$ with the diagonal $R[G]$-action. We describe the *standard resolution* $F(G)_\bullet$ of R by free $R[G]$-modules:

$$0 \xleftarrow{\partial_0} R \xleftarrow{\varepsilon} F(G)_0 := R[G] \xleftarrow{\partial_1} F(G)_1 := R[G^2] \xleftarrow{\partial_2} \cdots,$$

where we put (the 'hat' means that we leave out that element):

$$\partial_n := \sum_{i=0}^{n}(-1)^i d_i \quad \text{and} \quad d_i(g_0,\dots,g_n) := (g_0,\dots,\hat{g}_i,\dots,g_n).$$

The map ε is the usual augmentation map defined by sending $g \in G$ to $1 \in R$. By 'standard resolution' we refer to the straight maps. We have included the bended arrow ∂_0, which is 0 by definition, because it will be needed in the definition of group cohomology (Definition 3.13). In Exercise 3.35 you are asked to check that the standard resolution is indeed a resolution, i.e. that the above complex is exact.

Example: Bar Resolution of a Group

We continue to treat the standard resolution R by $R[G]$-modules, but we will write it differently. [21] calls the following the *unnormalised bar resolution* of G. We shall simply say *bar resolution*. If we let $h_r := g_{r-1}^{-1}g_r$, then we get the identity

$$(g_0, g_1, g_2, \dots, g_n) = g_0.(1, h_1, h_1h_2, \dots, h_1h_2\dots h_n) =: g_0.[h_1|h_2|\dots h_n].$$

The symbols $[h_1|h_2|\dots|h_n]$ with arbitrary $h_i \in G$ hence form an $R[G]$-basis of $F(G)_n$, and one has $F(G)_n = R[G] \otimes_R$ (free R-module on $[h_1|h_2|\dots|h_n]$). One computes the action of ∂_n on this basis and gets $\partial_n = \sum_{i=0}^{n}(-1)^i d_i$ where

$$d_i([h_1|\dots|h_n]) = \begin{cases} h_1[h_2|\dots|h_n] & i = 0 \\ [h_1|\dots|h_ih_{i+1}|\dots|h_n] & 0 < i < n \\ [h_1|\dots|h_{n-1}] & i = n. \end{cases}$$

We will from now on, if confusion is unlikely, simply write $(h_1,\dots,h_n)$ instead of $[h_1|\dots|h_n]$.

Example: Resolution of a Cyclic Group

Let $G = \langle T \rangle$ be an infinite cyclic group (i.e. a group isomorphic to $(\mathbb{Z}, +)$). Here is a very simple resolution of R by free $R[G]$-modules:

$$0 \to R[G] \xrightarrow{T-1} R[G] \xrightarrow{\varepsilon} R \to 0. \tag{11}$$

Let now $G = \langle \sigma \rangle$ be a finite cyclic group of order n and let $N_\sigma := \sum_{i=0}^{n-1}\sigma^i$. Here is a resolution of R by free $R[G]$-modules:

$$\cdots \to R[G] \xrightarrow{N_\sigma} R[G] \xrightarrow{1-\sigma} R[G] \xrightarrow{N_\sigma} R[G] \xrightarrow{1-\sigma} R[G] \to \cdots$$
$$\cdots \to R[G] \xrightarrow{1-\sigma} R[G] \xrightarrow{\varepsilon} R \to 0. \tag{12}$$

In Exercise 3.36 you are asked to verify the exactness of these two sequences.

Group Cohomology

A standard reference for group cohomology is [4].

Definition 3.13 Let R be a ring, G a group and M a left $R[G]$-module. Recall that $F(G)_\bullet$ denotes the standard resolution of R by free $R[G]$-modules.

(a) Let M be a left $R[G]$-module. When we apply the functor $\mathrm{Hom}_{R[G]}(\cdot, M)$ to the standard resolution $F(G)_\bullet$ cut off at 0 (i.e. $F(G)_1 \xrightarrow{\partial_1} F(G)_0 \xrightarrow{\partial_0} 0$), we get the cochain complex $\mathrm{Hom}_{R[G]}(F(G)_\bullet, M)$:

$$\to \mathrm{Hom}_{R[G]}(F(G)_{n-1}, M) \xrightarrow{\partial^n} \mathrm{Hom}_{R[G]}(F(G)_n, M) \xrightarrow{\partial^{n+1}} \mathrm{Hom}_{R[G]}(F(G)_{n+1}, M) \to .$$

Define the n-th cohomology group of G with values in the G-module M as

$$\mathrm{H}^n(G, M) := \mathrm{H}^n(\mathrm{Hom}_{R[G]}(F(G)_\bullet, M)).$$

(b) Let M be a right $R[G]$-module. When we apply the functor $M \otimes_{R[G]} \cdot$ to the standard resolution $F(G)_\bullet$ cut off at 0 we get the chain complex $M \otimes_{R[G]} F(G)_\bullet$:

$$\to M \otimes_{R[G]} F(G)_{n+1} \xrightarrow{\partial_{n+1}} M \otimes_{R[G]} F(G)_n \xrightarrow{\partial_n} M \otimes_{R[G]} F(G)_{n-1} \to .$$

Define the n-th homology group of G with values in the G-module M as

$$\mathrm{H}_n(G, M) := \mathrm{H}_n(M \otimes_{R[G]} F(G)_\bullet).$$

In this lecture, we shall only use group cohomology. As a motivation for looking at group cohomology in this lecture, we can already point out that

$$\mathrm{H}^1(\Gamma_1(N), V_{k-2}(R)) \cong \mathcal{M}_k(\Gamma_1(N), R),$$

provided that 6 is invertible in R (see Theorem 5.9). The reader is invited to compute explicit descriptions of H^0, H_0 and H^1 in Exercise 3.37.

3.3 *Theory: Cohomological Techniques*

The cohomology of groups fits into a general machinery, namely that of derived functor cohomology. Derived functors are universal cohomological δ-functors and many properties of them can be derived in a purely formal way from the universality. What this means will be explained in this section. We omit all proofs.

Definition 3.14 Let $\mathscr{C}$ and $\mathscr{D}$ be (abelian) categories (for instance, $\mathscr{C}$ the right cochain complexes of $R-$ modules and $\mathscr{D} = R-$ modules). A *positive covariant cohomological δ-functor* between $\mathscr{C}$ and $\mathscr{D}$ is a collection of functors $\mathrm{H}^n : \mathscr{C} \to \mathscr{D}$ for $n \geq 0$ together with *connecting morphisms*

$$\delta^n : \mathrm{H}^n(C) \to \mathrm{H}^{n+1}(A)$$

which are defined for every short exact sequence $0 \to A \to B \to C \to 0$ in $\mathscr{C}$ such that the following hold:

(a) (Positivity) H^n is the zero functor if $n < 0$.
(b) For every short exact sequence $0 \to A \to B \to C \to 0$ in $\mathscr{C}$ there is the *long exact sequence* in $\mathscr{D}$:

$$\cdots \to \mathrm{H}^{n-1}(C) \xrightarrow{\delta^{n-1}} \mathrm{H}^n(A) \to \mathrm{H}^n(B) \to \mathrm{H}^n(C) \xrightarrow{\delta^n} \mathrm{H}^{n+1}(A) \to \cdots,$$

where the maps $\mathrm{H}^n(A) \to \mathrm{H}^n(B) \to \mathrm{H}^n(C)$ are those that are induced from the homomorphisms in the exact sequence $0 \to A \to B \to C \to 0$.
(c) For every commutative diagram in $\mathscr{C}$

$$\begin{array}{ccccccccc} 0 & \longrightarrow & A & \longrightarrow & B & \longrightarrow & C & \longrightarrow & 0 \\ & & f\downarrow & & g\downarrow & & h\downarrow & & \\ 0 & \longrightarrow & A' & \longrightarrow & B' & \longrightarrow & C' & \longrightarrow & 0 \end{array}$$

with exact rows the following diagram in $\mathscr{D}$ commutes, too:

$$\begin{array}{ccccccccc} \mathrm{H}^{n-1}(C) & \xrightarrow{\delta^{n-1}} & \mathrm{H}^n(A) & \longrightarrow & \mathrm{H}^n(B) & \longrightarrow & \mathrm{H}^n(C) & \xrightarrow{\delta^n} & \mathrm{H}^{n+1}(A) \\ \mathrm{H}^{n-1}(h)\downarrow & & \mathrm{H}^n(f)\downarrow & & \mathrm{H}^n(g)\downarrow & & \mathrm{H}^n(h)\downarrow & & \mathrm{H}^{n+1}(f)\downarrow \\ \mathrm{H}^{n-1}(C') & \xrightarrow{\delta^{n-1}} & \mathrm{H}^n(A') & \longrightarrow & \mathrm{H}^n(B') & \longrightarrow & \mathrm{H}^n(C') & \xrightarrow{\delta^n} & \mathrm{H}^{n+1}(A') \end{array}$$

Theorem 3.15 *Let R be a ring (not necessarily commutative). Let $\mathscr{C}$ stand for the category of cochain complexes of left R-modules. Then the cohomology functors*

$$\mathrm{H}^n : \mathscr{C} \to R-\text{modules}, \quad C^\bullet \mapsto \mathrm{H}^n(C^\bullet)$$

form a cohomological δ-functor.

Proof This theorem is proved by some 'diagram chasing' starting from the snake lemma. See Chap. 1 of [21] for details. □

It is not difficult to conclude that group cohomology also forms a cohomological δ-functor.

Proposition 3.16 *Let R be a commutative ring and G a group.*

(a) *The functor from* $R[G]-\text{modules}$ *to cochain complexes of* $R[G]-\text{modules}$ *which associates to a left* $R[G]$*-module* M *the cochain complex* $\text{Hom}_{R[G]}(F(G)_\bullet, M)$ *with* $F(G)_\bullet$ *the bar resolution of* R *by free* $R[G]$*-modules is exact, i.e. it takes an exact sequence* $0 \to A \to B \to C \to 0$ *of* $R[G]$*-modules to the exact sequence*

$$0 \to \text{Hom}_{R[G]}(F(G)_\bullet, A) \to \text{Hom}_{R[G]}(F(G)_\bullet, B) \to \text{Hom}_{R[G]}(F(G)_\bullet, C) \to 0$$

of cochain complexes.

(b) *The functors*

$$\text{H}^n(G, \cdot) : R[G]-\text{modules} \to R-\text{modules}, \quad M \mapsto \text{H}^n(G, M)$$

form a positive cohomological δ*-functor.*

Proof Exercise 3.38. □

We will now come to universal δ-functors. Important examples of such (among them group cohomology) are obtained from injective resolutions. Although the following discussion is valid in any abelian category (with enough injectives), we restrict to $R-\text{modules}$ for a not necessarily commutative ring R.

Definition 3.17 Let R be a not necessarily commutative ring and let $M \in \text{obj}(R-\text{modules})$.

A *projective resolution* of M is a resolution

$$\cdots \to P_2 \xrightarrow{\partial_2} P_1 \xrightarrow{\partial_1} P_0 \to M \to 0,$$

i.e. an exact chain complex, in which all the P_n for $n \geq 0$ are projective R-modules.

An *injective resolution* of M is a resolution

$$0 \to M \to I^0 \xrightarrow{\partial^1} I^1 \xrightarrow{\partial^2} I^2 \to \cdots,$$

i.e. an exact cochain complex, in which all the I^n for $n \geq 0$ are injective R-modules.

We state the following lemma as a fact. It is easy for projective resolutions and requires work for injective ones (see e.g. [13]).

Lemma 3.18 *Injective and projective resolutions exist in the category of* R*-modules, where* R *is any ring (not necessarily commutative).*

Note that applying a left-exact covariant functor $\mathcal{F}$ to an injective resolution

$$0 \to M \to I^0 \to I^1 \to I^2 \to \cdots$$

of M gives rise to a cochain complex

$$0 \to \mathscr{F}(M) \to \mathscr{F}(I^0) \to \mathscr{F}(I^1) \to \mathscr{F}(I^2) \to \cdots,$$

of which only the part $0 \to \mathscr{F}(M) \to \mathscr{F}(I^0) \to \mathscr{F}(I^1)$ need to be exact. This means that the H^0 of the (cut off at 0) cochain complex $\mathscr{F}(I^0) \to \mathscr{F}(I^1) \to \mathscr{F}(I^2) \to \cdots$ is equal to $\mathscr{F}(M)$.

Definition 3.19 Let R be a not necessarily commutative ring.

(a) Let $\mathscr{F}$ be a left-exact covariant functor on the category of R-modules (mapping for instance to $\mathbb{Z}$ – modules).
The *right derived functors* $R^n\mathscr{F}(\cdot)$ of $\mathscr{F}$ are the functors on the category of R – modules defined as follows. For $M \in \mathrm{obj}(R - \text{modules})$ choose an injective resolution $0 \to M \to I^0 \to I^1 \to \cdots$ and let

$$R^n\mathscr{F}(M) := \mathrm{H}^n\left(\mathscr{F}(I^0) \to \mathscr{F}(I^1) \to \mathscr{F}(I^2) \to \cdots\right).$$

(b) Let $\mathscr{G}$ be a left-exact contravariant functor on the category of R-modules.
The *right derived functors* $R^n\mathscr{G}(\cdot)$ of $\mathscr{G}$ are the functors on the category of R – modules defined as follows. For $M \in \mathrm{obj}(R - \text{modules})$ choose a projective resolution $\cdots \to P_1 \to P_0 \to M \to 0$ and let

$$R^n\mathscr{G}(M) := \mathrm{H}^n\left(\mathscr{G}(P_0) \to \mathscr{G}(P_1) \to \mathscr{G}(P_2) \to \cdots\right).$$

We state the following lemma without a proof. It is a simple consequence of the injectivity, respectively, projectivity of the modules in the resolution.

Lemma 3.20 *The right derived functors do not depend on the choice of the resolution and they form a cohomological δ-functor.*

Of course, one can also define left derived functors of right-exact functors. An important example is the Tor-functor which is obtained by deriving the tensor product functor in a way dual to Ext (see below). As already mentioned, the importance of right and left derived functors comes from their universality.

Definition 3.21 (a) Let $(\mathrm{H}^n)_n$ and $(T^n)_n$ be cohomological δ-functors. A *morphism of cohomological δ-functors* is a collection of natural transformations $\eta^n : \mathrm{H}^n \Rightarrow T^n$ that commute with the connecting homomorphisms δ, i.e. for every short exact sequence $0 \to A \to B \to C \to 0$ and every n the diagram

$$\begin{array}{ccc} \mathrm{H}^n(C) & \xrightarrow{\delta} & \mathrm{H}^{n+1}(A) \\ \downarrow{\scriptstyle \eta^n_C} & & \downarrow{\scriptstyle \eta^{n+1}_A} \\ T^n(C) & \xrightarrow{\delta} & T^{n+1}(A) \end{array}$$

commutes.

(b) The cohomological δ-functor $(\mathrm{H}^n)_n$ is *universal* if for every cohomological δ-functor $(T^n)_n$ and every natural transformation $\eta^0 : \mathrm{H}^0(\cdot) \Rightarrow T^0(\cdot)$ there is a unique natural transformation $\eta^n : \mathrm{H}^n(\cdot) \Rightarrow T^n(\cdot)$ for all $n \geq 1$ such that the η^n form a morphism of cohomological δ-functors between $(\mathrm{H}^n)_n$ and $(T^n)_n$.

For the proof of the following central result, we refer to [21], Chap. 2.

Theorem 3.22 *Let R be a not necessarily commutative ring and let $\mathscr{F}$ be a left-exact covariant or contravariant functor on the category of R-modules (mapping for instance to $\mathbb{Z}$ –* modules*).*

The right derived functors $(R^n\mathscr{F}(\cdot))_n$ *of $\mathscr{F}$ form a* <u>*universal*</u> *cohomological δ-functor.*

Example 3.23 (a) Let R be a commutative ring and G a group. The functor

$$(\cdot)^G : R[G] - \text{modules} \to R - \text{modules}, \quad M \mapsto M^G$$

is left-exact and covariant, hence we can form its right derived functors $R^n(\cdot)^G$. Since we have the special case $(R^0(\cdot)^G)(M) = M^G$, universality gives a morphism of cohomological δ-functors $R^n(\cdot)^G \Rightarrow H^n(G, \cdot)$. We shall see that this is an isomorphism.

(b) Let R be a not necessarily commutative ring. We have seen that the functors $\mathrm{Hom}_R(\cdot, M)$ and $\mathrm{Hom}_R(M, \cdot)$ are left-exact. We write

$$\mathrm{Ext}^n_R(\cdot, M) := R^n\mathrm{Hom}_R(\cdot, M) \quad \text{and} \quad \mathrm{Ext}^n_R(M, \cdot) := R^n\mathrm{Hom}_R(M, \cdot).$$

See Theorem 3.24 below.

(c) Many cohomology theories in (algebraic) geometry are also of a right derived functor nature. For instance, let X be a topological space and consider the category of sheaves of abelian groups on X. The global sections functor $\mathscr{F} \mapsto \mathscr{F}(X) = \mathrm{H}^0(X, \mathscr{F})$ is left-exact and its right derived functors $R^n(\mathrm{H}^0(X, \cdot))$ can be formed. They are usually denoted by $\mathrm{H}^n(X, \cdot)$ and they define 'sheaf cohomology' on X. Etale cohomology is an elaboration of this based on a generalisation of topological spaces.

Universal Properties of Group Cohomology

Theorem 3.24 *Let R be a not necessarily commutative ring. The* Ext*-functor is* balanced*. This means that for any two R-modules M, N there are isomorphisms*

$$(\mathrm{Ext}^n_R(\cdot, N))(M) \cong (\mathrm{Ext}^n_R(M, \cdot))(N) =: \mathrm{Ext}^n_R(M, N).$$

Proof [21], Theorem 2.7.6. □

Corollary 3.25 *Let R be a commutative ring and G a group. For every $R[G]$-module M there are isomorphisms*

$$H^n(G, M) \cong \mathrm{Ext}^n_{R[G]}(R, M) \cong (R^n(\cdot)^G)(M)$$

and the functors $(H^n(G, \cdot))_n$ form a universal cohomological δ-functor. Moreover, apart from the standard resolution of R by free $R[G]$-modules, any resolution of R by projective $R[G]$-modules may be used to compute $H^n(G, M)$.

Proof We may compute $\mathrm{Ext}^n_{R[G]}(\cdot, M)(R)$ by any resolution of R by projective $R[G]$-modules. Our standard resolution is such a resolution, since any free module is projective. Hence, $H^n(G, M) \cong \mathrm{Ext}^n_{R[G]}(\cdot, M)(R)$. The key is now that Ext is balanced (Theorem 3.24), since it gives $H^n(G, M) \cong \mathrm{Ext}^n_{R[G]}(R, \cdot)(M) \cong R^n(\cdot)^G(M) \cong \mathrm{Ext}^n_{R[G]}(R, M)$. As the Ext-functor is universal (being a right derived functor), also $H^n(G, \cdot)$ is universal. For the last statement we recall that right derived functors do not depend on the chosen projective respectively injective resolution. □

You are invited to look at Exercise 3.39 now.

3.4 Theory: Generalities on Group Cohomology

We now apply the universality of the δ-functor of group cohomology. Let $\phi : H \to G$ be a group homomorphism and A an $R[G]$-module. Via ϕ we may consider A also as an $R[H]$-module and $\mathrm{res}^0 : \mathrm{H}^0(G, \cdot) \to \mathrm{H}^0(H, \cdot)$ is a natural transformation. By the universality of $\mathrm{H}^\bullet(G, \cdot)$ we get natural transformations

$$\mathrm{res}^n : \mathrm{H}^n(G, \cdot) \to \mathrm{H}^n(H, \cdot).$$

These maps are called *restrictions*. See Exercise 3.40 for a description in terms of cochains. Very often ϕ is just the embedding map of a subgroup.

Assume now that H is a normal subgroup of G and A is an $R[G]$-module. Then we can consider $\phi : G \to G/H$ and the restriction above gives natural transformations $\mathrm{res}^n : \mathrm{H}^n(G/H, (\cdot)^H) \to \mathrm{H}^n(G, (\cdot)^H)$. We define the *inflation maps* to be

$$\mathrm{infl}^n : \mathrm{H}^n(G/H, A^H) \xrightarrow{\mathrm{res}^n} \mathrm{H}^n(G, A^H) \longrightarrow \mathrm{H}^n(G, A)$$

where the last arrow is induced from the natural inclusion $A^H \hookrightarrow A$.

Under the same assumptions, conjugation by $g \in G$ preserves H and we have the isomorphism $H^0(H, A) = A^H \xrightarrow{a \mapsto ga} A^H = H^0(H, A)$. Hence by universality we obtain natural maps $H^n(H, A) \to H^n(H, A)$ for every $g \in G$. One even gets an $R[G]$-action on $\mathrm{H}^n(H, A)$. As $h \in H$ is clearly the identity on $\mathrm{H}^0(H, A)$, the above action is in fact also an $R[G/H]$-action.

Let now $H \leq G$ be a subgroup of finite index. Then the norm $N_{G/H} := \sum_{\{g_i\}} \in R[G]$ with $\{g_i\}$ a system of representatives of G/H gives a natural transformation $\mathrm{cores}^0 : \mathrm{H}^0(H, \cdot) \to \mathrm{H}^0(G, \cdot)$ where $\cdot$ is an $R[G]$-module. By universality we obtain

$$\mathrm{cores}^n : \mathrm{H}^n(H, \cdot) \to \mathrm{H}^n(G, \cdot),$$

the *corestriction (transfer)* maps.

The inflation map, the $R[G/H]$-action and the corestriction can be explicitly described in terms of cochains of the bar resolution (see Exercise 3.40).

It is clear that $\mathrm{cores}^0 \circ \mathrm{res}^0$ is multiplication by the index $(G : H)$. By universality, also $\mathrm{cores}^n \circ \mathrm{res}^n$ is multiplication by the index $(G : H)$. Hence we have proved the first part of the following proposition.

Proposition 3.26 *(a) Let $H < G$ be a subgroup of finite index $(G : H)$. For all i and all $R[G]$-modules M one has the equality*

$$\mathrm{cores}_H^G \circ \mathrm{res}_H^G = (G : H)$$

on all $\mathrm{H}^i(G, M)$.

(b) Let G be a finite group of order n and R a ring in which n is invertible. Then $\mathrm{H}^i(G, M) = 0$ for all $i \geq 1$ and all $R[G]$-modules M.

Proof Part (b) is an easy consequence with $H = 1$, since

$$\mathrm{H}^i(G, M) \xrightarrow{\mathrm{res}_H^G} \mathrm{H}^i(1, M) \xrightarrow{\mathrm{cores}_H^G} \mathrm{H}^i(G, M)$$

is the zero map (as $\mathrm{H}^i(1, M) = 0$ for $i \geq 1$), but it also is multiplication by n. □

The following exact sequence turns out to be very important for our purposes.

Theorem 3.27 (Hochschild–Serre) *Let $H \leq G$ be a normal subgroup and A an $R[G]$-module. There is the exact sequence:*

$$0 \to \mathrm{H}^1(G/H, A^H) \xrightarrow{\mathrm{infl}} \mathrm{H}^1(G, A) \xrightarrow{\mathrm{res}} \mathrm{H}^1(G, A)^{G/H} \to \mathrm{H}^2(G/H, A^H) \xrightarrow{\mathrm{infl}} \mathrm{H}^2(G, A).$$

Proof We only sketch the proof for those who know spectral sequences. It is, however, possible to verify the exactness on cochains explicitly (after having defined the missing map appropriately). Grothendieck's theorem on spectral sequences ([21], 6.8.2) associates to the composition of functors

$$(A \mapsto A^H \mapsto (A^H)^{G/H}) = (A \mapsto A^G)$$

the spectral sequence

$$E_2^{p,q} : H^p(G/H, H^q(H, A)) \Rightarrow H^{p+q}(G, A).$$

The statement of the theorem is then just the 5-term sequence that one can associate with every spectral sequence of this type. □

Coinduced Modules and Shapiro's Lemma

Let $H < G$ be a subgroup and A be a left $R[H]$-module. The $R[G]$-module

$$\mathrm{Coind}_H^G(A) := \mathrm{Hom}_{R[H]}(R[G], A)$$

is called the *coinduction* or the *coinduced module* from H to G of A. We make $\mathrm{Coind}_H^G(A)$ into a left $R[G]$-module by

$$(g.\phi)(g') = \phi(g'g) \;\; \forall\, g, g' \in G,\; \phi \in \mathrm{Hom}_{R[H]}(R[G], A).$$

Proposition 3.28 (Shapiro's Lemma) *For all $n \geq 0$, the map*

$$\mathrm{Sh} : \mathrm{H}^n(G, \mathrm{Coind}_H^G(A)) \to \mathrm{H}^n(H, A)$$

given on cochains is given by

$$c \mapsto ((h_1, \ldots, h_n) \to (c(h_1, \ldots, h_n))(1_G))$$

is an isomorphism.

Proof Exercise 3.41. □

Mackey's Formula and Stabilisers

If $H \leq G$ are groups and V is an $R[G]$-module, we denote by $\mathrm{Res}_H^G(V)$ the module V considered as an $R[H]$-module if we want to stress that the module is obtained by restriction. In later sections, we will often silently restrict modules to subgroups.

Proposition 3.29 *Let R be a ring, G be a group and H, K subgroups of G. Let furthermore V be an $R[H]$-module.* Mackey's formula *is the isomorphism*

$$\mathrm{Res}_K^G \mathrm{Coind}_H^G V \cong \prod_{g \in H\backslash G/K} \mathrm{Coind}_{K\cap g^{-1}Hg}^{K}\, {}^g(\mathrm{Res}_{H\cap gKg^{-1}}^H V).$$

Here ${}^g(\mathrm{Res}_{H\cap gKg^{-1}}^H V)$ denotes the $R[K \cap g^{-1}Hg]$-module obtained from V via the conjugated action $g^{-1}hg._g v := h.v$ for $v \in V$ and $h \in H$ such that $g^{-1}hg \in K$.

Proof We consider the commutative diagram

$$\begin{array}{ccc} \mathrm{Res}_K^G \mathrm{Hom}_{R[H]}(R[G], V) & \longrightarrow & \prod_{g\in H\backslash G/K} \mathrm{Hom}_{R[K\cap g^{-1}Hg]}(R[K], {}^g(\mathrm{Res}_{H\cap gKg^{-1}}^H V)) \\ & \searrow & \downarrow \sim \\ & & \prod_{g\in H\backslash G/K} \mathrm{Hom}_{R[H\cap gKg^{-1}]}(R[gKg^{-1}], \mathrm{Res}_{H\cap gKg^{-1}}^H V)). \end{array}$$

The vertical arrow is just given by conjugation and is clearly an isomorphism. The diagonal map is the product of the natural restrictions. From the bijection

$$(H \cap gKg^{-1})\backslash gKg^{-1} \xrightarrow{gkg^{-1} \mapsto Hgk} H\backslash HgK$$

it is clear that also the diagonal map is an isomorphism, proving the proposition. □

From Shapiro's Lemma 3.28 we directly get the following.

Corollary 3.30 *In the situation of Proposition 3.29 one has*

$$\begin{aligned} \mathrm{H}^i(K, \mathrm{Coind}_H^G V) &\cong \prod_{g \in H\backslash G/K} \mathrm{H}^i(K \cap g^{-1}Hg, {}^g(\mathrm{Res}^H_{H \cap gKg^{-1}} V)) \\ &\cong \prod_{g \in H\backslash G/K} \mathrm{H}^i(H \cap gKg^{-1}, \mathrm{Res}^H_{H \cap gKg^{-1}} V) \end{aligned}$$

for all $i \in \mathbb{N}$.

3.5 Theoretical Exercises

Exercise 3.31 Check the statements made in Example 3.6.

Exercise 3.32 Verify the statements of Example 3.8.

Exercise 3.33 Prove Lemma 3.10.

Hint: take a free R-module F which surjects onto P, i.e. $\pi : F \twoheadrightarrow P$, and use the definition of P being projective to show that the surjection admits a split $s : P \to F$, meaning that $\pi \circ s$ is the identity on P. This is then equivalent to the assertion.

Exercise 3.34 Let $\phi^\bullet : C^\bullet \to D^\bullet$ be a morphism of cochain complexes.

(a) Show that $\ker(\phi^\bullet)$ is a cochain complex and is a subcomplex of $C^\bullet$ in a natural way.
(b) Show that $\mathrm{im}(\phi^\bullet)$ is a cochain complex and is a subcomplex of $D^\bullet$ in a natural way.
(c) Show that $\mathrm{coker}(\phi^\bullet)$ is a cochain complex and is a quotient of $D^\bullet$ in a natural way.
(d) Show that $\phi^\bullet$ induces homomorphisms $\mathrm{H}^n(C^\bullet) \xrightarrow{\mathrm{H}^n(\phi^\bullet)} \mathrm{H}^n(D^\bullet)$ for all $n \in \mathbb{N}$.

Exercise 3.35 Check the exactness of the standard resolution of a group G.

Exercise 3.36 Check the exactness of the resolutions (11) and (12) for an infinite and a finite cyclic group, respectively.

Exercise 3.37 Let R, G, M be as in the definition of group (co-)homology.

(a) Prove $\mathrm{H}^0(G, M) \cong M^G$, the G-invariants of M.
(b) Prove $\mathrm{H}_0(G, M) \cong M_G$, the G-coinvariants of M.
(c) Prove the explicit descriptions:

$$\mathrm{Z}^1(G, M) = \{f : G \to M \text{ map} \mid f(gh) = g.f(h) + f(g)\ \forall g, h \in G\},$$
$$\mathrm{B}^1(G, M) = \{f : G \to M \text{ map} \mid \exists m \in M : f(g) = (1 - g)m\ \forall g \in G\},$$
$$\mathrm{H}^1(G, M) = \mathrm{Z}^1(G, M)/B^1(G, M).$$

In particular, if the action of G on M is trivial, the boundaries $B^1(G, M)$ are zero, and one has:

$$\mathrm{H}^1(G, M) = \mathrm{Hom}_{\text{group}}(G, M).$$

Exercise 3.38 Prove Proposition 3.16.

Hint: for (a), use that free modules are projective. (b) follows from (a) together with Theorem 3.15 or, alternatively, by direct calculation. See also [4, III.6.1].

Exercise 3.39 Let R be a commutative ring.

(a) Let $G = \langle T \rangle$ be a free cyclic group and M any $R[G]$-module. Prove

$$\mathrm{H}^0(G, M) = M^G, \quad \mathrm{H}^1(G, M) = M/(1 - T)M \quad \text{and} \quad \mathrm{H}^i(G, M) = 0$$

for all $i \geq 2$.
(b) For a finite cyclic group $G = \langle \sigma \rangle$ of order n and any $R[G]$-module M prove that

$$\mathrm{H}^0(G, M) \cong M^G, \qquad \mathrm{H}^1(G, M) \cong \{m \in M \mid N_\sigma m = 0\}/(1 - \sigma)M,$$
$$\mathrm{H}^2(G, M) \cong M^G/N_\sigma M, \quad \mathrm{H}^i(G, M) \cong \mathrm{H}^{i+2}(G, M) \text{ for all } i \geq 1.$$

Exercise 3.40 Let R be a commutative ring.

(a) Let $\phi : H \to G$ be a group homomorphism and A an $R[G]$-module. Prove that the restriction maps $\mathrm{res}^n : H^n(G, A) \to H^n(H, A)$ are given in terms of cochains of the bar resolution by composing the cochains by ϕ.
(b) Let H be a normal subgroup of G. Describe the inflation maps in terms of cochains of the bar resolution.
(c) Let H be a normal subgroup of G and A an $R[G]$-module. Describe the $R[G/H]$-action on $H^n(H, A)$ in terms of cochains of the bar resolution.
(d) Let now $H \leq G$ be a subgroup of finite index. Describe the corestriction maps in terms of cochains of the bar resolution.

Exercise 3.41 Prove Shapiro's lemma, i.e. Proposition 3.28.

Hint: see [21, (6.3.2)] for an abstract proof; see also [4, III.6.2] for the underlying map.

4 Cohomology of $\mathrm{PSL}_2(\mathbb{Z})$

In this section, we shall calculate the cohomology of the group $\mathrm{PSL}_2(\mathbb{Z})$ and important properties thereof. This will be at the basis of our treatment of Manin symbols in the following section. The key in this is the description of $\mathrm{PSL}_2(\mathbb{Z})$ as a free product of two cyclic groups.

4.1 Theory: The Standard Fundamental Domain for $\mathrm{PSL}_2(\mathbb{Z})$

We define the matrices of $\mathrm{SL}_2(\mathbb{Z})$

$$\sigma := \begin{pmatrix} 0 & -1 \\ 1 & 0 \end{pmatrix}, \quad \tau := \begin{pmatrix} -1 & 1 \\ -1 & 0 \end{pmatrix}, \quad T = \begin{pmatrix} 1 & 1 \\ 0 & 1 \end{pmatrix} = \tau\sigma.$$

By the definition of the action of $\mathrm{SL}_2(\mathbb{Z})$ on $\mathbb{H}$ in Eq. 1, we have for all $z \in \mathbb{H}$:

$$\sigma.z = \frac{-1}{z}, \quad \tau.z := 1 - \frac{1}{z}, \quad T.z = z + 1.$$

These matrices have the following conceptual meaning:

$$\langle \pm\sigma \rangle = \mathrm{Stab}_{\mathrm{SL}_2(\mathbb{Z})}(i), \ \langle \pm\tau \rangle = \mathrm{Stab}_{\mathrm{SL}_2(\mathbb{Z})}(\zeta_6) \ \text{ and } \ \langle \pm T \rangle = \mathrm{Stab}_{\mathrm{SL}_2(\mathbb{Z})}(\infty)$$

with $\zeta_6 = e^{2\pi i/6}$. From now on we will often represent classes of matrices in $\mathrm{PSL}_2(\mathbb{Z})$ by matrices in $\mathrm{SL}_2(\mathbb{Z})$. The orders of σ and τ in $\mathrm{PSL}_2(\mathbb{Z})$ are 2 and 3, respectively. These statements are checked by calculation. Exercise 4.19 is recommended at this point.

Even though in this section our interest concerns the full group $\mathrm{SL}_2(\mathbb{Z})$, we give the definition of fundamental domain for general subgroups of $\mathrm{SL}_2(\mathbb{Z})$ of finite index.

Definition 4.1 Let $\Gamma \leq \mathrm{SL}_2(\mathbb{Z})$ be a subgroup of finite index. A *fundamental domain* for the action of Γ on $\mathbb{H}$ is a subset $\mathcal{F} \subset \mathbb{H}$ such that the following hold:

(i) $\mathcal{F}$ is open.
(ii) For every $z \in \mathbb{H}$, there is $\gamma \in \Gamma$ such that $\gamma.z \in \overline{\mathcal{F}}$.
(iii) If $\gamma.z \in \mathcal{F}$ for $z \in \mathcal{F}$ and $\gamma \in \Gamma$, then one has $\gamma = \pm\begin{pmatrix} 1 & 0 \\ 0 & 1 \end{pmatrix}$.

In other words, a fundamental domain is an open set, which is small enough not to contain any two points that are equivalent under the operation by Γ, and which is big enough that every point in the upper half plane is equivalent to some point in the closure of the fundamental domain.

Proposition 4.2 *The set*

$$\mathscr{F} := \{z \in \mathbb{H} \mid |z| > 1 \text{ and } -\frac{1}{2} < \mathrm{Re}(z) < \frac{1}{2}\}$$

is a fundamental domain for the action of $\mathrm{SL}_2(\mathbb{Z})$ *on* $\mathbb{H}$.

It is clear that $\mathscr{F}$ is open. For (ii), we use the following lemma.

Lemma 4.3 *Let* $z \in \mathbb{H}$. *The orbit* $\mathrm{SL}_2(\mathbb{Z}).z$ *contains a point* $\gamma.z$ *with maximal imaginary part (geometrically also called 'height'), i.e.*

$$\mathrm{Im}(\gamma.z) \geq \mathrm{Im}(g.z) \;\; \forall g \in \mathrm{SL}_2(\mathbb{Z}).$$

A point $z \in \mathbb{H}$ *is of maximal height if* $|cz+d| \geq 1$ *for all coprime* $c, d \in \mathbb{Z}$.

Proof We have the simple formula $\mathrm{Im}(\gamma.z) = \frac{\mathrm{Im}(z)}{|cz+d|^2}$. It implies

$$\mathrm{Im}(z) \leq \mathrm{Im}(\gamma.z) \Leftrightarrow |cz+d| \leq 1.$$

For fixed $z = x + iy$ with $x, y \in \mathbb{R}$, consider the inequality

$$1 \geq |cz+d|^2 = (cx+d)^2 + c^2y^2.$$

This expression admits only finitely many solutions $c, d \in \mathbb{Z}$. Among these finitely many, we may choose a coprime pair (c, d) with minimal $|cz+d|$. Due to the coprimeness, there are $a, b \in \mathbb{Z}$ such that the matrix $M := \begin{pmatrix} a & b \\ c & d \end{pmatrix}$ belongs to $\mathrm{SL}_2(\mathbb{Z})$. It is now clear that $M.z$ has maximal height. □

We next use a simple trick to show (ii) in Definition 4.1 for $\mathscr{F}$. Let $z \in \mathbb{H}$. By Lemma 4.3, we choose $\gamma \in \mathrm{SL}_2(\mathbb{Z})$ such that $\gamma.z$ has maximal height. We now 'transport' $\gamma.z$ via an appropriate translation T^n in such a way that $-1/2 \leq \mathrm{Re}(T^n\gamma.z) < 1/2$. The height is obviously left invariant. Now we have $|T^n\gamma.z| \geq 1$ because otherwise the height of $T^n\gamma.z$ would not be maximal. For, if $|T^n\gamma.z + 0| < 1$ then applying σ (corresponding to reflection on the unit circle) would make the height strictly bigger. More precisely, we have the following result.

Lemma 4.4 *Every point of maximal height in* $\mathbb{H}$ *can be translated into the closure of the fundamental domain* $\overline{\mathscr{F}}$. *Conversely,* $\overline{\mathscr{F}}$ *only contains points of maximal height.*

Proof The first part was proved in the preceding discussion. The second one follows from the calculation

$$\begin{aligned} |cz+d|^2 =&(cx+d)^2 + c^2y^2 = c^2|z|^2 + 2cdx + d^2 \\ &\geq c^2|z|^2 - |cd| + d^2 \geq c^2 - |cd| + d^2 \geq (|c| - |d|)^2 + |cd| \geq 1 \end{aligned} \tag{13}$$

for all coprime integers c, d and $z = x + iy \in \mathbb{H}$ with $x, y \in \mathbb{R}$. □

Proof (*End of the proof of Proposition* 4.2.) Let $z \in \mathscr{F}$ and $\gamma := \left(\begin{smallmatrix} a & b \\ c & d \end{smallmatrix}\right) \in \mathrm{SL}_2(\mathbb{Z})$ such that $\gamma.z \in \mathscr{F}$. By Lemma 4.4, z and $\gamma.z$ both have maximal height, whence $|cz + d| = 1$. Hence the inequalities in Eq. 13 are equalities, implying $c = 0$. Thus, $\gamma = \pm T^n$ for some $n \in \mathbb{Z}$. But only $n = 0$ is compatible with the assumption $\gamma.z \in \mathscr{F}$. This proves (iii) in Definition 4.1 for $\mathscr{F}$. □

Proposition 4.5 *The group* $\mathrm{SL}_2(\mathbb{Z})$ *is generated by the matrices* σ *and* τ.

Proof Let $\Gamma := \langle \sigma, \tau \rangle$ be the subgroup of $\mathrm{SL}_2(\mathbb{Z})$ generated by σ and T.

We prove that for any $z \in \mathbb{H}$ there is $\gamma \in \Gamma$ such that $\gamma.z \in \overline{\mathscr{F}}$. For that, note that the orbit $\Gamma.z$ contains a point $\gamma.z$ for $\gamma \in \Gamma$ of maximal height as it is a subset of $\mathrm{SL}_2(\mathbb{Z}).z$, for which we have seen that statement. As Γ contains $T = \tau\sigma$, we can translate $\gamma.z$ so as to have real part in between $-\frac{1}{2}$ and $\frac{1}{2}$. As Γ also contains σ, the absolute value of the new point has to be at least 1 because other σ would make the height bigger.

In order to conclude, choose any point $z \in \mathscr{F}$ and let $M \in \mathrm{SL}_2(\mathbb{Z})$. We consider the point $M.z$ and 'transport' it back into $\mathscr{F}$ via a matrix $\gamma \in \Gamma$. We thus have $(\gamma M).z \in \mathscr{F}$. As $\mathscr{F}$ is a fundamental domain for $\mathrm{SL}_2(\mathbb{Z})$, it follows $\gamma M = \pm 1$, showing $M \in \Gamma$. □

An alternative algorithmic proof is provided in Algorithm 5.10 below.

4.2 *Theory:* $\mathbf{PSL_2(\mathbb{Z})}$ *as a Free Product*

We now apply the knowledge about the (existence of the) fundamental domain for $\mathrm{PSL}_2(\mathbb{Z})$ to derive that $\mathrm{PSL}_2(\mathbb{Z})$ is a free product.

Definition 4.6 Let G and H be two groups. The *free product* $G * H$ of G and H is the group having as elements all the possible *words*, i.e. sequences of symbols, $a_1 a_2 \ldots a_n$ with $a_i \in G - \{1\}$ or $a_i \in H - \{1\}$ such that elements from G and H alternate (i.e. if $a_i \in G$, then $a_{i+1} \in H$ and vice versa) together with the empty word, which we denote by 1. The integer n is called the *length* of the group element (word) $w = a_1 a_2 \ldots a_n$ and denoted by $l(w)$. We put $l(1) = 0$ for the empty word.

The group operation in $G * H$ is concatenation of words followed by 'reduction' (in order to obtain a new word obeying to the rules). The reduction rules are: for all words v, w, all $g_1, g_2 \in G$ and all $h_1, h_2 \in H$:

- $v1w = vw$,
- $vg_1g_2w = v(g_1g_2)w$ (i.e. the multiplication of g_1 and g_2 in G is carried out),
- $vh_1h_2w = v(h_1h_2)w$ (i.e. the multiplication of h_1 and h_2 in H is carried out).

In Exercise 4.18 you are asked to verify that $G * H$ is indeed a group and to prove a universal property. Alternatively, if G is given by the set of generators $\mathscr{G}_G$ together with relations $\mathscr{R}_G$ and similarly for the group H, then the free product $G * H$ can be described as the group generated by $\mathscr{G}_G \cup \mathscr{G}_H$ with relations $\mathscr{R}_G \cup \mathscr{R}_H$.

Theorem 4.7 *Let $\mathscr{P}$ be the free product $\langle\sigma\rangle * \langle\tau\rangle$ of the cyclic groups $\langle\sigma\rangle$ of order* 2 *and $\langle\tau\rangle$ of order* 3.

Then $\mathscr{P}$ is isomorphic to $\mathrm{PSL}_2(\mathbb{Z})$. *In particular, as an abstract group,* $\mathrm{PSL}_2(\mathbb{Z})$ *can be represented by generators and relations as* $\langle\sigma, \tau \mid \sigma^2 = \tau^3 = 1\rangle$.

In the proof, we will need the following statement, which we separate because it is entirely computational.

Lemma 4.8 *Let $\gamma \in \mathscr{P}$ be* 1 *or any word starting in σ on the left, i.e. $\sigma\tau^{e_1}\sigma\tau^{e_2}\ldots$. Then* $\mathrm{Im}(\tau^2\gamma.i) < 1$.

Proof For $\gamma = 1$, the statement is clear. Suppose $\gamma = \sigma\tau^{e_1}\sigma\tau^{e_2}\sigma\ldots\tau^{e_{r-1}}\sigma\tau^{e_r}$ with $r \geq 0$, $e_i \in \{1, 2\}$ for $i = 1, \ldots, r$. We prove more generally

$$\mathrm{Im}(\tau^2\gamma.i) = \mathrm{Im}(\tau^2(\gamma\sigma).i) > \mathrm{Im}(\tau^2(\gamma\sigma)\tau^e.i) = \mathrm{Im}(\tau^2(\gamma\sigma)\tau^e\sigma.i)$$

for any $e = 1, 2$. This means that extending the word to the right by $\sigma\tau^e$, the imaginary part goes strictly down for both $e = 1, 2$.

We first do some matrix calculations. Let us say that an integer matrix $\left(\begin{smallmatrix} a & b \\ c & d \end{smallmatrix}\right)$ satisfies (*) if $(c+d)^2 > \max(c^2, d^2)$. The matrix $\tau^2\sigma = \left(\begin{smallmatrix} -1 & 0 \\ -1 & -1 \end{smallmatrix}\right)$ clearly satisfies (*). Let us assume that $\gamma = \left(\begin{smallmatrix} a & b \\ c & d \end{smallmatrix}\right)$ satisfies (*). We show that $\gamma\tau\sigma = \left(\begin{smallmatrix} * & * \\ c & c+d \end{smallmatrix}\right)$ and $\gamma\tau^2\sigma = \left(\begin{smallmatrix} * & * \\ -c-d & -d \end{smallmatrix}\right)$ also satisfy (*). The first one follows once we know $(2c+d)^2 > \max(c^2, (c+d)^2)$. This can be seen like this:

$$(2c+d)^2 = (c^2 + 2cd) + 2c^2 + (c+d)^2 > 2c^2 + (c+d)^2 \geq \max(c^2, (c+d)^2),$$

where we used that (*) implies $(c+d)^2 > d^2$ and, thus, $c^2 + 2cd > 0$. The second inequality is obtained by exchanging the roles of c and d.

We thus see that $\tau^2\gamma = \left(\begin{smallmatrix} a & b \\ c & d \end{smallmatrix}\right)$ satisfies (*) for all words γ starting and ending in σ. Finally, we have for all such γ:

$$\begin{aligned}
\mathrm{Im}(\tau^2\gamma i) &= \frac{1}{|ci+d|^2} &&= \frac{1}{c^2+d^2}, \\
\mathrm{Im}(\tau^2\gamma\tau i) &= \mathrm{Im}(\tau^2\gamma(i+1)) = \frac{1}{|c(i+1)+d|^2} &&= \frac{1}{(c+d)^2+c^2}, \\
\mathrm{Im}(\tau^2\gamma\tau^2 i) &= \mathrm{Im}(\tau^2\gamma\frac{1+i}{2}) = \frac{1/2}{|c(i/2+1/2)+d|^2} &&= \frac{2}{(c+2d)^2+c^2}.
\end{aligned}$$

Now (*) implies the desired inequalities of the imaginary parts. □

Proof (*Proof of Theorem* 4.7) As $\mathrm{SL}_2(\mathbb{Z})$ is generated by σ and τ due to Proposition 4.5, the universal property of the free product gives us a surjection of groups $\mathscr{P} \twoheadrightarrow \mathrm{PSL}_2(\mathbb{Z})$.

Let B be the geodesic path from ζ_6 to i, i.e. the arc between ζ_6 and i in positive orientation (counter clockwise) on the circle of radius 1 around the origin, lying entirely on the closure $\overline{\mathscr{F}}$ of the standard fundamental domain from Proposition 4.2. Define the map

$$\mathrm{PSL}_2(\mathbb{Z}) \xrightarrow{\phi} \{\text{Paths in } \mathbb{H}\}$$

which sends $\gamma \in \mathrm{PSL}_2(\mathbb{Z})$ to $\gamma.B$, i.e. the image of B under γ. The proof of the theorem is finished by showing that the composite

$$\mathscr{P} \twoheadrightarrow \mathrm{PSL}_2(\mathbb{Z}) \xrightarrow{\phi} \{\text{Paths in } \mathbb{H}\}$$

is injective, as then the first map must be an isomorphism.

This composition is injective because its image is a tree, that is, a graph without circles. By drawing it, one convinces oneself very quickly hereof. We, however, give a formal argument, which can also be nicely visualised on the geometric realisation of the graph as going down further and further in every step.

In order to prepare for the proof, let us first suppose that $\gamma_1.B$ and $\gamma_2.B$ for some $\gamma_1, \gamma_2 \in \mathrm{PSL}_2(\mathbb{Z})$ meet in a point which is not the endpoint of either of the two paths. Then $\gamma.B$ intersects B in some interior point for $\gamma := \gamma_1^{-1}\gamma_2$. This intersection point lies on the boundary of the fundamental domain $\mathscr{F}$. Consequently, by (iii) in Definition 4.1, $\gamma = \pm 1$ and $\gamma_1.B = \gamma_2.B$. This implies that if $\mathrm{Im}(\gamma_1.i) \neq \mathrm{Im}(\gamma_2.i)$ where $i = \sqrt{-1}$, then $\gamma_1.B$ and $\gamma_2.B$ do not meet in any interior point and are thus distinct paths.

It is obvious that $B, \sigma.B, \tau.B$ are distinct paths. They share the property that their point that is conjugate to i has imaginary part 1 (in fact, the points conjugate to i in the paths are $i, i, i+1$, respectively).

By Lemma 4.8, for γ equal to 1 or any word in $\mathscr{P}$ starting with σ on the left, we obtain that $\tau^2\gamma.B$ is distinct from $B, \sigma.B, \tau.B$ because it lies 'lower'. In particular, $\tau^2\gamma.B \neq B$. As $\tau^2\gamma.B \neq \tau.B$, we also find $\tau\gamma.B \neq B$. Finally, if $\gamma.B = B$ and $\gamma = \sigma\tau^e\gamma'$ with $e \in \{1, 2\}$ and γ' starting in σ or $\gamma' = 1$, then $\tau^e\gamma'.B = \sigma.B$, which has already been excluded. We have thus found that for any non-trivial word $\gamma \in \mathscr{P}$, the conjugate $\gamma.B$ is distinct from B. This proves the desired injectivity. □

4.3 Theory: Mayer–Vietoris for $\mathrm{PSL}_2(\mathbb{Z})$

Motivated by the description $\mathrm{PSL}_2(\mathbb{Z}) = C_2 * C_3$, we now consider the cohomology of a group G which is the free product of two finite groups G_1 and G_2, i.e. $G = G_1 * G_2$.

Proposition 4.9 *Let R be a commutative ring. The sequence*

$$0 \to R[G] \xrightarrow{\alpha} R[G/G_1] \oplus R[G/G_2] \xrightarrow{\varepsilon} R \to 0$$

with $\alpha(g) = (gG_1, -gG_2)$ and $\varepsilon(gG_1, 0) = 1 = \varepsilon(0, gG_2)$ is exact.

Proof This proof is an even more elementary version of an elementary proof that I found in [3]. Clearly, ε is surjective and also $\varepsilon \circ \alpha = 0$.

Next we compute exactness at the centre. We first claim that for every element $g \in G$ we have

$$g - 1 = \sum_j \alpha_j g_j (h_j - 1) \in R[G/G_1]$$

for certain $\alpha_j \in R$ and certain $g_j \in G$, $h_j \in G_2$ and analogously with the roles of G_1 and G_2 exchanged. To see this, we write $g = a_1 a_2 \dots a_n$ with a_i alternatingly in G_1 and G_2 (we do not need the uniqueness of this expression). If $n = 1$, there is nothing to do. If $n > 1$, we have

$$a_1 a_2 \dots a_n - 1 = a_1 a_2 \dots a_{n-1}(a_n - 1) + (a_1 a_2 \dots a_{n-1} - 1)$$

and we obtain the claim by induction. Consequently, we have for all $\lambda = \sum_i r_i g_i G_1$ and all $\mu = \sum_k \tilde{r}_k \tilde{g}_k G_2$ with $r_i, \tilde{r}_k \in R$ and $g_i, \tilde{g}_k \in G$

$$\lambda - \sum_i r_i 1_G G_1 = \sum_j \alpha_j g_j (h_j - 1) \in R[G/G_1]$$

and

$$\mu - \sum_k \tilde{r}_k 1_G G_2 = \sum_l \tilde{\alpha}_l \tilde{g}_l (\tilde{h}_l - 1) \in R[G/G_2]$$

for certain $\alpha_j, \tilde{\alpha}_l \in R$, certain $g_j, \tilde{g}_l \in G$ and certain $h_j \in G_2$, $\tilde{h}_l \in G_1$. Suppose now that with λ and μ as above we have

$$\varepsilon(\lambda, \mu) = \sum_i r_i + \sum_k \tilde{r}_k = 0.$$

Then we directly get

$$\alpha\Big(\sum_j \alpha_j g_j (h_j - 1) - \sum_l \tilde{\alpha}_l \tilde{g}_l (\tilde{h}_l - 1) + \sum_i r_i 1_G\Big) = (\lambda, \mu)$$

and hence the exactness at the centre.

It remains to prove that α is injective. Now we use the freeness of the product. Let $\lambda = \sum_w a_w w \in R[G]$ be an element in the kernel of α. Hence, $\sum_w a_w w G_1 = 0$ and $\sum_w a_w w G_2 = 0$. Let us assume that $\lambda \neq 0$. It is clear that λ cannot just be a multiple of $1 \in G$, as otherwise it would not be in the kernel of α. Now pick the $g \in G$ with $a_g \neq 0$ having maximal length $l(g)$ (among all the $l(w)$ with $a_w \neq 0$). It follows that $l(g) > 0$. Assume without loss of generality that the representation of g ends in a non-zero element of G_1. Further, since $a_g \neq 0$ and $0 = \sum_w a_w w G_2$, there must be an $h \in G$ with $g \neq h$, $gG_2 = hG_2$ and $a_h \neq 0$. As g does not end in G_2, we must have $h = gy$ for some $0 \neq y \in G_2$. Thus, $l(h) > l(g)$, contradicting the maximality and proving the proposition. □

Recall that we usually denote the restriction of a module to a subgroup by the same symbol. For example, in the next proposition we will write $\mathrm{H}^1(G_1, M)$ instead of $\mathrm{H}^1(G_1, \mathrm{Res}^G_{G_1}(M))$.

Proposition 4.10 (Mayer–Vietoris) *Let $G = G_1 * G_2$ be a free product. Let M be a left $R[G]$-module. Then the Mayer–Vietoris sequence gives the exact sequences*

$$0 \to M^G \to M^{G_1} \oplus M^{G_2} \to M \to \mathrm{H}^1(G, M) \xrightarrow{\mathrm{res}} \mathrm{H}^1(G_1, M) \oplus \mathrm{H}^1(G_2, M) \to 0.$$

and for all $i \geq 2$ an isomorphism

$$\mathrm{H}^i(G, M) \cong \mathrm{H}^i(G_1, M) \oplus \mathrm{H}^i(G_2, M).$$

Proof We see that all terms in the exact sequence of Proposition 4.9 are free R-modules. We now apply the functor $\mathrm{Hom}_R(\cdot, M)$ to this exact sequence and obtain the exact sequence of $R[G]$-modules

$$0 \to M \to \underbrace{\mathrm{Hom}_{R[G_1]}(R[G], M)}_{\cong \mathrm{Coind}^G_{G_1}(M)} \oplus \underbrace{\mathrm{Hom}_{R[G_2]}(R[G], M)}_{\cong \mathrm{Coind}^G_{G_2}(M)} \to \underbrace{\mathrm{Hom}_R(R[G], M)}_{\cong \mathrm{Coind}^G_1(M)} \to 0.$$

The central terms, as well as the term on the right, can be identified with coinduced modules. Hence, the statements on cohomology follow by taking the long exact sequence of cohomology and invoking Shapiro's Lemma 3.28. □

We now apply the Mayer–Vietoris sequence (Proposition 4.10) to $\mathrm{PSL}_2(\mathbb{Z})$ and get that for any ring R and any left $R[\mathrm{PSL}_2(\mathbb{Z})]$-module M the sequence

$$0 \to M^{\mathrm{PSL}_2(\mathbb{Z})} \to M^{\langle\sigma\rangle} \oplus M^{\langle\tau\rangle} \to M \xrightarrow{m \mapsto f_m} \mathrm{H}^1(\mathrm{PSL}_2(\mathbb{Z}), M) \xrightarrow{\mathrm{res}} \mathrm{H}^1(\langle\sigma\rangle, M) \oplus \mathrm{H}^1(\langle\tau\rangle, M) \to 0 \quad (14)$$

is exact and for all $i \geq 2$ one has isomorphisms

$$\mathrm{H}^i(\mathrm{PSL}_2(\mathbb{Z}), M) \cong \mathrm{H}^i(\langle\sigma\rangle, M) \oplus \mathrm{H}^i(\langle\tau\rangle, M). \quad (15)$$

The 1-cocycle f_m can be explicitly described as the cocycle given by $f_m(\sigma) = (1 - \sigma)m$ and $f_m(\tau) = 0$ (see Exercise 4.21).

Lemma 4.11 *Let $\Gamma \leq \mathrm{PSL}_2(\mathbb{Z})$ be a subgroup of finite index and let $x \in \mathbb{H} \cup \mathbb{P}^1(\mathbb{Q})$ be any point. Recall that $\mathrm{PSL}_2(\mathbb{Z})_x$ denotes the stabiliser of x for the $\mathrm{PSL}_2(\mathbb{Z})$-action.*

(a) The map

$$\Gamma\backslash \mathrm{PSL}_2(\mathbb{Z})/\mathrm{PSL}_2(\mathbb{Z})_x \xrightarrow{g\mapsto gx} \Gamma\backslash \mathrm{PSL}_2(\mathbb{Z})x$$

is a bijection.

(b) For $g \in \mathrm{PSL}_2(\mathbb{Z})$ the stabiliser of gx for the Γ-action is

$$\Gamma_{gx} = \Gamma \cap g\mathrm{PSL}_2(\mathbb{Z})_x g^{-1}.$$

(c) For all $i \in \mathbb{N}$, and all $R[\Gamma]$-modules, Mackey's formula (Proposition 3.29) gives an isomorphism

$$\mathrm{H}^i(\mathrm{PSL}_2(\mathbb{Z})_x, \mathrm{Coind}_\Gamma^{\mathrm{PSL}_2(\mathbb{Z})} V) \cong \prod_{y\in\Gamma\backslash \mathrm{PSL}_2(\mathbb{Z})x} \mathrm{H}^i(\Gamma_y, V).$$

Proof (a) and (b) are clear and (c) follows directly from Mackey's formula. □

Corollary 4.12 *Let R be a ring and $\Gamma \leq \mathrm{PSL}_2(\mathbb{Z})$ be a subgroup of finite index such that all the orders of all stabiliser groups Γ_x for $x \in \mathbb{H}$ are invertible in R. Then for all $R[\Gamma]$-modules V one has*

$$\mathrm{H}^1(\Gamma, V) \cong M/(M^{\langle\sigma\rangle} + M^{\langle\tau\rangle})$$

with $M = \mathrm{Coind}_\Gamma^{\mathrm{PSL}_2(\mathbb{Z})}(V)$ and

$$\mathrm{H}^i(\Gamma, V) = 0$$

for all $i \geq 2$.

Proof By Lemma 4.11(b), all non-trivial stabiliser groups for the action of Γ on $\mathbb{H}$ are of the form $g\langle\sigma\rangle g^{-1} \cap \Gamma$ or $g\langle\tau\rangle g^{-1} \cap \Gamma$ for some $g \in \mathrm{PSL}_2(\mathbb{Z})$. Due to the invertibility assumption we get from Proposition 3.26 that the groups on the right in the equation in Lemma 4.11(c) are zero. Hence, by Shapiro's lemma (Proposition 3.28) we have

$$\mathrm{H}^i(\Gamma, V) \cong \mathrm{H}^i(\mathrm{PSL}_2(\mathbb{Z}), M)$$

for all $i \geq 0$, so that by Eqs. (14) and (15) we obtain the proposition. □

By Exercise 4.19, the assumptions of the proposition are for instance always satisfied if R is a field of characteristic not 2 or 3. Look at Exercise 4.20 to see for which N the assumptions hold for $\Gamma_1(N)$ and $\Gamma_0(N)$ over an arbitrary ring (e.g. the integers).

4.4 Theory: Parabolic Group Cohomology

Before going on, we include a description of the cusps as $\mathrm{PSL}_2(\mathbb{Z})$-orbits that is very useful for the sequel.

Lemma 4.13 *The cusps $\mathbb{P}^1(\mathbb{Q})$ lie in a single $\mathrm{PSL}_2(\mathbb{Z})$-orbit. The stabiliser group of ∞ for the $\mathrm{PSL}_2(\mathbb{Z})$-action is $\langle T \rangle$ and the map*

$$\mathrm{PSL}_2(\mathbb{Z})/\langle T \rangle \xrightarrow{g\langle T \rangle \mapsto g\infty} \mathbb{P}^1(\mathbb{Q})$$

is a $\mathrm{PSL}_2(\mathbb{Z})$-equivariant bijection.

Proof The claim on the stabiliser follows from a simple direct computation. This makes the map well-defined and injective. The surjectivity is equivalent to the claim that the cusps lie in a single $\mathrm{PSL}_2(\mathbb{Z})$-orbit and simply follows from the fact that any pair of coprime integers (a, c) appears as the first column of a matrix in $\mathrm{SL}_2(\mathbb{Z})$. □

Let R be a ring, $\Gamma \leq \mathrm{PSL}_2(\mathbb{Z})$ a subgroup of finite index. One defines the *parabolic cohomology group for the left $R[\Gamma]$-module V* as the kernel of the restriction map in

$$0 \to \mathrm{H}^1_{\mathrm{par}}(\Gamma, V) \to \mathrm{H}^1(\Gamma, V) \xrightarrow{\mathrm{res}} \prod_{g \in \Gamma \backslash \mathrm{PSL}_2(\mathbb{Z})/\langle T \rangle} \mathrm{H}^1(\Gamma \cap \langle gTg^{-1} \rangle, V). \quad (16)$$

Proposition 4.14 *Let R be a ring and $\Gamma \leq \mathrm{PSL}_2(\mathbb{Z})$ be a subgroup of finite index such that all the orders of all stabiliser groups Γ_x for $x \in \mathbb{H}$ are invertible in R. Let V be a left $R[\Gamma]$-module. Write for short $G = \mathrm{PSL}_2(\mathbb{Z})$ and $M = \mathrm{Hom}_{R[\Gamma]}(R[G], V)$. Then the following diagram is commutative, its vertical maps are isomorphisms and its rows are exact:*

$$\begin{array}{ccccccccc}
0 \to & \mathrm{H}^1_{\mathrm{par}}(\Gamma, V) & \longrightarrow & \mathrm{H}^1(\Gamma, V) & \xrightarrow{\mathrm{res}} & \prod_{g \in \Gamma \backslash \mathrm{PSL}_2(\mathbb{Z})/\langle T \rangle} \mathrm{H}^1(\Gamma \cap \langle gTg^{-1} \rangle, V) & \to & V_\Gamma & \to 0 \\
& \uparrow \scriptstyle{Shapiro} & & \uparrow \scriptstyle{Shapiro} & & \uparrow \scriptstyle{Mackey} & & \| & \\
0 \to & \mathrm{H}^1_{\mathrm{par}}(G, M) & \longrightarrow & \mathrm{H}^1(G, M) & \xrightarrow{\mathrm{res}} & \mathrm{H}^1(\langle T \rangle, M) & \longrightarrow & V_\Gamma & \to 0 \\
& \| & & \uparrow \scriptstyle{m \mapsto f_m} & & \downarrow \scriptstyle{c \mapsto c(T)} & & \uparrow \scriptstyle{\phi} & \\
0 \to & \mathrm{H}^1_{\mathrm{par}}(G, M) & \to & M/(M^{\langle\sigma\rangle} + M^{\langle\tau\rangle}) & \xrightarrow{m \mapsto (1-\sigma)m} & M/(1-T)M & \longrightarrow & M_G & \to 0
\end{array}$$

The map $\phi : M_G \to V_\Gamma$ is given as $f \mapsto \sum_{g \in \Gamma \backslash G} f(g)$.

Proof The commutativity of the diagram is checked in Exercise 4.22. By Exercise 3.39 we have $\mathrm{H}^1(\langle T \rangle, M) \cong M/(1-T)M$. Due to the assumptions we may apply Corollary 4.12. The cokernel of $M/(M^{\langle\sigma\rangle} + M^{\langle\tau\rangle}) \xrightarrow{m \mapsto (1-\sigma)m} M/(1-T)M$ is immediately seen to be $M/((1-\sigma)M + (1-T)M)$, which is equal to M_G, as T and σ generate $\mathrm{PSL}_2(\mathbb{Z})$. Hence, the lower row is an exact sequence.

We now check that the map ϕ is well-defined. For this we verify that the image of $f(g)$ in V_Γ only depends on the coset $\Gamma\backslash G$:

$$f(g) - f(\gamma g) = f(g) - \gamma f(g) = (1-\gamma)f(g) = 0 \in V_\Gamma.$$

Hence, for any $h \in G$ we get

$$\phi((1-h).f) = \sum_{g \in \Gamma\backslash \mathrm{PSL}_2(\mathbb{Z})} (f(g) - f(gh)) = 0,$$

as gh runs over all cosets. Thus, ϕ is well-defined. To show that ϕ is an isomorphism, we give an inverse ψ to ϕ by

$$\psi : V_\Gamma \to \mathrm{Hom}_{R[\Gamma]}(R[G], V)_G, \quad v \mapsto e_v \text{ with } e_v(g) = \begin{cases} gv, & \text{for } g \in \Gamma \\ 0, & \text{for } g \notin \Gamma. \end{cases}$$

It is clear that $\phi \circ \psi$ is the identity. The map ϕ is an isomorphism, as ψ is surjective. In order to see this, fix a system of representatives $\{1 = g_1, g_2, \ldots, g_n\}$ for $\Gamma\backslash\mathrm{PSL}_2(\mathbb{Z})$. We first have $f = \sum_{i=1}^n g_i^{-1}.e_{f(g_i)}$ because for all $h \in G$ we find

$$f(h) = g_j^{-1}.e_{f(g_j)}(h) = e_{f(g_j)}(hg_j^{-1}) = hg_j^{-1}.f(g_j) = .f(hg_j^{-1}g_j) = f(h),$$

where $1 \le j \le n$ is the unique index such that $h \in \Gamma g_j$. Thus

$$f = \sum_{i=1}^n e_{f(g_i)} - \sum_{i=2}^n (1 - g_i^{-1}).e_{f(g_i)} \in \mathrm{im}(\psi),$$

as needed.

More conceptually, one can first identify the coinduced module $\mathrm{Coind}_\Gamma^{\mathrm{PSL}_2(\mathbb{Z})}(V)$ with the induced one $\mathrm{Ind}_\Gamma^{\mathrm{PSL}_2(\mathbb{Z})}(V) = R[G] \otimes_{R[\Gamma]} V$. We claim that the G-coinvariants are isomorphic to $R \otimes_{R[\Gamma]} V \cong V_\Gamma$. As R-modules we have $R[G] = I_G \oplus R1_G$ since $r \mapsto r1_G$ defines a splitting of the augmentation map. Here I_G is the augmentation ideal defined in Exercise 1.29. Consequently, $R[G] \otimes_{R[\Gamma]} V \cong (I_G \otimes_{R[\Gamma]} V) \oplus R \otimes_{R[\Gamma]} V$. The claim follows, since $I_G(R[G] \otimes_{R[\Gamma]} V) \cong I_G \otimes_{R[\Gamma]} V$.

Since all the terms in the upper and the middle row are isomorphic to the respective terms in the lower row, all rows are exact. □

4.5 Theory: Dimension Computations

This seems to be a good place to compute the dimension of $\mathrm{H}^1(\Gamma, V_{k-2}(K))$ and $\mathrm{H}^1_{\mathrm{par}}(\Gamma, V_{k-2}(K))$ over a field K under certain conditions. The results will be important for the proof of the Eichler–Shimura theorem.

Lemma 4.15 *Let R be a ring and let $n \geq 1$ be an integer, $t = \left(\begin{smallmatrix} 1 & N \\ 0 & 1 \end{smallmatrix}\right)$ and $t' = \left(\begin{smallmatrix} 1 & 0 \\ N & 1 \end{smallmatrix}\right)$.*

(a) If $n!N$ is not a zero divisor in R, then for the t-invariants we have

$$V_n(R)^{\langle t \rangle} = \langle X^n \rangle$$

and for the t'-invariants

$$V_n(R)^{\langle t' \rangle} = \langle Y^n \rangle.$$

(b) If $n!N$ is invertible in R, then the coinvariants are given by

$$V_n(R)_{\langle t \rangle} = V_n(R)/\langle Y^n, XY^{n-1}, \dots, X^{n-1}Y \rangle$$

respectively

$$V_n(R)_{\langle t' \rangle} = V_n(R)/\langle X^n, X^{n-1}Y, \dots, XY^{n-1} \rangle.$$

(c) If $n!N$ is not a zero divisor in R, then the R-module of $\Gamma(N)$-invariants $V_n(R)^{\Gamma(N)}$ is zero. In particular, if R is a field of characteristic 0 and Γ is any congruence subgroup, then $V_n(R)^{\Gamma}$ is zero.

(d) If $n!N$ is invertible in R, then the R-module of $\Gamma(N)$-coinvariants $V_n(R)_{\Gamma(N)}$ is zero. In particular, if R is a field of characteristic 0 and Γ is any congruence subgroup, then $V_n(R)_{\Gamma}$ is zero.

Proof (a) The action of t is $t.(X^{n-i}Y^i) = X^{n-i}(NX+Y)^i$ and consequently

$$(t-1).(X^{n-i}Y^i) = \Big(\sum_{j=0}^{i} \tbinom{i}{j} N^{i-j} X^{i-j} Y^j\Big) X^{n-i} - X^{n-i}Y^i = \sum_{j=0}^{i-1} r_{i,j} X^{n-j} Y^j$$

with $r_{i,j} = N^{i-j}\binom{i}{j}$, which is not a zero divisor, respectively invertible, by assumption. For $x = \sum_{i=0}^{n} a_i X^{n-i} Y^i$ we have

$$\begin{aligned}(t-1).x &= \sum_{i=0}^{n} a_i \sum_{j=0}^{i-1} r_{i,j} X^{n-j} Y^j = \sum_{j=0}^{n-1} X^{n-j} Y^j \Big(\sum_{i=j+1}^{n} a_i r_{i,j}\Big) \\ &= XY^{n-1} a_n r_{n,n-1} + X^2 Y^{n-2}(a_n r_{n,n-2} + a_{n-1} r_{n-1,n-2}) + \dots .\end{aligned}$$

If $(t-1).x = 0$, we conclude for $j = n-1$ that $a_n = 0$. Next, for $j = n-2$ it follows that $a_{n-1} = 0$, and so on, until $a_1 = 0$. This proves the statement on the t-invariants. The one on the t'-invariants follows from symmetry.

(b) The claims on the coinvariants are proved in a very similar and straightforward way.

(c) and (d) As $\Gamma(N)$ contains the matrices t and t', this follows from Parts (a) and (b). □

Proposition 4.16 *Let K be a field of characteristic* 0 *and $\Gamma \leq \mathrm{SL}_2(\mathbb{Z})$ be a congruence subgroup of finite index μ such that $\Gamma_y = \{1\}$ for all $y \in \mathbb{H}$ (e.g. $\Gamma = \Gamma_1(N)$ with $N \geq 4$). We can and do consider Γ as a subgroup of* $\mathrm{PSL}_2(\mathbb{Z})$.

Then

$$\dim_K \mathrm{H}^1(\Gamma, V_{k-2}(K)) = (k-1)\frac{\mu}{6} + \delta_{k,2}$$

and

$$\dim_K \mathrm{H}^1_{\mathrm{par}}(\Gamma, V_{k-2}(K)) = (k-1)\frac{\mu}{6} - \nu_\infty + 2\delta_{k,2},$$

where ν_∞ is the number of cusps of Γ, i.e. the cardinality of $\Gamma\backslash\mathbb{P}^1(\mathbb{Q})$, and $\delta_{k,2} = \begin{cases} 1 & \text{if } k = 2 \\ 0 & \text{otherwise.} \end{cases}$

Proof Let $M = \mathrm{Coind}_\Gamma^{\mathrm{PSL}_2(\mathbb{Z})}(V_{k-2}(K))$. This module has dimension $(k-1)\mu$. From the Mayer–Vietoris exact sequence

$$0 \to M^{\mathrm{PSL}_2(\mathbb{Z})} \to M^{\langle\sigma\rangle} \oplus M^{\langle\tau\rangle} \to M \to \mathrm{H}^1(\mathrm{PSL}_2(\mathbb{Z}), M) \to 0,$$

we obtain

$$\dim \mathrm{H}^1(\Gamma, V_{k-2}(K)) = \dim M + \dim M^{\mathrm{PSL}_2(\mathbb{Z})} - \dim \mathrm{H}^0(\langle\sigma\rangle, M) - \dim \mathrm{H}^0(\langle\tau\rangle, M).$$

Recall the left $\mathrm{PSL}_2(\mathbb{Z})$-action on $\mathrm{Hom}_{K[\Gamma]}(K[\mathrm{PSL}_2(\mathbb{Z})], V_{k-2}(K))$, which is given by $(g.\phi)(h) = \phi(hg)$. It follows immediately that every function in the K-vector space $\mathrm{Hom}_{K[\Gamma]}(K[\mathrm{PSL}_2(\mathbb{Z})], V_{k-2}(K))^{\mathrm{PSL}_2(\mathbb{Z})}$ is constant and equal to its value at 1. The Γ-invariance, however, imposes additionally that this constant lies in $V_{k-2}(K)^\Gamma$. Hence, by Lemma 4.15, $\dim M^{\mathrm{PSL}_2(\mathbb{Z})} = \delta_{k,2}$. The term $\mathrm{H}^0(\langle\sigma\rangle, M)$ is handled by Mackey's formula:

$$\dim \mathrm{H}^0(\langle\sigma\rangle, M) = \sum_{x \in \Gamma\backslash\mathrm{PSL}_2(\mathbb{Z}).i} \dim V_{k-2}(K)^{\Gamma_x} = (k-1)\#(\Gamma\backslash\mathrm{PSL}_2(\mathbb{Z}).i) = (k-1)\frac{\mu}{2},$$

since all Γ_x are trivial by assumption and there are hence precisely $\mu/2$ points in Y_Γ lying over i in $Y_{\mathrm{SL}_2(\mathbb{Z})}$. By the same argument we get

$$\dim \mathrm{H}^0(\langle\tau\rangle, M) = \frac{\mu}{3}.$$

Putting these together gives the first formula:

$$\dim_K \mathrm{H}^1(\Gamma, V_{k-2}(K)) = (k-1)(\mu - \frac{\mu}{2} - \frac{\mu}{3}) + \delta_{k,2} = (k-1)\frac{\mu}{6} + \delta_{k,2}.$$

The second formula can be read off from the diagram in Proposition 4.14. It gives directly

$$\dim \mathrm{H}^1_{\mathrm{par}}(\Gamma, V_{k-2}(K)) = \dim \mathrm{H}^1(\Gamma, V_{k-2}(K)) + \dim V_{k-2}(K)_\Gamma - \sum_{g \in \Gamma\backslash \mathrm{PSL}_2(\mathbb{Z})/\langle T\rangle} \dim \mathrm{H}^1(\Gamma \cap \langle gTg^{-1}\rangle, V_{k-2}(K)).$$

All the groups $\Gamma \cap \langle gTg^{-1}\rangle$ are of the form $\langle T^n\rangle$ for some $n \geq 1$. Since they are cyclic, we have

$$\dim \mathrm{H}^1(\Gamma \cap \langle gTg^{-1}\rangle, V_{k-2}(K)) = \dim V_{k-2}(K)_{\langle T^n\rangle} = 1$$

by Lemma 4.15. As the set $\Gamma\backslash \mathrm{PSL}_2(\mathbb{Z})/\langle T\rangle$ is the set of cusps of Γ, we conclude

$$\sum_{g \in \Gamma\backslash \mathrm{PSL}_2(\mathbb{Z})/\langle T\rangle} \dim \mathrm{H}^1(\Gamma \cap \langle gTg^{-1}\rangle, V_{k-2}(K)) = \nu_\infty.$$

Moreover, also by Lemma 4.15, $\dim V_{k-2}(K)_\Gamma = \delta_{k,2}$. Putting everything together yields the formula

$$\dim \mathrm{H}^1_{\mathrm{par}}(\Gamma, V_{k-2}(K)) = (k-1)\frac{\mu}{6} + 2\delta_{k,2} - \nu_\infty,$$

as claimed. □

Remark 4.17 One can derive a formula for the dimension even if Γ is not torsion-free. One only needs to compute the dimensions $V_{k-2}(K)^{\langle\sigma\rangle}$ and $V_{k-2}(K)^{\langle\tau\rangle}$ and to modify the above proof slightly.

4.6 Theoretical Exercises

Exercise 4.18 (a) Verify that $G * H$ is a group.
(b) Prove the universal property represented by the commutative diagram

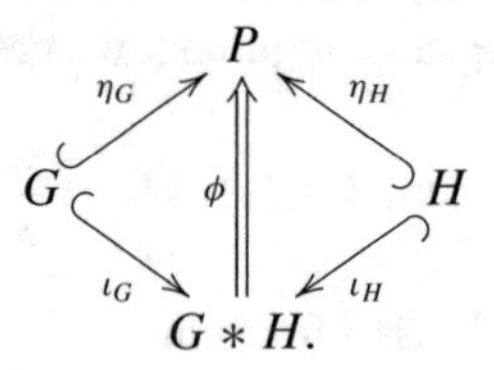

More precisely, let $\iota_G : G \to G * H$ and $\iota_H : H \to G * H$ be the natural inclusions. Let P be any group together with group injections $\eta_G : G \to P$ and

$\eta_H : H \to P$, then there is a unique group homomorphism $\phi : G * H \to P$ such that $\eta_G = \phi \circ \iota_G$ and $\eta_H = \phi \circ \iota_H$.

Exercise 4.19 (a) Let $M \in \mathrm{SL}_n(\mathbb{Z})$ be an element of finite order m. Determine the primes that may divide m. [Hint: Look at the characteristic polynomial of M.]
(b) Determine all conjugacy classes of elements of finite order in $\mathrm{PSL}_2(\mathbb{Z})$.

Exercise 4.20 (a) Determine the $N \geq 1$ for which $\Gamma_1(N)$ has no element of finite order apart from the identity. [Hint: You should get $N \geq 4$.]
(b) Determine the $N \geq 1$ for which $\Gamma_0(N)$ has no element of order 4. Also determine the cases in which there is no element of order 6.

Exercise 4.21 (a) Prove that the explicit description of f_m in the Mayer–Vietoris sequence (Eq. 14) satisfies the properties required for the 0-th connecting homomorphism in Definition 3.21.
Hint: Prove that if f_m is a boundary, then $m \in M^{\langle\sigma\rangle} + M^{\langle\tau\rangle}$. Moreover, prove that a 1-cocycle in $\mathrm{H}^1(\mathrm{PSL}_2(\mathbb{Z}), M)$ which becomes a coboundary when restricted to either $\langle\sigma\rangle$ or $\langle\tau\rangle$ can be changed by a coboundary to be of the form f_m for some $m \in M$.
(b) Let $0 \to A \to B \to C \to 0$ be an exact sequence of G-modules for some group G. Let $c \in C^G$ and write it as a class $b + A \in B/A \cong C$. As it is G-invariant, we have $0 = (1-g)c = (1-g)(b+A)$, whence $(1-g)b \in A$ for all $g \in G$. Define the 1-cocycle $\delta^0(c)$ as the map $G \to A$ sending g to $(1-g)b \in A$.
Prove that δ^0 satisfies the properties required for the 0-th connecting homomorphism in Definition 3.21.
Note that the connecting homomorphisms are not unique (one can, e.g. replace them by their negatives).
(c) As an alternative approach to (a), you may apply (b) to the exact sequence from which the Mayer–Vietoris sequence is derived as the associated long cohomology sequence in Proposition 4.10.

Exercise 4.22 Verify the commutativity of the diagram in Proposition 4.14.

4.7 Computer Exercises

Computer Exercise 4.23 Let $N \geq 1$. Compute a list of the elements of $\mathbb{P}^1(\mathbb{Z}/N\mathbb{Z})$. Compute a list of the cusps of $\Gamma_0(N)$ and $\Gamma_1(N)$ (cf. [20], p. 60). I recommend to use the decomposition of $\mathbb{P}^1(\mathbb{Z}/N\mathbb{Z})$ into $\mathbb{P}^1(\mathbb{Z}/p^n\mathbb{Z})$.

Computer Exercise 4.24 Let K be some field. Let $\chi : (\mathbb{Z}/N\mathbb{Z})^\times \to K^\times$ be a Dirichlet character of modulus N. For given N and K, compute the group of all Dirichlet characters. Every Dirichlet character should be implemented as a map $\phi : \mathbb{Z} \to K^\times$ such that $\phi(a) = 0$ for all $a \in \mathbb{Z}$ with $(a, N) \neq 1$ and $\phi(a) = \chi(a \bmod N)$ otherwise.

5 Modular Symbols and Manin Symbols

5.1 Theory: Manin Symbols

This section is an extended version of a specialisation of parts of my article [23] to the group $\mathrm{PSL}_2(\mathbb{Z})$. Manin symbols provide an alternative description of modular symbols. See Definition 5.6 below. We shall use this description for the comparison with group cohomology and for implementing the modular symbols formalism. We stay in the general setting over a ring R.

Proposition 5.1 *The sequence of* R*-modules*

$$0 \to R[\mathrm{PSL}_2(\mathbb{Z})]N_\sigma + R[\mathrm{PSL}_2(\mathbb{Z})]N_\tau \to R[\mathrm{PSL}_2(\mathbb{Z})] \xrightarrow{g \mapsto g(1-\sigma)\infty} R[\mathbb{P}^1(\mathbb{Q})] \xrightarrow{g\infty \mapsto 1} R \to 0$$

is exact. Here we are considering $R[\mathrm{PSL}_2(\mathbb{Z})]$ *as a right* $R[\mathrm{PSL}_2(\mathbb{Z})]$*-module.*

Proof Let H be a finite subgroup of a group G and let $H\backslash G = \{g_i \mid i \in I\}$ stand for a fixed system of representatives of the cosets. We write $R[H\backslash G]$ for the free R-module on the set of representatives. The map

$$\mathrm{Hom}_R(R[H], R[H\backslash G]) \to R[G], \quad f \mapsto \sum_{h \in H} h.f(h)$$

is an isomorphism. Indeed, suppose that for $f \in \mathrm{Hom}_R(R[H], R[H\backslash G])$ we have

$$0 = \sum_{h\in H} h.(f(h)) = \sum_{h\in H} h.(\sum_{i\in I} a_{h,i} g_i) = \sum_{h\in H}(\sum_{i\in I} a_{h,i} h g_i),$$

then $a_{h,i} = 0$ for all $h \in H$ and all $i \in I$ (since the elements hg_i are all distinct), whence $f = 0$. For the surjectivity, note that all elements in $R[G]$ can be written as (finite) sums of the form $\sum_{h\in H}\sum_{i\in I} a_{h,i} h g_i$ because any element in G is of the form hg_i for a unique $h \in H$ and a unique $i \in I$.

This yields via Shapiro's lemma that

$$H^i(\langle\sigma\rangle, R[\mathrm{PSL}_2(\mathbb{Z})]) = H^i(\langle 1\rangle, R[\langle\sigma\rangle\backslash\mathrm{PSL}_2(\mathbb{Z})]) = 0$$

for all $i \geq 1$, and similarly for $\langle\tau\rangle$. The resolution for a finite cyclic group (12) gives

$$\begin{aligned}
R[\mathrm{PSL}_2(\mathbb{Z})]N_\sigma &= \ker_{R[\mathrm{PSL}_2(\mathbb{Z})]}(1-\sigma) = R[\mathrm{PSL}_2(\mathbb{Z})]^{\langle\sigma\rangle},\\
R[\mathrm{PSL}_2(\mathbb{Z})]N_\tau &= \ker_{R[\mathrm{PSL}_2(\mathbb{Z})]}(1-\tau) = R[\mathrm{PSL}_2(\mathbb{Z})]^{\langle\tau\rangle},\\
R[\mathrm{PSL}_2(\mathbb{Z})](1-\sigma) &= \ker_{R[\mathrm{PSL}_2(\mathbb{Z})]} N_\sigma \quad \text{and}\\
R[\mathrm{PSL}_2(\mathbb{Z})](1-\tau) &= \ker_{R[\mathrm{PSL}_2(\mathbb{Z})]} N_\tau.
\end{aligned}$$

By Proposition 4.9, we have the exact sequence

$$0 \to R[\mathrm{PSL}_2(\mathbb{Z})] \to R[\mathrm{PSL}_2(\mathbb{Z})]_{\langle\sigma\rangle} \oplus R[\mathrm{PSL}_2(\mathbb{Z})]_{\langle\tau\rangle} \to R \to 0.$$

The injectivity of the first map in the exact sequence (which we recall is a consequence of $\mathrm{PSL}_2(\mathbb{Z}) = \langle\sigma\rangle * \langle\tau\rangle$) leads to

$$R[\mathrm{PSL}_2(\mathbb{Z})](1-\sigma) \cap R[\mathrm{PSL}_2(\mathbb{Z})](1-\tau) = 0.$$

Sending g to $g\infty$ yields a bijection between $R[\mathrm{PSL}_2(\mathbb{Z})]/R[\mathrm{PSL}_2(\mathbb{Z})](1-T)$ and $R[\mathbb{P}^1(\mathbb{Q})]$. In order to prove the exactness at $R[\mathrm{PSL}_2(\mathbb{Z})]$, we show that the equality $x(1-\sigma) = y(1-T)$ for $x, y \in R[\mathrm{PSL}_2(\mathbb{Z})]$ yields that x belongs to $R[\mathrm{PSL}_2(\mathbb{Z})]^{\langle\sigma\rangle} + R[\mathrm{PSL}_2(\mathbb{Z})]^{\langle\tau\rangle}$.

Note that $x(1-\sigma) = y(1-T) = y(1-\tau) - yT(1-\sigma)$ because of the equality $\tau = T\sigma$. This yields $x(1-\sigma) + yT(1-\sigma) = y(1-\tau)$. This expression, however, is equal to zero. Hence, there exists a $z \in R[\mathrm{PSL}_2(\mathbb{Z})]$ satisfying $y = zN_\tau$. We have $N_\tau T = N_\tau \sigma$ because of $T = \tau\sigma$. Consequently, we get

$$y(1-T) = zN_\tau(1-T) = zN_\tau(1-\sigma) = y(1-\sigma).$$

The equality $x(1-\sigma) = y(1-\sigma)$ implies that $x - y$ belongs to $R[\mathrm{PSL}_2(\mathbb{Z})]^{\langle\sigma\rangle}$. Since $y \in R[\mathrm{PSL}_2(\mathbb{Z})]^{\langle\tau\rangle}$, we see get that $x = (x-y) + y$ lies in $R[\mathrm{PSL}_2(\mathbb{Z})]^{\langle\sigma\rangle} + R[\mathrm{PSL}_2(\mathbb{Z})]^{\langle\tau\rangle}$, as required.

It remains to prove the exactness at $R[\mathbb{P}^1(\mathbb{Q})]$. The kernel of $R[\mathrm{PSL}_2(\mathbb{Z})] \xrightarrow{g \mapsto 1} R$ is the augmentation ideal, which is generated by all elements of the $1 - g$ for $g \in \mathrm{PSL}_2(\mathbb{Z})$. Noticing further that we can write

$$1 - \alpha\beta = \alpha.(1-\beta) + (1-\alpha)$$

for $\alpha, \beta \in \mathrm{PSL}_2(\mathbb{Z})$, the fact that σ and $T = \tau\sigma$ generate $\mathrm{PSL}_2(\mathbb{Z})$ implies that the kernel of $R[\mathrm{PSL}_2(\mathbb{Z})] \xrightarrow{g \mapsto 1} R$ equals

$$R[\mathrm{PSL}_2(\mathbb{Z})](1-\sigma) + R[\mathrm{PSL}_2(\mathbb{Z})](1-T)$$

inside $R[\mathrm{PSL}_2(\mathbb{Z})]$ It suffices to take the quotient by $R[\mathrm{PSL}_2(\mathbb{Z})](1-T)$ to obtain the desired exactness. □

Lemma 5.2 *The sequence of R-modules*

$$0 \to \mathcal{M}_R \xrightarrow{\{\alpha,\beta\} \mapsto \beta - \alpha} R[\mathbb{P}^1(\mathbb{Q})] \xrightarrow{\alpha \mapsto 1} R \to 0$$

is exact.

Proof Note that, using the relations defining $\mathcal{M}_R$, any element in $\mathcal{M}_R$ can be written as $\sum_{\alpha \neq \infty} r_\alpha \{\infty, \alpha\}$ with $r_\alpha \in R$. This element is mapped to $\sum_{\alpha \neq \infty} r_\alpha \alpha -$

$(\sum_{\alpha \neq \infty} r_\alpha)\infty$. If this expression equals zero, all coefficients r_α have to be zero. This shows the injectivity of the first map.

Let $\sum_\alpha r_\alpha \alpha \in R[\mathbb{P}^1(\mathbb{Q})]$ be an element in the kernel of the second map. Then $\sum_\alpha r_\alpha = 0$, so that we can write

$$\sum_\alpha r_\alpha \alpha = \sum_{\alpha \neq \infty} r_\alpha \alpha - (\sum_{\alpha \neq \infty} r_\alpha)\infty$$

to obtain an element in the image of the first map. □

Proposition 5.3 *The homomorphism of R-modules*

$$R[\mathrm{PSL}_2(\mathbb{Z})] \xrightarrow{\phi} \mathcal{M}_R, \quad g \mapsto \{g.0, g.\infty\}$$

is surjective with kernel $R[\mathrm{PSL}_2(\mathbb{Z})]N_\sigma + R[\mathrm{PSL}_2(\mathbb{Z})]N_\tau$.

Proof This follows from Proposition 5.1 and Lemma 5.2. □

We have now provided all the input required to prove the description of modular symbols in terms of Manin symbols. For this we need the notion of an induced module. In homology it plays the role that the coinduced module plays in cohomology.

Definition 5.4 Let R be a ring, G a group, $H \leq G$ a subgroup and V a left $R[H]$-module. The *induced module* of V from H to G is defined as

$$\mathrm{Ind}_H^G(V) := R[G] \otimes_{R[H]} V,$$

where we view $R[G]$ as a right $R[H]$-module via the natural action. The induced module is a left $R[G]$-module via the natural left action of G on $R[G]$.

In case of H having a finite index in G (as in our standard example $\Gamma_1(N) \leq \mathrm{PSL}_2(\mathbb{Z})$), the induced module is isomorphic to the coinduced one:

Lemma 5.5 *Let R be a ring, G a group,* $H \leq G$ *a subgroup of finite index and V a left* $R[H]$*-module.*

(a) $\mathrm{Ind}_H^G(V)$ *and* $\mathrm{Coind}_H^G(V)$ *are isomorphic as left* $R[G]$*-modules.*
(b) Equip $(R[G] \otimes_R V)$ *with the diagonal left H-action* $h.(g \otimes v) = hg \otimes h.v$ *and the right G-action* $(g \otimes v).\tilde{g} = g\tilde{g} \otimes v$*. Consider the induced module* $\mathrm{Ind}_H^G(V)$ *as a right* $R[G]$*-module by inverting the left action in the definition. Then*

$$\mathrm{Ind}_H^G(V) \to (R[G] \otimes_R V)_H, \quad g \otimes v \mapsto g^{-1} \otimes v$$

is an isomorphism of right $R[G]$*-modules.*

Proof Exercise 5.11. □

Definition 5.6 Let $\Gamma \subseteq \mathrm{PSL}_2(\mathbb{Z})$ a finite index subgroup, V a left $R[\Gamma]$-module and $M = \mathrm{Ind}_\Gamma^{\mathrm{PSL}_2(\mathbb{Z})}(V)$, which we identify with the right $R[\mathrm{PSL}_2(\mathbb{Z})]$-module $(R[\mathrm{PSL}_2(\mathbb{Z})] \otimes_R V)_\Gamma$ as in Lemma 5.5(b).

Elements in $M/(MN_\sigma + MN_\tau)$ are called *Manin symbols* over R (for the subgroup $\Gamma \subseteq \mathrm{PSL}_2(\mathbb{Z})$ and the left $R[\Gamma]$-module V).

Theorem 5.7 *In the setting of Definition 5.6, the following statements hold:*

(a) The homomorphism ϕ from Proposition 5.3 induces the exact sequence of R-modules

$$0 \to MN_\sigma + MN_\tau \to M \to \mathcal{M}_R(\Gamma, V) \to 0,$$

and the homomorphism $M \to \mathcal{M}_R(\Gamma, V)$ is given by $g \otimes v \mapsto \{g.0, g.\infty\} \otimes v$. In other words, this map induces an isomorphism between Manin symbols over R (for the subgroup $\Gamma \subseteq \mathrm{PSL}_2(\mathbb{Z})$ and the left $R[\Gamma]$-module V) and the modular symbols module $\mathcal{M}_R(\Gamma, V)$.

(b) The homomorphism $R[\mathrm{PSL}_2(\mathbb{Z})] \to R[\mathbb{P}^1(\mathbb{Q})]$ sending g to $g.\infty$ induces the exact sequence of R-modules

$$0 \to M(1 - T) \to M \to \mathcal{B}_R(\Gamma, V) \to 0.$$

(c) The identifications of (a) and (b) imply the isomorphism

$$\mathcal{C}\mathcal{M}_R(\Gamma, V) \cong \ker\big(M/(MN_\sigma + MN_\tau) \xrightarrow{m \mapsto m(1-\sigma)} M/M(1 - T)\big).$$

Proof (a) Proposition 5.3 gives the exact sequence

$$0 \to R[\mathrm{PSL}_2(\mathbb{Z})]N_\sigma + R[\mathrm{PSL}_2(\mathbb{Z})]N_\tau \to R[\mathrm{PSL}_2(\mathbb{Z})] \to \mathcal{M}_R \to 0,$$

which we tensor with V over R, yielding the exact sequence of left $R[\Gamma]$-modules

$$0 \to (R[\mathrm{PSL}_2(\mathbb{Z})] \otimes_R V)N_\sigma + (R[\mathrm{PSL}_2(\mathbb{Z})] \otimes_R V)N_\tau \\ \to (R[\mathrm{PSL}_2(\mathbb{Z})] \otimes_R V) \to \mathcal{M}_R(V) \to 0.$$

Passing to left Γ-coinvariants yields (a) because MN_σ and MN_τ are the images of $(R[\mathrm{PSL}_2(\mathbb{Z})] \otimes_R V)N_\sigma$ and $(R[\mathrm{PSL}_2(\mathbb{Z})] \otimes_R V)N_\tau$ inside M, respectively. (b) is clear from the definition and (c) has already been observed in the proof of Proposition 5.1. □

In the literature on Manin symbols one usually finds a more explicit version of the induced module. This is the contents of the following proposition. It establishes the link with the main theorem on Manin symbols in [20], namely Theorem 8.4.

Since in the following proposition left and right actions are involved, we sometimes indicate left (co-)invariants by using left subscripts (resp. superscripts) and right (co-)invariants by right ones.

Proposition 5.8 *Let $\chi : (\mathbb{Z}/N\mathbb{Z})^\times \to R^\times$ be a character such that $\chi(-1) = (-1)^k$. Consider the R-module*

$$X := R[\Gamma_1(N)\backslash \mathrm{SL}_2(\mathbb{Z})] \otimes_R V_{k-2}(R) \otimes_R R^\chi$$

equipped with the right $\mathrm{SL}_2(\mathbb{Z})$-action $(\Gamma_1(N)h \otimes V \otimes r)g = (\Gamma_1(N)hg \otimes g^{-1}v \otimes r)$ and with the left $\Gamma_1(N)\backslash\Gamma_0(N)$-action $g(\Gamma_1(N)h \otimes v \otimes r) = (\Gamma_1(N)gh \otimes v \otimes \chi(g)r)$.

Then

$$X \cong \mathrm{Ind}_{\Gamma_1(N)}^{\mathrm{SL}_2(\mathbb{Z})}(V_k^\chi(R))$$

as a right $R[\mathrm{SL}_2(\mathbb{Z})]$-module and a left $R[\Gamma_1(N)\backslash\Gamma_0(N)]$-module. Moreover,

$$_{\Gamma_1(N)\backslash\Gamma_0(N)}X \cong \mathrm{Ind}_{\Gamma_0(N)}^{\mathrm{SL}_2(\mathbb{Z})}(V_k^\chi(R)).$$

If $N \geq 3$, then the latter module is isomorphic to $\mathrm{Ind}_{\Gamma_0(N)/\{\pm 1\}}^{\mathrm{PSL}_2(\mathbb{Z})}(V_k^\chi(R))$.

Proof Mapping $g \otimes v \otimes r$ to $g \otimes g^{-1}v \otimes r$ defines an isomorphism of right $R[\mathrm{SL}_2(\mathbb{Z})]$-modules and of left $R[\Gamma_1(N)\backslash\Gamma_0(N)]$-modules

$$_{\Gamma_1(N)}(R[\mathrm{SL}_2(\mathbb{Z})] \otimes_R V_{k-2}(R) \otimes_R R^\chi) \to X.$$

As we have seen above, the left-hand side module is naturally isomorphic to the induced module $\mathrm{Ind}_{\Gamma_1(N)}^{\mathrm{SL}_2(\mathbb{Z})}(V_k^\chi(R))$ (equipped with its right $R[\mathrm{SL}_2(\mathbb{Z})]$-action described before). This establishes the first statement. The second one follows from $_{\Gamma_1(N)\backslash\Gamma_0(N)}(_{\Gamma_1(N)}M) = {}_{\Gamma_0(N)}M$ for any $\Gamma_0(N)$-module M. The third statement is due to the fact that $_{\langle -1\rangle}(R[\mathrm{SL}_2(\mathbb{Z})] \otimes_R V_{k-2}^\chi(R))$ is naturally isomorphic to $R[\mathrm{PSL}_2(\mathbb{Z})] \otimes_R V_{k-2}^\chi(R)$, since -1 acts trivially on the second factor, as the assumption assures that $-1 \in \Gamma_0(N)$ but $-1 \notin \Gamma_1(N)$. □

For one more description of the induced module $\mathrm{Ind}_{\Gamma_0(N)/\{\pm 1\}}^{\mathrm{PSL}_2(\mathbb{Z})}(V_k^\chi(R))$ see Exercise 5.12. It is this description that uses up the least memory in an implementation. Now all the prerequisites have been provided for implementing Manin symbols (say for $\Gamma_0(N)$ and a character). This is the task of Computer Exercise 5.14.

5.2 *Theory: Manin Symbols and Group Cohomology*

Let $\Gamma \leq \mathrm{PSL}_2(\mathbb{Z})$ be a subgroup of finite index, and V a left $R[\Gamma]$-module for a ring R.

Theorem 5.9 *Suppose that the orders of all stabiliser subgroups of Γ for the action on $\mathbb{H}$ are invertible in R. Then we have isomorphisms:*

$$\mathrm{H}^1(\Gamma, V) \cong \mathcal{M}_R(\Gamma, V)$$

and

$$\mathrm{H}^1_{\mathrm{par}}(\Gamma, V) \cong \mathscr{CM}_R(\Gamma, V).$$

Proof As before, set $M = \mathrm{Ind}_{\Gamma}^{\mathrm{PSL}_2(\mathbb{Z})}(V)$ and recall that this module is isomorphic to $\mathrm{Coind}_{\Gamma}^{\mathrm{PSL}_2(\mathbb{Z})}(V)$. To see the first statement, in view of Theorem 5.7 and the corollary of the Mayer–Vietoris exact sequence (Corollary 4.12), it suffices to show $M^{\langle\sigma\rangle} = MN_\sigma$ and $M^{\langle\tau\rangle} = MN_\tau$. By the resolution of R for a cyclic group in (12), the quotient $M^{\langle\sigma\rangle}/MN_\sigma$ is equal to $H^2(\langle\sigma\rangle, M)$, but this one is zero by the application of Mackey's formula done in Lemma 4.11(c). The same argument works with τ instead of σ.

The passage to the parabolic/cuspidal subspaces is immediate because the boundary map with source M has the same explicit description in both cases (see Theorem 5.7(c) and Proposition 4.14). □

5.3 Algorithms and Implementations: Conversion Between Manin and Modular Symbols

We now use the Euclidean Algorithm to represent any element $g \in \mathrm{PSL}_2(\mathbb{Z})$ in terms of σ and T.

Algorithm 5.10 Input: A matrix $M = \begin{pmatrix} a & b \\ c & d \end{pmatrix}$ with integer entries and determinant 1.
Output: A list of matrices $[A_1, A_2, \ldots, A_n]$ where all $A_i \in \{T^n | n \in \mathbb{Z}\} \cup \{\sigma\}$ and σ and T^n alternate.

(1) create an empty list `output`.
(2) if $|c| > |a|$ then
(3) append σ to `output`.
(4) $M := \sigma M$.
(5) end if;
(6) while $c \neq 0$ do
(7) $q := a \text{ div } c$.
(8) append T^q to `output`.
(9) append σ to `output`.
(10) $M := \sigma T^{-q} M$.
(11) end while;
(12) if $M \notin \{\begin{pmatrix} 1 & 0 \\ 0 & 1 \end{pmatrix}, \begin{pmatrix} -1 & 0 \\ 0 & -1 \end{pmatrix}\}$ then [At this point $M \in \{\begin{pmatrix} 1 & * \\ 0 & 1 \end{pmatrix}, \begin{pmatrix} -1 & * \\ 0 & -1 \end{pmatrix}\}$.]
(13) append M to `output`.
(14) end if;
(15) return `output`.

This algorithm gives a constructive proof of the fact (Proposition 4.5) that $\mathrm{PSL}_2(\mathbb{Z})$ is generated by σ and T, and hence also by σ and τ. Note, however, that the algorithm does not necessarily give the shortest such representation. See Exercise 5.13 for a relation to continued fractions.

We can use the algorithm to make a conversion between modular symbols and Manin symbols, as follows. Suppose we are given the modular symbols $\{\alpha, \infty\}$ (this is no loss of generality, as we can represent $\{\alpha, \beta\} = \{\alpha, \infty\} - \{\beta, \infty\}$). Suppose α is given as $g\infty$ with some $g \in \mathrm{SL}_2(\mathbb{Z})$ (i.e. representing the cusp as a fraction $\frac{a}{c}$ with $(a, c) = 1$, then we can find b, d by the Euclidean Algorithm such that $g = \left(\begin{smallmatrix} a & b \\ c & d \end{smallmatrix}\right) \in \mathrm{SL}_2(\mathbb{Z})$ satisfies the requirements). We now use Algorithm 5.10 to represent g as $\sigma T^{a_1} \sigma T^{a_2} \sigma \ldots T^{a_n} \sigma$ (for example). Then we have

$$\{\alpha, \infty\} = \sigma T^{a_1} \sigma T^{a_2} \sigma \ldots T^{a_n} \{0, \infty\} + \sigma T^{a_1} \sigma T^{a_2} \sigma \ldots T^{a_{n-1}} \{0, \infty\} + \cdots + \sigma T^{a_1} \{0, \infty\} + \{0, \infty\}.$$

If g does not end in σ but T^{a_n}, then we must drop T^{a_n} from the above formula (since T stabilises ∞). If g starts in T^{a_1} (instead of σ), then we must drop the last summand.

Since we are in weight 2 (i.e. trivial module V), the space of Manin symbols is a quotient of $R[\mathrm{PSL}_2(\mathbb{Z})]/\Gamma$ (see Definition 5.6). The Manin symbol corresponding to the above example chosen for the modular symbol $\{\alpha, \infty\}$ is then simply represented by the formal sum

$$\sigma T^{a_1} \sigma T^{a_2} \sigma \ldots T^{a_n} + \sigma T^{a_1} \sigma T^{a_2} \sigma \ldots T^{a_{n-1}} + \cdots + \sigma T^{a_1} + 1. \tag{17}$$

If the module V is not trivial, a modular symbol would typically look like $\{\alpha, \infty\} \otimes v$ for $v \in V$ and the corresponding Manin symbol would be the formal sum in (17) tensored with v.

In Computer Exercise 5.15 you are asked to implement a conversion between Manin and modular symbols.

5.4 Theoretical Exercises

Exercise 5.11 Prove Lemma 5.5.

Exercise 5.12 Assume the set-up of Proposition 5.8. Describe a right $\mathrm{PSL}_2(\mathbb{Z})$-action on

$$Y := R[\mathbb{P}^1(\mathbb{Z}/N\mathbb{Z})] \otimes_R V_{k-2}(R) \otimes_R R^\chi$$

and an isomorphism

$${}_{\Gamma_1(N)\backslash\Gamma_0(N)}X \to Y$$

of right $\mathrm{PSL}_2(\mathbb{Z})$-modules.

Exercise 5.13 Provide a relationship between Algorithm 5.10 and continued fractions.

5.5 Computer Exercises

Computer Exercise 5.14 Use the description of Exercise 5.12 and your results from Computer Exercises 4.23 and 4.24 to implement Manin symbols for $\Gamma_0(N)$ and a character over a field. As a first approach you may use the trivial character only.

Computer Exercise 5.15 (a) Write an algorithm to represent any element of the group $\mathrm{PSL}_2(\mathbb{Z})$ in terms of σ and T.
(b) Write an algorithm that represents any modular symbol $\{\alpha, \beta\}$ as a Manin symbol (inside the vector space created in Computer Exercise 5.14).

6 Eichler–Shimura

This section is devoted to proving the theorem by Eichler and Shimura that is at the basis of the modular symbols algorithm and its group cohomological variant. The standard reference for the Eichler–Shimura theorem is [19, §8.2]. In the entire section, let $k \geq 2$ be an integer.

6.1 Theory: Petersson Scalar Product

Recall the standard fundamental domain for $\mathrm{SL}_2(\mathbb{Z})$

$$\mathscr{F} = \{z = x + iy \in \mathbb{H} \mid |z| > 1, |x| < \frac{1}{2}\}$$

from Proposition 4.2. Every subgroup $\Gamma \leq \mathrm{SL}_2(\mathbb{Z})$ of finite index has a fundamental domain, for example, $\bigcup_{\gamma \in \overline{\Gamma}\backslash \mathrm{PSL}_2(\mathbb{Z})} \gamma \mathscr{F}$ for any choice of system of representatives of the cosets $\overline{\Gamma}\backslash \mathrm{PSL}_2(\mathbb{Z})$, where we put $\overline{\Gamma} = \Gamma/(\langle \pm 1 \rangle \cap \Gamma)$.

Lemma 6.1 *(a) Let $\gamma \in \mathrm{GL}_2(\mathbb{R})^+$ be a real matrix with positive determinant. Let $f \in \mathrm{M}_k(\Gamma\,;\,\mathbb{C})$ and $g \in \mathrm{S}_k(\Gamma\,;\,\mathbb{C})$. We have with $z \in \mathbb{H}$*

$$f(\gamma z)\overline{g(\gamma z)}(\gamma z - \gamma \overline{z})^k = \det(\gamma)^{2-k} f|_\gamma(z)\overline{g|_\gamma(z)}(z - \overline{z})^k$$

for all $\gamma \in \mathrm{SL}_2(\mathbb{R})$. The function $G(z) := f(z)\overline{g(z)}(z - \overline{z})^k$ is bounded on $\mathbb{H}$.
(b) We have $d\gamma z = \frac{\det(\gamma)}{(cz+d)^2} dz$ for all $\gamma \in \mathrm{GL}_2(\mathbb{R})^+$.
(c) The differential form $\frac{dz \wedge d\overline{z}}{(z-\overline{z})^2}$ is $\mathrm{GL}_2(\mathbb{R})^+$-invariant. In terms of $z = x + iy$ we have $\frac{dz \wedge d\overline{z}}{(z-\overline{z})^2} = \frac{i}{2}\frac{dx \wedge dy}{y^2}$.
(d) Let $\Gamma \leq \mathrm{SL}_2(\mathbb{Z})$ be a subgroup with finite index $\mu = (\mathrm{PSL}_2(\mathbb{Z}) : \overline{\Gamma})$. The volume of any fundamental domain $\mathscr{F}_\Gamma$ for Γ with respect to the differential form $\frac{2 dz \wedge d\overline{z}}{i(z-\overline{z})^2}$, i.e.

$$\mathrm{vol}(\mathscr{F}_\Gamma) = \int_{\mathscr{F}_\Gamma} \frac{2dz \wedge d\overline{z}}{i(z-\overline{z})^2},$$

is equal to $\mu\frac{\pi}{3}$.

Proof (a) The first statement is computed as follows:

$$\begin{aligned}
&f(\gamma z)\overline{g(\gamma z)}(\gamma z - \gamma\overline{z})^k \\
&= \det(\gamma)^{2(1-k)}(f|_\gamma(z)(cz+d)^k)\overline{(g|_\gamma(z)(cz+d)^k)}(\frac{az+b}{cz+d} - \frac{a\overline{z}+b}{c\overline{z}+d})^k \\
&= \det(\gamma)^{2-2k} f|_\gamma(z)\overline{g|_\gamma(z)}((az+b)(c\overline{z}+d) - (a\overline{z}+b)(cz+d))^k \\
&= \det(\gamma)^{2-k} f|_\gamma(z)\overline{g|_\gamma(z)}(z-\overline{z})^k,
\end{aligned}$$

where we write $\gamma = \left(\begin{smallmatrix} a & b \\ c & d \end{smallmatrix}\right)$. By the preceding computation, the function $G(z)$ is invariant under $\gamma \in \Gamma$. Hence, it suffices to check that $|G(z)|$ is bounded on the closure of any fundamental domain $\mathscr{F}_\Gamma$ for Γ. For this, it is enough to verify for every γ in a system of representatives of $\Gamma\backslash\mathrm{SL}_2(\mathbb{Z})$ that any of the functions $G(\gamma z)$ is bounded on the closure of the standard fundamental domain $\mathscr{F}$. By the preceding computation, we also have $G(\gamma z) = f|_\gamma(z)\overline{g|_\gamma(z)}(z-\overline{z})^k$ for $\gamma \in \mathrm{SL}_2(\mathbb{Z})$. Note that $f(z)g(z)$ is a cusp form in $\mathrm{S}_{2k}(\Gamma\,;\,\mathbb{C})$, in particular, for every $\gamma \in \mathrm{SL}_2(\mathbb{Z})$ the function $f|_\gamma(z)g|_\gamma(z)$ has a Fourier expansion in ∞ of the form $\sum_{n=1}^\infty a_n e^{2\pi i z n}$. This series converges absolutely and uniformly on compact subsets of $\mathbb{H}$, in particular, for any $C > 1$

$$K_\gamma := \sum_{n=1}^\infty |a_n e^{2\pi i (x+iC)n}| = \sum_{n=1}^\infty |a_n| e^{-2\pi C n}$$

is a positive real number, depending on γ (in a system of representatives $\Gamma\backslash\mathrm{SL}_2(\mathbb{Z})$). We have with $z = x + iy$ and $y \geq C$

$$\begin{aligned}
|G(\gamma z)| \leq (2y)^k \sum_{n=1}^\infty |a_n| e^{-2\pi y n} &= (2y)^k e^{-2\pi y} \sum_{n=1}^\infty |a_n| e^{-2\pi y(n-1)} \\
&\leq (2y)^k e^{-2\pi y} \sum_{n=1}^\infty |a_n| e^{-2\pi C(n-1)} \\
&\leq (2y)^k e^{-2\pi y} K_\gamma e^{2\pi C}.
\end{aligned}$$

This tends to 0 if y tends to ∞. Consequently, the function $G(\gamma z)$ is bounded on the closure of the standard fundamental domain, as desired.

(b) Again writing $\gamma = \left(\begin{smallmatrix} a & b \\ c & d \end{smallmatrix}\right)$ we have

$$\frac{d\gamma z}{dz} = \frac{d\frac{az+b}{cz+d}}{dz} = \frac{1}{(cz+d)^2}(a(cz+d) - (az+b)c) = \frac{\det(\gamma)}{(cz+d)^2},$$

which gives the claim.

(c) This is again a simple computation:

$$(\gamma z - \gamma \overline{z})^{-2} d\gamma z \wedge d\gamma \overline{z} = \det(\gamma)^2 (\frac{az+b}{cz+d} - \frac{a\overline{z}+b}{c\overline{z}+d})^{-2} (cz+d)^{-2} (c\overline{z}+d)^{-2} dz \wedge d\overline{z}$$
$$= (z-\overline{z})^{-2} dz \wedge d\overline{z},$$

using (b). The last statement is

$$\frac{dz \wedge d\overline{z}}{(z-\overline{z})^2} = \frac{(dx+idy)\wedge(dx-idy)}{(2iy)^2} = \frac{-2idx \wedge dy}{-4y^2} = \frac{idx \wedge dy}{2y^2}.$$

(d) Due to the Γ-invariance, it suffices to show

$$\int_{\mathscr{F}} \frac{dz \wedge d\overline{z}}{(z-\overline{z})^2} = \frac{i\pi}{6}.$$

Let $\omega = -\frac{dz}{z-\overline{z}}$. The total derivative of ω is

$$d\omega = ((z-\overline{z})^{-2} dz - (z-\overline{z})^{-2} d\overline{z}) \wedge dz = \frac{dz \wedge d\overline{z}}{(z-\overline{z})^2}.$$

Hence, Stokes' theorem yields

$$\int_{\mathscr{F}} \frac{dz \wedge d\overline{z}}{(z-\overline{z})^2} = -\int_{\partial\mathscr{F}} \frac{dz}{z-\overline{z}},$$

where $\partial\mathscr{F}$ is the positively oriented border of $\mathscr{F}$, which we describe concretely as the path A from ∞ to ζ_3 on the vertical line, followed by the path C from ζ_3 to ζ_6 on the unit circle and finally followed by $-TA$. Hence with $z = x+iy$ we have

$$\int_{\mathscr{F}} \frac{dz \wedge d\overline{z}}{(z-\overline{z})^2} = -\frac{1}{2i}\Big(\int_A \frac{dz}{y} - \int_{TA} \frac{dz}{y} + \int_C \frac{dz}{y}\Big) = -\frac{1}{2i}\int_C \frac{dz}{y},$$

since $dz = dTz$. Using the obvious parametrisation of C we obtain

$$-\frac{1}{2i}\int_C \frac{dz}{y} = -\frac{1}{2i}\int_{2\pi/3}^{2\pi/6} \frac{1}{\mathrm{Im}(e^{i\phi})} \frac{de^{i\phi}}{d\phi} d\phi = -\frac{1}{2}\int_{2\pi/3}^{2\pi/6} \frac{e^{i\phi}}{\mathrm{Im}(e^{i\phi})} d\phi$$
$$= -\frac{1}{2}\int_{2\pi/3}^{2\pi/6} (\frac{\cos(\phi)}{\sin(\phi)} + i) d\phi = -\frac{i}{2}(\frac{2\pi}{6} - \frac{2\pi}{3}) = \frac{i\pi}{6},$$

since sin is symmetric around $\pi/2$ and cos is antisymmetric, so that the integral over $\frac{\cos(\phi)}{\sin(\phi)}$ cancels. □

Definition 6.2 Let $\Gamma \leq \mathrm{SL}_2(\mathbb{Z})$ be a subgroup of finite index and $\mu := (\mathrm{PSL}_2(\mathbb{Z}) : \overline{\Gamma})$ be the index of $\overline{\Gamma} = \Gamma/(\langle \pm 1 \rangle \cap \Gamma)$ in $\mathrm{PSL}_2(\mathbb{Z})$. We define the *Petersson pairing* as

$$\begin{aligned}\mathrm{M}_k(\Gamma\,;\,\mathbb{C}) \times \mathrm{S}_k(\Gamma\,;\,\mathbb{C}) \to & \mathbb{C} \\ (f,g) \mapsto & \frac{-1}{(2i)^{k-1}\mu}\int_{\mathscr{F}_\Gamma} f(z)\overline{g(z)}(z-\overline{z})^k \frac{dz\wedge d\overline{z}}{(z-\overline{z})^2} \\ = & \frac{1}{\mu}\int_{\mathscr{F}_\Gamma} f(z)\overline{g(z)}y^{k-2}dx\wedge dy =: (f,g),\end{aligned}$$

where $\mathscr{F}_\Gamma$ is any fundamental domain for Γ.

Proposition 6.3 *(a) The integral in the Petersson pairing converges. It does not depend on the choice of the fundamental domain $\mathscr{F}_\Gamma$.*

(b) The Petersson pairing is a sesqui-linear pairing (linear in the first and anti-linear in the second variable).

(c) The restriction of the Petersson pairing to $\mathrm{S}_k(\Gamma\,;\,\mathbb{C})$ is a positive definite scalar product (the Petersson scalar product*).*

(d) If f, g are modular (cusp) forms for the group Γ and $\Gamma' \le \Gamma$ is a subgroup of finite index, then the Petersson pairing of f and g with respect to Γ gives the same value as the one with respect to Γ'.

Proof (a) By Lemma 6.1 the integral converges, since the function

$$G(z) := f(z)\overline{g(z)}(z-\overline{z})^k$$

is bounded on $\mathscr{F}_\Gamma$ and the volume of $\mathscr{F}_\Gamma$ for the measure in question is finite. The integral does not depend on the choice of the fundamental domain by the invariance of $G(z)$ under Γ.

(b) is clear.

(c) We have

$$(f,f) = \frac{1}{\mu}\int_{\mathscr{F}_\Gamma} |f(z)|^2 y^{k-2} dx\wedge dy,$$

which is clearly non-negative. It is 0 if and only if f is the zero function, showing that the product is positive definite.

(d) If $\mathscr{F}_\Gamma$ is a fundamental domain for Γ, then $\bigcup_{\gamma\in\Gamma'\backslash\Gamma}\gamma\mathscr{F}_\Gamma$ is a fundamental domain for Γ' (for any choice of representatives of $\Gamma'\backslash\Gamma$). But on every $\gamma\mathscr{F}_\Gamma$ the integral takes the same value. □

Proposition 6.4 *Let $f, g \in \mathrm{S}_k(\Gamma\,;\,\mathbb{C})$. We have*

$$(f,g) = \frac{-1}{(2i)^{k-1}\mu}\sum_{\gamma\in\overline{\Gamma}\backslash\mathrm{PSL}_2(\mathbb{Z})}\int_{\zeta_3}^{i}\int_{\infty}^{0} f|_\gamma(z)\overline{g|_\gamma(z)}(z-\overline{z})^{k-2}dz d\overline{z}.$$

Proof Let us write for short $G_\gamma(z,\overline{z}) = f|_\gamma(z)\overline{g|_\gamma(z)}(z-\overline{z})^k$ for $\gamma \in \mathrm{SL}_2(\mathbb{Z})$. Then

$$-(2i)^{k-1}\mu(f,g) = \int_{\bigcup_\gamma \gamma\mathscr{F}} G(z,\overline{z})\frac{dz\wedge d\overline{z}}{(z-\overline{z})^2} = \sum_\gamma \int_{\mathscr{F}} G_\gamma(z,\overline{z})\frac{dz\wedge d\overline{z}}{(z-\overline{z})^2}$$

by Lemma 6.1, where the union resp. sum runs over a fixed system of coset representatives of $\overline{\Gamma}\backslash\mathrm{PSL}_2(\mathbb{Z})$; by our observations, everything is independent of this choice. Consider the differential form

$$\omega_\gamma := \Big(\int_\infty^z f|_\gamma(u)(u-\overline{z})^{k-2}du\Big)\overline{g|_\gamma(z)}d\overline{z}.$$

Note that the integral converges since f is a cusp form. The total derivative of ω_γ is $d\omega_\gamma = G_\gamma(z,\overline{z})\frac{dz\wedge d\overline{z}}{(z-\overline{z})^2}$. Consequently, Stokes' theorem gives

$$\sum_\gamma \int_{\mathscr{F}} G_\gamma(z,\overline{z})\frac{dz\wedge d\overline{z}}{(z-\overline{z})^2} = \sum_\gamma \int_{\partial\mathscr{F}} \Big(\int_\infty^z f|_\gamma(u)(u-\overline{z})^{k-2}du\Big)\overline{g|_\gamma(z)}d\overline{z},$$

where as above $\partial\mathscr{F}$ is the positively oriented border of the standard fundamental domain $\mathscr{F}$, which we describe as the path A along the vertical line from ∞ to ζ_3, followed by the path B from ζ_3 to i along the unit circle, followed by $-\sigma B$ and by $-TA$.

We now make a small calculation. Let for this C be any (piecewise continuously differentiable) path in $\mathbb{H}$ and $M \in \mathrm{SL}_2(\mathbb{Z})$:

$$\begin{aligned}
&\int_{MC}\int_\infty^z f|_\gamma(u)\overline{g|_\gamma(z)}(u-\overline{z})^{k-2}dud\overline{z}\\
&= \int_C\int_\infty^{Mz} f|_\gamma(u)\overline{g|_\gamma(Mz)}(u-M\overline{z})^{k-2}du\frac{dM\overline{z}}{d\overline{z}}d\overline{z}\\
&= \int_C\int_{M^{-1}\infty}^z f|_{\gamma M}(u)\overline{g|_{\gamma M}(z)}(u-\overline{z})^{k-2}dud\overline{z}\\
&= \int_C\int_\infty^z f|_{\gamma M}(u)\overline{g|_{\gamma M}(z)}(u-\overline{z})^{k-2}dud\overline{z} - \int_C\int_\infty^{M^{-1}\infty} f|_{\gamma M}(u)\overline{g|_{\gamma M}(z)}(u-\overline{z})^{k-2}dud\overline{z}.
\end{aligned}$$

This gives

$$\begin{aligned}
&\int_{C-MC}\int_\infty^z f|_\gamma(u)\overline{g|_\gamma(z)}(u-\overline{z})^{k-2}dud\overline{z} =\\
&\qquad \int_C\int_\infty^z (G_\gamma(u,\overline{z}) - G_{\gamma M}(u,\overline{z}))dud\overline{z} + \int_C\int_\infty^{M^{-1}\infty} G_{\gamma M}(u,\overline{z})dud\overline{z}.
\end{aligned}$$

Continuing with the main calculation, we have

$$
\begin{aligned}
&-(2i)^{k-1}\mu(f,g)\\
=&\sum_{\gamma}\Big[\int_A\int_{\infty}^{z}(G_\gamma(u,\overline{z})-G_{\gamma T}(u,\overline{z}))du d\overline{z}+\int_A\int_{\infty}^{T^{-1}\infty}G_{\gamma T}(u,\overline{z})du d\overline{z}\Big]\\
+&\sum_{\gamma}\Big[\int_B\int_{\infty}^{z}(G_\gamma(u,\overline{z})-G_{\gamma\sigma}(u,\overline{z}))du d\overline{z}+\int_B\int_{\infty}^{\sigma^{-1}\infty}G_{\gamma\sigma}(u,\overline{z})du d\overline{z}\Big]\\
=&\sum_{\gamma}\int_B\int_{\infty}^{0}G_{\gamma\sigma}(u,\overline{z})du d\overline{z},
\end{aligned}
$$

using $T^{-1}\infty=\infty, \sigma^{-1}\infty=0$ and the fact that the γT and $\gamma\sigma$ are just permutations of the cosets. □

6.2 Theory: The Eichler–Shimura Map

Let $\Gamma \leq \mathrm{SL}_2(\mathbb{Z})$ be a subgroup of finite index. We fix some $z_0 \in \mathbb{H}$. For $f \in \mathrm{M}_k(\Gamma\,;\,\mathbb{C})$ with $k \geq 2$ and γ, δ in $\mathbb{Z}^{2\times 2}$ with positive determinant, let

$$I_f(\gamma z_0, \delta z_0) := \int_{\gamma z_0}^{\delta z_0} f(z)(Xz+Y)^{k-2}dz \in V_{k-2}(\mathbb{C}).$$

The integral is to be taken coefficient-wise. Note that it is independent of the chosen path since we are integrating holomorphic functions.

Lemma 6.5 *For any $z_0 \in \mathbb{H}$ and any matrices $\gamma, \delta \in \mathbb{Z}^{2\times 2}$ with positive determinant we have*

$$I_f(z_0, \gamma\delta z_0) = I_f(z_0, \gamma z_0) + I_f(\gamma z_0, \gamma\delta z_0)$$

and

$$I_f(\gamma z_0, \gamma\delta z_0) = \det(\gamma)^{2-k}(\gamma.\big(I_{f|_\gamma}(z_0, \delta z_0)\big)) = (\det(\gamma)^{-1}\gamma).\big(I_{f|_\gamma}(z_0, \delta z_0)\big).$$

Proof The first statement is clear. Write $\gamma = \left(\begin{smallmatrix} a & b \\ c & d \end{smallmatrix}\right)$. Recall that by Lemma 6.1(b), we have $d\gamma z = \frac{\det(\gamma)}{(cz+d)^2}dz$. We compute further

$$
\begin{aligned}
I_f(\gamma z_0, \gamma\delta z_0) &= \int_{\gamma z_0}^{\gamma\delta z_0} f(z)(Xz+Y)^{k-2}dz \\
&= \int_{z_0}^{\delta z_0} f(\gamma z)(X\gamma z+Y)^{k-2}\frac{d\gamma z}{dz}dz \\
&= \det(\gamma)^{2-k}\int_{z_0}^{\delta z_0} f|_\gamma(z)(cz+d)^{k-2}(X\frac{az+b}{cz+d}+Y)^{k-2}dz \\
&= \det(\gamma)^{2-k}\int_{z_0}^{\delta z_0} f|_\gamma(z)(X(az+b)+Y(cz+d))^{k-2}dz \\
&= \det(\gamma)^{2-k}\int_{z_0}^{\delta z_0} f|_\gamma(z)((Xa+Yc)z+(Xb+Yd))^{k-2}dz \\
&= \det(\gamma)^{2-k}\int_{z_0}^{\delta z_0} f|_\gamma(z)(\gamma.(Xz+Y))^{k-2}dz \\
&= \det(\gamma)^{2-k}\cdot\gamma.\Big(\int_{z_0}^{\delta z_0} f|_\gamma(z)(Xz+Y)^{k-2}dz\Big) \\
&= \det(\gamma)^{2-k}\cdot\gamma.\big(I_{f|_\gamma}(z_0,\delta z_0)\big).
\end{aligned}
$$

We recall that for a polynomial $P(X, Y)$ we have the action

$$(g.P)(X,Y) = P((X,Y)\left(\begin{smallmatrix} a & b \\ c & d\end{smallmatrix}\right)) = P(Xa+Yc, Xb+Yd).$$

□

Definition 6.6 The space of *antiholomorphic cusp forms* $\overline{\mathrm{S}_k(\Gamma\,;\,\mathbb{C})}$ consists of the functions $z \mapsto \overline{f}(z) := \overline{f(z)}$ with $f \in \mathrm{S}_k(\Gamma\,;\,\mathbb{C})$.

We can consider an antiholomorphic cusp form as a power series in $\overline{z}$. For instance, if $f(z) = \sum_{n=1}^{\infty} a_n e^{2\pi i n z}$, then $\overline{f(z)} = \sum_{n=1}^{\infty} \overline{a_n} e^{2\pi i n(-\overline{z})} = \tilde{f}(-\overline{z})$, where $\tilde{f}(z) = \sum_{n=1}^{\infty} \overline{a_n} e^{2\pi i n z}$. Note that

$$
\begin{aligned}
\int_\alpha \overline{F(z)}d\overline{z} = \int_0^1 \overline{F(\alpha(t))}\frac{d\overline{\alpha}}{dt}dt &= \int_0^1 \overline{F(\alpha(t))}\,\overline{\frac{d\alpha}{dt}}dt \\
&= \overline{\int_0^1 F(\alpha(t))\frac{d\alpha}{dt}dt} = \overline{\int_\alpha F(z)dz} \quad (18)
\end{aligned}
$$

for any piecewise analytic path $\alpha : [0, 1] \to \mathbb{C}$ and any integrable complex valued function F. This means for $f \in \mathrm{S}_k(\Gamma\,;\,\mathbb{C})$:

$$\overline{I_f(\gamma z_0, \delta z_0)} = \int_{\gamma z_0}^{\delta z_0} \overline{f(z)}(X\overline{z}+Y)^{k-2}d\overline{z} \in V_{k-2}(\mathbb{C}).$$

Proposition 6.7 *Let $k \geq 2$ and $\Gamma \leq \mathrm{SL}_2(\mathbb{Z})$ be a subgroup of finite index and fix $z_0, z_1 \in \mathbb{H}$.*

(a) The Eichler–Shimura map

$$\mathrm{M}_k(\Gamma\,;\,\mathbb{C}) \oplus \overline{\mathrm{S}_k(\Gamma\,;\,\mathbb{C})} \to \mathrm{H}^1(\Gamma, V_{k-2}(\mathbb{C})),$$
$$(f, \overline{g}) \mapsto (\gamma \mapsto I_f(z_0, \gamma z_0) + \overline{I_g(z_1, \gamma z_1)})$$

is a well-defined homomorphism of $\mathbb{C}$*-vector spaces. It does not depend on the choice of* z_0 *and* z_1*.*

(b) The induced Eichler–Shimura map

$$\mathrm{M}_k(\Gamma\,;\,\mathbb{C}) \oplus \overline{\mathrm{S}_k(\Gamma\,;\,\mathbb{C})} \to \mathrm{H}^1(\mathrm{SL}_2(\mathbb{Z}), \mathrm{Hom}_{\mathbb{C}[\Gamma]}(\mathbb{C}[\mathrm{SL}_2(\mathbb{Z})], V_{k-2}(\mathbb{C}))),$$
$$(f, \overline{g}) \mapsto (a \mapsto (b \mapsto I_f(bz_0, baz_0) + \overline{I_g(bz_1, baz_1)}))$$

is a well-defined homomorphism of $\mathbb{C}$*-vector spaces. It does not depend on the choice of* z_0 *and* z_1*. Via the map from Shapiro's lemma, this homomorphism coincides with the one from (a).*

Proof (a) For checking that the map is well-defined, it suffices to compute that $\gamma \mapsto I_f(z_0, \gamma z_0)$ is a 1-cocycle:

$$I_f(z_0, \gamma\delta z_0) = I_f(z_0, \gamma z_0) + I_f(\gamma z_0, \gamma\delta z_0) = I_f(z_0, \gamma z_0) + \gamma.I_f(z_0, \delta z_0),$$

using Lemma 6.5 and $f|_\gamma = f$ since $\gamma \in \Gamma$.

The independence of the base point is seen as follows. Let $\tilde{z}_0$ be any base point.

$$I_f(\tilde{z}_0, \gamma\tilde{z}_0) = I_f(\tilde{z}_0, z_0) + I_f(z_0, \gamma z_0) + I_f(\gamma z_0, \gamma\tilde{z}_0) = I_f(z_0, \gamma z_0) + (1-\gamma)I_f(\tilde{z}_0, z_0).$$

The difference of the cocycles $(\gamma \mapsto I_f(\tilde{z}_0, \gamma\tilde{z}_0))$ and $(\gamma \mapsto I_f(z_0, \gamma z_0))$ is hence the coboundary $(\gamma \mapsto (1-\gamma)I_f(\tilde{z}_0, z_0))$.

(b) We first check that the map $(b \mapsto I_f(bz_0, baz_0) + \overline{I_g(bz_1, baz_1)})$ is indeed in the coinduced module $\mathrm{Hom}_{\mathbb{C}[\Gamma]}(\mathbb{C}[\mathrm{SL}_2(\mathbb{Z})], V_{k-2}(\mathbb{C}))$. For that let $\gamma \in \Gamma$. We have

$$I_f(\gamma bz_0, \gamma baz_0) = \gamma.(I_f(bz_0, baz_0))$$

by Lemma 6.5, as desired. The map $\phi(a) := (b \mapsto I_f(bz_0, baz_0) + \overline{I_g(bz_1, baz_1)})$ is a cocycle:

$$\phi(a_1a_2)(b) = I_f(bz_0, ba_1a_2z_0) + \overline{I_g(bz_1, ba_1a_2z_1)} =$$
$$I_f(bz_0, ba_1z_0) + I_f(ba_1z_0, ba_1a_2z_0) + \overline{I_g(bz_1, ba_1z_1)} + \overline{I_g(ba_1z_1, ba_1a_2z_1)}$$
$$= \phi(a_1)(b) + \phi(a_2)(ba_1) = \phi(a_1)(b) + (a_1.(\phi(a_2)))(b),$$

by the definition of the left action of $\mathrm{SL}_2(\mathbb{Z})$ on the coinduced module. Note that the map in Shapiro's lemma in our situation is given by

$$\phi \mapsto (\gamma \mapsto \phi(\gamma)(1) = I_f(z_0, \gamma z_0)) + \overline{I_g(z_1, \gamma z_1)}),$$

which shows that the maps from (a) and (b) coincide. The independence from the base point in (b) now follows from the independence in (a). □

Next we identify the cohomology of $\mathrm{SL}_2(\mathbb{Z})$ with the one of $\mathrm{PSL}_2(\mathbb{Z})$.

Proposition 6.8 *Let $\Gamma \leq \mathrm{SL}_2(\mathbb{Z})$ be a subgroup of finite index and let R be a ring in which 2 is invertible. Let V be a left $R[\Gamma]$-module. Assume that either $-1 \notin \Gamma$ or $-1 \in \Gamma$ acts trivially on V. Then the inflation map*

$$\mathrm{H}^1(\mathrm{PSL}_2(\mathbb{Z}), \mathrm{Hom}_{R[\overline{\Gamma}]}(R[\mathrm{PSL}_2(\mathbb{Z})], V)) \xrightarrow{\text{infl}} \mathrm{H}^1(\mathrm{SL}_2(\mathbb{Z}), \mathrm{Hom}_{R[\Gamma]}(R[\mathrm{SL}_2(\mathbb{Z})], V))$$

is an isomorphism. We shall identify these two R-modules from now on.

Proof If $-1 \notin \Gamma$, then $\Gamma \cong \overline{\Gamma}$ and $\mathrm{Hom}_{R[\Gamma]}(R[\mathrm{SL}_2(\mathbb{Z})], V)^{\langle -1 \rangle}$ consists of all the functions satisfying $f(g) = f(-g)$ for all $g \in \mathrm{SL}_2(\mathbb{Z})$, which are precisely the functions in $\mathrm{Hom}_{R[\overline{\Gamma}]}(R[\mathrm{PSL}_2(\mathbb{Z})], V)$.

If $-1 \in \Gamma$ and -1 acts trivially on V, then $f(-g) = (-1).f(g) = f(g)$ and so -1 already acts trivially on $\mathrm{Hom}_{R[\Gamma]}(R[\mathrm{SL}_2(\mathbb{Z})], V)$. This $R[\mathrm{SL}_2(\mathbb{Z})]$-module is then naturally isomorphic to $\mathrm{Hom}_{R[\overline{\Gamma}]}(R[\mathrm{PSL}_2(\mathbb{Z})], V)$ since any function is uniquely determined on its classes modulo $\langle -1 \rangle$.

Due to the invertibility of 2, the Hochschild–Serre exact sequence (Theorem 3.27) shows that inflation indeed gives the desired isomorphism because the third term $\mathrm{H}^1(\langle \pm 1 \rangle, \mathrm{Hom}_{R[\Gamma]}(R[\mathrm{SL}_2(\mathbb{Z})], V))$ in the inflation-restriction sequence is zero (see Proposition 3.26). □

Proposition 6.9 *The kernel of the Eichler–Shimura map composed with the restriction*

$$\mathrm{M}_k(\Gamma\,;\,\mathbb{C}) \oplus \overline{\mathrm{S}_k(\Gamma\,;\,\mathbb{C})} \to \mathrm{H}^1(\Gamma, V_{k-2}(\mathbb{C})) \to \prod_{c \in \Gamma \backslash \mathbb{P}^1(\mathbb{Q})} \mathrm{H}^1(\Gamma_c, V_{k-2}(\mathbb{C}))$$

is equal to $\mathrm{S}_k(\Gamma\,;\,\mathbb{C}) \oplus \overline{\mathrm{S}_k(\Gamma\,;\,\mathbb{C})}$. In particular, the image of $\mathrm{S}_k(\Gamma\,;\,\mathbb{C}) \oplus \overline{\mathrm{S}_k(\Gamma\,;\,\mathbb{C})}$ under the Eichler–Shimura map lies in the parabolic cohomology $\mathrm{H}^1_{\mathrm{par}}(\Gamma, V_{k-2}(\mathbb{C}))$.

Proof In order to simplify the notation of the proof, we shall only prove the case of a modular form $f \in \mathrm{M}_k(\Gamma\,;\,\mathbb{C})$. The statement for antihomolorphic forms is proved in the same way. The composition maps the modular form f to the 1-cocycle (for $\gamma \in \Gamma_c$)

$$\gamma \mapsto \int_{z_0}^{\gamma z_0} f(z)(Xz + Y)^{k-2} dz$$

with a fixed base point $z_0 \in \mathbb{H}$. The aim is now to move the base point to the cusps. We cannot just replace z_0 by ∞, as then the integral might not converge any more (it converges on cusp forms). Let $c = M\infty$ be any cusp with $M = \left(\begin{smallmatrix} a & b \\ c & d \end{smallmatrix}\right) \in \mathrm{SL}_2(\mathbb{Z})$.

We then have $\Gamma_c = \langle MTM^{-1}\rangle \cap \Gamma = \langle MT^r M^{-1}\rangle$ for some $r \geq 1$. Since f is holomorphic in the cusps, we have

$$f|_M(z) = \sum_{n=0}^{\infty} a_n e^{2\pi i n/rz} = a_0 + g(z)$$

and thus

$$f(z) = a_0|_{M^{-1}}(z) + g|_{M^{-1}}(z) = \frac{a_0}{(-cz+a)^k} + g|_{M^{-1}}(z).$$

Now we compute the cocycle evaluated at $\gamma = MT^r M^{-1}$:

$$\int_{z_0}^{\gamma z_0} f(z)(Xz+Y)^{k-2}dz = a_0 \int_{z_0}^{\gamma z_0} \frac{(Xz+Y)^{k-2}}{(-cz+a)^k}dz + \int_{z_0}^{\gamma z_0} g|_{M^{-1}}(z)(Xz+Y)^{k-2}dz.$$

Before we continue by evaluating the right summand, we remark that the integral

$$I_{g|_{M^{-1}}}(z_0, M\infty) = \int_{z_0}^{M\infty} g|_{M^{-1}}(z)(Xz+Y)^{k-2}dz = M.\int_{M^{-1}z_0}^{\infty} g(z)(Xz+Y)^{k-2}dz$$

converges. We have

$$\begin{aligned}\int_{z_0}^{\gamma z_0} g|_{M^{-1}}(z)(Xz+Y)^{k-2}dz &= (\int_{z_0}^{M\infty} + \int_{\gamma M\infty}^{\gamma z_0})g|_{M^{-1}}(z)(Xz+Y)^{k-2}dz \\ &= (1-\gamma).\int_{z_0}^{M\infty} g|_{M^{-1}}(z)(Xz+Y)^{k-2}dz\end{aligned}$$

since $g|_{M^{-1}\gamma}(z) = g|_{T^r M^{-1}}(z) = g|_{M^{-1}}(z)$. The 1-cocycle $\gamma \mapsto \int_{z_0}^{\gamma z_0} g|_{M^{-1}}(z)(Xz+Y)^{k-2}dz$ is thus a 1-coboundary. Consequently, the class of the image of f is equal to the class of the 1-cocycle

$$\gamma \mapsto a_0 \int_{z_0}^{\gamma z_0} \frac{(Xz+Y)^{k-2}}{(-cz+a)^k}dz.$$

We have the isomorphism (as always for cyclic groups)

$$\mathrm{H}^1(\Gamma_c, V_{k-2}(\mathbb{C})) \xrightarrow{\phi \mapsto \phi(MT^r M^{-1})} V_{k-2}(\mathbb{C})_{\Gamma_c}.$$

Furthermore, we have the isomorphism

$$V_{k-2}(\mathbb{C})_{\Gamma_c} \xrightarrow{P \mapsto M^{-1}P} V_{k-2}(\mathbb{C})_{\langle T^r\rangle} \xrightarrow{P \mapsto P(0,1)} \mathbb{C}$$

with polynomials $P(X, Y)$. Note that the last map is an isomorphism by the explicit description of $V_{k-2}(\mathbb{C})_{\langle T^r \rangle}$. Under the composition the image of the cocycle coming from the modular form f is

$$a_0 M^{-1}. \int_{z_0}^{\gamma z_0} \frac{(Xz+Y)^{k-2}}{(-cz+a)^k} dz(0,1) = a_0 \int_{z_0}^{\gamma z_0} \frac{(Xz+Y)^{k-2}}{(-cz+a)^k} dz(-c,a)$$

$$= a_0 \int_{z_0}^{\gamma z_0} \frac{1}{(-cz+a)^2} dz = a_0 \int_{M^{-1}z_0}^{T^r M^{-1} z_0} dz = a_0(M^{-1}z_0 + r - M^{-1}z_0) = r a_0,$$

as $(0,1)M^{-1} = (0,1)\left(\begin{smallmatrix} d & -b \\ -c & a \end{smallmatrix}\right) = (-c, a)$. This expression is zero if and only if $a_0 = 0$, i.e. if and only if f vanishes at the cusp c.

A similar argument works for antiholomorphic cusp forms. □

6.3 Theory: Cup Product and Petersson Scalar Product

This part owes much to the treatment of the Petersson scalar product by Haberland in [14] (see also [5, §12]).

Definition 6.10 Let G be a group and M and N be two left $R[G]$-modules. We equip $M \otimes_R N$ with the diagonal left $R[G]$-action. Let $m, n \geq 0$. Then we define the *cup product*

$$\cup : \mathrm{H}^n(G, M) \otimes_R \mathrm{H}^m(G, N) \to \mathrm{H}^{n+m}(G, M \otimes_R N)$$

by

$$\phi \cup \psi := ((g_1, \ldots, g_n, g_{n+1}, \ldots, g_{n+m}) \mapsto \phi(g_1, \ldots, g_n) \otimes (g_1 \cdots g_n).\psi(g_{n+1}, \ldots, g_{n+m})$$

on cochains of the bar resolution.

This description can be derived easily from the natural one on the standard resolution. For instance, [4, §5.3] gives the above formula up to a sign (which does not matter in our application anyway because we work in fixed degree). In Exercise 6.17 it is checked that the cup product is well-defined.

Lemma 6.11 *Keep the notation of Definition 6.10 and let $\phi \in \mathrm{H}^n(G, M)$ and $\psi \in \mathrm{H}^m(G, N)$. Then*

$$\phi \cup \psi = (-1)^{mn} \psi \cup \phi$$

via the natural isomorphism $M \otimes_R N \cong N \otimes_R M$.

Proof Exercise 6.18. □

We are now going to formulate a pairing on cohomology, which will turn out to be a version of the Petersson scalar product. We could introduce compactly supported cohomology for writing it in more conceptual terms, but have decided not to do this in order not to increase the amount of new material even more.

Definition 6.12 Let M be an $R[\mathrm{PSL}_2(\mathbb{Z})]$-module. The *parabolic* 1*-cocycles* are defined as

$$\mathrm{Z}^1_{\mathrm{par}}(\Gamma, M) = \ker(\mathrm{Z}^1(\Gamma, M) \xrightarrow{\mathrm{res}} \prod_{g \in \Gamma \backslash \mathrm{PSL}_2(\mathbb{Z})/\langle T \rangle} \mathrm{Z}^1(\Gamma \cap \langle gTg^{-1} \rangle, M)).$$

Proposition 6.13 *Let R be a ring in which 6 is invertible. Let M and N be left $R[\mathrm{PSL}_2(\mathbb{Z})]$-modules together with a $R[\mathrm{PSL}_2(\mathbb{Z})]$-module homomorphism $\pi : M \otimes_R N \to R$ where we equip $M \otimes_R N$ with the diagonal action. Write G for $\mathrm{PSL}_2(\mathbb{Z})$. We define a pairing*

$$\langle , \rangle : \mathrm{Z}^1(G, M) \times \mathrm{Z}^1(G, N) \to R$$

as follows: Let (ϕ, ψ) be a pair of 1-cocycles. Form their cup product $\rho := \pi_*(\phi \cup \psi)$ in $\mathrm{Z}^2(G, R)$ via $\mathrm{Z}^2(G, M \otimes_R N) \xrightarrow{\pi_*} \mathrm{Z}^2(G, R)$. As $\mathrm{H}^2(G, R)$ is zero (Corollary 4.12), ρ must be a 2-coboundary, i.e. there is $a : G \to R$ (depending on (ϕ, ψ)) such that

$$\rho(g, h) = \pi(\phi(g) \otimes g.\psi(h)) = a(h) - a(gh) + a(g).$$

We define the pairing by

$$\langle \phi, \psi \rangle := a(T).$$

(a) The pairing is well-defined and bilinear. It can be expressed as

$$\langle \phi, \psi \rangle = -\rho(\tau, \sigma) + \frac{1}{2}\rho(\sigma, \sigma) + \frac{1}{3}(\rho(\tau, \tau) + \rho(\tau, \tau^2)).$$

(b) If $\phi \in \mathrm{Z}^1_{\mathrm{par}}(G, M)$, then $\rho(\tau, \sigma) = \rho(\sigma, \sigma)$ and

$$\langle \phi, \psi \rangle = -\frac{1}{2}\rho(\sigma, \sigma) + \frac{1}{3}(\rho(\tau, \tau) + \rho(\tau, \tau^2)).$$

Moreover, $\langle \phi, \psi \rangle$ only depends on the class of ψ in $\mathrm{H}^1(G, N)$.

(c) If $\psi \in \mathrm{Z}^1_{\mathrm{par}}(G, N)$, then $\rho(\tau, \sigma) = \rho(\tau, \tau^2)$ and

$$\langle \phi, \psi \rangle = \frac{1}{2}\rho(\sigma, \sigma) + \frac{1}{3}\rho(\tau, \tau) - \frac{2}{3}\rho(\tau, \tau^2).$$

Moreover, $\langle \phi, \psi \rangle$ only depends on the class of ϕ in $\mathrm{H}^1(G, M)$.

(d) If $\phi \in \mathrm{Z}^1_{\mathrm{par}}(G, M)$ *and* $\psi \in \mathrm{Z}^1_{\mathrm{par}}(G, N)$*, then* $\rho(\sigma, \sigma) = \rho(\tau, \tau^2)$ *and*

$$\langle \phi, \psi \rangle = -\frac{1}{6}\rho(\sigma, \sigma) + \frac{1}{3}\rho(\tau, \tau).$$

Proof (a) We first have

$$0 = \pi(\phi(1) \otimes \psi(1)) = \rho(1, 1) = a(1) - a(1) + a(1) = a(1),$$

since ϕ and ψ are 1-cocycles. Recall that the value of a 1-cocycle at 1 is always 0 due to $\phi(1) = \phi(1 \cdot 1) = \phi(1) + \phi(1)$. Furthermore, we have

$$\begin{aligned}
\rho(\tau, \sigma) &= a(\sigma) - a(T) + a(\tau)\\
\rho(\sigma, \sigma) &= a(\sigma) - a(1) + a(\sigma) = 2a(\sigma)\\
\rho(\tau, \tau^2) &= a(\tau^2) - a(1) + a(\tau) = a(\tau) + a(\tau^2)\\
\rho(\tau, \tau) &= a(\tau) - a(\tau^2) + a(\tau) = 2a(\tau) - a(\tau^2)
\end{aligned}$$

Hence, we get $a(T) = -\rho(\tau, \sigma) + a(\sigma) + a(\tau)$ and $a(\sigma) = \frac{1}{2}\rho(\sigma, \sigma)$ as well as $a(\tau) = \frac{1}{3}(\rho(\tau, \tau) + \rho(\tau, \tau^2))$, from which the claimed formula follows. The formula also shows the independence of the choice of a and the bilinearity.

(b) Now assume $\phi(T) = 0$. Using $T = \tau\sigma$ we obtain

$$\begin{aligned}
\rho(\tau, \sigma) &= \pi(\phi(\tau) \otimes \tau\psi(\sigma)) = -\pi(\tau.\phi(\sigma) \otimes \tau\psi(\sigma))\\
&= -\pi(\phi(\sigma) \otimes \psi(\sigma)) = \pi((\phi(\sigma) \otimes \sigma\psi(\sigma))) = \rho(\sigma, \sigma)
\end{aligned}$$

because $0 = \phi(T) = \phi(\tau\sigma) = \tau.\phi(\sigma) + \phi(\tau)$ and $0 = \psi(1) = \psi(\sigma^2) = \sigma.\psi(\sigma) + \psi(\sigma)$. This yields the formula.

We now show that the pairing does not depend on the choice of 1-cocycle in the class of ψ. To see this, let $\psi(g) = (g - 1)n$ with $n \in N$ be a 1-coboundary. Put $b(g) := \pi(-\phi(g) \otimes gn)$. Then, using $\phi(gh) = g(\phi(h)) + \phi(g)$, one immediately checks the equality

$$\rho(g, h) = \pi(\phi(g) \otimes g(h - 1)n) = g.b(h) - b(gh) + b(g).$$

Hence, (ϕ, ψ) is mapped to $b(T) = \pi(-\phi(T) \otimes Tn) = \pi(0 \otimes Tn) = 0$.

(c) Let now $\psi(T) = 0$. Then $0 = \psi(T) = \psi(\tau\sigma) = \tau\psi(\sigma) + \psi(\tau)$ and $0 = \psi(\tau^3) = \tau\psi(\tau^2) + \psi(\tau)$, whence $\tau\psi(\tau^2) = \tau\psi(\sigma)$. Consequently,

$$\rho(\tau, \sigma) = \pi(\phi(\tau) \otimes \tau\psi(\sigma)) = \pi(\phi(\tau) \otimes \tau\psi(\tau^2)) = \rho(\tau, \tau^2),$$

implying the formula.

The pairing does not depend on the choice of 1-cocycle in the class of ϕ. Let $\phi(g) = (g - 1)m$ be a 1-coboundary and put $c(g) := \pi(m \otimes \psi(g))$. Then the equality

$$\rho(g,h) = \pi((g-1)m \otimes g\psi(h)) = g.c(h) - c(gh) + c(g)$$

holds. Hence, (ϕ, ψ) is mapped to $c(T) = \pi(m \otimes \psi(T)) = \pi(m \otimes 0) = 0$.

(d) Suppose now that $\phi(T) = 0 = \psi(T)$, then by what we have just seen

$$\rho(\tau, \sigma) = \rho(\sigma, \sigma) = \rho(\tau, \tau^2).$$

This implies the claimed formula. □

Our next aim is to specialise this pairing to the cocycles coming from modular forms under the Eichler–Shimura map. We must first define a pairing on the modules used in the cohomology groups.

On the modules $\mathrm{Sym}^{k-2}(R^2)$ we now define the *symplectic pairing* over any ring R in which $(k-2)!$ is invertible. Let $n = k-2$ for simplicity. The pairing for $n = 0$ is just the multiplication on R. We now define the pairing for $n = 1$ as

$$R^2 \times R^2 \to R, \quad \left(\begin{smallmatrix} a \\ c \end{smallmatrix}\right) \bullet \left(\begin{smallmatrix} b \\ d \end{smallmatrix}\right) := \det\left(\begin{smallmatrix} a & b \\ c & d \end{smallmatrix}\right).$$

For any $g \in \mathrm{SL}_2(\mathbb{Z})$ we have

$$g\left(\begin{smallmatrix} a \\ c \end{smallmatrix}\right) \bullet g\left(\begin{smallmatrix} b \\ d \end{smallmatrix}\right) = \det g \left(\begin{smallmatrix} a & b \\ c & d \end{smallmatrix}\right) = \det\left(\begin{smallmatrix} a & b \\ c & d \end{smallmatrix}\right) = \left(\begin{smallmatrix} a \\ c \end{smallmatrix}\right) \bullet \left(\begin{smallmatrix} b \\ d \end{smallmatrix}\right).$$

As the next step, we define a pairing on the n-th tensor power of R^2

$$(R^2 \otimes_R \cdots \otimes_R R^2) \times (R^2 \otimes_R \cdots \otimes_R R^2) \to R$$

by

$$(\left(\begin{smallmatrix} a_1 \\ c_1 \end{smallmatrix}\right) \otimes \cdots \otimes \left(\begin{smallmatrix} a_n \\ c_n \end{smallmatrix}\right)) \bullet (\left(\begin{smallmatrix} b_1 \\ d_1 \end{smallmatrix}\right) \otimes \cdots \otimes \left(\begin{smallmatrix} b_n \\ d_n \end{smallmatrix}\right)) := \prod_{i=1}^{n} \left(\begin{smallmatrix} a_i \\ c_i \end{smallmatrix}\right) \bullet \left(\begin{smallmatrix} b_i \\ d_i \end{smallmatrix}\right).$$

This pairing is still invariant under the $\mathrm{SL}_2(\mathbb{Z})$-action.

Now we use the assumption on the invertibility of $n!$ in order to embed $\mathrm{Sym}^n(R^2)$ as an $R[S_n]$-module in the n-th tensor power, where the action of the symmetric group S_n is on the indices. We have that the map (in fact, $1/n!$ times the norm)

$$\mathrm{Sym}^n(R^2) \to R^2 \otimes_R \cdots \otimes_R R^2, \quad [\left(\begin{smallmatrix} a_1 \\ c_1 \end{smallmatrix}\right) \otimes \cdots \otimes \left(\begin{smallmatrix} a_n \\ c_n \end{smallmatrix}\right)] \mapsto \frac{1}{n!} \sum_{\sigma \in S_n} \begin{pmatrix} a_{\sigma(1)} \\ c_{\sigma(1)} \end{pmatrix} \otimes \cdots \otimes \begin{pmatrix} a_{\sigma(n)} \\ c_{\sigma(n)} \end{pmatrix}$$

is injective (one can use Tate cohomology groups to see this) as the order of S_n is invertible in the ring.

Finally, we define the pairing on $\mathrm{Sym}^n(R^2)$ as the restriction of the pairing on the n-th tensor power to the image of $\mathrm{Sym}^n(R^2)$ under the embedding that we just described. This pairing is, of course, still $\mathrm{SL}_2(\mathbb{Z})$-invariant.

We point to the important special case

$$\binom{a}{c}^{\otimes(k-2)} \bullet \binom{b}{d}^{\otimes(k-2)} = (ad - bc)^{k-2}.$$

Hence, after the identification $\mathrm{Sym}^{k-2}(R^2) \cong V_{k-2}(R)$ from Exercise 1.31, the resulting pairing on $V_{k-2}(R)$ has the property

$$(aX + cY)^{k-2} \bullet (bX + dY)^{k-2} \mapsto (ad - bc)^{k-2}.$$

This pairing extends to a paring on coinduced modules

$$\pi : \mathrm{Hom}_{R[\Gamma]}(R[\mathrm{PSL}_2(\mathbb{Z})], V_{k-2}(R)) \otimes_R \mathrm{Hom}_{R[\Gamma]}(R[\mathrm{PSL}_2(\mathbb{Z})], V_{k-2}(R)) \to R \tag{19}$$

by mapping (α, β) to $\sum_{\gamma \in \Gamma \backslash \mathrm{PSL}_2(\mathbb{Z})} \alpha(\gamma) \bullet \beta(\gamma)$.

Proposition 6.14 *Let $k \geq 2$. Assume $-1 \notin \Gamma$ (whence we view Γ as a subgroup of $\mathrm{PSL}_2(\mathbb{Z})$). Let $f, g \in \mathrm{S}_k(\Gamma\,;\,\mathbb{C})$ be cusp forms. Denote by ϕ_f the 1-cocycle associated with f under the Eichler–Shimura map for the base point $z_0 = \infty$, i.e.*

$$\phi_f(a) = (b \mapsto I_f(b\infty, ba\infty)) \in \mathrm{Z}^1(\mathrm{PSL}_2(\mathbb{Z}), \mathrm{Coind}_\Gamma^{\mathrm{PSL}_2(\mathbb{Z})}(V_{k-2}(\mathbb{C}))).$$

Further denote

$$\overline{\phi_f}(a) = (b \mapsto \overline{I_f(b\infty, ba\infty)}) \in \mathrm{Z}^1(\mathrm{PSL}_2(\mathbb{Z}), \mathrm{Coind}_\Gamma^{\mathrm{PSL}_2(\mathbb{Z})}(V_{k-2}(\mathbb{C}))).$$

Similarly, denote by ψ_g the 1-cocycle associated with g for the base point $z_1 = \zeta_6$. Define a bilinear pairing as in Proposition 6.13

$$\langle , \rangle : \left(\mathrm{Z}^1(\mathrm{PSL}_2(\mathbb{Z}), \mathrm{Coind}_\Gamma^{\mathrm{PSL}_2(\mathbb{Z})}(V_{k-2}(\mathbb{C})))\right)^2 \to \mathbb{C}$$

with the product on the coinduced modules described in (19). *Then the equation*

$$\langle \phi_f, \overline{\psi_g} \rangle = (2i)^{k-1} \mu(f, g)$$

holds where (f, g) denotes the Petersson scalar product and μ the index of Γ in $\mathrm{PSL}_2(\mathbb{Z})$.

Proof Note that the choice of base point ∞ is on the one hand well-defined (the integral converges, as it is taken over a cusp form) and on the other hand it ensures that $\phi_f(T) = \overline{\phi_f}(T) = 0$. But note that ψ_g is not a parabolic cocycle in general since the chosen base point is not ∞ even though g is also a cusp form.

Now consider $\langle \phi_f, \overline{\psi_g} \rangle$. Let $\rho(a, b) := \pi(\phi_f(a) \otimes a\overline{\psi_g}(b))$, where π is from (19). We describe $\rho(a, b)$:

$$\begin{aligned}
\rho(a,b) &= \sum_{\gamma} \Big(\int_{\gamma\infty}^{\gamma a\infty} f(z)(Xz+Y)^{k-2} dz \Big) \bullet \Big(\int_{\gamma a \zeta_6}^{\gamma ab\zeta_6} \overline{g(z)(X\overline{z}+Y)^{k-2} dz} \Big) \\
&= \sum_{\gamma} \int_{\gamma a\zeta_6}^{\gamma ab\zeta_6} \int_{\gamma\infty}^{\gamma a\infty} f(z)\overline{g(z)}\big((Xz+Y)^{k-2} \bullet (X\overline{z}+Y)^{k-2}\big) dz d\overline{z} \\
&= \sum_{\gamma} \int_{\gamma a\zeta_6}^{\gamma ab\zeta_6} \int_{\gamma\infty}^{\gamma a\infty} f(z)\overline{g(z)}(z-\overline{z})^{k-2} dz d\overline{z} \\
&= \sum_{\gamma} \int_{a\zeta_6}^{ab\zeta_6} \int_{\infty}^{a\infty} f|_{\gamma}(z)\overline{g|_{\gamma}(z)}(z-\overline{z})^{k-2} dz d\overline{z}.
\end{aligned}$$

where the sums run over a system of representatives of $\Gamma\backslash\mathrm{PSL}_2(\mathbb{Z})$. We obtain

$$\begin{aligned}
&\rho(\sigma,\sigma) \\
&= \sum_{\gamma} \int_{\sigma\zeta_6}^{\sigma^2\zeta_6} \int_{\infty}^{\sigma\infty} f|_{\gamma}(z)\overline{g|_{\gamma}(z)}(z-\overline{z})^{k-2} dz d\overline{z} \\
&= \sum_{\gamma} \int_{\zeta_3}^{\zeta_6} \int_{\infty}^{0} f|_{\gamma}(z)\overline{g|_{\gamma}(z)}(z-\overline{z})^{k-2} dz d\overline{z}, \\
&= \sum_{\gamma} \Big[\int_{\zeta_3}^{i} \int_{\infty}^{0} f|_{\gamma}(z)\overline{g|_{\gamma}(z)}(z-\overline{z})^{k-2} dz d\overline{z} + \int_{\sigma\zeta_3}^{\sigma i} \int_{\sigma\infty}^{\sigma 0} f|_{\gamma}(z)\overline{g|_{\gamma}(z)}(z-\overline{z})^{k-2} dz d\overline{z} \Big] \\
&= \sum_{\gamma} \Big[\int_{\zeta_3}^{i} \int_{\infty}^{0} f|_{\gamma}(z)\overline{g|_{\gamma}(z)}(z-\overline{z})^{k-2} dz d\overline{z} + \int_{\zeta_3}^{i} \int_{\infty}^{0} f|_{\gamma\sigma}(z)\overline{g|_{\gamma\sigma}(z)}(z-\overline{z})^{k-2} dz d\overline{z} \Big] \\
&= 2\sum_{\gamma} \int_{\zeta_3}^{i} \int_{\infty}^{0} f|_{\gamma}(z)\overline{g|_{\gamma}(z)}(z-\overline{z})^{k-2} dz d\overline{z},
\end{aligned}$$

and

$$\begin{aligned}
\rho(\tau,\tau) &= \sum_{\gamma} \int_{\tau\zeta_6}^{\tau^2\zeta_6} \int_{\infty}^{\tau\infty} f|_{\gamma}(z)\overline{g|_{\gamma}(z)}(z-\overline{z})^{k-2} dz d\overline{z} = 0 \\
\rho(\tau,\tau^2) &= \sum_{\gamma} \int_{\tau\zeta_6}^{\tau^3\zeta_6} \int_{\infty}^{\tau\infty} f|_{\gamma}(z)\overline{g|_{\gamma}(z)}(z-\overline{z})^{k-2} dz d\overline{z} = 0,
\end{aligned}$$

since τ stabilises ζ_6. It now suffices to compare with the formulas computed before (Propositions 6.13 and 6.4) to obtain the claimed formula. □

6.4 Theory: The Eichler–Shimura Theorem

We can now, finally, prove that the Eichler–Shimura map is an isomorphism. It should be pointed out again that the cohomology groups can be replaced by modular symbols according to Theorem 5.9.

Theorem 6.15 (Eichler–Shimura) *Let $N \geq 4$ and $k \geq 2$. The Eichler–Shimura map and the induced Eichler–Shimura map (Proposition 6.7) are isomorphisms for $\Gamma = \Gamma_1(N)$. The image of $\mathrm{S}_k(\Gamma_1(N)\,;\,\mathbb{C}) \oplus \overline{\mathrm{S}_k(\Gamma_1(N)\,;\,\mathbb{C})}$ is isomorphic to the parabolic subspace.*

Proof We first assert that the dimensions of both sides of the Eichler–Shimura map agree and also that twice the dimension of the space of cusp forms equals the dimension of the parabolic subspace. The dimension of the cohomology group and its parabolic subspace was computed in Proposition 4.16. For the dimension of the left-hand side we refer to [20, §6.2].

Suppose that (f, g) are in the kernel of the Eichler–Shimura map. Then by Proposition 6.9 it follows that f and g are both cuspidal. Hence, it suffices to prove that the restriction of the Eichler–Shimura map to $\mathrm{S}_k(\Gamma_1(N)\,;\,\mathbb{C}) \oplus \overline{\mathrm{S}_k(\Gamma_1(N)\,;\,\mathbb{C})}$ is injective. In order to do this we choose $z_0 = z_1 = \infty$ as base points for the Eichler–Shimura map, which is possible as the integrals converge on cusp forms (as in Proposition 6.7 one sees that this choice of base point does not change the cohomology class). As in Proposition 6.14, we write ϕ_f for the 1-cocycle associated with a cusp form f for the base point ∞.

We now make use of the pairing from Proposition 6.14 on

$$\mathrm{Z}^1(\mathrm{PSL}_2(\mathbb{Z}), \mathrm{Coind}_\Gamma^{\mathrm{PSL}_2(\mathbb{Z})}(V_{k-2}(\mathbb{C}))),$$

where we put $\Gamma := \Gamma_1(N)$ for short. This pairing induces a $\mathbb{C}$-valued pairing $\langle\ ,\ \rangle$ on

$$\mathrm{H}^1_{\mathrm{par}}(\mathrm{PSL}_2(\mathbb{Z}), \mathrm{Coind}_\Gamma^{\mathrm{PSL}_2(\mathbb{Z})}(V_{k-2}(\mathbb{C}))).$$

Next observe that the map

$$\mathrm{S}_k(\Gamma_1(N)\,;\,\mathbb{C}) \oplus \overline{\mathrm{S}_k(\Gamma_1(N)\,;\,\mathbb{C})} \xrightarrow{(f,\overline{g})\mapsto(f+g,\overline{f}-\overline{g})} \mathrm{S}_k(\Gamma_1(N)\,;\,\mathbb{C}) \oplus \overline{\mathrm{S}_k(\Gamma_1(N)\,;\,\mathbb{C})}$$

is an isomorphism. Let $f, g \in \mathrm{S}_k(\Gamma_1(N)\,;\,\mathbb{C})$ be cusp forms and assume now that $(f+g, \overline{f}-\overline{g})$ is sent to the zero-class in $\mathrm{H}^1_{\mathrm{par}}(\mathrm{PSL}_2(\mathbb{Z}), \mathrm{Coind}_\Gamma^{\mathrm{PSL}_2(\mathbb{Z})}(V_{k-2}(\mathbb{C})))$. In that cohomology space, we thus have

$$0 = \phi_f + \phi_g + \overline{\phi_f} - \overline{\phi_g} = (\phi_f + \overline{\phi_f}) + (\phi_g - \overline{\phi_g}) = 2\,\mathrm{Re}(\phi_f) + 2i\,\mathrm{Im}(\phi_g).$$

We conclude that the cohomology classes of $\phi_f + \overline{\phi_f}$ and $\phi_g - \overline{\phi_g}$ are both zero.

Now we apply the pairing as follows:

$$0 = \langle \phi_f, \phi_f + \overline{\phi_f} \rangle = \langle \phi_f, \phi_f \rangle + \langle \phi_f, \overline{\phi_f} \rangle = (2i)^{k-1} \mu(f, f)$$

where we used $\langle \phi_f, \phi_f \rangle = 0$ because of Lemma 6.11 (since the pairing is given by the cup product), as well as Proposition 6.14. Hence, $(f, f) = 0$ and, thus, $f = 0$ since the Petersson scalar product is positive definite. Similar arguments with $0 = \langle \phi_g, \phi_g - \overline{\phi_g} \rangle$ show $g = 0$. This proves the injectivity. □

Remark 6.16 The Eichler–Shimura map is, in fact, an isomorphism for all subgroups Γ of $\mathrm{SL}_2(\mathbb{Z})$ of finite index. The proof is the same, but must use more involved dimension formulae for the cohomology group (see Remark 4.17) and modular forms.

In Corollary 7.30, we will see that there also is an Eichler–Shimura isomorphism with a Dirichlet character.

6.5 Theoretical Exercises

Exercise 6.17 Check that the cup product is well-defined.

Hint: this is a standard exercise that can be found in many textbooks (e.g. [4]).

Exercise 6.18 Prove Lemma 6.11.

Hint: [4, (5.3.6)].

7 Hecke Operators

In this section, we introduce Hecke operators on group cohomology using the double cosets approach and we prove that the Eichler–Shimura isomorphism is compatible with the Hecke action on group cohomology and modular forms.

7.1 Theory: Hecke Rings

Definition 7.1 Let $N, n \in \mathbb{N}$. We define

$$\begin{aligned}
\Delta_0^n(N) &= \{\left(\begin{smallmatrix} a & b \\ c & d \end{smallmatrix}\right) \in M_2(\mathbb{Z}) | \left(\begin{smallmatrix} a & b \\ c & d \end{smallmatrix}\right) \equiv \left(\begin{smallmatrix} * & * \\ 0 & * \end{smallmatrix}\right) \mod N, (a, N) = 1, \det\left(\begin{smallmatrix} a & b \\ c & d \end{smallmatrix}\right) = n\},\\
\Delta_1^n(N) &= \{\left(\begin{smallmatrix} a & b \\ c & d \end{smallmatrix}\right) \in M_2(\mathbb{Z}) | \left(\begin{smallmatrix} a & b \\ c & d \end{smallmatrix}\right) \equiv \left(\begin{smallmatrix} 1 & * \\ 0 & * \end{smallmatrix}\right) \mod N, \det\left(\begin{smallmatrix} a & b \\ c & d \end{smallmatrix}\right) = n\},\\
\Delta_0(N) &= \bigcup_{n \in \mathbb{N}} \Delta_0^n(N),\\
\Delta_1(N) &= \bigcup_{n \in \mathbb{N}} \Delta_1^n(N).
\end{aligned}$$

From now on, let $(\Delta, \Gamma) = (\Delta_1(N), \Gamma_1(N))$ or $(\Delta, \Gamma) = (\Delta_0(N), \Gamma_0(N))$.

Lemma 7.2 *Let $\alpha \in \Delta$. We put*

$$\Gamma_\alpha = \Gamma \cap \alpha^{-1}\Gamma\alpha \text{ and } \Gamma^\alpha = \Gamma \cap \alpha\Gamma\alpha^{-1}.$$

Then Γ_α has finite index in Γ and $\alpha^{-1}\Gamma\alpha$ (one says that Γ and $\alpha^{-1}\Gamma\alpha$ are commensurable), and also Γ^α has finite index in Γ and $\alpha\Gamma\alpha^{-1}$ (hence, Γ and $\alpha\Gamma\alpha^{-1}$ are commensurable).

Proof Let $n = \det\alpha$. One checks by matrix calculation that

$$\alpha^{-1}\Gamma(Nn)\alpha \subset \Gamma(N).$$

Thus,

$$\Gamma(Nn) \subset \alpha^{-1}\Gamma(N)\alpha \subset \alpha^{-1}\Gamma\alpha.$$

Hence, we have $\Gamma(Nn) \subset \Gamma_\alpha$ and the first claim follows. For the second claim, one proceeds similarly. □

Example 7.3 Let $\Gamma = \Gamma_0(N)$ and p a prime. The most important case for the sequel is $\alpha = \left(\begin{smallmatrix} 1 & 0 \\ 0 & p \end{smallmatrix}\right)$. An elementary calculation shows $\Gamma^\alpha = \Gamma_0(Np)$.

Definition 7.4 Let $\alpha \in \Delta$. We consider the diagram

$$\begin{array}{ccc} \Gamma_\alpha\backslash\mathbb{H} & \xrightarrow[\tau\mapsto\alpha\tau]{\alpha} & \Gamma^\alpha\backslash\mathbb{H} \\ \downarrow{\scriptstyle \pi_\alpha} & & \downarrow{\scriptstyle \pi^\alpha} \\ \Gamma\backslash\mathbb{H} & & \Gamma\backslash\mathbb{H}, \end{array}$$

in which π^α and π_α are the natural projections. One checks that this is well-defined by using $\alpha\Gamma_\alpha\alpha^{-1} = \Gamma^\alpha$.

The *group of divisors* $\mathrm{Div}(S)$ on a Riemann surface S consists of all formal $\mathbb{Z}$-linear combinations of points of S. For a morphism $\pi : S \to T$ of Riemann surfaces, define the *pull-back* $\pi^* : \mathrm{Div}(T) \to \mathrm{Div}(S)$ and the *push-forward* $\pi_* : \mathrm{Div}(S) \to \mathrm{Div}(T)$ uniquely by the rules $\pi^*(t) = \sum_{s\in S:\pi(s)=t} s$ and $\pi_*(s) = \pi(s)$ for points $t \in T$ and $s \in S$.

The *modular correspondence* or *Hecke correspondence* τ_α is defined as

$$\tau_\alpha : \mathrm{Div}(Y_\Gamma) \xrightarrow{\pi_\alpha^*} \mathrm{Div}(Y_{\Gamma_\alpha}) \xrightarrow{\alpha_*} \mathrm{Div}(Y_{\Gamma^\alpha}) \xrightarrow{\pi_*^\alpha} \mathrm{Div}(Y_\Gamma).$$

These modular correspondences will be described more explicitly in a moment. First a lemma:

Lemma 7.5 *Let $\alpha_i \in \Gamma$ for $i \in I$ with some index set I. Then we have*

$$\Gamma = \bigsqcup_{i\in I} \Gamma_\alpha\alpha_i \Leftrightarrow \Gamma\alpha\Gamma = \bigsqcup_{i\in I} \Gamma\alpha\alpha_i.$$

Proof This is proved by quite a straight forward calculation. □

Corollary 7.6 *Let $\alpha \in \Delta$ and $\Gamma\alpha\Gamma = \bigsqcup_{i\in I} \Gamma\alpha\alpha_i$. Then the Hecke correspondence $\tau_\alpha : \mathrm{Div}(Y_\Gamma) \to \mathrm{Div}(Y_\Gamma)$ is given by $\tau \mapsto \sum_{i\in I} \alpha\alpha_i\tau$ for representatives $\tau \in \mathbb{H}$.*

Proof It suffices to check the definition using Lemma 7.5. □

Remark 7.7 We have $\Delta^n = \bigcup_{\alpha\in\Delta,\det\alpha=n} \Gamma\alpha\Gamma$ and one can choose finitely many α_i for $i \in I$ such that $\Delta^n = \bigsqcup_{i\in I} \Gamma\alpha_i\Gamma$.

Definition 7.8 Let $\Delta^n = \bigsqcup_{i\in I} \Gamma\alpha_i\Gamma$. The Hecke operator T_n on $\mathrm{Div}(Y_\Gamma)$ is defined as

$$T_n = \sum_{i\in I} \tau_{\alpha_i}.$$

Let us recall from Eq. (2) the matrix $\sigma_a \in \Gamma_0(N)$ (for $(a, N) = 1$) which satisfies

$$\sigma_a \equiv \left(\begin{smallmatrix} a^{-1} & 0 \\ 0 & a \end{smallmatrix}\right) \mod N.$$

Proposition 7.9 *(a) We have the decomposition*

$$\Delta_0^n(N) = \bigsqcup_a \bigsqcup_b \Gamma_0(N) \left(\begin{smallmatrix} a & b \\ 0 & d \end{smallmatrix}\right),$$

where a runs through the positive integers with $a \mid n$ and $(a, N) = 1$ and b runs through the integers such that $0 \le b < d := n/a$.

(b) We have the decomposition

$$\Delta_1^n(N) = \bigsqcup_a \bigsqcup_b \Gamma_1(N)\sigma_a \left(\begin{smallmatrix} a & b \\ 0 & d \end{smallmatrix}\right)$$

with a, b, d as in (a).

Proof This proof is elementary. □

Note that due to $\sigma_a \in \Gamma_0(N)$, the matrices $\sigma_a \left(\begin{smallmatrix} a & b \\ 0 & d \end{smallmatrix}\right)$ used in part (b) also work in part (a). One can thus use the same representatives regardless if one works with $\Gamma_0(N)$ or $\Gamma_1(N)$. Note also that for $n = \ell$ a prime, these representatives are exactly the elements of $\mathcal{R}_\ell$ from Eq. (3).

Next, we turn to the important description of the Hecke algebra as a double coset algebra.

Definition 7.10 The *Hecke ring* $R(\Delta, \Gamma)$ is the free abelian group on the double cosets $\Gamma\alpha\Gamma$ for $\alpha \in \Delta$.

As our next aim we would like to define a multiplication, which then also justifies the name 'ring'. First let $\Gamma\alpha\Gamma = \bigsqcup_{i=1}^n \Gamma\alpha_i$ and $\Gamma\beta\Gamma = \bigsqcup_{j=1}^m \Gamma\beta_j$. We just start computing.

$$\Gamma\alpha\Gamma \cdot \Gamma\beta\Gamma = \bigcup_j \Gamma\alpha\Gamma\beta_j = \bigcup_{i,j} \Gamma\alpha_i\beta_j.$$

This union is not necessarily disjoint. The left-hand side can be written as a disjoint union of double cosets $\bigsqcup_{k=1}^{r} \Gamma\gamma_k\Gamma$. Each of these double cosets is again of the form

$$\Gamma\gamma_k\Gamma = \bigsqcup_{l=1}^{n_k} \Gamma\gamma_{k,l}.$$

We obtain in summary

$$\Gamma\alpha\Gamma \cdot \Gamma\beta\Gamma = \bigcup_{i,j} \Gamma\alpha_i\beta_j = \bigsqcup_k \bigsqcup_l \Gamma\gamma_{k,l}.$$

We will now introduce a piece of notation for the multiplicity with which every coset on the right appears in the centre of the above equality. For fixed k we define for every l

$$m_{k,l} = \#\{(i, j) | \Gamma\gamma_{k,l} = \Gamma\alpha_i\beta_j\}.$$

The important point is the following lemma.

Lemma 7.11 *The number $m_{k,l}$ is independent of l. We put $m_k := m_{k,l}$.*

Proof The proof is combinatorial and quite straight forward. □

Definition 7.12 We define the multiplication on $R(\Delta, \Gamma)$ by

$$\Gamma\alpha\Gamma \cdot \Gamma\beta\Gamma = \sum_{k=1}^{n} m_k \Gamma\gamma_k\Gamma,$$

using the preceding notation.

In Exercise 7.34 you are asked to check that the Hecke ring is indeed a ring. The definition of the multiplication makes sense, as it gives for Hecke correspondences:

$$\tau_\alpha \circ \tau_\beta = \sum_{k=1}^{n} m_k \tau_{\gamma_k}.$$

Definition 7.13 For $\alpha \in \Delta$ let $\tau_\alpha = \Gamma\alpha\Gamma$. We define (as above)

$$T_n = \sum_\alpha \tau_\alpha \in R(\Delta, \Gamma),$$

where the sum runs over a set of α such that $\Delta^n = \bigsqcup_\alpha \Gamma\alpha\Gamma$. For $a \mid d$ and $(d, N) = 1$ we let

$$T(a,d) = \Gamma\sigma_a \begin{pmatrix} a & 0 \\ 0 & d \end{pmatrix} \Gamma \in R(\Delta, \Gamma).$$

From Exercise 7.35, we obtain the following important corollary.

Corollary 7.14 *We have $T_m T_n = T_n T_m$ and hence $R(\Delta, \Gamma)$ is a commutative ring.*

7.2 Theory: Hecke Operators on Modular Forms

In this section, we again let $(\Delta, \Gamma) = (\Delta_0(N), \Gamma_0(N))$ or $(\Delta_1(N), \Gamma_1(N))$. We now define an action of the Hecke ring $R(\Delta, \Gamma)$ on modular forms.

Definition 7.15 Let $\alpha \in \Delta$. Suppose $\Gamma\alpha\Gamma = \bigsqcup_{i=1}^n \Gamma\alpha_i$ and let $f \in M_k(\Gamma)$. We put

$$f.\tau_\alpha := \sum_{i=1}^n f|_{\alpha_i}.$$

Lemma 7.16 *The function $f.\tau_\alpha$ again lies in $M_k(\Gamma)$.*

Proof For $\gamma \in \Gamma$ we check the transformation rule:

$$\sum_i f|_{\alpha_i}|_\gamma = \sum_i f|_{\alpha_i\gamma} = \sum_i f|_{\alpha_i},$$

since the cosets $\Gamma(\alpha_i\gamma)$ are a permutation of the cosets $\Gamma\alpha_i$. The holomorphicity of $f.\tau_\alpha$ is clear and the holomorphicity in the cusps is not difficult. □

This thus gives the desired operation of $R(\Delta, \Gamma)$ on $M_k(\Gamma)$.

Proposition 7.17 *Let $(\Delta, \Gamma) = (\Delta_0(N), \Gamma_0(N))$ and $f \in M_k(\Gamma)$. The following formulae hold:*

(a) $(f.T_m)(\tau) = \frac{1}{m}\sum_{a|m,(a,N)=1}\sum_{b=0}^{\frac{m}{a}-1} a^k f(\frac{a\tau+b}{m/a})$,
(b) $a_n(f.T_m) = \sum_{a|(m,n),(a,N)=1} a^{k-1} a_{\frac{mn}{a^2}}$.

Similar formulae hold for $(\Delta_1(N), \Gamma_1(N))$, if one includes a Dirichlet character at the right places.

Proof (a) follows directly from Proposition 7.9.

(b) is a simple calculation using $\sum_{b=0}^{d-1} e^{2\pi i \frac{b}{d} n} = \begin{cases} 0, & \text{if } d \nmid n \\ d, & \text{if } d \mid n. \end{cases}$ □

Remark 7.18 The Hecke ring $R(\Delta, \Gamma)$ also acts on $S_k(\Gamma)$.

Corollary 7.19 *Let $(\Delta, \Gamma) = (\Delta_0(N), \Gamma_0(N))$. For the action of the Hecke operators on $M_k(\Gamma)$ and $S_k(\Gamma)$ the following formulae hold:*

(a) $T_n T_m = T_{nm}$ *for* $(n, m) = 1$,
(b) $T_{p^{r+1}} = T_p T_{p^r} - p^{k-1} T_{p^{r-1}}$, *if* $p \nmid N$, *and*
(c) $T_{p^{r+1}} = T_p T_{p^r}$, *if* $p \mid N$.

Here, p always denotes a prime number. Similar formulae hold for $(\Delta_1(N), \Gamma_1(N))$, *if one includes a Dirichlet character at the right places.*

Proof These formulae follow from Exercise 7.35 and the definition of the action. □

Even though it is not directly relevant for our purposes, we include Euler products, which allow us to express the formulae from the corollary in a very elegant way.

Proposition 7.20 (Euler product) *The action of the Hecke operators* T_n *on modular forms satisfies the formal identity:*

$$\sum_{n=1}^{\infty} T_n n^{-s} = \prod_{p \nmid N} (1 - T_p p^{-s} + p^{k-1-2s})^{-1} \cdot \prod_{p \mid N} (1 - T_p p^{-s})^{-1}.$$

That the identity is formal means that we can arbitrarily permute terms in sums and products without considering questions of convergence.

Proof The proof is carried out in three steps.

1st step: Let $g : \mathbb{Z} \to \mathbb{C}$ be any function. Then we have the formal identity

$$\prod_{p \text{ prime}} \sum_{r=0}^{\infty} g(p^r) = \sum_{n=1}^{\infty} \prod_{p^r \| n} g(p^r).$$

For its proof, let first S be a finite set of prime numbers. Then we have the formal identity:

$$\prod_{p \in S} \sum_{r=0}^{\infty} g(p^r) = \sum_{n=1, n \text{ only has prime factors in } S}^{\infty} \prod_{p^r \| n} g(p^r),$$

which one proves by multiplying out the left-hand side (Attention! Here one permutes the terms!). We finish the first step by letting S run through arbitrarily large sets.

2nd step: For $p \nmid N$ we have

$$\left(\sum_{r=0}^{\infty} T_{p^r} p^{-rs}\right)(1 - T_p p^{-s} + p^{k-1-2s}) = 1$$

and for $p \mid N$:

$$\left(\sum_{r=0}^{\infty} T_{p^r} p^{-rs}\right)(1 - T_p p^{-s}) = 1.$$

The proof of the second step consists of multiplying out these expressions and to identify a 'telescope'.

3rd step: The proposition now follows by using the first step with $g(p^r) = T_{p^r} p^{-rs}$ and plugging in the formulae from the second step. □

7.3 Theory: Hecke Operators on Group Cohomology

In this section, we again let $(\Delta, \Gamma) = (\Delta_0(N), \Gamma_0(N))$ or $(\Delta_1(N), \Gamma_1(N))$. Let R be a ring and V a left $R[\Gamma]$-module which extends to a semi-group action by the semi-group consisting of all α^ι for $\alpha \in \Delta^n$ for all n. Recall that $\left(\begin{smallmatrix} a & b \\ c & d \end{smallmatrix}\right)^\iota = \left(\begin{smallmatrix} d & -b \\ -c & a \end{smallmatrix}\right)$.

We now give the definition of the Hecke operator τ_α on $\mathrm{Div}(\Gamma\backslash\mathbb{H})$ (see, for instance, [9] or [22]).

Definition 7.21 Let $\alpha \in \Delta$. The *Hecke operator* τ_α acting on group cohomology is the composite

$$\mathrm{H}^1(\Gamma, V) \xrightarrow{\mathrm{res}} \mathrm{H}^1(\Gamma^\alpha, V) \xrightarrow{\mathrm{conj}_\alpha} \mathrm{H}^1(\Gamma_\alpha, V) \xrightarrow{\mathrm{cores}} \mathrm{H}^1(\Gamma, V).$$

The first map is the *restriction*, and the third one is the *corestriction*. We explicitly describe the second map on cocycles:

$$\mathrm{conj}_\alpha : \mathrm{H}^1(\Gamma^\alpha, V) \to \mathrm{H}^1(\Gamma_\alpha, V), \quad c \mapsto \big(g_\alpha \mapsto \alpha^\iota.c(\alpha g_\alpha \alpha^{-1})\big).$$

There is a similar description on the parabolic subspace and the two are compatible, see Exercise 7.36.

Proposition 7.22 *Let $\alpha \in \Delta$. Suppose that $\Gamma\alpha\Gamma = \bigcup_{i=1}^n \Gamma\delta_i$ is a disjoint union. Then the Hecke operator τ_α acts on $\mathrm{H}^1(\Gamma, V)$ and $\mathrm{H}^1_{\mathrm{par}}(\Gamma, V)$ by sending the cocycle c to $\tau_\alpha c$ defined by*

$$(\tau_\alpha c)(g) = \sum_{i=1}^n \delta_i^\iota c(\delta_i g \delta_{\sigma_g(i)}^{-1})$$

for $g \in \Gamma$. Here $\sigma_g(i)$ is the index such that $\delta_i g \delta_{\sigma_g(i)}^{-1} \in \Gamma$.

Proof We only have to describe the corestriction explicitly. For that we use that $\Gamma = \bigcup_{i=1}^n \Gamma_\alpha g_i$ with $\alpha g_i = \delta_i$. Furthermore, by Exercise 7.37 the corestriction of a cocycle $u \in \mathrm{H}^1(\Gamma_\alpha, V)$ is the cocycle $\mathrm{cores}(u)$ uniquely given by

$$\mathrm{cores}(u)(g) = \sum_{i=1}^n g_i^{-1} u(g_i g g_{\sigma_g(i)}^{-1}) \tag{20}$$

for $g \in \Gamma$. Combining with the explicit description of the map conj_α yields the result. □

Definition 7.23 For a positive integer n, the *Hecke operator* T_n is defined as $\sum_\alpha \tau_\alpha$, where the sum runs through a system of representatives of the double cosets $\Gamma\backslash\Delta^n/\Gamma$.

Let a be an integer coprime to N. The *diamond operator* $\langle a\rangle$ is defined as τ_α for the matrix $\sigma_a \in \Gamma_0(N)$, defined in Eq. 2 (if the Γ-action on V extends to an action of the semi-group generated by Γ and α^ι; note that $\alpha \in \Delta_0^1$, but in general not in Δ_1^1).

It is clear that the Hecke and diamond operators satisfy the 'usual' Euler product.

Proposition 7.24 *The Eichler–Shimura isomorphism is compatible with the Hecke operators.*

Proof We recall the definition of Shimura's main involution: $\left(\begin{smallmatrix} a & b \\ c & d\end{smallmatrix}\right)^\iota = \left(\begin{smallmatrix} d & -b \\ -c & a\end{smallmatrix}\right)$. In other words, for matrices with a non-zero determinant, we have

$$\left(\begin{smallmatrix} a & b \\ c & d\end{smallmatrix}\right)^\iota = (\det\left(\begin{smallmatrix} a & b \\ c & d\end{smallmatrix}\right)) \cdot \left(\begin{smallmatrix} a & b \\ c & d\end{smallmatrix}\right)^{-1}.$$

Let now $f \in \mathrm{M}_k(\Gamma\,;\,\mathbb{C})$ be a modular form, $\gamma \in \Gamma$ and $z_0 \in \mathbb{H}$. For any matrix g with non-zero determinant, Lemma 6.5 yields

$$I_{f|_g}(z_0, \gamma z_0) = g^\iota I_f(gz_0, g\gamma z_0).$$

Let $\alpha \in \Delta$. We show the compatibility of the Hecke operator τ_α with the map

$$f \mapsto (\gamma \mapsto I_f(z_0, \gamma z_0))$$

between $\mathrm{M}_k(\Gamma\,;\,\mathbb{C})$ and $\mathrm{H}^1(\Gamma, V_{k-2}(\mathbb{C}))$. The same arguments will also work, when $I_f(z_0, \gamma z_0)$ is replaced by $J_{\overline{g}}(z_1, \gamma z_1))$ with antiholomorphic cusp forms $\overline{g}$.

Consider a coset decomposition $\Gamma\alpha\Gamma = \bigsqcup_i \Gamma\delta_i$. We use notation as in Proposition 7.22 and compute:

$$\begin{aligned}
&I_{\tau_\alpha f}(z_0, \gamma z_0)\\
=&I_{\sum_i f|_{\delta_i}}(z_0, \gamma z_0) = \sum_i I_{f|_{\delta_i}}(z_0, \gamma z_0) = \sum_i \delta_i^\iota I_f(\delta_i z_0, \delta_i\gamma z_0)\\
=&\sum_i \delta_i^\iota\big(I_f(\delta_i z_0, z_0) + I_f(z_0, \delta_i\gamma\delta_{\sigma_\gamma(i)}^{-1} z_0) + I_f(\delta_i\gamma\delta_{\sigma_\gamma(i)}^{-1} z_0, \delta_i\gamma\delta_{\sigma_\gamma(i)}^{-1}\delta_{\sigma_\gamma(i)} z_0)\big)\\
=&\sum_i \delta_i^\iota I_f(z_0, \delta_i\gamma\delta_{\sigma_\gamma(i)}^{-1} z_0) + \sum_i \delta_i^\iota I_f(\delta_i z_0, z_0) - \sum_i \delta_i^\iota\delta_i\gamma\delta_{\sigma_\gamma(i)}^{-1} I_f(\delta_{\sigma_\gamma(i)} z_0, z_0)\\
=&\sum_i \delta_i^\iota I_f(z_0, \delta_i\gamma\delta_{\sigma_\gamma(i)}^{-1} z_0) + (1-\gamma)\sum_i \delta_i^\iota I_f(\delta_i z_0, z_0),
\end{aligned}$$

since $\delta_i^\iota\delta_i\gamma\delta_{\sigma_\gamma(i)}^{-1} = \gamma\delta_{\sigma_\gamma(i)}^\iota$. Up to coboundaries, the cocycle $\gamma \mapsto I_{\tau_\alpha f}(z_0, \gamma z_0)$ is thus equal to the cocycle $\gamma \mapsto \sum_i \delta_i^\iota I_f(z_0, \delta_i\gamma\delta_{\sigma_\gamma(i)}^{-1} z_0)$, which by Proposition 7.22 is equal to τ_α applied to the cocycle $\gamma \mapsto I_f(z_0, \gamma z_0)$, as required. □

Remark 7.25 The conceptual reason why the above proposition is correct, is, of course, that the Hecke operators come from Hecke correspondences.

7.4 Theory: Hecke Operators and Shapiro's Lemma

We now prove that the Hecke operators are compatible with Shapiro's lemma. This was first proved by Ash and Stevens [1]. We need to say what the action of $\alpha \in \Delta$ on the coinduced module $\mathrm{Hom}_{R[\Gamma]}(R[\mathrm{SL}_2(\mathbb{Z})], V)$ should be. Here we are assuming that V carries an action by the semi-group Δ^ι (that is, ι applied to all elements of Δ).

Let U_N be the image of Δ^ι in $\mathrm{Mat}_2(\mathbb{Z}/N\mathbb{Z})$. The natural map

$$\Gamma\backslash\mathrm{SL}_2(\mathbb{Z}) \to U_N\backslash\mathrm{Mat}_2(\mathbb{Z}/N\mathbb{Z})$$

is injective. Its image consists of those $U_N g$ such that

$$(0,1)g = (u,v) \text{ with } \langle u,v\rangle = \mathbb{Z}/N\mathbb{Z}. \tag{21}$$

If that is so, then we say for short that g satisfies (21). Note that this condition does not depend on the choice of g in $U_N g$. Define the $R[\Delta^\iota]$-module $\mathcal{C}(N,V)$ as

$$\{f \in \mathrm{Hom}_R(R[U_N\backslash\mathrm{Mat}_2(\mathbb{Z}/N\mathbb{Z})], V) \mid f(g) = 0 \text{ if } g \text{ does not satisfy (21)}\}$$

with the action of $\delta \in \Delta^\iota$ given by $(\delta.f)(g) = \delta.(f(g\delta))$. The module $\mathcal{C}(N,V)$ is isomorphic to the coinduced module $\mathrm{Hom}_{R[\Gamma]}(R[\mathrm{SL}_2(\mathbb{Z})], V)$ as an $R[\Gamma]$-module by

$$\mathrm{Hom}_{R[\Gamma]}(R[\mathrm{SL}_2(\mathbb{Z})], V) \to \mathcal{C}(N,V),$$

$$f \mapsto \begin{cases} (g \mapsto gf(g^{-1})) & \text{for any } g \in \mathrm{SL}_2(\mathbb{Z}), \\ 0 & \text{if } g \text{ does not satisfy (21).} \end{cases}$$

One might wonder why we introduce the module $\mathcal{C}(N,V)$ instead of working with $\mathrm{Hom}_{R[\Gamma]}(R[\mathrm{SL}_2(\mathbb{Z})], V)$. The point is that we cannot directly act on the latter with a matrix of determinant different from 1. Hence we need a way to naturally extend the action. We do this by embedding $\Gamma\backslash\mathrm{SL}_2(\mathbb{Z})$ into $U_N\backslash\mathrm{Mat}_2(\mathbb{Z}/N\mathbb{Z})$. Of course, we then want to work on the image of this embedding, which is exactly described by (21). The module $\mathcal{C}(N,V)$ is then immediately written down in view of the identification between $\mathrm{Hom}_{R[\Gamma]}(R[\mathrm{SL}_2(\mathbb{Z})], V)$ and $\mathrm{Hom}_R(R[\Gamma\backslash\mathrm{SL}_2(\mathbb{Z})], V)$ given by sending f to $(g \mapsto g.f(g^{-1}))$ (which is clearly independent of the choice of g in the coset Γg).

Proposition 7.26 *The Hecke operators are compatible with Shapiro's Lemma. More precisely, for all $n \in \mathbb{N}$ the following diagram commutes:*

$$\begin{array}{ccc} \mathrm{H}^1(\Gamma, V) & \xrightarrow{T_n} & \mathrm{H}^1(\Gamma, V) \\ \uparrow{\scriptstyle \mathit{Shapiro}} & & \uparrow{\scriptstyle \mathit{Shapiro}} \\ \mathrm{H}^1(\mathrm{SL}_2(\mathbb{Z}), \mathscr{C}(N, V)) & \xrightarrow{T_n} & \mathrm{H}^1(\mathrm{SL}_2(\mathbb{Z}), \mathscr{C}(N, V)). \end{array}$$

Proof Let $j \in \{0, 1\}$ indicate whether we work with Γ_0 or Γ_1. Let δ_i, for $i = 1, \ldots, r$ be the representatives of $\mathrm{SL}_2(\mathbb{Z})\backslash\Delta_j^n(1)$ provided by Proposition 7.9. Say, that they are ordered such that δ_i for $i = 1, \ldots, s$ with $s \le r$ are representatives for $\Gamma\backslash\Delta_j^n(N)$. This explicitly means that the lower row of δ_i^ι is $(0, a)$ with $(a, N) = 1$ (or even $(0, 1)$ if $j = 1$) for $i = 1, \ldots, s$. If $s < i \le r$, then the lower row is (u, v) with $\langle u, v\rangle \lneq \mathbb{Z}/N\mathbb{Z}$.

Let $c \in \mathrm{H}^1(\mathrm{SL}_2(\mathbb{Z}), \mathscr{C}(N, V))$ be a 1-cochain. Then, as required, we find

$$\begin{aligned} \mathrm{Shapiro}(T_n(c))(\gamma) &= \sum_{i=1}^{r} (\delta_i^\iota.c(\delta^i\gamma\delta_{\sigma_\gamma(i)}^{-1}))(\left(\begin{smallmatrix}1&0\\0&1\end{smallmatrix}\right)) = \sum_{i=1}^{r} \delta_i^\iota(c(\delta^i\gamma\delta_{\sigma_\gamma(i)}^{-1})(\delta_i^\iota)) \\ &= \sum_{i=1}^{s} (\delta_i^\iota.c(\delta^i\gamma\delta_{\sigma_\gamma(i)}^{-1}))(\left(\begin{smallmatrix}1&0\\0&1\end{smallmatrix}\right))) = T_n(\mathrm{Shapiro}(c))(\gamma), \end{aligned}$$

where the second equality is due to the definition of the action and the third one holds since $c(\delta^i\gamma\delta_{\sigma_\gamma(i)}^{-1})$ lies in $\mathscr{C}(N, V)$ and thus evaluates to 0 on δ_i^ι for $i > s$. □

Remark 7.27 A very similar description exists involving $\mathrm{PSL}_2(\mathbb{Z})$.

Remark 7.28 It is possible to give an explicit description of Hecke operators on Manin symbols from Theorem 5.7 by using Heilbronn matrices and variations as, for instance, done in [16].

Remark 7.29 One can show that the isomorphisms from Theorem 5.9 are compatible with Hecke operators.

7.5 Theory: Eichler–Shimura Revisited

In this section, we present some corollaries and extensions of the Eichler–Shimura theorem. We first come to modular symbols with a character and, thus, also to modular symbols for $\Gamma_0(N)$.

Corollary 7.30 (Eichler–Shimura) *Let $N \ge 1$, $k \ge 2$ and $\chi : (\mathbb{Z}/N\mathbb{Z})^\times \to \mathbb{C}^\times$ be a Dirichlet character. Then the Eichler–Shimura map gives isomorphisms*

$$\mathrm{M}_k(N, \chi\,;\, \mathbb{C}) \oplus \overline{\mathrm{S}_k(N, \chi\,;\, \mathbb{C})} \to \mathrm{H}^1(\Gamma_0(N), V_{k-2}^{\iota,\chi}(\mathbb{C})),$$

and

$$\mathrm{S}_k(N,\chi\,;\,\mathbb{C}) \oplus \overline{\mathrm{S}_k(N,\chi\,;\,\mathbb{C})} \to \mathrm{H}^1_{\mathrm{par}}(\Gamma_0(N), V^{\iota,\chi}_{k-2}(\mathbb{C})),$$

which are compatible with the Hecke operators.

Proof Recall that the σ_a form a system of coset representatives for $\Gamma_0(N)/\Gamma_1(N) =: \Delta$ and that the group Δ acts on $\mathrm{H}^1(\Gamma_0(N), V)$ by sending a cocycle c to the cocycle δc (for $\delta \in \Delta$) which is defined by

$$\gamma \mapsto \delta.c(\delta^{-1}\gamma\delta).$$

With $\delta = \sigma_a^{-1} = \sigma_a^{\iota}$, this reads

$$\gamma \mapsto \sigma_a^{\iota}.c(\sigma_a\gamma\sigma_a^{-1}) = \tau_{\sigma_a} c = \langle a\rangle c.$$

Hence, $\sigma_a \in \Delta$-action acts through the inverse of the diamond operators.

We now appeal to the Hochschild–Serre exact sequence, using that the cohomology groups (from index 1 onwards) vanish if the group order is finite and invertible. We get the isomorphism

$$\mathrm{H}^1(\Gamma_0(N), V^{\iota,\chi}_{k-2}(\mathbb{C})) \xrightarrow{\mathrm{res}} \mathrm{H}^1(\Gamma_1(N), V^{\iota,\chi}_{k-2}(\mathbb{C}))^{\Delta}.$$

Moreover, the Eichler–Shimura isomorphism is an isomorphism of Hecke modules

$$\mathrm{M}_k(\Gamma_1(N)\,;\,\mathbb{C}) \oplus \overline{\mathrm{S}_k(\Gamma_1(N)\,;\,\mathbb{C})} \to \mathrm{H}^1(\Gamma_1(N), V^{\iota,\chi}_{k-2}(\mathbb{C})),$$

since for matrices in $\Delta_1(N)$ acting through the Shimura main involution the modules $V^{\iota,\chi}_{k-2}(\mathbb{C})$ and $V_{k-2}(\mathbb{C})$ coincide. Note that it is necessary to take $V^{\iota,\chi}_{k-2}(\mathbb{C})$ because the action on group cohomology involves the Shimura main involution. Moreover, with this choice, the Eichler–Shimura isomorphism is Δ-equivariant.

To finish the proof, it suffices to take Δ-invariants on both sides, i.e. to take invariants for the action of the diamond operators. The result on the parabolic subspace is proved in the same way.

Since Hecke and diamond operators commute, the Hecke action is compatible with the decomposition into χ-isotypical components. □

Next we consider the action of complex conjugation.

Corollary 7.31 *Let* $\Gamma = \Gamma_1(N)$. *The maps*

$$\mathrm{S}_k(\Gamma\,;\,\mathbb{C}) \to \mathrm{H}^1_{\mathrm{par}}(\Gamma, V_{k-2}(\mathbb{R})), \quad f \mapsto (\gamma \mapsto \mathrm{Re}(I_f(z_0, \gamma z_0)))$$

and

$$\mathrm{S}_k(\Gamma\,;\,\mathbb{C}) \to \mathrm{H}^1_{\mathrm{par}}(\Gamma, V_{k-2}(\mathbb{R})), \quad f \mapsto (\gamma \mapsto \mathrm{Im}(I_f(z_0, \gamma z_0)))$$

are isomorphisms (of real vector spaces) compatible with the Hecke operators. A similar result holds in the presence of a Dirichlet character.

Proof We consider the composite

$$\mathrm{S}_k(\Gamma\,;\,\mathbb{C}) \xrightarrow{f\mapsto \frac{1}{2}(f+\overline{f})} \mathrm{S}_k(\Gamma\,;\,\mathbb{C}) \oplus \overline{\mathrm{S}_k(\Gamma\,;\,\mathbb{C})} \xrightarrow{\text{Eichler–Shimura}} \mathrm{H}^1_{\mathrm{par}}(\Gamma, V_{k-2}(\mathbb{C})).$$

It is clearly injective. As $J_{\overline{f}}(z_0, \gamma z_0) = \overline{I_f(z_0, \gamma z_0)}$, the composite map coincides with the first map in the statement. Its image is thus already contained in the real vector space $\mathrm{H}^1_{\mathrm{par}}(\Gamma, V_{k-2}(\mathbb{R}))$. Since the real dimensions coincide, the map is an isomorphism. In order to prove the second isomorphism, we use $f \mapsto \frac{1}{2i}(f - \overline{f})$ and proceed as before. □

We now treat the + and the −-space for the involution attached to the matrix $\eta = \left(\begin{smallmatrix} -1 & 0 \\ 0 & 1 \end{smallmatrix}\right)$ from Eq. (9). The action of η on $\mathrm{H}^1(\Gamma, V)$ is the action of the Hecke operator τ_η; strictly speaking, this operator is not defined because the determinant is negative, however we use the same definition. To be precise we have

$$\tau_\eta : \mathrm{H}^1(\Gamma, V) \to \mathrm{H}^1(\Gamma, V), \quad c \mapsto (\gamma \mapsto \eta^\iota.c(\eta\gamma\eta)),$$

provided, of course, that η^ι acts on V (compatibly with the Γ-action).

We also want to define an involution τ_η on $\mathrm{S}_k(\Gamma\,;\,\mathbb{C}) \oplus \overline{\mathrm{S}_k(\Gamma\,;\,\mathbb{C})}$. For that recall that if $f(z) = \sum a_n e^{2\pi i n z}$, then $\tilde{f}(z) := \sum \overline{a_n} e^{2\pi i n z}$ is again a cusp form in $\mathrm{S}_k(\Gamma\,;\,\mathbb{C})$ since we only applied a field automorphism (complex conjugation) to the coefficients (think of cusp forms as maps from the Hecke algebra over $\mathbb{Q}$ to $\mathbb{C}$). We define τ_η as the composite

$$\tau_\eta : \mathrm{S}_k(\Gamma\,;\,\mathbb{C}) \xrightarrow{f\mapsto(-1)^{k-1}\tilde{f}} \mathrm{S}_k(\Gamma\,;\,\mathbb{C}) \xrightarrow{\tilde{f}\mapsto\overline{\tilde{f}}} \overline{\mathrm{S}_k(\Gamma\,;\,\mathbb{C})}.$$

Similarly, we also define $\tau_\eta : \overline{\mathrm{S}_k(\Gamma\,;\,\mathbb{C})} \to \mathrm{S}_k(\Gamma\,;\,\mathbb{C})$ and obtain in consequence an involution τ_η on $\mathrm{S}_k(\Gamma\,;\,\mathbb{C}) \oplus \overline{\mathrm{S}_k(\Gamma\,;\,\mathbb{C})}$. We consider the function $(-1)^{k-1}\overline{\tilde{f}(z)}$ as a function of $\overline{z}$. We have

$$\begin{aligned}\tau_\eta(f)(\overline{z}) = (-1)^{k-1}\overline{\tilde{f}(z)} = (-1)^{k-1}\overline{\sum_n \overline{a_n} e^{2\pi i n z}} &= (-1)^{k-1}\sum_n a_n e^{2\pi i n(-\overline{z})} \\ &= (-1)^{k-1} f(-\overline{z}) = f|_\eta(\overline{z}).\end{aligned}$$

Proposition 7.32 *The Eichler–Shimura map commutes with* τ_η.

Proof Let $f \in \mathrm{S}_k(\Gamma\,;\,\mathbb{C})$ (for simplicity). We have to check whether τ_η of the cocycle attached to f is the same as the cocycle attached to $\tau_\eta(f)$. We evaluate the latter at a general $\gamma \in \Gamma$ and compute:

$$
\begin{aligned}
J_{(-1)^{k-1}\overline{f}}(\infty, \gamma\infty) &= (-1)^{k-1} \int_{\infty}^{\gamma\infty} f(-\overline{z})(X\overline{z} + Y)^{k-2} d\overline{z} \\
&= - \int_{\infty}^{\gamma\infty} f(-\overline{z})(X(-\overline{z}) - Y)^{k-2} d\overline{z} \\
&= \int_{\gamma\infty}^{\infty} f(-\overline{z})(X(-\overline{z}) - Y)^{k-2} d\overline{z} \\
&= \int_{0}^{\infty} f(-\overline{(\gamma\infty + it)})(X(-\overline{(\gamma\infty + it)}) - Y)^{k-2} (-i) dt \\
&= - \int_{0}^{\infty} f(-\gamma\infty + it)(X(-\gamma\infty + it) - Y)^{k-2} i dt \\
&= \int_{\infty}^{-\gamma\infty} f(z)(Xz - Y)^{k-2} dz \\
&= \eta^{\iota}.I_f(\infty, -\gamma\infty) = \eta^{\iota}.I_f(\infty, \eta\gamma\eta\infty).
\end{aligned}
$$

This proves the claim. □

Corollary 7.33 *Let $\Gamma = \Gamma_1(N)$. The maps*

$$
\mathrm{S}_k(\Gamma\,;\,\mathbb{C}) \to \mathrm{H}^1_{\mathrm{par}}(\Gamma, V_{k-2}(\mathbb{C}))^+, \quad f \mapsto (1 + \tau_\eta).(\gamma \mapsto I_f(z_0, \gamma z_0))
$$

and

$$
\mathrm{S}_k(\Gamma\,;\,\mathbb{C}) \to \mathrm{H}^1_{\mathrm{par}}(\Gamma, V_{k-2}(\mathbb{C}))^-, \quad f \mapsto (1 - \tau_\eta).(\gamma \mapsto I_f(z_0, \gamma z_0))
$$

are isomorphisms compatible with the Hecke operators, where the $+$ (respectively the $-$) indicate the subspace invariant (respectively anti-invariant) for the involution τ_η. A similar result holds in the presence of a Dirichlet character.

Proof Both maps are clearly injective (consider them as being given by $f \mapsto f + \tau_\eta f$ followed by the Eichler–Shimura map) and so dimension considerations show that they are isomorphisms. □

7.6 Theoretical Exercises

Exercise 7.34 Check that $R(\Delta, \Gamma)$ is a ring (associativity and distributivity).

Exercise 7.35 Show the formula

$$
T_m T_n = \sum_{d|(m,n),(d,N)=1} d T(d,d) T_{\frac{mn}{d^2}}.
$$

Also show that $R(\Delta, \Gamma)$ is generated by T_p and $T(p, p)$ for p running through all prime numbers.

Exercise 7.36 Check that the Hecke operator τ_α from Definition 7.21 restricts to $\mathrm{H}^1_{\mathrm{par}}(\Gamma, V)$.

Exercise 7.37 Prove Eq. 20.

7.7 Computer Exercises

Computer Exercise 7.38 Implement Hecke operators.

References

1. A. Ash and G. Stevens. *Modular forms in characteristic l and special values of their L-functions*, Duke Math. J. **53** (1986), no. 3, 849–868.
2. M. F. Atiyah and I. G. Macdonald. *Introduction to commutative algebra*, Addison-Wesley Publishing Co., Reading, Mass.-London-Don Mills, Ont., 1969 ix+128 pp.
3. R. Bieri. *Homological dimension of discrete groups*. Queen Mary College Mathematics Notes, London, 1976.
4. K. S. Brown. *Cohomology of groups*, Springer, New York, 1982.
5. H. Cohen, F. Strömberg. *Modular forms. A classical approach.* Graduate Studies in Mathematics, 179. American Mathematical Society, Providence, RI, 2017. xii+700 pp.
6. J. E. Cremona. *Algorithms for modular elliptic curves*. Second edition. Cambridge University Press, Cambridge, 1997.
7. H. Darmon, F. Diamond, R. Taylor. *Fermat's last theorem.* Elliptic curves, modular forms & Fermat's last theorem (Hong Kong, 1993), 2–140, Int. Press, Cambridge, MA, 1997.
8. P. Deligne, J. P. Serre. *Formes modulaires de poids 1.* Ann. Sci. Ecole Norm. Sup. (4) 7 (1974), 507–530.
9. F. Diamond and J. Im. *Modular forms and modular curves*, in *Seminar on Fermat's Last Theorem (Toronto, ON, 1993–1994)*, 39–133, Amer. Math. Soc., Providence, RI, 1995.
10. Diamond, Fred; Shurman, Jerry. *A first course in modular forms.* Graduate Text in Mathematics, 228. Springer, 2005.
11. W. Eberly. *Decomposition of algebras over finite fields and number fields.* Comput. Complexity 1 (1991), no. 2, 183–210.
12. B. Edixhoven, J.-M. Couveignes. *Computational aspects of modular forms and Galois representations*. Ann. of Math. Stud., 176, Princeton Univ. Press, Princeton, NJ, 2011.
13. D. Eisenbud. *Commutative algebra with a view toward algebraic geometry*, Graduate Texts in Mathematics, **150**, Springer, New York, 1995.
14. K. Haberland. *Perioden von Modulformen einer Variabler and Gruppencohomologie. I, II, III.* Math. Nachr. 112 (1983), 245–282, 283–295, 297–315.
15. C. Khare, J. P. Wintenberger. *Serre's modularity conjecture. I.* Invent. Math. 178 (2009), no. 3, 485–504.
16. L. Merel. *Universal Fourier expansions of modular forms*, in *On Artin's conjecture for odd 2-dimensional representations*, 59–94, Lecture Notes in Math., 1585, Springer, Berlin, 1994.
17. J.-P. Serre. *Sur les représentations modulaires de degré* 2 *de* $\mathrm{Gal}(\overline{\mathbb{Q}}/\mathbb{Q})$. Duke Mathematical Journal **54**, No. 1 (1987), 179–230.
18. K. Ribet. *On l-adic representations attached to modular forms. II.* Glasgow Math. J. 27 (1985), 185–194.

19. G. Shimura. *Introduction to the Arithmetic Theory of Automorphic Forms*. Princeton University Press, 1994.
20. W. A. Stein. *Modular forms, a computational approach.* With an appendix by Paul E. Gunnells. Graduate Studies in Mathematics, 79. American Mathematical Society, Providence, RI, 2007.
21. C. A. Weibel. *An introduction to homological algebra*, Cambridge Univ. Press, Cambridge, 1994.
22. G. Wiese. *On the faithfulness of parabolic cohomology as a Hecke module over a finite field.* J. Reine Angew. Math. 606 (2007), 79–103.
23. G. Wiese. *On modular symbols and the cohomology of Hecke triangle surfaces.* Int. J. Number Theory 5 (2009), no. 1, 89–108.

Computational Number Theory in Relation with L-Functions

Henri Cohen

Abstract We give a number of theoretical and practical methods related to the computation of L-functions, both in the local case (counting points on varieties over finite fields, involving in particular a detailed study of Gauss and Jacobi sums), and in the global case (for instance, Dirichlet L-functions, involving in particular the study of inverse Mellin transforms); we also give a number of little-known but very useful numerical methods, usually but not always related to the computation of L-functions.

1 L-Functions

This course is divided into five parts. In the first part (Sects. 1 and 2), we introduce the notion of L-function, give a number of results and conjectures concerning them, and explain some of the computational problems in this theory. In the second part (Sects. 3–6), we give a number of computational methods for obtaining the Dirichlet series coefficients of the L-function, so is *arithmetic* in nature. In the third part (Sect. 7), we give a number of *analytic* tools necessary for working with L-functions. In the fourth part (Sects. 8 and 9), we give a number of very useful numerical methods which are not sufficiently well known, most of which being also related to the computation of L-functions. The fifth part (Sects. 10 and 11) gives the `Pari/GP` commands corresponding to most of the algorithms and examples given in the course. A final Sect. 12 gives as an appendix some basic definitions and results used in the course which may be less familiar to the reader.

1.1 Introduction

The theory of L-functions is one of the most exciting subjects in number theory. It includes, for instance, two of the crowning achievements of twentieth-century

---

H. Cohen (✉)
Université de Bordeaux, CNRS, INRIA, IMB, UMR 5251, 33400 Talence, France
e-mail: Henri.Cohen@math.u-bordeaux.fr

I. Inam and E. Büyükasik (eds.), *Notes from the International Autumn School on Computational Number Theory*, Tutorials, Schools, and Workshops in the Mathematical Sciences, https://doi.org/10.1007/978-3-030-12558-5_3

mathematics, first the proof of the Weil conjectures and of the Ramanujan conjecture by Deligne in the early 1970s, using the extensive development of modern algebraic geometry initiated by Weil himself and pursued by Grothendieck and followers in the famous EGA and SGA treatises, and second the proof of the Shimura–Taniyama–Weil conjecture by Wiles et al., implying among other things the proof of Fermat's last theorem. It also includes two of the seven 1-million dollar Clay problems for the twenty-first century, first the Riemann hypothesis, and second the Birch–Swinnerton-Dyer conjecture which in my opinion is the most beautiful, if not the most important, conjecture in number theory, or even in the whole of mathematics, together with similar conjectures such as the Beilinson–Bloch conjecture.

There are two kinds of L-functions: local L-functions and global L-functions. Since the proof of the Weil conjectures, local L-functions are rather well understood from a theoretical standpoint, but somewhat less from a computational standpoint. Much less is known on global L-functions, even theoretically, so here the computational standpoint is much more important since it may give some insight on the theoretical side.

Before giving a definition of L-functions, we look in some detail at a large number of special cases of global L-functions.

1.2 *The Prototype: The Riemann Zeta Function* $\zeta(s)$

The simplest of all (global) L-function is the Riemann zeta function $\zeta(s)$ defined by

$$\zeta(s) = \sum_{n\geq 1} \frac{1}{n^s} .$$

This is an example of a *Dirichlet series* (more generally $\sum_{n\geq 1} a(n)/n^s$, or even more generally $\sum_{n\geq 1} 1/\lambda_n^s$, but we will not consider the latter). As such, it has a half-plane of absolute convergence, here $\Re(s) > 1$.

The properties of this function, studied initially by Bernoulli and Euler, are as follows, given historically:

1. (Bernoulli, Euler): it has *special values*. When $s = 2, 4, \ldots$ is a strictly positive even integer, $\zeta(s)$ is equal to π^s times a *rational number*. π is here a *period*, and is of course the usual π used for measuring circles. These rational numbers have elementary *generating functions*, and are equal up to easy terms to the so-called *Bernoulli numbers*. For example, $\zeta(2) = \pi^2/6$, $\zeta(4) = \pi^4/90$, etc. This was conjectured by Bernoulli and proved by Euler. Note that the proof in 1735 of the so-called *Basel problem*:

$$\zeta(2) = 1 + \frac{1}{2^2} + \frac{1}{3^2} + \frac{1}{4^2} + \cdots = \frac{\pi^2}{6}$$

is one of the crowning achievements of mathematics of that time.

2. (Euler): it has an *Euler product*: for $\Re(s) > 1$ one has the identity

$$\zeta(s) = \prod_{p \in P} \frac{1}{1 - 1/p^s} ,$$

where P is the set of prime numbers. This is exactly equivalent to the so-called fundamental theorem of arithmetic. Note in passing (this does not seem interesting here but will be important later) that if we consider $1 - 1/p^s$ as a polynomial in $1/p^s = T$, its reciprocal roots all have the same modulus, here 1, this being of course trivial.

3. (Riemann, but already "guessed" by Euler in special cases): it has an *analytic continuation* to a meromorphic function in the whole complex plane, with a single pole, at $s = 1$, with residue 1, and a *functional equation* $\Lambda(1 - s) = \Lambda(s)$, where $\Lambda(s) = \Gamma_{\mathbb{R}}(s)\zeta(s)$, with $\Gamma_{\mathbb{R}}(s) = \pi^{-s/2}\Gamma(s/2)$, and Γ is the gamma function (see appendix).

4. As a consequence of the functional equation, we have $\zeta(s) = 0$ when $s = -2$, -4,..., $\zeta(0) = -1/2$, but we also have *special values* at $s = -1$, $s = -3$,... which are symmetrical to those at $s = 2, 4$,... (for instance, $\zeta(-1) = -1/12$, $\zeta(-3) = 1/120$, etc.). This is the part which was guessed by Euler.

Roughly speaking, one can say that a global L-function is a function having properties similar to *all* the above. We will of course be completely precise below. Two things should be added immediately: first, the existence of special values will not be a part of the definition but, at least conjecturally, a consequence. Second, all the global L-functions that we will consider should *conjecturally* satisfy a Riemann hypothesis: when suitably normalized, and excluding "trivial" zeros, all the zeros of the function should be on the line $\Re(s) = 1/2$, axis of symmetry of the functional equation. Note that even for the simplest L-function, $\zeta(s)$, this is not proved.

1.3 Dedekind Zeta Functions

The Riemann zeta function is perhaps too simple an example to get the correct feeling about global L-functions, so we generalize:

Let K be a number field (a finite extension of $\mathbb{Q}$) of degree d. We can define its *Dedekind zeta function* $\zeta_K(s)$ for $\Re(s) > 1$ by

$$\zeta_K(s) = \sum_{\mathfrak{a}} \frac{1}{\mathcal{N}(\mathfrak{a})^s} = \sum_{n \geq 1} \frac{i(n)}{n^s} ,$$

where $\mathfrak{a}$ ranges over all (nonzero) integral ideals of the ring of integers $\mathbb{Z}_K$ of K, $\mathcal{N}(\mathfrak{a}) = [\mathbb{Z}_K : \mathfrak{a}]$ is the norm of $\mathfrak{a}$, and $i(n)$ denotes the number of integral ideals of norm n.

This function has very similar properties to those of $\zeta(s)$ (which is the special case $K = \mathbb{Q}$). We give them in a more logical order:

1. It can be analytically continued to the whole complex plane into a meromorphic function having a single pole, at $s = 1$, with known residue, and it has a functional equation $\Lambda_K(1-s) = \Lambda_K(s)$, where
$$\Lambda_K(s) = |D_K|^{s/2}\Gamma_{\mathbb{R}}(s)^{r_1+r_2}\Gamma_{\mathbb{R}}(s+1)^{r_2}\ ,$$
where $(r_1, 2r_2)$ are the number of real and complex embeddings of K and D_K its discriminant.
2. It has an Euler product $\zeta_K(s) = \prod_{\mathfrak{p}} 1/(1 - 1/\mathcal{N}(\mathfrak{p})^s)$, where the product is over all prime ideals of $\mathbb{Z}_K$. Note that this can also be written
$$\zeta_K(s) = \prod_{p\in P}\prod_{\mathfrak{p}|p}\frac{1}{1 - 1/p^{f(\mathfrak{p}/p)s}}\ ,$$
where $f(\mathfrak{p}/p) = [\mathbb{Z}_K/\mathfrak{p} : \mathbb{Z}/p\mathbb{Z}]$ is the so-called *residual index* of $\mathfrak{p}$ above p. Once again, note that if we set as usual $1/p^s = T$, the reciprocal roots of $1 - T^{f(\mathfrak{p}/p)}$ all have modulus 1.
3. It has *special values*, but only when K is a *totally real* number field ($r_2 = 0$, $r_1 = d$): in that case $\zeta_K(s)$ is a *rational number* if s is a negative odd integer, or equivalently by the functional equation, it is a rational multiple of $\sqrt{|D_K|}\pi^{ds}$ if s is a positive even integer.

An important new phenomenon occurs: recall that $\sum_{\mathfrak{p}|p} e(\mathfrak{p}/p)f(\mathfrak{p}/p) = d$, where $e(\mathfrak{p}/p)$ is the so-called *ramification index*, which is equivalent to the defining equality $p\mathbb{Z}_K = \prod_{\mathfrak{p}|p}\mathfrak{p}^{e(\mathfrak{p}/p)}$. In particular $\sum_{\mathfrak{p}|p} f(\mathfrak{p}/p) = d$ if and only if $e(\mathfrak{p}/p) = 1$ for all $\mathfrak{p}$, which means that p is *unramified* in $K/\mathbb{Q}$; one can prove that this is equivalent to $p \nmid D_K$. Thus, the *local L-function* $L_{K,p}(T) = \prod_{\mathfrak{p}|p}(1 - T^{f(\mathfrak{p}/p)})$ has degree in T exactly equal to d for all but a finite number of primes p, which are exactly those which divide the discriminant D_K, and for those "bad" primes the degree is strictly less than d. In addition, note that the number of $\Gamma_{\mathbb{R}}$ factors in the *completed* function $\Lambda_K(s)$ is equal to $r_1 + 2r_2$, hence once again equal to d.

Examples:

1. Let D be the discriminant of a quadratic field, and let $K = \mathbb{Q}(\sqrt{D})$. In that case, $\zeta_K(s)$ *factors* as $\zeta_K(s) = \zeta(s)L(\chi_D, s)$, where $\chi_D = \left(\frac{D}{.}\right)$ is the Legendre–Kronecker symbol, and $L(\chi_D, s) = \sum_{n\geq 1}\chi_D(n)/n^s$. Thus, the local L-function at a prime p is given by
$$L_{K,p}(T) = (1-T)(1-\chi_D(p)T) = 1 - a_pT + \chi_D(p)T^2\ ,$$
with $a_p = 1 + \chi_D(p)$. Note that a_p is equal to the number of solutions in $\mathbb{F}_p$ of the equation $x^2 = D$.

2. Let us consider two special cases of (1): first $K = \mathbb{Q}(\sqrt{5})$. Since it is a real quadratic field, it has special values, for instance,

$$\zeta_K(-1) = \frac{1}{30}\ , \quad \zeta_K(-3) = \frac{1}{60}\ , \quad \zeta_K(2) = \frac{2\sqrt{5}\pi^4}{375}\ , \quad \zeta_K(4) = \frac{4\sqrt{5}\pi^8}{84375}\ .$$

In addition, note that its *gamma factor* is $5^{s/2}\Gamma_{\mathbb{R}}(s)^2$.
Second, consider $K = \mathbb{Q}(\sqrt{-23})$. Since it is not a totally real field, $\zeta_K(s)$ does not have special values. However, because of the factorization $\zeta_K(s) = \zeta(s)L(\chi_D, s)$, we can look *separately* at the special values of $\zeta(s)$, which we have already seen (negative odd integers and positive even integers), and of $L(\chi_D, s)$. It is easy to prove that the special values of this latter function occurs at negative *even* integers and positive *odd* integers, which have empty intersection which those of $\zeta(s)$ and explains why $\zeta_K(s)$ itself has none. For instance,

$$L(\chi_D, -2) = -48\ , \quad L(\chi_D, -4) = 6816\ , \quad L(\chi_D, 3) = \frac{96\sqrt{23}\pi^3}{12167}\ .$$

In addition, note that its gamma factor is

$$23^{s/2}\Gamma_{\mathbb{R}}(s)\Gamma_{\mathbb{R}}(s+1) = 23^{s/2}\Gamma_{\mathbb{C}}(s)\ ,$$

where we set by definition

$$\Gamma_{\mathbb{C}}(s) = \Gamma_{\mathbb{R}}(s)\Gamma_{\mathbb{R}}(s+1) = 2\cdot(2\pi)^{-s}\Gamma(s)$$

by the duplication formula for the gamma function.
3. Let K be the unique cubic field up to isomorphism of discriminant -23, defined, for instance, by a root of the equation $x^3 - x - 1 = 0$. We have $(r_1, 2r_2) = (1, 2)$ and $D_K = -23$. Here, one can prove (it is less trivial) that $\zeta_K(s) = \zeta(s)L(\rho, s)$, where $L(\rho, s)$ is a holomorphic function. Using both properties of ζ_K and ζ, this L-function has the following properties:

   - It extends to an entire function on $\mathbb{C}$ with a functional equation $\Lambda(\rho, 1-s) = \Lambda(\rho, s)$, with

   $$\Lambda(\rho, s) = 23^{s/2}\Gamma_{\mathbb{R}}(s)\Gamma_{\mathbb{R}}(s+1)L(\rho, s) = 23^{s/2}\Gamma_{\mathbb{C}}(s)L(\rho, s)\ .$$

   Note that this is the *same* gamma factor as for $\mathbb{Q}(\sqrt{-23})$. However the functions are fundamentally different, since $\zeta_{\mathbb{Q}(\sqrt{-23})}(s)$ has a pole at $s = 1$, while $L(\rho, s)$ is an entire function.
   - It is immediate to show that if we let $L_{\rho,p}(T) = L_{K,p}(T)/(1-T)$ be the local L function for $L(\rho, s)$, we have $L_{\rho,p}(T) = 1 - a_pT + \chi_{-23}(p)T^2$, with $a_p = 1$ if $p = 23$, $a_p = 0$ if $\left(\frac{-23}{p}\right) = -1$, and $a_p = 1$ or 2 if $\left(\frac{-23}{p}\right) = 1$.

Remark 1.1 In all of the above examples, the function $\zeta_K(s)$ is *divisible* by the Riemann zeta function $\zeta(s)$, i.e., the function $\zeta_K(s)/\zeta(s)$ is an *entire function*. This is known for some number fields K, but is *not* known in general, even in degree $d=5$ for instance: it is a consequence of the more precise *Artin conjecture* on the holomorphy of Artin L-functions.

1.4 Further Examples in Weight 0

It is now time to give examples not coming from number fields. Define $a_1(n)$ by the formal equality

$$q\prod_{n\ge1}(1-q^n)(1-q^{23n})=\sum_{n\ge1}a_1(n)q^n=q-q^2-q^3+q^6+q^8-\cdots\;,$$

and set $L_1(s)=\sum_{n\ge1}a_1(n)/n^s$. The theory of modular forms (here of the Dedekind eta function) tells us that $L_1(s)$ will satisfy exactly the same properties as $L(\rho,s)$ with ρ as above.

Define $a_2(n)$ by the formal equality

$$\frac{1}{2}\left(\sum_{(m,n)\in\mathbb{Z}\times\mathbb{Z}}q^{m^2+mn+6n^2}-q^{2m^2+mn+3n^2}\right)=\sum_{n\ge1}a_2(n)q^n\;,$$

and set $L_2(s)=\sum_{n\ge1}a_2(n)/n^s$. The theory of modular forms (here of theta functions) tells us that $L_2(s)$ will satisfy exactly the same properties as $L(\rho,s)$.

And indeed, it is an interesting *theorem* that

$$L_1(s)=L_2(s)=L(\rho,s)\;:$$

The "moral" of this story is the following, which can be made mathematically precise: if two L-functions are holomorphic, have the same gamma factor (including in this case the $23^{s/2}$), then (conjecturally in general) they belong to a finite-dimensional vector space. Thus, in particular, if this vector space is 1-dimensional and the L-functions are suitably normalized (usually with $a(1)=1$), this implies as here that they are equal.

1.5 Examples in Weight 1

Although we have not yet defined the notion of weight, let me give two further examples.

Define $a_3(n)$ by the formal equality

$$q\prod_{n\ge1}(1-q^n)^2(1-q^{11n})^2=\sum_{n\ge1}a_3(n)q^n=q-2q^2-q^3+2q^4+\cdots\;,$$

and set $L_3(s)=\sum_{n\ge1}a_3(n)/n^s$. The theory of modular forms (again of the Dedekind eta function) tells us that $L_3(s)$ will satisfy the following properties, analogous but more general than those satisfied by $L_1(s)=L_2(s)=L(\rho,s)$:

- It has an analytic continuation to the whole complex plane, and if we set

$$\Lambda_3(s)=11^{s/2}\Gamma_{\mathbb{R}}(s)\Gamma_{\mathbb{R}}(s+1)L_3(s)=11^{s/2}\Gamma_{\mathbb{C}}(s)L_3(s)\;,$$

  we have the functional equation $\Lambda_3(2-s)=\Lambda_3(s)$. Note the crucial difference that here $1-s$ is replaced by $2-s$.
- There exists an Euler product $L_3(s)=\prod_{p\in P}1/L_{3,p}(1/p^s)$ similar to the preceding ones in that $L_{3,p}(T)$ is for all but a finite number of p a second-degree polynomial in T. More precisely, if $p=11$ we have $L_{3,p}(T)=1-T$, while for $p\neq11$ we have $L_{3,p}(T)=1-a_pT+pT^2$, for some a_p such that $|a_p|<2\sqrt{p}$. This is expressed more vividly by saying that for $p\neq11$ we have $L_{3,p}(T)=(1-\alpha_pT)(1-\beta_pT)$, where the reciprocal roots α_p and β_p have modulus exactly equal to $p^{1/2}$. Note again the crucial difference with "weight 0" in that the coefficient of T^2 is equal to p instead of ±1, hence that $|\alpha_p|=|\beta_p|=p^{1/2}$ instead of 1.

As a second example, consider the equation $y^2+y=x^3-x^2-10x-20$ (an elliptic curve E), and denote by $N_q(E)$ the number of projective points of this curve over the finite field $\mathbb{F}_q$ (it is clear that there is a unique point at infinity, so if you want $N_q(E)$ is one plus the number of affine points). There is a universal recipe to construct an L-function out of a variety which we will recall below, but here let us simplify: for p prime, set $a_p=p+1-N_p(E)$ and

$$L_4(s)=\prod_{p\in P}1/(1-a_pp^{-s}+\chi(p)p^{1-2s})\;,$$

where $\chi(p)=1$ for $p\neq11$ and $\chi(11)=0$. It is not difficult to show that $L_4(s)$ satisfies exactly the same properties as $L_3(s)$ (using, for instance, the elementary theory of modular curves), so by the moral explained above, it should not come as a surprise that in fact $L_3(s)=L_4(s)$.

1.6 Definition of a Global L-Function

With all these examples at hand, it is quite natural to give the following definition of an L-function, which is not the most general but will be sufficient for us.

Definition 1.2 Let d be a nonnegative integer. We say that a Dirichlet series $L(s) = \sum_{n\geq 1} a(n)n^{-s}$ with $a(1) = 1$ is an L-function of *degree* d and *weight* 0 if the following conditions are satisfied:

1. (Ramanujan bound): We have $a(n) = O(n^{\varepsilon})$ for all $\varepsilon > 0$, so that in particular the Dirichlet series converges absolutely and uniformly in any half-plane $\Re(s) \geq \sigma > 1$.
2. (Meromorphy and Functional equation): The function $L(s)$ can be extended to $\mathbb{C}$ to a meromorphic function of order 1 (see appendix) having a finite number of poles; furthermore there exist complex numbers λ_i with nonnegative real part and an integer N called the *conductor* such that if we set
$$\gamma(s) = N^{s/2} \prod_{1\leq i\leq d} \Gamma_{\mathbb{R}}(s + \lambda_i) \quad \text{and} \quad \Lambda(s) = \gamma(s)L(s) ,$$
we have the *functional equation*
$$\Lambda(s) = \omega\overline{\Lambda(1 - \overline{s})}$$
for some complex number ω, called the *root number*, which will necessarily be of modulus 1.
3. (Euler Product): For $\Re(s) > 1$, we have an Euler product
$$L(s) = \prod_{p\in P} 1/L_p(1/p^s) \quad \text{with} \quad L_p(T) = \prod_{1\leq j\leq d} (1 - \alpha_{p,j}T) ,$$
and the reciprocal roots $\alpha_{p,j}$ are called the *Satake parameters*.
4. (Local Riemann hypothesis): for $p \nmid N$ we have $|\alpha_{p,j}| = 1$, and for $p \mid N$ we have either $\alpha_{p,j} = 0$ or $|\alpha_{p,j}| = p^{-m/2}$ for some m such that $1 \leq m \leq d$.

Remarks 1.3 1. More generally Selberg has defined a more general class of L-functions, which first allows $\Gamma(\mu_i s + \lambda_i)$ with μ_i positive real in the gamma factors and second allows weaker assumptions on N and the Satake parameters.
2. Note that d is *both* the number of $\Gamma_{\mathbb{R}}$ factors, *and* the degree in T of the Euler factors $L_p(T)$, at least for $p \nmid N$, while the degree decreases for the "bad" primes p which divide N.
3. The Ramanujan bound (1) is easily seen to be a consequence of the conditions that we have imposed on the Satake parameters: in Selberg's more general definition this is not the case.

It is important to generalize this definition in the following trivial way:

Definition 1.4 Let w be a nonnegative integer. A function $L(s)$ is said to be an L-function of degree d and *motivic weight* w if $L(s + w/2)$ is an L-function of degree d and weight 0 as above (with the slight additional technical condition that the nonzero Satake parameters $\alpha_{p,j}$ for $p \mid N$ satisfy $|\alpha_{p,j}| = p^{-m/2}$ with $1 \leq m \leq w$).

For an L-function of weight w, it is clear that the functional equation is $\Lambda(s) = \omega\overline{\Lambda(k-\overline{s})}$ with $k = w+1$, and that the Satake parameters will satisfy $|\alpha_{p,j}| = p^{w/2}$ for $p \nmid N$, and for $p \mid N$ we have either $\alpha_{p,j} = 0$ or $|\alpha_{p,j}| = p^{(w-m)/2}$ for some integer m such that $1 \le m \le w$.

Thus, the first examples that we have given are all of weight 0, and the last two (which are in fact equal) are of weight 1. For those who know the theory of modular forms, note that the motivic weight (that we denote by w) is one less than the weight k of the modular form.

2 Origins of L-Functions

As can already be seen in the above examples, it is possible to construct L-functions in many different ways. In the present section, we look at three different ways for constructing L-functions: the first is by the theory of modular forms or more generally of *automorphic forms* (of which we have seen a few examples above), the second is by using Weil's construction of local L-functions attached to varieties, and more generally to *motives*, and third, as a special but much simpler case of this, by the theory of *hypergeometric motives*.

2.1 L-Functions Coming from Modular Forms

The basic notion that we need here is that of *Mellin transform*: if $f(t)$ is a nice function tending to zero exponentially fast at infinity, we can define its Mellin transform $\Lambda(f;s) = \int_0^\infty t^s f(t)\,dt/t$, the integral being written in this way because dt/t is the invariant Haar measure on the locally compact group $\mathbb{R}_{>0}$. If we set $g(t) = t^{-k}f(1/t)$ and assume that g also tends to zero exponentially fast at infinity, it is immediate to see by a change of variable that $\Lambda(g;s) = \Lambda(f;k-s)$. This is exactly the type of functional equation needed for an L-function.

The other fundamental property of L-functions that we need is the existence of an Euler product of a specific type. This will come from the theory of *Hecke operators*.

A crash course in modular forms (see for instance, [6] for a complete introduction): we use the notation $q = e^{2\pi i\tau}$, for $\tau \in \mathbb{C}$ such that $\Im(\tau) > 0$, so that $|q| < 1$. A function $f(\tau) = \sum_{n\ge1} a(n)q^n$ is said to be a modular cusp form of (positive, even) weight k if $f(-1/\tau) = \tau^k f(\tau)$ for all $\Im(\tau) > 0$. Note that because of the notation q we also have $f(\tau+1) = f(\tau)$, hence it is easy to deduce that $f((a\tau+b)/(c\tau+d)) = (c\tau+d)^k f(\tau)$ if $\left(\begin{smallmatrix} a & b \\ c & d \end{smallmatrix}\right)$ is an integer matrix of determinant 1. We define the L-function attached to f as $L(f;s) = \sum_{n\ge1} a(n)/n^s$, and the Mellin transform $\Lambda(f;s)$ of the function $f(it)$ is on the one hand equal to $(2\pi)^{-s}\Gamma(s)L(f;s) = (1/2)\Gamma_{\mathbb{C}}(s)L(f;s)$, and on the other hand as we have seen above satisfies the functional equation $\Lambda(k-s) = (-1)^{k/2}\Lambda(s)$.

One can easily show the fundamental fact that the vector space of modular forms of given weight k is *finite dimensional*, and compute its dimension explicitly.

If $f(\tau) = \sum_{n\geq 1} a(n)q^n$ is a modular form and p is a prime number, one defines $T(p)(f)$ by $T(p)(f) = \sum_{n\geq 1} b(n)q^n$ with $b(n) = a(pn) + p^{k-1}a(n/p)$, where $a(n/p)$ is by convention 0 when $p \nmid n$, or equivalently

$$T(p)(f)(\tau) = p^{k-1}f(p\tau) + \frac{1}{p}\sum_{0\leq j<p} f\left(\frac{\tau+j}{p}\right).$$

Then $T(p)f$ is also a modular cusp form, so $T(p)$ is an operator on the space of modular forms, and it is easy to show that the $T(p)$ commute and are diagonalizable, so they are simultaneously diagonalizable hence there exists a basis of common *eigenforms* for all the $T(p)$. Since one can show that for such an eigenform one has $a(1) \neq 0$, we can normalize them by asking that $a(1) = 1$, and we then obtain a canonical basis.

If $f(\tau) = \sum_{n\geq 1} a(n)q^n$ is such a *normalized eigenform*, it follows that the corresponding L function $\sum_{n\geq 1} a(n)/n^s$ will indeed have an Euler product, and using the elementary properties of the operators $T(p)$ that it will in fact be of the form:

$$L(f; s) = \prod_{p\in P} \frac{1}{1 - a(p)p^{-s} + p^{k-1-2s}}.$$

As a final remark, note that the analytic continuation and functional equation of this L-function is an *elementary consequence* of the definition of a modular form. This is totally different from the motivic cases that we will see below, where this analytic continuation is in general completely *conjectural*.

The above describes briefly the theory of modular forms on the modular group $\mathrm{PSL}_2(\mathbb{Z})$. One can generalize (nontrivially) this theory to *subgroups* of the modular group, the most important being $\Gamma_0(N)$ (matrices as above with $N \mid c$), to other *Fuchsian groups*, to forms in several variables, and even more generally to *reductive groups*.

2.2 Local L-Functions of Algebraic Varieties

The second very important source of L-functions comes from algebraic geometry. Let V be some algebraic object. In modern terms, V may be a *motive*, whatever that may mean for the moment, but assume, for instance, that V is an algebraic variety, in other words that for each suitable field K, $V(K)$ is the set of common zeros of a family of polynomials in several variables. If K is a *finite* field $\mathbb{F}_q$ (recall that we must then have $q = p^n$ for some prime p and that $\mathbb{F}_q$ exists and is unique up to isomorphism), then $V(\mathbb{F}_q)$ will also be finite.

After studying a number of special cases, such as elliptic curves (due to Hasse), and quasi-diagonal hypersurfaces in $\mathbb{P}^d$, in 1949 Weil was led to make a number of more precise conjectures concerning the number of *projective* points $|V(\mathbb{F}_q)|$, assuming that V is a *smooth projective* variety, and proved these conjectures in the special case of curves (the proof is already quite deep).

The first *Weil conjecture* says that (for p fixed) the number $|V(\mathbb{F}_{p^n})|$ of projective points of V over the finite field $\mathbb{F}_{p^n}$ satisfies a (nonhomogeneous) linear recurrence with constant coefficients. For instance, if V is an *elliptic curve* defined over $\mathbb{Q}$ (such as $y^2 = x^3 + x + 1$) and if we set $a(p^n) = p^n + 1 - |V(\mathbb{F}_{p^n})|$, then

$$a(p^{n+1}) = a(p)a(p^n) - \chi(p)pa(p^{n-1}) ,$$

where $\chi(p) = 1$ unless p divides the so-called *conductor* of the elliptic curve, in which case $\chi(p) = 0$ (this is not quite true because we must choose a suitable model for V, but it suffices for us).

Exercise 2.1 Using the above recursion for $a(p^n)$, find the corresponding recursion for $v_n = |V(\mathbb{F}_{p^n})|$.

Exercise 2.2 1. Given a prime p and $n \geq 1$, write a computer program which runs through all the elements of $\mathbb{F}_{p^n}$, represented in a suitable way.
2. For the elliptic curve $y^2 = x^3 + x + 1$, compute (on a computer) $a(5)$ and $a(5^2)$, and check the recursion.
3. Similarly, compute $a(31)$ and $a(31^2)$, and check the recursion (here $\chi(31) = 0$).

This first Weil conjecture was proved by Dwork in the early 1960s. It is better reformulated in terms of *local L-functions* as follows: define the Hasse–Weil zeta function of V as the *formal power series* in T given by the formula

$$Z_p(V; T) = \exp\left(\sum_{n\geq 1} \frac{|V(\mathbb{F}_{p^n})|}{n} T^n\right) .$$

There should be no difficulty in understanding this: setting for simplicity $v_n = |V(\mathbb{F}_{p^n})|$, we have

$$\begin{aligned} Z_p(V; T) &= \exp(v_1 T + v_2 T^2/2 + v_3 T^3/3 + \cdots) \\ &= 1 + v_1 T + (v_1^2 + v_2)T^2/2 + (v_1^3 + 3v_1 v_2 + 2v_3)T^3/6 + \cdots \end{aligned}$$

For instance, if V is projective d-space $\mathbb{P}^d$, we have $|V(\mathbb{F}_q)| = q^d + q^{d-1} + \cdots + 1$, and since $\sum_{n\geq 1} p^{nj} T^n/n = -\log(1 - p^j T)$, we deduce that $Z_p(\mathbb{P}^d; T) = 1/((1 - T)(1 - pT) \cdots (1 - p^d T))$.

In terms of this language, the existence of the recurrence relation is equivalent to the fact that $Z_p(V; T)$ is a *rational function* of T, and as already mentioned, this was proved by Dwork in 1960.

The second conjecture of Weil states that this rational function is of the form

$$Z_p(V;T)=\prod_{0\le i\le 2d}P_{i,p}(V;T)^{(-1)^{i+1}}=\frac{P_{1,p}(V;T)\cdots P_{2d-1,p}(V;T)}{P_{0,p}(V;T)P_{2,p}(V;T)\cdots P_{2d,p}(V;T)}\;,$$

where $d=\dim(V)$, and the $P_{i,p}$ are polynomials in T. Furthermore, a basic result in algebraic geometry called Poincaré duality implies that $Z_p(V;1/(p^dT))=\pm p^{de/2}T^eZ_p(V;T)$, where e is the degree of the rational function (called the Euler characteristic of V), which means that there is a relation between $P_{i,p}$ and $P_{2d-i,p}$. In addition the $P_{i,p}$ have integer coefficients, and $P_{0,p}(T)=1-T$, $P_{2d,p}(T)=1-p^dT$. For instance, for *curves*, this means that $Z_p(V;T)=P_1(V;T)/((1-T)(1-pT))$, the polynomial P_1 is of even degree $2g$ (g is the so-called *genus* of the curve) and satisfies $p^{dg}P_1(V;1/(p^dT))=\pm P_1(V;T)$.

For knowledgeable readers, in highbrow language, the polynomial $P_{i,p}$ is the reverse characteristic polynomial of the Frobenius endomorphism acting on the ith ℓ-adic cohomology group $H^i(V;\mathbb{Q}_\ell)$ for any $\ell\neq p$.

The third, most important and most difficult of the Weil conjectures is the local *Riemann hypothesis*, which says that the reciprocal roots of $P_{i,p}$ have modulus exactly equal to $p^{i/2}$, in other words that

$$P_{i,p}(V;T)=\prod_j(1-\alpha_{i,j}T)\quad\text{with}\quad|\alpha_{i,j}|=p^{i/2}\;.$$

This last is the most important in applications.

The Weil conjectures were completely proved by Deligne in the early 1970s following a strategy already put forward by Weil, and is considered as one of the two or three major accomplishments of mathematics of the second half of the twentieth century.

Exercise 2.3 (*You need to know some algebraic number theory for this*) Let $P\in\mathbb{Z}[X]$ be a monic irreducible polynomial and $K=\mathbb{Q}(\theta)$, where θ is a root of P be the corresponding number field. Assume that $p^2\nmid\operatorname{disc}(P)$. Show that the Hasse–Weil zeta function at p of the 0-dimensional variety defined by $P=0$ is the Euler factor at p of the Dedekind zeta function $\zeta_K(s)$ attached to K, where p^{-s} is replaced by T.

2.3 Global L-Function Attached to a Variety

We are now ready to "globalize" the above construction, and build *global* L-functions attached to a variety.

Let V be an algebraic variety defined over $\mathbb{Q}$, say. We assume that V is "nice", meaning, for instance, that we choose V to be projective, smooth, and absolutely irreducible. For all but a finite number of primes p we can consider V as a smooth variety over $\mathbb{F}_p$, so for each i we can set $L_i(V;s)=\prod_p 1/P_{i,p}(V;p^{-s})$, where the product is over all the "good" primes, and the $P_{i,p}$ are as above. The factor $1/P_{i,p}(V;p^{-s})$ is as usual called the Euler factor at p. These functions L_i can be called the global L-functions attached to V.

This naïve definition is insufficient to construct interesting objects. First and most importantly, we have omitted a finite number of Euler factors at the so-called "bad primes", which include in particular those for which V is not smooth over $\mathbb{F}_p$, and although there do exist cohomological recipes to define them, as far as the author is aware these recipes do not really give practical algorithms. (In highbrow language, these recipes are based on the computation of ℓ-adic cohomology groups, for which the known algorithms are useless in practice; in the simplest case of Artin L-functions, one must determine the action of Frobenius on the vector space fixed by the inertia group, which can be done reasonably easily.)

Another much less important reason is the fact that most of the L_i are uninteresting or related. For instance, in the case of elliptic curves seen above, we have (up to a finite number of Euler factors) $L_0(V;s) = \zeta(s)$ and $L_2(V;s) = \zeta(s-1)$, so the only interesting L-function, called *the* L-function of the elliptic curve, is the function $L_1(V;s) = \prod_p(1 - a(p)p^{-s} + \chi(p)p^{1-2s})^{-1}$ (if the model of the curve is chosen to be *minimal*, this happens to be the correct definition, including for the "bad" primes). For varieties of higher dimension d, as we have mentioned as part of the Weil conjecture the functions L_i and L_{2d-i} are related by Poincaré duality, and L_0 and L_{2d} are translates of the Riemann zeta function (as above), so only the L_i for $1 \le i \le d$ need to be studied.

2.4 Hypergeometric Motives

Still another way to construct L-functions is through the use of *hypergeometric motives*, due to Katz and Rodriguez-Villegas. Although this construction is a special case of the construction of L-functions of varieties studied above, the corresponding variety is *hidden* (although it can be recovered if desired), and the computations are in some sense much simpler.

Let me give a short and unmotivated introduction to the subject: let $\gamma = (\gamma_n)_{n\ge1}$ be a finite sequence of (positive or negative) integers satisfying the essential condition $\sum_n n\gamma_n = 0$. For any finite field $\mathbb{F}_q$ with $q = p^f$ and any character χ of $\mathbb{F}_q^*$, recall that the Gauss sum $\mathfrak{g}(\chi)$ is defined by

$$\mathfrak{g}(\chi) = \sum_{x\in\mathbb{F}_q^*} \chi(x)\exp(2\pi i\,\mathrm{Tr}_{\mathbb{F}_q/\mathbb{F}_p}(x)/p)\ ,$$

see Sect. 4.1 below. We set

$$Q_q(\gamma;\chi) = \prod_{n\ge1} \mathfrak{g}(\chi^n)^{\gamma_n}$$

and for any $t \in \mathbb{F}_q \setminus \{0,1\}$

$$a_q(\gamma;t) = \frac{1}{1-q}\left(1 + \sum_{\chi\neq\varepsilon} \chi(Mt)Q_q(\gamma;\chi)\right),$$

where ε is the trivial character and $M = \prod_n n^{n\gamma_n}$ is a normalizing constant (this is not quite the exact formula but it will suffice for our purposes). The theorem of Katz is that for $t \neq 0, 1$ the quantity $a_q(\gamma;t)$ is the *trace of Frobenius* on some *motive defined over* $\mathbb{Q}$. In the language of L-functions, this means the following: define as usual the local L-function at p by the formal power series

$$L_p(\gamma;t;T) = \exp\left(\sum_{f\geq 1} a_{p^f}(\gamma;t)\frac{T^f}{f}\right).$$

Then L_p is a rational function of T, satisfies the local Riemann hypothesis, and if we set

$$L(\gamma;t;s) = \prod_p L_p(\gamma;t;p^{-s})^{-1},$$

then L once completed at the "bad" primes should be a global L-function of the standard type described above.

Let me give one of the simplest examples of a hypergeometric motive, and show how one can recover the underlying algebraic variety. We choose $\gamma_1 = 4$, $\gamma_2 = -2$, $\gamma_n = 0$ for $n > 2$, which does satisfy the condition $\sum_n n\gamma_n = 0$ (we could choose the simpler values $\gamma_1 = 2$, $\gamma_2 = -1$, but this would give a 0-dimensional variety, i.e., a number field, so less representative of the general case). We thus have $Q_q(\gamma,\chi) = \mathfrak{g}(\chi)^4/\mathfrak{g}(\chi^2)^2$ and $M = 1/4$. By the results on Jacobi sums that we will see below (Proposition 4.9), if χ^2 is not the trivial character ε we have $Q_q(\gamma,\chi) = J(\chi,\chi)^2$, where $J(\chi,\chi) = \sum_{x\in\mathbb{F}_q\setminus\{0,1\}} \chi(x)\chi(1-x)$. As mentioned above, we did not give the precise formula, here it simply corresponds to setting $Q_q(\gamma,\chi) = J(\chi,\chi)^2$, including when $\chi^2 = \varepsilon$. Thus

$$a_q(\gamma;t) = \frac{1}{1-q}\left(1 + \sum_{\chi\neq\varepsilon} \chi(t/4)J(\chi,\chi)^2\right).$$

If by a temporary abuse of notation[1] we define $J(\varepsilon,\varepsilon)$ by the same formula as above, we have $J(\varepsilon,\varepsilon) = (q-2)^2$ hence

[1]The definition of J given below is a sum over all $x \in \mathbb{F}_q$, so that $J(\varepsilon,\varepsilon) = q^2$ and not $(q-2)^2$.

$$a_q(\gamma;t)=\frac{1}{1-q}\left(1-(q-2)^2+\sum_{\chi}\chi(t/4)J(\chi,\chi)^2\right).$$

Now

$$\sum_{\chi}\chi(t/4)J(\chi,\chi)^2=\sum_{x,y\in\mathbb{F}_q\setminus\{0,1\}}\sum_{\chi}\chi(t/4)\chi(x)\chi(1-x)\chi(y)\chi(1-y)\;.$$

The point of writing it this way is that because of orthogonality of characters (Exercise 4.4 below) the sum on χ vanishes unless the argument is equal to 1 in which case it is equal to $q-1$, so that

$$\sum_{\chi}\chi(t/4)J(\chi,\chi)^2=(q-1)N_q(t)\;,\quad\text{where}\quad N_q(t)=\sum_{\substack{x,y\in\mathbb{F}_q\setminus\{0,1\}\\(t/4)x(1-x)y(1-y)=1}}1$$

is the number of *affine* points over $\mathbb{F}_q$ of the algebraic variety defined by $(t/4)x(1-x)y(1-y)=1$ (which automatically implies x and y are different from 0 and 1). We have thus shown that

$$a_q(\gamma;t)=\frac{1}{1-q}(1-(q-2)^2+(q-1)N_q(t))=q-3-N_q(t)\;.$$

Exercise 2.4 By making the change of variables $X=(4/t)(1-1/x)$, $Y=(4/t)(y-1)(1-1/x)$, show that

$$a_q(\gamma;t)=q+1-|E(\mathbb{F}_q)|\;,$$

where $|E(\mathbb{F}_q)|$ is the number of projective points over $\mathbb{F}_q$ of the elliptic curve $Y^2+XY=X(X-4/t)^2$. Thus, the global L-function attached to the hypergeometric motive defined by γ is equal to the L-function attached to the elliptic curve E.

Since we will see below fast methods for computing expressions such as $\sum_{\chi}\chi(t/4)J(\chi,\chi)^2$, these will consequently give fast methods for computing $|E(\mathbb{F}_q)|$ for an arbitrary elliptic curve E.

Exercise 2.5 1. In a similar way, study the hypergeometric motive corresponding to $\gamma_1=3$, $\gamma_3=-1$, and $\gamma_n=0$ otherwise, assuming that the correct formula for Q_q corresponds as above to the replacement of quotients of Gauss sums by Jacobi sums for all characters χ, not only those allowed by Proposition 4.9. To find the elliptic curve, use the change of variable $X=-xy$, $Y=x^2y$.

2. Deduce that the global L-function of this hypergeometric motive is equal to the L-function attached to the elliptic curve $y^2=x^3+x^2+4x+4$ and to the L-function attached to the modular form $q\prod_{n\geq1}(1-q^{2n})^2(1-q^{10n})^2$.

2.5 Other Sources of L-Functions

There exist many other sources of L-functions in addition to those that we have already mentioned, that we will not expand upon:

- Hecke L-functions, attached to Hecke Grössencharacters.
- Artin L-functions, of which we have met a couple of examples in Sect. 1.
- Functorial constructions of L-functions such as Rankin–Selberg L-functions, symmetric squares and more generally symmetric powers.
- L-functions attached to Galois representations.
- General automorphic L-functions.

Of course these are not disjoint sets, and as already mentioned, when some L-functions lie in an intersection, this usually corresponds to an interesting arithmetic property. Probably, the most general such correspondence is the *Langlands program*.

2.6 Results and Conjectures on $L(V; s)$

The problem with global L-functions is that most of their properties are only *conjectural*. We mention these conjectures in the case of global L-functions attached to algebraic varieties:

1. The function L_i is only defined through its Euler product, and thanks to the last of Weil's conjectures, the local Riemann hypothesis, proved by Deligne, it converges absolutely for $\Re(s) > 1 + i/2$. Note that, with the definitions introduced above, L_i is an L-function of degree d_i, the common degree of $P_{i,p}$ for all but a finite number of p, and of motivic weight exactly $w = i$ since the Satake parameters satisfy $|\alpha_{i,p}| = p^{i/2}$, again by the local Riemann hypothesis.
2. A first conjecture is that L_i should have an *analytic continuation* to the whole complex plane with a *finite number* of *known* poles with *known* polar part.
3. A second conjecture, which can in fact be considered as part of the first, is that this extended L-function should satisfy a *functional equation* when s is changed into $i + 1 - s$. More precisely, when completed with the Euler factors at the "bad" primes as mentioned (but not explained) above, then if we set

$$\Lambda_i(V; s) = N^{s/2} \prod_{1 \le j \le d_i} \Gamma_{\mathbb{R}}(s + \mu_j) L_i(V; s)$$

then $\Lambda_i(V; i + 1 - s) = \omega \overline{\Lambda_i(V^*; s)}$ for some variety V^* in some sense "dual" to V and a complex number ω of modulus 1. In the above, N is some integer divisible exactly by all the "bad" primes, i.e., essentially (but not exactly) the primes for which V reduced modulo p is not smooth, and the μ_j are in this case (varieties) *integers* which can be computed in terms of the *Hodge numbers* $h^{p,q}$

of the variety thanks to a recipe due to Serre [15]. The number i is called the *motivic weight*, and it is important to note that the "weight" k usually attached to an L-function with functional equation $s \mapsto k - s$ is equal to $k = i + 1$, i.e., to *one more* than the motivic weight.
In many cases, the L-function is self-dual, in which case the functional equation is simply of the form $\Lambda_i(V; i + 1 - s) = \pm\Lambda_i(V; s)$.
4. The function Λ_i should satisfy the generalized Riemann hypothesis (GRH): all its zeros in $\mathbb{C}$ are on the vertical line $\Re(s) = (i + 1)/2$. Equivalently, the zeros of L_i are on the one hand real zeros at some integers coming from the poles of the gamma factors, and all the others satisfy $\Re(s) = (i + 1)/2$.
5. The function Λ_i should have *special values*: for the integer values of s (called special points) which are those for which neither the gamma factor at s nor at $i + 1 - s$ has a pole, it should be computable "explicitly": it should be equal to a *period* (integral of an algebraic function on an algebraic cycle) times an algebraic number. This has been stated (conjecturally) in great detail by Deligne in the 1970s.

It is conjectured that *all* L-functions of degree d_i and weight i as defined at the beginning should satisfy all the above properties, not only the L-functions coming from varieties.

I now give the status of these conjectures.

1. The first conjecture (analytic continuation) is known only for a very restricted class of L-functions: first L-functions of degree 1, which can be shown to be Dirichlet L-functions, L-functions of Hecke characters, L-functions attached to modular forms as shown above, and more generally to *automorphic forms*. For L-functions attached to varieties, one knows this *only* when one can prove that the corresponding L-function comes from an automorphic form: this is how Wiles proves the analytic continuation of the L-function attached to an elliptic curve defined over $\mathbb{Q}$, a very deep and difficult result, with Deligne's proof of the Weil conjectures one of the most important result of the end of the twentieth century. More results of this type are known for certain higher dimensional varieties such as certain *Calabi–Yau manifolds*. Note, however, that for such simple objects as most *Artin L-functions* (degree 0, in which case only *meromorphic* continuation is known) or abelian surfaces, this is not known, although the work of Brumer–Kramer–Poor–Yuen, as well as more recent work of G. Boxer, F. Calegari, T. Gee, and V. Pilloni on the *paramodular conjecture* may someday lead to a proof in this last case.
2. The second conjecture on the existence of a functional equation is of course intimately linked to the first, and the work of Wiles et al. also proves the existence of this functional equation. But in addition, in the case of Artin L-functions for which only meromorphy (possibly with infinitely many poles) is known thanks to a theorem of Brauer, this same theorem implies the functional equation which is thus known in this case. Also, as mentioned, the Euler factors which we must include for the "bad" primes in order to have a clean functional equation are often quite difficult to compute.

3. The (global) Riemann hypothesis is not known for *any* global L-function of the type mentioned above, not even for the simplest one, the Riemann zeta function $\zeta(s)$. Note that it *is* known for other kinds of L-functions such as *Selberg zeta functions*, but these are functions of order 2, so are not in the class considered above.
4. Concerning *special values*: many cases are known, and many conjectured. This is probably one of the most *fun* conjectures since everything can be computed explicitly to thousands of decimals if desired. For instance, for modular forms it is a theorem of Manin, for symmetric squares of modular forms it is a theorem of Rankin, and for higher symmetric powers one has very precise conjectures of Deligne, which check perfectly on a computer, but none of them are proved. For the Riemann zeta function or Dirichlet L-functions, of course all these results such as $\zeta(2) = \pi^2/6$ date back essentially to Euler.
   In the case of an elliptic curve E over $\mathbb{Q}$, the only special point is $s = 1$, and in this case the whole subject revolves around the *Birch and Swinnerton-Dyer conjecture* (BSD) which predicts the behavior of $L_1(E; s)$ around $s = 1$. The only known results, already quite deep, due to Kolyvagin and Gross–Zagier, deal with the case where the *rank* of the elliptic curve is 0 or 1.

There exist a number of other very important conjectures linked to the behavior of L-functions at integer points which are not necessarily special, such as the Bloch, Beilinson, Kato, Lichtenbaum, or Zagier conjectures, but it would carry us too far afield to describe them in general. However, in the next subsections, we will give three completely explicit numerical examples of these conjectures, so that the reader can convince himself both that they are easy to check numerically, and that the results are spectacular.

2.7 *An Explicit Numerical Example of BSD*

Let us now be a little more precise. Even if this subsection involves notions not introduced in these notes, we ask the reader to be patient since the numerical work only involves standard notions.

Let E be an elliptic curve defined over $\mathbb{Q}$. Elliptic curves have a natural *abelian group* structure, and it is a theorem of Mordell that the group of rational points on E is *finitely generated*, i.e., $E(\mathbb{Q}) \simeq \mathbb{Z}^r \oplus E_{\text{tors}}(\mathbb{Q})$, where $E_{\text{tors}}(\mathbb{Q})$ is a finite group, and r is called the *rank* of the curve.

On the analytic side, we have mentioned that E has an L-function $L(E, s)$ (denoted L_1 above), and the deep theorem of Wiles et al. says that it has an analytic continuation to the whole of $\mathbb{C}$ into an entire function with a functional equation linking $L(E, s)$ to $L(E, 2 - s)$. The only special point in the above sense is $s = 1$, and a weak form of the Birch and Swinnerton-Dyer conjecture states that the order of vanishing v of $L(E, s)$ at $s = 1$ should be equal to r.

This has been proved for $r = 0$ (by Kolyvagin) and for $r = 1$ (by Gross–Zagier–Kolyvagin), and nothing is known for $r \geq 2$. However, this is not quite true: if $r = 2$ then we cannot have $v = 0$ or 1 by the previous results, so $v \geq 2$. On the other hand, for any given elliptic curve it is easy to check numerically that $L''(E, 1) \neq 0$, so to check that $v = 2$. Similarly, if $r = 3$ we again cannot have $v = 0$ or 1. But for any given elliptic curve one can compute the *sign* of the functional equation linking $L(E, s)$ to $L(E, 2 - s)$, and this will show that if $r = 3$ all derivatives $L^{(k)}(E, s)$ for k even will vanish. Thus we cannot have $v = 2$, and once again for any E it is easy to check that $L'''(E, 1) \neq 0$, hence to check that $v = 3$.

Unfortunately, this argument does not work for $r \geq 4$. Assume for instance $r = 4$. The same reasoning will show that $L(E, 1) = 0$ (by Kolyvagin), that $L'(E, 1) = L'''(E, 1) = 0$ (because the sign of the functional equation will be $+$), and that $L''''(E, 1) \neq 0$ by direct computation. The BSD conjecture tells us that $L''(E, 1) = 0$, but this is not known for a single curve.

Let us give the simplest numerical example, based on an elliptic curve with $r = 4$. I emphasize that no knowledge of elliptic curves is needed for this.

For every prime p, consider the congruence

$$y^2 + xy \equiv x^3 - x^2 - 79x + 289 \pmod{p} ,$$

and denote by $N(p)$ the number of pairs $(x, y) \in (\mathbb{Z}/p\mathbb{Z})^2$ satisfying it. We define an arithmetic function $a(n)$ in the following way:

1. $a(1) = 1$.
2. If p is prime, we set $a(p) = p - N(p)$.
3. For $k \geq 2$ and p is prime, we define $a(p^k)$ by induction:

$$a(p^k) = a(p)a(p^{k-1}) - \chi(p)p \cdot a(p^{k-2}) ,$$

   where $\chi(p) = 1$ unless $p = 2$ or $p = 117223$, in which case $\chi(p) = 0$.
4. For arbitrary n, we extend by multiplicativity: if $n = \prod_i p_i^{k_i}$ then $a(n) = \prod_i a(p_i^{k_1})$.

Remarks 2.6 • The number 117223 is simply a prime factor of the discriminant of the cubic equation obtained by completing the square in the equation of the above elliptic curve.

• Even though the definition of $a(n)$ looks complicated, it is *very* easy to compute (see below), for instance, only a few seconds for a million terms. In addition $a(n)$ is quite small: for $n = 1, 2, \ldots$ we have

$$a(n) = 1, -1, -3, 1, -4, 3, -5, -1, 6, 4, -6, -3, -6, 5, \ldots$$

On the analytic side, define a function $f(x)$ for $x > 0$ by

$$f(x) = \int_1^{\infty} e^{-xt} \log(t)^2 \, dt \,.$$

Note that it is very easy to compute this integral to thousands of digits if desired and also note that f tends to 0 exponentially fast as $x \to \infty$ (more precisely $f(x) \sim 2e^{-x}/x^3$).

In this specific situation, the BSD conjecture tells us that $S = 0$, where

$$S = \sum_{n \geq 1} a(n) f\left(\frac{2\pi n}{\sqrt{234446}}\right) .$$

It takes only a few seconds to compute *thousands* of digits of S, and we can indeed check that S is extremely close to 0, but as of now nobody knows how to prove that $S = 0$.

2.8 An Explicit Numerical Example of Beilinson–Bloch

This subsection is entirely due to V. Golyshev (personal communication) whom I heartily thank.

Let $u > 1$ be a real parameter. Consider the elliptic curve $E(u)$ with affine equation

$$y^2 = x(x+1)(x+u^2) \,.$$

As usual one can define its L-function $L(E(u), s)$ using a general recipe. The BSD conjecture deals with the value of $L(E(u), s)$ (and its derivatives) at $s = 1$. The Beilinson–Bloch conjectures deal with values at other integer values of s, in the present case we consider $L(E(u), 2)$. Once again it is very easy to compute thousands of decimals of this quantity if desired.

On the other hand, for $u > 1$ consider the function

$$g(u) = 2\pi \int_0^1 \frac{\operatorname{asin}(t)}{\sqrt{1 - t^2/u^2}} \frac{dt}{t} + \pi^2 \operatorname{acosh}(u) = \frac{\pi^2}{2} \left(2\log(4u) - \sum_{n \geq 1} \frac{\binom{2n}{n}^2}{n} (4u)^{-2n} \right) .$$

The conjecture says that when u is an integer, $L(E(u), 2)/g(u)$ should be a *rational number*. In fact, if we let $N(u)$ be the *conductor* of $E(u)$ (notion that I have not defined), then it seems that when $u \neq 4$ and $u \neq 8$ we even have $F(u) = N(u)L(E(u), 2)/g(u) \in \mathbb{Z}$.

Once again, this is a conjecture which can immediately be tested on modern computer algebra systems such as `Pari/GP`. For instance, for $u = 2, 3, \ldots$ we find *numerically* to thousands of decimal digits (remember that nothing is proved)

$$F(u) = 1, 2, 4/11, 8, 32, 8, 4/3, 8, 32, 64, 8, 96, 256, 48, 16, 16, 192, \ldots$$

Exercise 2.7 Check numerically that the conjecture seems still to be true when $4u \in \mathbb{Z}$, i.e., if u is a rational number with denominator 2 or 4. On the other hand, it is definitely wrong, for instance, if $3u \in \mathbb{Z}$ (and $u \notin \mathbb{Z}$), i.e., when the denominator is 3. It is possible that there is a replacement formula, but Bloch and Golyshev tell me that this is unlikely.

2.9 An Explicit Numerical Example of Mahler Measures

This example is entirely due to W. Zudilin (personal communication), whom I heartily thank. The reader does not need any knowledge of Mahler measures since we are again going to give the example as an equality between values of L-functions and integrals. Note that this can also be considered an isolated example of the Bloch–Beilinson conjecture.

Consider the elliptic curve E with equation $y^2 = x^3 - x^2 - 4x + 4$, of conductor 24. Its associated L-function $L(E, s)$ can easily be shown to be equal to the L-function associated to the modular form

$$q \prod_{n \geq 1} (1 - q^{2n})(1 - q^{4n})(1 - q^{6n})(1 - q^{12n})$$

(we do not need this for this example, but this will give us two ways to create the L-function in `Pari/GP`). We have the conjectural identity due to Zudilin:

$$L(E, 3) = \frac{\pi^2}{36} \left(\pi G + \int_0^1 \operatorname{asin}(x) \operatorname{asin}(1 - x) \frac{dx}{x} \right),$$

where $G = \sum_{n \geq 0} (-1)^n/(2n + 1)^2 = 0.91596559 \cdots$ is Catalan's constant.

At the end of this course, the reader will find three complete `Pari/GP` scripts which implement the BSD, Beilinson–Bloch, and Mahler measure examples that we have just given.

2.10 Computational Goals

Now that we have a handle on what L-functions are, we come to the computational and algorithmic problems, which are the main focus of these notes. This involves many different aspects, all interesting in their own right.

In a first type of situation, we assume that we are "given" the L-function, in other words that we are given a reasonably "efficient" algorithm to compute the coefficients $a(n)$ of the Dirichlet series (or the Euler factors), and that we know the gamma factor $\gamma(s)$. The main computational goals are then the following:

1. Compute $L(s)$ for "reasonable" values of s: for example, compute $\zeta(3)$. More sophisticated, but much more interesting: check the Birch–Swinnerton-Dyer conjecture, the Beilinson–Bloch conjecture, and the conjectures of Deligne concerning special values of symmetric powers L-functions of modular forms.
2. Check the numerical validity of the functional equation, and in passing, if unknown, compute the numerical value of the *root number* ω occurring in the functional equation.
3. Compute $L(s)$ for $s = 1/2 + it$ for rather large real values of t (in the case of weight 0, more generally for $s = (w + 1)/2 + it$), and/or make a plot of the corresponding Z function (see below).
4. Compute all the zeros of $L(s)$ on the critical line up to a given height, and check the corresponding Riemann hypothesis.
5. Compute the residue of $L(s)$ at $s = 1$ (typically): for instance, if L is the Dedekind zeta function of a number field, this gives the product hR.
6. Compute the *order* of the zeros of $L(s)$ at integer points (if it has one), and the leading term in the Taylor expansion: for instance, for the L-function of an elliptic curve and $s = 1$, this gives the *analytic rank* of an elliptic curve, together with the Birch and Swinnerton-Dyer data.

Unfortunately, we are not always given an L-function completely explicitly. We can lack more or less partial information on the L-function:

1. One of the most frequent situations is that one knows the Euler factors for the "good" primes, as well as the corresponding part of the conductor, and that one is lacking both the Euler factors for the bad primes and the bad part of the conductor. The goal is then to find numerically the missing factors and missing parts.
2. A more difficult but much more interesting problem is when essentially nothing is known on the L-function except $\gamma(s)$, in other words the $\Gamma_{\mathbb{R}}$ factors and the constant N, essentially equal to the conductor. It is quite amazing that nonetheless one can quite often tell whether an L-function with the given data can exist, and give some of the initial Dirichlet coefficients (even when several L-functions may be possible).
3. Even more difficult is when essentially nothing is known except the degree d and the constant N, and one looks for possible $\Gamma_{\mathbb{R}}$ factors: this is the case in the search for Maass forms over $\mathrm{SL}_n(\mathbb{Z})$, which has been conducted very successfully for $n = 2$, 3, and 4.

We will not consider these more difficult problems.

2.11 Available Software for L-Functions

Many people working on the subject have their own software. I mention the available public data.

• M. Rubinstein's `C++` program `lcalc`, which can compute values of L-functions, make large tables of zeros, and so on. The program uses `C++` language `double`, so is limited to 15 decimal digits, but is highly optimized, hence very fast, and used in most situations. Also optimized for large values of the imaginary part using Riemann–Siegel. Available in `Sage`.

• T. Dokchitser's program `computel`, initially written in `GP/Pari`, rewritten for `magma`, and also available in `Sage`. Similar to Rubinstein's, but allows arbitrary precision, hence slower, and has no built-in zero finder, although this is not too difficult to write. It is not optimized for large imaginary parts.

• Since June 2015, `Pari/GP` has a complete package for computing with L-functions, written by B. Allombert, K. Belabas, P. Molin, and myself, based on the ideas of T. Dokchitser for the computation of inverse Mellin transforms (see below) but put on a more solid footing, and on the ideas of P. Molin for computing the L-function values themselves, which avoid computing generalized incomplete gamma functions (see also below). Note the related complete `Pari/GP` package for computing with modular forms, available since July 2018.

• Last but not least, not a program but a huge *database* of L-functions, modular forms, number fields, etc., which is the result of a collaborative effort of approximately 30–40 people headed by D. Farmer. This database can, of course, be queried in many different ways, it is possible and useful to navigate between related pages, and it also contains `knowls`, bits of knowledge which give the main definitions. In addition to the stored data, the site can compute additional required information on the fly using the software mentioned above, i.e., `Pari`, `Sage`, `magma`, and `lcalc`) Available at:

`http://www.lmfdb.org`

3 Arithmetic Methods: Computing $a(n)$

We now come to the second part of this course: the computation of the Dirichlet series coefficients $a(n)$ and/or of the Euler factors, which is usually the same problem. Of course, this depends entirely on how the L-function is *given*: in view of what we have seen, it can be given, for instance, (but not only) as the L-function attached to a modular form, to a variety, or to a hypergeometric motive. Since there are so many relations between these L-functions (we have seen several identities above), we will not separate the way in which they are given, but treat everything at once.

In view of the preceding section, an important computational problem is the computation of $|V(\mathbb{F}_q)|$ for a variety V. This may, of course, be done by a naïve point count: if V is defined by polynomials in n variables, we can range through the q^n possibilities for the n variables and count the number of common zeros. In other words, there always exists a trivial algorithm requiring q^n steps. We, of course, want something better.

3.1 General Elliptic Curves

Let us first look at the special case of *elliptic curves*, i.e., a projective curve V with affine equation $y^2 = x^3 + ax + b$ such that $p \nmid 6(4a^3 + 27b^2)$, which is almost the general equation for an *elliptic curve*. For simplicity assume that $q = p$, but it is immediate to generalize. If you know the definition of the Legendre symbol, you know that the number of solutions in $\mathbb{F}_p$ to the equation $y^2 = n$ is equal to $1 + \left(\frac{n}{p}\right)$. If you do not, since $\mathbb{F}_p$ is a field, it is clear that this number is equal to 0, 1, or 2, and so one can *define* $\left(\frac{n}{p}\right)$ as one less, so -1, 0, or 1. Thus, since it is immediate to see that there is a single projective point at infinity, we have

$$|V(\mathbb{F}_p)| = 1 + \sum_{x \in \mathbb{F}_p} \left(1 + \left(\frac{x^3 + ax + b}{p}\right)\right) = p + 1 - a(p)\;, \quad \text{with}$$

$$a(p) = -\sum_{0 \le x \le p-1} \left(\frac{x^3 + ax + b}{p}\right)\;.$$

Now a Legendre symbol can be computed very efficiently using the *quadratic reciprocity law*. Thus, considering that it can be computed in constant time (which is not quite true but almost), this gives a $O(p)$ algorithm for computing $a(p)$, already much faster than the trivial $O(p^2)$ algorithm consisting in looking at all pairs (x, y).

To do better, we have to use an additional and crucial property of an elliptic curve: it is an *abelian group*. Using this combined with the so-called Hasse bounds $|a(p)| < 2\sqrt{p}$ (a special case of the Weil conjectures), and the so-called *baby-step giant-step algorithm* due to Shanks, one can obtain a $O(p^{1/4})$ algorithm, which is very fast for all practical purposes.

However, a remarkable discovery due to Schoof in the early 1980s is that there exists a practical algorithm for computing $a(p)$ which is *polynomial in* $\log(p)$, for instance, $O(\log^6(p))$. The idea is to compute $a(p)$ modulo ℓ for small primes ℓ using *ℓ-division polynomials*, and then use the Chinese remainder theorem and the bound $|a(p)| < 2\sqrt{p}$ to recover $a(p)$. Several important improvements have been made on this basic algorithm, in particular by Atkin and Elkies, and the resulting SEA algorithm (which is implemented in many computer packages) is able to compute $a(p)$ for p with several thousand decimal digits. Note, however, that in practical ranges (say $p < 10^{12}$), the $O(p^{1/4})$ algorithm mentioned above is sufficient.

3.2 Elliptic Curves with Complex Multiplication

In certain special cases, it is possible to compute $|V(\mathbb{F}_q)|$ for an elliptic curve V much faster than with any of the above methods: when the elliptic curve V has *complex multiplication*. Let us consider the special cases $y^2 = x^3 - nx$ (the general case is more complicated but not really slower). By the general formula for $a(p)$, we have for $p \geq 3$:

$$\begin{aligned}
a(p) &= -\sum_{-(p-1)/2 \leq x \leq (p-1)/2} \left(\frac{x(x^2-n)}{p}\right) \\
&= -\sum_{1 \leq x \leq (p-1)/2} \left(\left(\frac{x(x^2-n)}{p}\right) + \left(\frac{-x(x^2-n)}{p}\right)\right) \\
&= -\left(1 + \left(\frac{-1}{p}\right)\right) \sum_{1 \leq x \leq (p-1)/2} \left(\frac{x(x^2-n)}{p}\right)
\end{aligned}$$

by the multiplicative property of the Legendre symbol. This already shows that if $\left(\frac{-1}{p}\right) = -1$, in other words $p \equiv 3 \pmod 4$, we have $a(p) = 0$. But we can also find a formula when $p \equiv 1 \pmod 4$: recall that in that case by a famous theorem due to Fermat, there exist integers u and v such that $p = u^2 + v^2$. If necessary by exchanging u and v, and/or changing the sign of u, we may assume that $u \equiv -1 \pmod 4$, in which case the decomposition is unique, up to the sign of v. It is then not difficult to prove the following theorem (see Sect. 8.5.2 of [3] for the proof):

Theorem 3.1 *Assume that* $p \equiv 1 \pmod 4$ *and* $p = u^2 + v^2$ *with* $u \equiv -1 \pmod 4$. *The number of projective points on the elliptic curve* $y^2 = x^3 - nx$ *(where* $p \nmid n$*) is equal to* $p + 1 - a(p)$*, where*

$$a(p) = 2\left(\frac{2}{p}\right) \begin{cases} -u & \text{if } n^{(p-1)/4} \equiv 1 \pmod p \\ u & \text{if } n^{(p-1)/4} \equiv -1 \pmod p \\ -v & \text{if } n^{(p-1)/4} \equiv -u/v \pmod p \\ v & \text{if } n^{(p-1)/4} \equiv u/v \pmod p \end{cases}$$

(note that one of these four cases must occur).

To apply this theorem from a computational standpoint, we note the following two *facts*:

(1) The quantity $n^{(p-1)/4} \bmod p$ can be computed efficiently by the *binary powering algorithm* (in $O(\log^3(p))$ operations). It is, however, possible to compute it more efficiently in $O(\log^2(p))$ operations using the *quartic reciprocity law*.

(2) The numbers u and v, such that $u^2 + v^2 = p$, can be computed efficiently (in $O(\log^2(p))$ operations) using *Cornacchia's algorithm* which is very easy to describe but not so easy to prove. It is a variant of Euclid's algorithm. It proceeds as follows:

• As a first step, we compute a square root of -1 modulo p, i.e., an x such that $x^2 \equiv -1 \pmod p$. This is done by choosing randomly a $z \in [1, p-1]$ and computing the Legendre symbol $\left(\frac{z}{p}\right)$ until it is equal to -1 (we can also simply try $z = 2, 3, \ldots$). Note that this is a fast computation. When this is the case, we have by definition

$z^{(p-1)/2} \equiv -1 \pmod{p}$, hence $x^2 \equiv -1 \pmod{p}$ for $x = z^{(p-1)/4} \bmod p$. Reducing x modulo p and possibly changing x into $p - x$, we normalize x so that $p/2 < x < p$.

• As a second step, we perform the Euclidean algorithm on the pair (p, x), writing $a_0 = p, a_1 = x$, and $a_{n-1} = q_n a_n + a_{n+1}$ with $0 \le a_{n+1} < a_n$, and we stop at the exact n for which $a_n^2 < p$. It can be proved (this is the difficult part) that for this specific n we have $a_n^2 + a_{n+1}^2 = p$, so up to exchange of u and v and/or change of signs, we can take $u = a_n$ and $v = a_{n+1}$.

Note that Cornacchia's algorithm can easily be generalized to solving efficiently $u^2 + dv^2 = p$ or $u^2 + dv^2 = 4p$ for any $d \ge 1$, see Sect. 1.5.2 of [2] (incidentally one can also solve this for $d < 0$, but it poses completely different problems since there may be infinitely many solutions).

The above theorem is given for the special elliptic curves $y^2 = x^3 - nx$, which have complex multiplication by the (ring of integers of the) field $\mathbb{Q}(i)$, but a similar theorem is valid for all curves with complex multiplication, see Sect. 8.5.2 of [3].

3.3 *Using Modular Forms of Weight* 2

By Wiles' celebrated theorem, the L-function of an elliptic curve is equal to the L-function of a modular form of weight 2 for $\Gamma_0(N)$, where N is the conductor of the curve. We do not need to give the precise definitions of these objects, but only a specific example.

Let V be the elliptic curve with affine equation $y^2 + y = x^3 - x^2$. It has conductor 11. It can be shown using classical modular form methods (i.e., without Wiles' theorem) that the global L-function $L(V; s) = \sum_{n \ge 1} a(n)/n^s$ is the same as that of the modular form of weight 2 over $\Gamma_0(11)$ given by

$$f(\tau) = q \prod_{m \ge 1} (1 - q^m)^2 (1 - q^{11m})^2 ,$$

with $q = \exp(2\pi i \tau)$. Even with no knowledge of modular forms, this simply means that if we formally expand the product on the right-hand side as

$$q \prod_{m \ge 1} (1 - q^m)^2 (1 - q^{11m})^2 = \sum_{n \ge 1} b(n) q^n ,$$

we have $b(n) = a(n)$ for all n, and in particular for $n = p$ prime. We have already seen this example above with a slightly different equation for the elliptic curve (which makes no difference for its L-function outside of the primes 2 and 3).

We see that this gives an alternate method for computing $a(p)$ by expanding the infinite product. Indeed, the function

$$\eta(\tau) = q^{1/24} \prod_{m\geq 1}(1 - q^m)$$

is a modular form of weight $1/2$ with known expansion:

$$\eta(\tau) = \sum_{n\geq 1} \left(\frac{12}{n}\right) q^{n^2/24} ,$$

and so using Fast Fourier Transform techniques for formal power series multiplication we can compute all the coefficients $a(n)$ simultaneously (as opposed to one by one) for $n \leq B$ in time $O(B \log^2(B))$. This amounts to computing each individual $a(n)$ in time $O(\log^2(n))$, so it seems to be competitive with the fast methods for elliptic curves with complex multiplication, but this is an illusion since we must store all B coefficients, so it can be used only for $B \leq 10^{12}$, say, far smaller than what can be reached using Schoof's algorithm, which is truly polynomial in $\log(p)$ for each fixed prime p.

3.4 *Higher Weight Modular Forms*

It is interesting to note that the dichotomy between elliptic curves with or without complex multiplication is also valid for modular forms of higher weight (again, whatever that means, you do not need to know the definitions). For instance, consider

$$\Delta(\tau) = \Delta_{24}(\tau) = \eta^{24}(\tau) = q \prod_{m\geq 1}(1 - q^m)^{24} := \sum_{n\geq 1} \tau(n) q^n .$$

The function $\tau(n)$ is a famous function called the *Ramanujan* τ *function*, and has many important properties, analogous to those of the $a(p)$ attached to an elliptic curve (i.e., to a modular form of weight 2).

There are several methods to compute $\tau(p)$ for p prime, say. One is to do as above, using FFT techniques. The running time is similar, but again we are limited to $B \leq 10^{12}$, say. A second more sophisticated method is to use the *Eichler–Selberg trace formula*, which enables the computation of an individual $\tau(p)$ in time $O(p^{1/2+\varepsilon})$ for all $\varepsilon > 0$. A third very deep method, developed by Edixhoven, Couveignes, et al., is a generalization of Schoof's algorithm. While in principle polynomial time in $\log(p)$, it is not yet practical compared to the preceding method.

For those who want to see the formula using the trace formula explicitly, we let $H(N)$ be the *Hurwitz class number* $H(N)$ (essentially the class number of imaginary quadratic orders counted with suitable multiplicity): if we set $H_3(N) = H(4N) + 2H(N)$ (note that $H(4N)$ can be computed in terms of $H(N)$), then for p prime

$$\tau(p) = 28p^6 - 28p^5 - 90p^4 - 35p^3 - 1$$
$$- 128 \sum_{1\le t<p^{1/2}} t^6(4t^4 - 9pt^2 + 7p^2)H_3(p - t^2) ,$$

which is the fastest *practical* formula that I know for computing $\tau(p)$.

On the contrary, consider

$$\Delta_{26}(\tau) = \eta^{26}(\tau) = q^{13/12} \prod_{m\ge1}(1 - q^m)^{26} := q^{13/12} \sum_{n\ge1} \tau_{26}(n)q^n .$$

This is what is called a modular form with complex multiplication. Whatever the definition, this means that the coefficients $\tau_{26}(p)$ can be computed in time polynomial in $\log(p)$ using a generalization of Cornacchia's algorithm, hence very fast.

Exercise 3.2 (*You need some extra knowledge for this*) In the literature, find an exact formula for $\tau_{26}(p)$ in terms of values of Hecke *Grössencharacters*, and program this formula. Use it to compute some values of $\tau_{26}(p)$ for p prime as large as you can go.

3.5 *Computing* $|V(\mathbb{F}_q)|$ *for Quasi-diagonal Hypersurfaces*

We now consider a completely different situation, where $|V(\mathbb{F}_q)|$ can be computed without too much difficulty.

As we have seen, in the case of elliptic curves V defined over $\mathbb{Q}$, the corresponding L-function is of *degree* 2, in other words is of the form $\prod_p 1/(1 - a(p)p^{-s} + b(p)p^{-2s})$, where $b(p) \neq 0$ for all but a finite number of p. L-functions of degree 1 such as the Riemann zeta function are essentially L-functions of Dirichlet characters, in other words simple "twists" of the Riemann zeta function. L-functions of degree 2 are believed to be always L-functions attached to modular forms, and $b(p) = \chi(p)p^{k-1}$ for a suitable integer k ($k = 2$ for elliptic curves), the *weight* (note that this is *one more* than the so-called *motivic weight*). Even though many unsolved questions remain, this case is also quite well understood. Much more mysterious are L-functions of higher degree, such as 3 or 4, and it is interesting to study natural mathematical objects leading to such functions. A case where this can be done reasonably easily is the case of diagonal or *quasi-diagonal hypersurfaces*. We study a special case:

Definition 3.3 Let $m \ge 2$, for $1 \le i \le m$ let $a_i \in \mathbb{F}_q^*$ be nonzero, and let $b \in \mathbb{F}_q$. The quasi-diagonal hypersurface defined by this data is the hypersurface in $\mathbb{P}^{m-1}$ defined by the projective equation

$$\sum_{1\le i\le m} a_i x_i^m - b \prod_{1\le i\le m} x_i = 0 .$$

When $b = 0$, it is a diagonal hypersurface.

Of course, we could study more general equations, for instance, where the degree is not equal to the number of variables, but we stick to this special case.

To compute the number of (projective) points on this hypersurface, we need an additional definition:

Definition 3.4 We let ω be a generator of the group of characters of $\mathbb{F}_q^*$, either with values in $\mathbb{C}$, or in the p-adic field $\mathbb{C}_p$ (do not worry if you are not familiar with this).

Indeed, by a well-known theorem of elementary algebra, the multiplicative group $\mathbb{F}_q^*$ of a finite field is *cyclic*, so its group of characters, which is *non-canonically isomorphic* to $\mathbb{F}_q^*$, is also cyclic, so ω indeed exists.

It is not difficult to prove the following theorem:

Theorem 3.5 *Assume that* $\gcd(m, q-1) = 1$ *and* $b \neq 0$, *and set* $B = \prod_{1 \le i \le m} (a_i/b)$. *If* V *is the above quasi-diagonal hypersurface, the number* $|V(\mathbb{F}_q)|$ *of* affine *points on* V *is given by*

$$|V(\mathbb{F}_q)| = q^{m-1} + (-1)^{m-1} + \sum_{1 \le n \le q-2} \omega^{-n}(B) J_m(\omega^n, \dots, \omega^n) ,$$

where J_m *is the m-variable Jacobi sum.*

We will study in great detail below the definition and properties of J_m.

Note that the number of *projective* points is simply $(|V(\mathbb{F}_q)| - 1)/(q-1)$.

There also exists a more general theorem with no restriction on $\gcd(m, q-1)$, which we do not give.

The occurrence of Jacobi sums is very natural and frequent in point counting results. It is, therefore, important to look at efficient ways to compute them, and this is what we do in the next section, where we also give complete definitions and basic results.

4 Gauss and Jacobi Sums

In this long section, we study in great detail Gauss and Jacobi sums. Most results are standard, and I would like to emphasize that almost all of them can be proved with little difficulty by easy algebraic manipulations.

4.1 Gauss Sums over $\mathbb{F}_q$

We can define and study Gauss and Jacobi sums in two different contexts: first, and most importantly, over finite fields $\mathbb{F}_q$, with $q = p^f$ a prime power (note that from now on we write $q = p^f$ and not $q = p^n$). Second, over the ring $\mathbb{Z}/N\mathbb{Z}$. The two notions coincide when $N = q = p$ is prime, but the methods and applications are quite different.

To give the definitions over $\mathbb{F}_q$ we need to recall some fundamental (and easy) results concerning finite fields.

Proposition 4.1 *Let p be a prime, $f \geq 1$, and $\mathbb{F}_q$ be the finite field with $q = p^f$ elements, which exists and is unique up to isomorphism.*

1. *The map ϕ such that $\phi(x) = x^p$ is a field isomorphism from $\mathbb{F}_q$ to itself leaving $\mathbb{F}_p$ fixed. It is called the* Frobenius map.
2. *The extension $\mathbb{F}_q/\mathbb{F}_p$ is a* normal *(i.e., separable and Galois) field extension, with Galois group which is cyclic of order f generated by ϕ.*

In particular, we can define the *trace* $\mathrm{Tr}_{\mathbb{F}_q/\mathbb{F}_p}$ and the *norm* $\mathscr{N}_{\mathbb{F}_q/\mathbb{F}_p}$, and we have the formulas (where from now on we omit $\mathbb{F}_q/\mathbb{F}_p$ for simplicity):

$$\mathrm{Tr}(x) = \sum_{0 \leq j \leq f-1} x^{p^j} \quad \text{and} \quad \mathscr{N}(x) = \prod_{0 \leq j \leq f-1} x^{p^j} = x^{(p^f-1)/(p-1)} = x^{(q-1)/(p-1)} .$$

Definition 4.2 Let χ be a character from $\mathbb{F}_q^*$ to an algebraically closed field C of characteristic 0. For $a \in \mathbb{F}_q$ we define the *Gauss sum* $\mathfrak{g}(\chi, a)$ by

$$\mathfrak{g}(\chi, a) = \sum_{x \in \mathbb{F}_q^*} \chi(x) \zeta_p^{\mathrm{Tr}(ax)} ,$$

where ζ_p is a fixed primitive pth root of unity in C. We also set $\mathfrak{g}(\chi) = \mathfrak{g}(\chi, 1)$.

Note that strictly speaking this definition depends on the choice of ζ_p. However, if ζ_p' is some other primitive pth root of unity we have $\zeta_p' = \zeta_p^k$ for some $k \in \mathbb{F}_p^*$, so

$$\sum_{x \in \mathbb{F}_q^*} \chi(x) \zeta_p'^{\mathrm{Tr}(ax)} = \mathfrak{g}(\chi, ka) .$$

In fact, it is trivial to see (this follows from the next proposition) that $\mathfrak{g}(\chi, ka) = \chi^{-1}(k)\mathfrak{g}(\chi, a)$.

Definition 4.3 We define ε to be the trivial character, i.e., such that $\varepsilon(x) = 1$ for all $x \in \mathbb{F}_q^*$. We extend characters χ to the whole of $\mathbb{F}_q$ by setting $\chi(0) = 0$ if $\chi \neq \varepsilon$ and $\varepsilon(0) = 1$.

Note that this apparently innocuous definition of $\varepsilon(0)$ is *crucial* because it simplifies many formulas. Note also that the definition of $\mathfrak{g}(\chi, a)$ is a sum over $x \in \mathbb{F}_q^*$ and not $x \in \mathbb{F}_q$, while for Jacobi sums we will use all of $\mathbb{F}_q$.

Exercise 4.4 1. Show that $\mathfrak{g}(\varepsilon, a) = -1$ if $a \in \mathbb{F}_q^*$ and $\mathfrak{g}(\varepsilon, 0) = q - 1$.

2. If $\chi \neq \varepsilon$, show that $\mathfrak{g}(\chi, 0) = 0$, in other words that

$$\sum_{x \in \mathbb{F}_q} \chi(x) = 0$$

(here it does not matter if we sum over $\mathbb{F}_q$ or $\mathbb{F}_q^*$).

3. Deduce that if $\chi_1 \neq \chi_2$ then

$$\sum_{x \in \mathbb{F}_q^*} \chi_1(x)\chi_2^{-1}(x) = 0 .$$

This relation is called for evident reasons *orthogonality of characters.*

4. Dually, show that if $x \neq 0, 1$ we have $\sum_\chi \chi(x) = 0$, where the sum is over all characters of $\mathbb{F}_q^*$.

Because of this exercise, if necessary we may assume that $\chi \neq \varepsilon$ and/or that $a \neq 0$.

Exercise 4.5 Let χ be a character of $\mathbb{F}_q^*$ of exact order n.

1. Show that $n \mid (q-1)$ and that $\chi(-1) = (-1)^{(q-1)/n}$. In particular, if n is odd and $p > 2$ we have $\chi(-1) = 1$.
2. Show that $\mathfrak{g}(\chi, a) \in \mathbb{Z}[\zeta_n, \zeta_p]$, where as usual ζ_m denotes a primitive mth root of unity.

Proposition 4.6 *1. If $a \neq 0$ we have*

$$\mathfrak{g}(\chi, a) = \chi^{-1}(a)\mathfrak{g}(\chi) .$$

2. We have

$$\mathfrak{g}(\chi^{-1}) = \chi(-1)\overline{\mathfrak{g}(\chi)} .$$

3. We have

$$\mathfrak{g}(\chi^p, a) = \chi^{1-p}(a)\mathfrak{g}(\chi, a) .$$

4. If $\chi \neq \varepsilon$ we have

$$|\mathfrak{g}(\chi)| = q^{1/2} .$$

4.2 *Jacobi Sums over* $\mathbb{F}_q$

Recall that we have extended characters of $\mathbb{F}_q^*$ by setting $\chi(0) = 0$ if $\chi \neq \varepsilon$ and $\varepsilon(0) = 1$.

Definition 4.7 For $1 \leq j \leq k$ let χ_j be characters of $\mathbb{F}_q^*$. We define the Jacobi sum

$$J_k(\chi_1, \ldots, \chi_k; a) = \sum_{x_1+\cdots+x_k=a} \chi_1(x_1) \cdots \chi_k(x_k)$$

and $J_k(\chi_1, \ldots, \chi_k) = J_k(\chi_1, \ldots, \chi_k; 1)$.

Note that, as mentioned above, we do not exclude the cases where some $x_i = 0$, using the convention of Definition 4.3 for $\chi(0)$.

The following easy lemma shows that it is only necessary to study $J_k(\chi_1, \ldots, \chi_k)$:

Lemma 4.8 *Set* $\chi = \chi_1 \cdots \chi_k$.

1. *If* $a \neq 0$ *we have*

$$J_k(\chi_1, \ldots, \chi_k; a) = \chi(a) J_k(\chi_1, \ldots, \chi_k) .$$

2. *If* $a = 0$, *abbreviating* $J_k(\chi_1, \ldots, \chi_k; 0)$ *to* $J_k(0)$ *we have*

$$J_k(0) = \begin{cases} q^{k-1} & \text{if } \chi_j = \varepsilon \text{ for all } j , \\ 0 & \text{if } \chi \neq \varepsilon , \\ \chi_k(-1)(q-1)J_{k-1}(\chi_1, \ldots, \chi_{k-1}) & \text{if } \chi = \varepsilon \text{ and } \chi_k \neq \varepsilon . \end{cases}$$

As we have seen, a Gauss sum $\mathfrak{g}(\chi)$ belongs to the rather large ring $\mathbb{Z}[\zeta_{q-1}, \zeta_p]$ (and in general not to a smaller ring). The advantage of Jacobi sums is that they belong to the smaller ring $\mathbb{Z}[\zeta_{q-1}]$, and as we are going to see, that they are closely related to Gauss sums. Thus, when working *algebraically*, it is almost always better to use Jacobi sums instead of Gauss sums. On the other hand, when working *analytically* (for instance in $\mathbb{C}$ or $\mathbb{C}_p$), it may be better to work with Gauss sums: we will see below the use of root numbers (suggested by Louboutin), and of the Gross–Koblitz formula.

Note that $J_1(\chi_1) = 1$. Outside of this trivial case, the close link between Gauss and Jacobi sums is given by the following easy proposition, whose apparently technical statement is only due to the trivial character ε: if none of the χ_j nor their product is trivial, we have the simple formula given by (3).

Proposition 4.9 *Denote by t the number of χ_j equal to the trivial character ε, and as above set $\chi = \chi_1 \ldots \chi_k$.*

1. *If* $t = k$ *then* $J_k(\chi_1, \ldots, \chi_k) = q^{k-1}$.
2. *If* $1 \leq t \leq k-1$ *then* $J_k(\chi_1, \ldots, \chi_k) = 0$.
3. *If* $t = 0$ *and* $\chi \neq \varepsilon$ *then*

$$J_k(\chi_1, \ldots, \chi_k) = \frac{\mathfrak{g}(\chi_1) \cdots \mathfrak{g}(\chi_k)}{\mathfrak{g}(\chi_1 \cdots \chi_k)} = \frac{\mathfrak{g}(\chi_1) \cdots \mathfrak{g}(\chi_k)}{\mathfrak{g}(\chi)} .$$

4. *If $t = 0$ and $\chi = \varepsilon$ then*

$$J_k(\chi_1, \dots, \chi_k) = -\frac{\mathfrak{g}(\chi_1)\cdots\mathfrak{g}(\chi_k)}{q}$$
$$= -\chi_k(-1)\frac{\mathfrak{g}(\chi_1)\cdots\mathfrak{g}(\chi_{k-1})}{\mathfrak{g}(\chi_1\cdots\chi_{k-1})} = -\chi_k(-1)J_{k-1}(\chi_1, \dots, \chi_{k-1}) .$$

In particular, in this case we have

$$\mathfrak{g}(\chi_1)\cdots\mathfrak{g}(\chi_k) = \chi_k(-1)qJ_{k-1}(\chi_1, \dots, \chi_{k-1}) .$$

Corollary 4.10 *With the same notation, assume that $k \geq 2$ and all the χ_j are nontrivial. Setting $\psi = \chi_1 \cdots \chi_{k-1}$, we have the following recursive formula:*

$$J_k(\chi_1, \dots, \chi_k) = \begin{cases} J_{k-1}(\chi_1, \dots, \chi_{k-1})J_2(\psi, \chi_k) & \text{if } \psi \neq \varepsilon , \\ \chi_{k-1}(-1)qJ_{k-2}(\chi_1, \dots, \chi_{k-2}) & \text{if } \psi = \varepsilon . \end{cases}$$

The point of this recursion is that the definition of a k-fold Jacobi sum J_k involves a sum over q^{k-1} values for $x_1, \dots, x_{k-1}$, the last variable x_k being determined by $x_k = 1 - x_1 - \cdots - x_{k-1}$, so neglecting the time to compute the $\chi_j(x_j)$ and their product (which is a reasonable assumption), using the definition takes time $O(q^{k-1})$. On the other hand, using the above recursion boils down at worst to computing $k - 1$ Jacobi sums J_2, for a total time of $O((k - 1)q)$. Nonetheless, we will see that in some cases it is still better to use directly Gauss sums and formula (3) of the proposition.

Since Jacobi sums J_2 are the simplest and the above recursion in fact shows that one can reduce to J_2, we will drop the subscript 2 and simply write $J(\chi_1, \chi_2)$. Note that

$$J(\chi_1, \chi_2) = \sum_{x \in \mathbb{F}_q} \chi_1(x)\chi_2(1 - x) ,$$

where the sum is over the whole of $\mathbb{F}_q$ and *not* $\mathbb{F}_q \setminus \{0, 1\}$ (which makes a difference only if one of the χ_i is trivial). More precisely it is clear that $J(\varepsilon, \varepsilon) = q^2$, and that if $\chi \neq \varepsilon$ we have $J(\chi, \varepsilon) = \sum_{x \in \mathbb{F}_q} \chi(x) = 0$, which are special cases of Proposition 4.9.

Exercise 4.11 Let $n \mid (q - 1)$ be the order of χ. Prove that $\mathfrak{g}(\chi)^n \in \mathbb{Z}[\zeta_n]$.

Exercise 4.12 Assume that none of the χ_j is equal to ε, but that their product χ is equal to ε. Prove that (using the same notation as in Lemma 4.8):

$$J_k(0) = \left(1 - \frac{1}{q}\right)\mathfrak{g}(\chi_1)\cdots\mathfrak{g}(\chi_k) .$$

Exercise 4.13 Prove the following reciprocity formula for Jacobi sums: if the χ_j are all nontrivial and $\chi = \chi_1 \cdots \chi_k$, we have

$$J_k(\chi_1^{-1},\dots,\chi_k^{-1})=\frac{q^{k-1-\delta}}{J_k(\chi_1,\dots,\chi_k)}\;,$$

where $\delta=1$ if $\chi=\varepsilon$, and otherwise $\delta=0$.

4.3 Applications of $J(\chi,\chi)$

In this short subsection, we give without proof a couple of applications of the special Jacobi sums $J(\chi,\chi)$. Once again the proofs are not difficult. We begin by the following result, which is a special case of the Hasse–Davenport relations that we will give below.

Lemma 4.14 *Assume that q is odd, and let ρ be the unique character of order 2 on $\mathbb{F}_q^*$. For any nontrivial character χ we have*

$$\chi(4)J(\chi,\chi)=J(\chi,\rho)\;.$$

Equivalently, if $\chi\neq\rho$ we have

$$\mathfrak{g}(\chi)\mathfrak{g}(\chi\rho)=\chi^{-1}(4)\mathfrak{g}(\rho)\mathfrak{g}(\chi^2)\;.$$

Exercise 4.15 1. Prove this lemma.
2. Show that $\mathfrak{g}(\rho)^2=(-1)^{(q-1)/2}q$.

Proposition 4.16 *1. Assume that $q\equiv 1\pmod 4$, let χ be one of the two characters of order 4 on $\mathbb{F}_q^*$, and write $J(\chi,\chi)=a+bi$. Then $q=a^2+b^2$, $2\mid b$, and $a\equiv -1\pmod 4$.*
2. *Assume that $q\equiv 1\pmod 3$, let χ be one of the two characters of order 3 on $\mathbb{F}_q^*$, and write $J(\chi,\chi)=a+b\rho$, where $\rho=\zeta_3$ is a primitive cube root of unity. Then $q=a^2-ab+b^2$, $3\mid b$, $a\equiv -1\pmod 3$, and $a+b\equiv q-2\pmod 9$.*
3. *Let $p\equiv 2\pmod 3$, $q=p^{2m}\equiv 1\pmod 3$, and let χ be one of the two characters of order 3 on $\mathbb{F}_q^*$. We have*

$$J(\chi,\chi)=(-1)^{m-1}p^m=(-1)^{m-1}q^{1/2}\;.$$

Corollary 4.17 *1. (Fermat.) Any prime $p\equiv 1\pmod 4$ is a sum of two squares.*
2. *Any prime $p\equiv 1\pmod 3$ is of the form a^2-ab+b^2 with $3\mid b$, or equivalently $4p=(2a-b)^2+27(b/3)^2$ is of the form c^2+27d^2.*
3. *(Gauss.) $p\equiv 1\pmod 3$ is itself of the form $p=u^2+27v^2$ if and only if 2 is a cube in $\mathbb{F}_p^*$.*

Exercise 4.18 Assuming the proposition, prove the corollary.

4.4 The Hasse–Davenport Relations

All the results that we have given up to now on Gauss and Jacobi sums have rather simple proofs, which is one of the reasons we have not given them. Perhaps surprisingly, there exist other important relations which are considerably more difficult to prove. Before giving them, it is instructive to explain how one can "guess" their existence, if one knows the classical theory of the gamma function $\Gamma(s)$ (of course skip this part if you do not know it, since it would only confuse you, or read the appendix).

Recall that $\Gamma(s)$ is defined (at least for $\Re(s) > 0$) by

$$\Gamma(s) = \int_0^\infty e^{-t} t^s dt/t ,$$

and the beta function $B(a, b)$ by $B(a, b) = \int_0^1 t^{a-1}(1-t)^{b-1}\,dt$. The function e^{-t} transforms sums into products, so is an *additive* character, analogous to ζ_p^t. The function t^s transforms products into products, so is a multiplicative character, analogous to $\chi(t)$ (dt/t is simply the Haar invariant measure on $\mathbb{R}_{>0}$). Thus $\Gamma(s)$ is a continuous analogue of the Gauss sum $\mathfrak{g}(\chi)$.

Similarly, since $J(\chi_1, \chi_2) = \sum_t \chi_1(t)\chi_2(1-t)$, we see the similarity with the function B. Thus, it does not come too much as a surprise that analogous formulas are valid on both sides. To begin with, it is not difficult to show that $B(a, b) = \Gamma(a)\Gamma(b)/\Gamma(a+b)$, exactly analogous to $J(\chi_1, \chi_2) = \mathfrak{g}(\chi_1)\mathfrak{g}(\chi_2)/\mathfrak{g}(\chi_1\chi_2)$. The analogue of $\Gamma(s)\Gamma(-s) = -\pi/(s\sin(s\pi))$ is

$$\mathfrak{g}(\chi)\mathfrak{g}(\chi^{-1}) = \chi(-1)q .$$

But, it is well known that the gamma function has a duplication formula $\Gamma(s)\Gamma(s+1/2) = 2^{1-2s}\Gamma(1/2)\Gamma(2s)$, and more generally a multiplication (or distribution) formula. This duplication formula is clearly the analogue of the formula

$$\mathfrak{g}(\chi)\mathfrak{g}(\chi\rho) = \chi^{-1}(4)\mathfrak{g}(\rho)\mathfrak{g}(\chi^2)$$

given above. The *Hasse–Davenport product relation* is the analogue of the distribution formula for the gamma function.

Theorem 4.19 *Let ρ be a character of exact order m dividing $q-1$. For any character χ of $\mathbb{F}_q^*$ we have*

$$\prod_{0 \le a < m} \mathfrak{g}(\chi\rho^a) = \chi^{-m}(m)k(p, f, m)q^{(m-1)/2}\mathfrak{g}(\chi^m) ,$$

where $k(p, f, m)$ is the fourth root of unity given by

$$k(p,f,m)=\begin{cases}\left(\dfrac{p}{m}\right)^f & \textit{if } m \textit{ is odd},\\ (-1)^{f+1}\left(\dfrac{(-1)^{m/2+1}m/2}{p}\right)^f\left(\dfrac{-1}{p}\right)^{f/2} & \textit{if } m \textit{ is even},\end{cases}$$

where $(-1)^{f/2}$ is to be understood as i^f when f is odd.

Remark 4.20 For some reason, in the literature this formula is usually stated in the weaker form where the constant $k(p,f,m)$ is not given explicitly.

Contrary to the proof of the distribution formula for the gamma function, the proof of this theorem is quite long. There are essentially two completely different proofs: one using classical algebraic number theory, and one using p-adic analysis. The latter is simpler and gives directly the value of $k(p,f,m)$. See Sect. 3.7.2 of [3] and Sect. 11.7.4 of [4] for both detailed proofs.

Gauss sums satisfy another type of nontrivial relation, also due to Hasse–Davenport, the so-called *lifting relation*, as follows:

Theorem 4.21 *Let $\mathbb{F}_{q^n}/\mathbb{F}_q$ be an extension of finite fields, let χ be a character of $\mathbb{F}_q^*$, and define the* lift *of χ to $\mathbb{F}_{q^n}$ by the formula $\chi^{(n)}=\chi\circ\mathcal{N}_{\mathbb{F}_{q^n}/\mathbb{F}_q}$. We have*

$$\mathfrak{g}(\chi^{(n)})=(-1)^{n-1}\mathfrak{g}(\chi)^n\ .$$

This relation is essential in the initial proof of the Weil conjectures for diagonal hypersurfaces done by Weil himself. This is not surprising, since we have seen in Theorem 3.5 that $|V(\mathbb{F}_q)|$ is closely related to Jacobi sums, hence also to Gauss sums.

5 Practical Computations of Gauss and Jacobi Sums

As above, let ω be a character of order exactly $q-1$, so that ω is a generator of the group of characters of $\mathbb{F}_q^*$. For notational simplicity, we will write $J(r_1,\dots,r_k)$ instead of $J(\omega^{r_1},\dots,\omega^{r_k})$. Let us consider the specific example of efficient computation of the quantity

$$S(q;z)=\sum_{0\le n\le q-2}\omega^{-n}(z)J_5(n,n,n,n,n)\ ,$$

which occurs in the computation of the Hasse–Weil zeta function of a quasi-diagonal threefold, see Theorem 3.5.

5.1 Elementary Methods

By the recursion of Corollary 4.10, we have *generically* (i.e., except for special values of n which will be considered separately):

$$J_5(n,n,n,n,n) = J(n,n)J(2n,n)J(3n,n)J(4n,n)\ .$$

Since $J(n,an) = \sum_x \omega^n(x)\omega^{an}(1-x)$, the cost of computing J_5 as written is $\widetilde{O}(q)$, where here and after we write $\widetilde{O}(q^\alpha)$ to mean $O(q^{\alpha+\varepsilon})$ for all $\varepsilon > 0$ (soft-O notation). Thus computing $S(q;z)$ by this direct method requires time $\widetilde{O}(q^2)$.

We can, however, do much better. Since the values of the characters are all in $\mathbb{Z}[\zeta_{q-1}]$, we work in this ring. In fact, even better, we work in the ring with zero divisors $R = \mathbb{Z}[X]/(X^{q-1}-1)$, together with the natural surjective map sending the class of X in R to ζ_{q-1}. Indeed, let g be the generator of $\mathbb{F}_q^*$ such that $\omega(g) = \zeta_{q-1}$. We have, again *generically*:

$$J(n,an) = \sum_{1\le u\le q-2} \omega^n(g^u)\omega^{an}(1-g^u) = \sum_{1\le u\le q-2} \zeta_{q-1}^{nu+an\log_g(1-g^u)}\ ,$$

where $\log_g$ is the *discrete logarithm* to base g defined modulo $q-1$, i.e., such that $g^{\log_g(x)} = x$. If $(q-1) \nmid n$ but $(q-1) \mid an$ we have $\omega^{an} = \varepsilon$ so we must add the contribution of $u = 0$, which is 1, and if $(q-1) \mid n$ we must add the contribution of $u = 0$ *and* of $x = 0$, which is 2 (recall the *essential* convention that $\chi(0) = 0$ if $\chi \ne \varepsilon$ and $\varepsilon(0) = 1$, see Definition 4.3).

In other words, if we set

$$P_a(X) = \sum_{1\le u\le q-2} X^{(u+a\log_g(1-g^u)) \bmod (q-1)} \in R\ ,$$

we have

$$J(n,an) = P_a(\zeta_{q-1}^n) + \begin{cases} 0 & \text{if } (q-1)\nmid an\ , \\ 1 & \text{if } (q-1)\mid an \text{ but } (q-1)\nmid n\ , \text{and} \\ 2 & \text{if } (q-1)\mid n\ . \end{cases}$$

Thus, if we set finally

$$P(X) = P_1(X)P_2(X)P_3(X)P_4(X) \bmod X^{q-1} \in R\ ,$$

we have (still generically) $J_5(n,n,n,n,n) = P(\zeta_{q-1}^n)$. Assume for the moment that this is true for all n (we will correct this below), let $\ell = \log_g(z)$, so that $\omega(z) = \omega(g^\ell) = \zeta_{q-1}^\ell$, and write

$$P(X) = \sum_{0\le j\le q-2} a_j X^j\ .$$

We thus have

$$\omega^{-n}(z)J_5(n,n,n,n,n)=\zeta_{q-1}^{-n\ell}\sum_{0\le j\le q-2}a_j\zeta_{q-1}^{nj}=\sum_{0\le j\le q-2}a_j\zeta_{q-1}^{n(j-\ell)},$$

hence

$$\begin{aligned}S(q;z)&=\sum_{0\le n\le q-2}\omega^{-n}(z)J_5(n,n,n,n,n)=\sum_{0\le j\le q-2}a_j\sum_{0\le n\le q-2}\zeta_{q-1}^{n(j-\ell)}\\&=(q-1)\sum_{0\le j\le q-2,\ j\equiv\ell\pmod{q-1}}a_j=(q-1)a_\ell\ .\end{aligned}$$

The result is thus immediate as soon as we know the coefficients of the polynomial P. Since there exist fast methods for computing discrete logarithms, this leads to a $\widetilde{O}(q)$ method for computing $S(q;z)$.

To obtain the correct formula, we need to adjust for the special n for which $J_5(n,n,n,n,n)$ is not equal to $J(n,n)J(n,2n)J(n,3n)J(n,4n)$, which are the same for which $(q-1)\mid an$ for some a such that $2\le a\le 4$, together with $a=5$. This is easy but boring, and should be skipped on first reading.

1. For $n=0$ we have $J_5(n,n,n,n,n)=q^4$, and on the other hand $P(1)=(J(0,0)-2)^4=(q-2)^4$, so the correction term is $q^4-(q-2)^4=8(q-1)(q^2-2q+2)$.
2. For $n=(q-1)/2$ (if q is odd) we have

$$J_5(n,n,n,n,n)=\mathfrak{g}(\omega^n)^5/\mathfrak{g}(\omega^{5n})=\mathfrak{g}(\omega^n)^4=\mathfrak{g}(\rho)^4$$

   since $5n\equiv n\pmod{q-1}$, where ρ is the character of order 2, and we have $\mathfrak{g}(\rho)^2=(-1)^{(q-1)/2}q$, so $J_5(n,n,n,n,n)=q^2$. On the other hand

$$\begin{aligned}P(\zeta_{q-1}^n)&=J(\rho,\rho)(J(\rho,2\rho)-1)J(\rho,\rho)(J(\rho,2\rho)-1)\\&=J(\rho,\rho)^2=\mathfrak{g}(\rho)^4/q^2=1\ ,\end{aligned}$$

   so the correction term is $\rho(z)(q^2-1)$.
3. For $n=\pm(q-1)/3$ (if $q\equiv 1\pmod 3$), writing $\chi_3=\omega^{(q-1)/3}$, which is one of the two cubic characters, we have

$$\begin{aligned}J_5(n,n,n,n,n)&=\mathfrak{g}(\omega^n)^5/\mathfrak{g}(\omega^{5n})=\mathfrak{g}(\omega^n)^5/\mathfrak{g}(\omega^{-n})\\&=\mathfrak{g}(\omega^n)^6/(\mathfrak{g}(\omega^{-n})\mathfrak{g}(\omega^n))=\mathfrak{g}(\omega^n)^6/q\\&=qJ(n,n)^2\end{aligned}$$

   (check all this). On the other hand

$$\begin{aligned}P(\zeta_{q-1}^n) &= J(n,n)J(n,2n)(J(n,3n)-1)J(n,4n)\\ &= \frac{\mathfrak{g}(\omega^n)^2}{\mathfrak{g}(\omega^{2n})}\frac{\mathfrak{g}(\omega^n)\mathfrak{g}(\omega^{2n})}{q}\frac{\mathfrak{g}(\omega^n)^2}{\mathfrak{g}(\omega^{2n})}\\ &= \frac{\mathfrak{g}(\omega^n)^5}{q\mathfrak{g}(\omega^{-n})} = \frac{\mathfrak{g}(\omega^n)^6}{q^2} = J(n,n)^2\ ,\end{aligned}$$

so the correction term is $2(q-1)\Re(\chi_3^{-1}(z)J(\chi_3,\chi_3)^2)$.

4. For $n = \pm(q-1)/4$ (if $q \equiv 1 \pmod 4$), writing $\chi_4 = \omega^{(q-1)/4}$, which is one of the two quartic characters, we have

$$J_5(n,n,n,n,n) = \mathfrak{g}(\omega^n)^5/\mathfrak{g}(\omega^{5n}) = \mathfrak{g}(\omega^n)^4 = \omega^n(-1)qJ_3(n,n,n)\ .$$

In addition, we have

$$J_3(n,n,n) = J(n,n)J(n,2n) = \omega^n(4)J(n,n)^2 = \rho(2)J(n,n)^2\ ,$$

so

$$J_5(n,n,n,n,n) = \mathfrak{g}(\omega^n)^4 = \omega^n(-1)q\rho(2)J(n,n)^2\ .$$

Note that

$$\chi_4(-1) = \chi_4^{-1}(-1) = \rho(2) = (-1)^{(q-1)/4}\ ,$$

(Exercise: prove it!), so that $\omega^n(-1)\rho(2) = 1$ and the above simplifies to $J_5(n,n,n,n,n) = qJ(n,n)^2$.

On the other hand,

$$\begin{aligned}P(\zeta_{q-1}^n) &= J(n,n)J(n,2n)J(n,3n)(J(n,4n)-1)\\ &= \frac{\mathfrak{g}(\omega^n)^2}{\mathfrak{g}(\omega^{2n})}\frac{\mathfrak{g}(\omega^n)\mathfrak{g}(\omega^{2n})}{\mathfrak{g}(\omega^{3n})}\frac{\mathfrak{g}(\omega^n)\mathfrak{g}(\omega^{3n})}{q}\\ &= \frac{\mathfrak{g}(\omega^n)^4}{q} = \omega^n(-1)\rho(2)J(n,n)^2 = J(n,n)^2\end{aligned}$$

as above, so the correction term is $2(q-1)\Re(\chi_4^{-1}(z)J(\chi_4,\chi_4)^2)$.

5. For $n = a(q-1)/5$ with $1 \le a \le 4$ (if $q \equiv 1 \pmod 5$), writing $\chi_5 = \omega^{(q-1)/5}$ we have $J_5(n,n,n,n,n) = -\mathfrak{g}(\chi_5^a)^5/q$, while abbreviating $\mathfrak{g}(\chi_5^{am})$ to $g(m)$ we have

$$\begin{aligned}P(\zeta_{q-1}^n) &= J(n,n)J(n,2n)J(n,3n)J(n,4n)\\ &= -\frac{g(n)^2}{g(2n)}\frac{g(n)g(2n)}{g(3n)}\frac{g(n)g(3n)}{g(4n)}\frac{g(n)g(4n)}{q}\\ &= -\frac{g(n)^5}{q}\ ,\end{aligned}$$

so there is no correction term.

Summarizing, we have shown the following:

Proposition 5.1 *Let* $S(q;z)=\sum_{0\le n\le q-2}\omega^{-n}(z)J_5(n,n,n,n,n)$. *Let* $\ell=\log_g(z)$ *and let* $P(X)=\sum_{0\le j\le q-2}a_jX^j$ *be the polynomial defined above. We have*

$$S(q;z)=(q-1)(T_1+T_2+T_3+T_4+a_\ell)\ ,$$

where $T_m=0$ *if* $m\nmid(q-1)$ *and otherwise*

$$T_1=8(q^2-2q+2)\ ,\quad T_2=\rho(z)(q+1)\ ,$$
$$T_3=2\Re(\chi_3^{-1}(z)J(\chi_3,\chi_3)^2)\ ,\quad \textit{and}\quad T_4=2\Re(\chi_4^{-1}(z)J(\chi_4,\chi_4)^2)\ ,$$

with the above notation.

Note that thanks to Proposition 4.16, these supplementary Jacobi sums $J(\chi_3,\chi_3)$ and $J(\chi_4,\chi_4)$ can be computed in logarithmic time using Cornacchia's algorithm (this is not quite true, one needs an additional slight computation, do you see why?).

Note also for future reference that the above proposition *proves* that $(q-1)\mid S(q,z)$, which is not clear from the definition.

5.2 Sample Implementations

For simplicity, assume that $q=p$ is prime. I have written simple implementations of the computation of $S(q;z)$. In the first implementation, I use the naïve formula expressing J_5 in terms of $J(n,an)$ and sum on n, except that I use the reciprocity formula which gives $J_5(-n,-n,-n,-n,-n)$ in terms of $J_5(n,n,n,n,n)$ to sum only over $(p-1)/2$ terms instead of $p-1$. Of course to avoid recomputation, I precompute a discrete logarithm table.

The timings for $p\approx 10^k$ for $k=2$, 3, and 4 are 0.03, 1.56, and 149 s respectively, compatible with $\widetilde{O}(q^2)$ time.

On the other hand, implementing in a straightforward manner the algorithm given by the above proposition gives timings for $p\approx 10^k$ for $k=2$, 3, 4, 5, 6, and 7 of 0, 0.02, 0.08, 0.85, 9.90, and 123 s respectively, of course much faster and compatible with $\widetilde{O}(q)$ time.

The main drawback of this method is that it requires $O(q)$ storage: it is thus applicable only for $q\le 10^8$, say, which is more than sufficient for many applications, but of course not for all. For instance, the case $p\approx 10^7$ mentioned above already required a few gigabytes of storage.

5.3 Using Theta Functions

A completely different way of computing Gauss and Jacobi sums has been suggested by S. Louboutin. It is related to the theory of L-functions of Dirichlet characters that we study below, and in our context is valid only for $q = p$ prime, not for prime powers, but in the context of Dirichlet characters it is valid in general (simply replace p by N and $\mathbb{F}_p$ by $\mathbb{Z}/N\mathbb{Z}$ in the following formulas when χ is a primitive character of conductor N, see below for definitions):

Definition 5.2 Let χ be a nontrivial character of $\mathbb{F}_p^*$, and let $e = 0$ or 1 be such that $\chi(-1) = (-1)^e$. The *theta function* associated to χ is the function defined on the upper half-plane by

$$\Theta(\chi, \tau) = 2\sum_{m\geq 1} m^e \chi(m) e^{i\pi m^2 \tau/p} .$$

The main property of this function, which is a direct consequence of the *Poisson summation formula*, and is equivalent to the functional equation of Dirichlet L-functions, is as follows:

Proposition 5.3 *We have the functional equation*

$$\Theta(\chi, -1/\tau) = \omega(\chi)(\tau/i)^{(2e+1)/2}\Theta(\chi^{-1}, \tau) ,$$

with the principal determination of the square root, and where $\omega(\chi) = \mathfrak{g}(\chi)/(i^e p^{1/2})$ is the so-called root number.

Corollary 5.4 *If $\chi(-1) = 1$ we have*

$$\mathfrak{g}(\chi) = p^{1/2} \frac{\sum_{m\geq 1} \chi(m) \exp(-\pi m^2/pt)}{t^{1/2}\sum_{m\geq 1} \chi^{-1}(m) \exp(-\pi m^2 t/p)}$$

and if $\chi(-1) = -1$ we have

$$\mathfrak{g}(\chi) = p^{1/2} i \frac{\sum_{m\geq 1} \chi(m) m \exp(-\pi m^2/pt)}{t^{3/2}\sum_{m\geq 1} \chi^{-1}(m) m \exp(-\pi n^2 t/p)}$$

for any t such that the denominator does not vanish.

Note that the optimal choice of t is $t = 1$, and (at least for p prime) it seems that the denominator never vanishes (there are counterexamples when p is not prime, but apparently only four, see [8]).

It follows from this corollary that $\mathfrak{g}(\chi)$ can be computed numerically as a complex number in $\widetilde{O}(p^{1/2})$ operations. Thus, if χ_1 and χ_2 are nontrivial characters such that $\chi_1\chi_2 \neq \varepsilon$ (otherwise $J(\chi_1, \chi_2)$ is trivial to compute), the formula $J(\chi_1, \chi_2) = \mathfrak{g}(\chi_1)\mathfrak{g}(\chi_2)/\mathfrak{g}(\chi_1\chi_2)$ allows the computation of J_2 *numerically* as a complex number in $\widetilde{O}(p^{1/2})$ operations.

To recover J itself as an algebraic number, we could either compute all its conjugates, but this would require more time than the direct computation of J, or possibly use the LLL algorithm, which although fast, would also require some time. In practice, to perform computations such as that of the sum $S(q; z)$ above, we only need J to sufficient accuracy: we perform all the elementary operations in $\mathbb{C}$, and since we know that at the end the result will be an integer for which we know an upper bound, we thus obtain a proven exact result.

More generally, we have generically $J_5(n, n, n, n, n) = \mathfrak{g}(\omega^n)^5/\mathfrak{g}(\omega^{5n})$, which can thus be computed in $\widetilde{O}(p^{1/2})$ operations. It follows that $S(p; z)$ can be computed in $\widetilde{O}(p^{3/2})$ operations, which is slower than the elementary method seen above. The main advantage is that we do not need much storage: more precisely, we want to compute $S(p; z)$ to sufficiently small accuracy that we can recognize it as an integer, so a priori up to an absolute error of 0.5. However, we have seen that $(p-1) \mid S(p; z)$: it is thus sufficient to have an absolute error less than $(p-1)/2$ thus at worse each of the $p-1$ terms in the sum to an absolute error less than $1/2$. Since generically $|J_5(n, n, n, n, n)| = p^2$, we need a relative error less than $1/(2p^2)$, so less than $1/(10p^2)$ on each Gauss sum. In practice, of course, this is overly pessimistic, but it does not matter. For $p \le 10^9$, this means that 19 decimal digits suffice.

The main term in the theta function computation (with $t = 1$) is $\exp(-\pi m^2/p)$, so we need $\exp(-\pi m^2/p) \le 1/(100p^2)$, say, in other words $\pi m^2/p \ge 4.7 + 2\log(p)$, so $m^2 \ge p(1.5 + 0.7\log(p))$.

This means that we will need the values of $\omega(m)$ only up to this limit, of the order of $O((p\log(p))^{1/2})$, considerably smaller than p. Thus, instead of computing a full discrete logarithm table, which takes some time but more importantly a lot of memory, we compute only discrete logarithms up to that limit, using specific algorithms for doing so which exist in the literature, some of which being quite easy.

A straightforward implementation of this method gives timings for $k = 2, 3, 4$, and 5 of 0.02, 0.40, 16.2, and 663 s respectively, compatible with $\widetilde{O}(p^{3/2})$ time. This is faster than the completely naïve method, but slower than the method explained above. Its advantage is that it requires much less memory. For p around 10^7, however, it is much too slow so this method is rather useless. We will see that its usefulness is mainly in the context where it was invented, i.e., for L-functions of Dirichlet characters.

5.4 *Using the Gross–Koblitz Formula*

This section is of a higher mathematical level than the preceding ones, but is very important since it gives the best method for computing Gauss (and Jacobi) sums. We refer to Sects. 11.6 and 11.7 of [4] for complete details, and urge the reader to try to understand what follows.

In the preceding sections, we have considered Gauss sums as belonging to a number of different rings: the ring $\mathbb{Z}[\zeta_{q-1}, \zeta_p]$ or the field $\mathbb{C}$ of complex numbers, and for Jacobi sums the ring $\mathbb{Z}[\zeta_{q-1}]$, but also the ring $\mathbb{Z}[X]/(X^{q-1}-1)$, and again the field $\mathbb{C}$.

In number theory, there exist other algebraically closed fields which are useful in many contexts, the fields $\mathbb{C}_\ell$ of ℓ-adic numbers, one for each prime number ℓ. These fields come with a topology and analysis which are rather special: one of the main things to remember is that a sequence of elements tends to 0 if and only the ℓ-adic valuation of the elements (the largest exponent of ℓ dividing them) tends to infinity. For instance, 2^m tends to 0 in $\mathbb{C}_2$, but in no other $\mathbb{C}_\ell$, and 15^m tends to 0 in $\mathbb{C}_3$ and in $\mathbb{C}_5$.

The most important subrings of $\mathbb{C}_\ell$ are the ring $\mathbb{Z}_\ell$ of ℓ-adic integers, the elements of which can be written as $x = a_0 + a_1\ell + \cdots + a_k\ell^k + \cdots$ with $a_j \in [0, \ell - 1]$, and its field of fractions $\mathbb{Q}_\ell$, which contains $\mathbb{Q}$, whose elements can be represented in a similar way as $x = a_{-m}\ell^{-m} + a_{-(m-1)}\ell^{-(m-1)} + \cdots + a_{-1}\ell^{-1} + a_0 + a_1\ell + \cdots$.

In dealing with Gauss and Jacobi sums over $\mathbb{F}_q$ with $q = p^f$, the only $\mathbb{C}_\ell$ which is of use for us is the one with $\ell = p$ (in highbrow language, we are going to use implicitly *crystalline* p-adic methods, while for $\ell \neq p$ it would be *étale* ℓ-adic methods).

Apart from this relatively strange topology, many definitions and results valid on $\mathbb{C}$ have analogues in $\mathbb{C}_p$. The main object that we will need in our context is the analogue of the gamma function, naturally called the p-adic gamma function, in the present case due to Morita (there is another one, see Sect. 11.5 of [4]), and denoted Γ_p. Its definition is in fact quite simple:

Definition 5.5 For $s \in \mathbb{Z}_p$ we define

$$\Gamma_p(s) = \lim_{m \to s} (-1)^m \prod_{\substack{0 \le k < m \\ p \nmid k}} k\ ,$$

where the limit is taken over any sequence of positive integers m tending to s for the p-adic topology.

It is, of course, necessary to show that this definition makes sense, but this is not difficult, and most of the important properties of $\Gamma_p(s)$, analogous to those of $\Gamma(s)$, can be deduced from it.

Exercise 5.6 Choose $p = 5$ and $s = -1/4$, so that p-adically $s = 1/(1-5) = 1 + 5 + 5^2 + 5^3 + \cdots$.

1. Compute the right-hand side of the above definition with small 5-adic accuracy for $m = 1$, $1 + 5$, and $1 + 5 + 5^2$.
2. It is in fact easy to compute that

$$\Gamma_5(-1/4) = 4 + 4 \cdot 5 + 5^3 + 3 \cdot 5^4 + 2 \cdot 5^5 + 2 \cdot 5^6 + 2 \cdot 5^7 + 4 \cdot 5^8 + \cdots$$

Using this, show that $\Gamma_5(-1/4)^2/16$ seems to be a 5-adic root of the polynomial $5X^2 + 4X + 1$. This is in fact true, see the Gross–Koblitz formula below.

We need a much deeper property of $\Gamma_p(s)$ known as the Gross–Koblitz formula: it is in fact an analogue of a formula for $\Gamma(s)$ known as the Chowla–Selberg formula, and it is also closely related to the Davenport–Hasse relations that we have seen above.

The proof of the Gross–Koblitz formula was initially given using tools of crystalline cohomology, but an elementary proof due to A. Robert now exists, see for instance, Sect. 11.7 of [4] once again.

The Gross–Koblitz formula tells us that certain products of p-adic gamma functions at *rational* arguments are in fact *algebraic numbers*, more precisely *Gauss sums* (explaining their importance for us). This is quite surprising since usually transcendental functions such as Γ_p take transcendental values.

To give a specific example, we have $\Gamma_5(1/4)^2 = -2 + \sqrt{-1}$, where $\sqrt{-1}$ is the square root in $\mathbb{Z}_5$ congruent to 3 modulo 5. In view of the elementary properties of the p-adic gamma function, this is equivalent to the result stated in the above exercise as $\Gamma_5(-1/4)^2 = -(16/5)(2 + \sqrt{-1})$.

Before stating the formula we need to collect a number of facts, both on classical algebraic number theory and on p-adic analysis. None are difficult to prove, see Chap. 4 of [3]. Recall that $q = p^f$.

• We let $K = \mathbb{Q}(\zeta_p)$ and $L = K(\zeta_{q-1}) = \mathbb{Q}(\zeta_{q-1}, \zeta_p) = \mathbb{Q}(\zeta_{p(q-1)})$, so that L/K is an extension of degree $\phi(q-1)$. There exists a unique prime ideal $\mathfrak{p}$ of K above p, and we have $\mathfrak{p} = (1 - \zeta_p)\mathbb{Z}_K$ and $\mathfrak{p}^{p-1} = p\mathbb{Z}_K$, and $\mathbb{Z}_K/\mathfrak{p} \simeq \mathbb{F}_p$. The prime ideal $\mathfrak{p}$ splits into a product of $g = \phi(q-1)/f$ prime ideals $\mathfrak{P}_j$ of degree f in the extension L/K, i.e., $\mathfrak{p}\mathbb{Z}_L = \mathfrak{P}_1 \cdots \mathfrak{P}_g$, and for any prime ideal $\mathfrak{P} = \mathfrak{P}_j$ we have $\mathbb{Z}_L/\mathfrak{P} \simeq \mathbb{F}_q$.

Exercise 5.7 Prove directly that for any f we have $f \mid \phi(p^f - 1)$.

• Fix one of the prime ideals $\mathfrak{P}$ as above. There exists a unique group isomorphism $\omega = \omega_{\mathfrak{P}}$ from $(\mathbb{Z}_L/\mathfrak{P})^*$ to the group of $(q-1)$st roots of unity in L, such that for all $x \in (\mathbb{Z}_L/\mathfrak{P})^*$ we have $\omega(x) \equiv x \pmod{\mathfrak{P}}$. It is called the *Teichmüller character*, and it can be considered as a character of order $q-1$ on $\mathbb{F}_q^* \simeq (\mathbb{Z}_L/\mathfrak{P})^*$. We can thus *instantiate* the definition of a Gauss sum over $\mathbb{F}_q$ by defining it as $\mathfrak{g}(\omega_{\mathfrak{P}}^{-r}) \in L$.

• Let ζ_p be a primitive pth root of unity in $\mathbb{C}_p$, fixed once and for all. There exists a unique $\pi \in \mathbb{Z}[\zeta_p]$ satisfying $\pi^{p-1} = -p$, $\pi \equiv 1 - \zeta_p \pmod{\pi^2}$, and we set $K_{\mathfrak{p}} = \mathbb{Q}_p(\pi) = \mathbb{Q}_p(\zeta_p)$, and $L_{\mathfrak{P}}$ the *completion* of L at $\mathfrak{P}$. The field extension $L_{\mathfrak{P}}/K_{\mathfrak{p}}$ is Galois, with Galois group isomorphic to $\mathbb{Z}/f\mathbb{Z}$ (which is the same as the Galois group of $\mathbb{F}_q/\mathbb{F}_p$, where $\mathbb{F}_p$ (resp., $\mathbb{F}_q$) is the so-called *residue field* of K (resp., L)).

• We set the following:

Definition 5.8 We define the *p-adic Gauss sum* by

$$\mathfrak{g}_q(r) = \sum_{x \in L_{\mathfrak{P}},\ x^{q-1}=1} x^{-r} \zeta_p^{\operatorname{Tr}_{L_{\mathfrak{P}}/K_{\mathfrak{p}}}(x)} \in L_{\mathfrak{P}}\ .$$

Note that this depends on the choice of ζ_p, or equivalently of π. Since $\mathfrak{g}_q(r)$ and $\mathfrak{g}(\omega_{\mathfrak{P}}^{-r})$ are algebraic numbers, it is clear that they are equal, although viewed in fields having different topologies. Thus, results about $\mathfrak{g}_q(r)$ translate immediately into results about $\mathfrak{g}(\omega_{\mathfrak{P}}^{-r})$, hence about general Gauss sums over finite fields.

The Gross–Koblitz formula is as follows:

Theorem 5.9 (Gross–Koblitz) *Denote by $s(r)$ the sum of digits in base p of the integer r mod $(q-1)$, i.e., of the unique integer r' such that $r' \equiv r \pmod{q-1}$ and $0 \le r' < q-1$. We have*

$$\mathfrak{g}_q(r) = -\pi^{s(r)} \prod_{0 \le i < f} \Gamma_p\left(\left\{\frac{p^{f-i}r}{q-1}\right\}\right) ,$$

where $\{x\}$ denotes the fractional part of x.

Let us show how this can be used to compute Gauss or Jacobi sums, and in particular our sum $S(q; z)$. Assume for simplicity that $f = 1$, in other words that $q = p$: the right-hand side is thus equal to $-\pi^{s(r)}\Gamma_p(\{pr/(p-1)\})$. Since we can always choose r such that $0 \le r < p-1$, we have $s(r) = r$ and $\{pr/(p-1)\} = \{r + r/(p-1)\} = r/(p-1)$, so the RHS is $-\pi^r\Gamma_p(r/(p-1))$. Now an easy property of Γ_p is that it is differentiable: recall that p is "small" in the p-adic topology, so $r/(p-1)$ is close to $-r$, more precisely $r/(p-1) = -r + pr/(p-1)$ (this is how we obtained it in the first place!). Thus in particular, if $p > 2$ we have the Taylor expansion

$$\begin{aligned}\Gamma_p(r/(p-1)) &= \Gamma_p(-r) + (pr/(p-1))\Gamma_p'(-r) + O(p^2)\\ &= \Gamma_p(-r) - pr\Gamma_p'(-r) + O(p^2) .\end{aligned}$$

Since $\mathfrak{g}_q(r)$ depends only on r modulo $p-1$, we will assume that $0 \le r < p-1$. In that case, it is easy to show from the definition that

$$\Gamma_p(-r) = 1/r! \quad \text{and} \quad \Gamma_p'(-r) = (-\gamma_p + H_r)/r! ,$$

where $H_r = \sum_{1 \le n \le r} 1/n$ is the harmonic sum, and $\gamma_p = -\Gamma_p'(0)$ is the p-adic analogue of Euler's constant.

Exercise 5.10 Prove these formulas, as well as the congruence for γ_p given below.

There exist infinite (p-adic) series enabling accurate computation of γ_p, but since we only need it modulo p, we use the easily proved congruence $\gamma_p \equiv ((p-1)! + 1)/p = W_p \pmod{p}$, the so-called *Wilson quotient*.

We will see below that, as a consequence of the Weil conjectures proved by Deligne, it is sufficient to compute $S(p; z)$ modulo p^2. Thus, in the following p-adic computation we only work modulo p^2.

The Gross–Koblitz formula tells us that for $0 \le r < p-1$ we have

$$\mathfrak{g}_q(r) = -\frac{\pi^r}{r!}(1 - pr(H_r - W_p) + O(p^2)) .$$

It follows that for $(p-1) \nmid 5r$ we have

$$J(-r,-r,-r,-r,-r) = \frac{\mathfrak{g}(\omega_{\mathfrak{P}})^5}{\mathfrak{g}(\omega_{\mathfrak{P}}^5)} = \frac{\mathfrak{g}_q(r)^5}{\mathfrak{g}_q(5r)} = \pi^{f(r)}(a + bp + O(p^2)) ,$$

where a and b will be computed below and

$$\begin{aligned} f(r) &= 5r - (5r \bmod p-1) = 5r - (5r - (p-1)\lfloor 5r/(p-1)\rfloor) \\ &= (p-1)\lfloor 5r/(p-1)\rfloor , \end{aligned}$$

so that $\pi^{f(r)} = (-p)^{\lfloor 5r/(p-1)\rfloor}$ since $\pi^{p-1} = -p$. Since we want the result modulo p^2, we consider three intervals together with special cases:

1. If $r > 2(p-1)/5$ but $(p-1) \nmid 5r$, we have
$$J(-r,-r,-r,-r,-r) \equiv 0 \pmod{p^2} .$$
2. If $(p-1)/5 < r < 2(p-1)/5$ we have
$$J(-r,-r,-r,-r,-r) \equiv (-p)\frac{(5r-(p-1))!}{r!^5} \pmod{p^2} .$$
3. If $0 < r < (p-1)/5$ we have $f(r) = 0$ and $0 \le 5r < (p-1)$ hence
$$\begin{aligned} J(-r,-r,-r,-r,-r) &= \frac{(5r)!}{r!^5}(1 - 5pr(H_r - W_p) + O(p^2))\cdot \\ &\quad \cdot (1 + 5pr(H_{5r} - W_p) + O(p^2)) \\ &\equiv \frac{(5r)!}{r!^5}(1 + 5pr(H_{5r} - H_r)) \pmod{p^2} . \end{aligned}$$
4. Finally, if $r = j(p-1)/5$ we have $J(-r,-r,-r,-r,-r) = p^4 \equiv 0 \pmod{p^2}$ if $j = 0$, and otherwise $J(-r,-r,-r,-r,-r) = -\mathfrak{g}_q(r)^5/p$, and since the p-adic valuation of $\mathfrak{g}_q(r)$ is equal to $r/(p-1) = j/5$, that of $J(-r,-r,-r,-r,-r)$ is equal to $j-1$, which is greater or equal to 2 as soon as $j \ge 3$. For $j = 2$, i.e., $r = 2(p-1)/5$, we thus have
$$J(-r,-r,-r,-r,-r) \equiv p\frac{1}{r!^5} \equiv (-p)\frac{(5r-(p-1))!}{r!^5} \pmod{p^2} ,$$

which is the same formula as for $(p-1)/5 < r \le 2(p-1)/5$. For $j = 1$, i.e., $r = (p-1)/5$, we thus have

$$J(-r,-r,-r,-r,-r) \equiv -\frac{1}{r!^5}(1 - 5pr(H_r - W_p)) \pmod{p^2}\ ,$$

while on the other hand

$$(5r)! = (p-1)! = -1 + pW_p \equiv -1 - p(p-1)W_p \equiv -1 - 5prW_p\ ,$$

and $H_{5r} = H_{p-1} \equiv 0 \pmod{p}$ (Wolstenholme's congruence, easy), so

$$\begin{aligned}\frac{(5r)!}{r!^5}(1 + 5pr(H_{5r} - H_r)) &\equiv -\frac{1}{r!^5}(1 - 5prH_r)(1 + 5prW_p)\\ &\equiv -\frac{1}{r!^5}(1 - 5pr(H_r - W_p)) \pmod{p^2}\ ,\end{aligned}$$

which is the same formula as for $0 < r < (p-1)/5$.

An important point to note is that we are working p-adically, but the final result $S(p;z)$ being an integer, it does not matter at the end. There is one small additional detail to take care of: we have

$$\begin{aligned}S(p;z) &= \sum_{0\le r\le p-2} \omega^{-r}(z)J(r,r,r,r,r)\\ &= \sum_{0\le r\le p-2} \omega^{r}(z)J(-r,-r,-r,-r,-r)\ ,\end{aligned}$$

so we must express $\omega^r(z)$ in the p-adic setting. Since $\omega = \omega_{\mathfrak{P}}$ is the *Teichmüller character*, in the p-adic setting it is easy to show that $\omega(z)$ is the p-adic limit of z^{p^k} as $k \to \infty$, in particular $\omega(z) \equiv z \pmod{p}$, but more precisely $\omega(z) \equiv z^p \pmod{p^2}$.

Exercise 5.11 Let $p \ge 3$. Assume that $z \in \mathbb{Z}_p \setminus p\mathbb{Z}_p$ (for instance that $z \in \mathbb{Z} \setminus p\mathbb{Z}$). Prove that z^{p^k} has a p-adic limit $\omega(z)$ when $k \to \infty$, that $\omega^{p-1}(z) = 1$, that $\omega(z) \equiv z \pmod{p}$, and $\omega(z) \equiv z^p \pmod{p^2}$.

We have thus proved the following.

Proposition 5.12 *We have*

$$\begin{aligned}S(p;z) \equiv &\sum_{0<r\le(p-1)/5} \frac{(5r)!}{r!^5}(1 + 5pr(H_{5r} - H_r))z^{pr}\\ &- p\sum_{(p-1)/5<r\le 2(p-1)/5} \frac{(5r-(p-1))!}{r!^5}z^r \pmod{p^2}\ .\end{aligned}$$

In particular

$$S(p;z) \equiv \sum_{0<r\le(p-1)/5} \frac{(5r)!}{r!^5} z^r \pmod{p}\ .$$

Remarks 5.13 1. Note that, as must be the case, all mention of p-adic numbers has disappeared from this formula. We used the p-adic setting only in the proof. It can be proved "directly", but with some difficulty.

2. We used the Taylor expansion only to order 2. It is, of course, possible to use it to any order, thus giving a generalization of the above proposition to any power of p.

The point of giving all these details is as follows: it is easy to show that $(p-1) \mid S(p;z)$ (in fact we have seen this in the elementary method above). We can thus easily compute $S(p;z)$ modulo $p^2(p-1)$. On the other hand, it is possible to prove (but not easy, it is part of the Weil conjectures proved by Deligne), that $|S(p;z) - p^4| < 4p^{5/2}$. It follows that as soon as $8p^{5/2} < p^2(p-1)$, in other words $p \ge 67$, the computation that we perform modulo p^2 is sufficient to determine $S(p;z)$ exactly. It is clear that the time to perform this computation is $\widetilde{O}(p)$, and in fact much faster than any that we have seen.

In fact, implementing in a reasonable way the algorithm given by the above proposition gives timings for $p \approx 10^k$ for $k = 2, 3, 4, 5, 6, 7$, and 8 of 0, 0.01, 0.03, 0.21, 2.13, 21.92, and 229.6 s, respectively, of course much faster and compatible with $\widetilde{O}(p)$ time. The great additional advantage is that we use very small memory. This is, therefore, the best known method.

Numerical example: Choose $p = 10^6 + 3$ and $z = 2$. In 2.13 s we find that $S(p;z) \equiv a \pmod{p^2}$ with $a = 356022712041$. Using the Chinese remainder formula

$$S(p;z) = p^4 + ((a - (1+a)p^2) \bmod ((p-1)p^2))\ ,$$

we immediately deduce that

$$S(p;z) = 1000012000056356142712140\ .$$

Here is a summary of the timings (in seconds) that we have mentioned:

k	2	3	4	5	6	7	8
Naïve	0.03	1.56	149	$*$	$*$	$*$	$*$
Theta	0.02	0.40	16.2	663	$*$	$*$	$*$
Mod $X^{q-1} - 1$	0	0.02	0.08	0.85	9.90	123	$*$
Gross–Koblitz	0	0.01	0.03	0.21	2.13	21.92	229.6

Time for computing $S(p;z)$ for $p \approx 10^k$

6 Gauss and Jacobi Sums over $\mathbb{Z}/N\mathbb{Z}$

Another context in which one encounters Gauss sums is over finite rings such as $\mathbb{Z}/N\mathbb{Z}$. The theory coincides with that over $\mathbb{F}_q$ when $q = p = N$ is prime, but is rather different otherwise. These other Gauss sums enter in the important theory of *Dirichlet characters*.

6.1 Definitions

We recall the following definition:

Definition 6.1 Let χ be a (multiplicative) character from the multiplicative group $(\mathbb{Z}/N\mathbb{Z})^*$ of invertible elements of $\mathbb{Z}/N\mathbb{Z}$ to the complex numbers $\mathbb{C}$. We denote by abuse of notation again by χ the map from $\mathbb{Z}$ to $\mathbb{C}$ defined by $\chi(x) = \chi(x \bmod N)$ when x is coprime to N, and $\chi(x) = 0$ if x is not coprime to N, and call it the Dirichlet character modulo N associated to χ.

It is clear that a Dirichlet character satisfies $\chi(xy) = \chi(x)\chi(y)$ for all x and y, that $\chi(x + N) = \chi(x)$, and that $\chi(x) = 0$ if and only if x is not coprime with N. Conversely, it immediate that these properties characterize Dirichlet characters.

A crucial notion (which has no equivalent in the context of characters of $\mathbb{F}_q^*$) is that of *primitivity*:

Assume that $M \mid N$. If χ is a Dirichlet character modulo M, we can transform it into a character χ_N modulo N by setting $\chi_N(x) = \chi(x)$ if x is coprime to N, and $\chi_N(x) = 0$ otherwise. We say that the characters χ and χ_N are *equivalent*. Conversely, if ψ is a character modulo N, it is not always true that one can find χ modulo M such that $\psi = \chi_N$. If it is possible, we say that ψ *can be defined modulo* M.

Definition 6.2 Let χ be a character modulo N. We say that χ is a *primitive character* if χ cannot be defined modulo M for any proper divisor M of N, i.e., for any $M \mid N$ such that $M \neq N$.

Exercise 6.3 Assume that $N \equiv 2 \pmod 4$. Show that there do not exist any primitive characters modulo N.

Exercise 6.4 Assume that $p^a \mid N$ with p prime. Show that if χ is a primitive character modulo N, the *order* of χ (the smallest k such that χ^k is a trivial character) is *divisible* by p^{a-1}.

As we will see, questions about general Dirichlet characters can always be reduced to questions about primitive characters, and the latter have much nicer properties.

Proposition 6.5 *Let χ be a character modulo N. There exists a divisor f of N called the* conductor *of χ (this f has nothing to do with the f used above such that $q = p^f$), having the following properties:*

1. *The character χ can be defined modulo f, in other words there exists a character ψ modulo f such that $\chi = \psi_N$ using the notation above.*
2. *f is the smallest divisor of N having this property.*
3. *The character ψ is a primitive character modulo f.*

There is also the notion of *trivial character modulo N*: however we must be careful here, and we set the following.

Definition 6.6 The trivial character modulo N is the Dirichlet character associated with the trivial character of $(\mathbb{Z}/N\mathbb{Z})^*$. It is usually denoted by χ_0 (but be careful, the index N is implicit, so χ_0 may represent different characters), and its values are as follows: $\chi_0(x) = 1$ if x is coprime to N, and $\chi_0(x) = 0$ if x is not coprime to N.

In particular, $\chi_0(0) = 0$ if $N \neq 1$. The character χ_0 can also be characterized as the only character modulo N of conductor 1.

Definition 6.7 Let χ be a character modulo N. The *Gauss sum* associated to χ and $a \in \mathbb{Z}$ is

$$\mathfrak{g}(\chi, a) = \sum_{x \bmod N} \chi(x)\zeta_N^{ax} ,$$

and we write simply $\mathfrak{g}(\chi)$ instead of $\mathfrak{g}(\chi, 1)$.

The most important results concerning these Gauss sums is the following.

Proposition 6.8 *Let χ be a character modulo N.*

1. *If a is coprime to N we have*

$$\mathfrak{g}(\chi, a) = \chi^{-1}(a)\mathfrak{g}(\chi) = \overline{\chi(a)}\mathfrak{g}(\chi) ,$$

   and more generally $\mathfrak{g}(\chi, ab) = \chi^{-1}(a)\mathfrak{g}(\chi, b) = \overline{\chi(a)}\mathfrak{g}(\chi, b)$.
2. *If χ is a* primitive *character, we have*

$$\mathfrak{g}(\chi, a) = \overline{\chi(a)}\mathfrak{g}(\chi)$$

   for all *a, in other words, in addition to (1), we have $\mathfrak{g}(\chi, a) = 0$ if a is not coprime to N.*
3. *If χ is a* primitive *character, we have $|\mathfrak{g}(\chi)|^2 = N$.*

Note that (1) is trivial, and that since $\chi(a)$ has modulus 1 when a is coprime to N, we can write indifferently $\chi^{-1}(a)$ or $\overline{\chi(a)}$. On the other hand, (2) is not completely trivial.

We leave to the reader the easy task of defining Jacobi sums and of proving the easy relations between Gauss and Jacobi sums.

6.2 Reduction to Prime Gauss Sums

A fundamental and little-known fact is that in the context of Gauss sums over $\mathbb{Z}/N\mathbb{Z}$ (as opposed to $\mathbb{F}_q$), one can in fact always reduce to prime N. First note (with proof) the following easy result:

Proposition 6.9 *Let $N = N_1N_2$ with N_1 and N_2 coprime, and let χ be a character modulo N.*

1. *There exist unique characters χ_i modulo N_i such that $\chi = \chi_1\chi_2$ in an evident sense, and if χ is primitive, the χ_i will also be primitive.*
2. *We have the identity (valid even if χ is not primitive):*

$$\mathfrak{g}(\chi) = \chi_1(N_2)\chi_2(N_1)\mathfrak{g}(\chi_1)\mathfrak{g}(\chi_2)\ .$$

Proof (1) Since N_1 and N_2 are coprime there exist u_1 and u_2 such that $u_1N_1 + u_2N_2 = 1$. We define $\chi_1(x) = \chi(xu_2N_2 + u_1N_1)$ and $\chi_2(x) = \chi(xu_1N_1 + u_2N_2)$. We leave to the reader to check (1) using these definitions.

(2) When x_i ranges modulo N_i, $x = x_1u_2N_2 + x_2u_1N_1$ ranges modulo N (check it, in particular that the values are distinct!), and $\chi(x) = \chi_1(x)\chi_2(x) = \chi_1(x_1)\chi_2(x_2)$. Furthermore,

$$\zeta_N = \exp(2\pi i/N) = \exp(2\pi i(u_1/N_2 + u_2/N_1)) = \zeta_{N_1}^{u_2}\zeta_{N_2}^{u_1}\ ,$$

hence

$$\begin{aligned}\mathfrak{g}(\chi) &= \sum_{x \bmod N} \chi(x)\zeta_N^x \\ &= \sum_{x_1 \bmod N_1,\ x_2 \bmod N_2} \chi_1(x_1)\chi_2(x_2)\zeta_{N_1}^{u_2x_1}\zeta_{N_2}^{u_1x_2} \\ &= \mathfrak{g}(\chi_1; u_2)\mathfrak{g}(\chi_2; u_1) = \chi_1^{-1}(u_2)\chi_2^{-1}(u_1)\mathfrak{g}(\chi_1)\mathfrak{g}(\chi_2)\ ,\end{aligned}$$

so the result follows since $N_2u_2 \equiv 1 \pmod{N_1}$ and $N_1u_1 \equiv 1 \pmod{N_2}$. □

Thanks to the above result, the computation of Gauss sums modulo N can be reduced to the computation of Gauss sums modulo prime powers.

Here a remarkable simplification occurs, due to Odoni: Gauss sums modulo p^a for $a \geq 2$ can be "explicitly computed", in the sense that there is a direct formula not involving a sum over p^a terms for computing them. Although the proof is not difficult, we do not give it, and refer instead to [5] which can be obtained from the author. We use the classical notation $\mathbf{e}(x)$ to mean $e^{2\pi ix}$. Furthermore, we use the p-adic logarithm $\log_p(m)$, but in a totally elementary manner since we will always have $m \equiv 1 \pmod{p}$ and the standard expansion $-\log_p(1-x) = \sum_{k\geq 1} x^k/k$, which we stop as soon as all the terms are divisible by p^n:

Theorem 6.10 (Odoni et al.) *Let χ be a* primitive *character modulo p^n.*

1. *Assume that $p \geq 3$ is prime and $n \geq 2$. Write $\chi(1+p) = \mathbf{e}(-b/p^{n-1})$ with $p \nmid b$. Define*
$$A(p) = \frac{p}{\log_p(1+p)} \quad \text{and} \quad B(p) = A(p)(1 - \log_p(A(p)))\ ,$$
except when $p^n = 3^3$, in which case we define $B(p) = 10$. Then
$$\mathfrak{g}(\chi) = p^{n/2}\mathbf{e}\left(\frac{bB(p)}{p^n}\right)\chi(b)\cdot\begin{cases} 1 & \text{if } n \geq 2 \text{ is even,} \\ \left(\frac{b}{p}\right) i^{p(p-1)/2} & \text{if } n \geq 3 \text{ is odd.}\end{cases}$$

2. *Let $p = 2$ and assume that $n \geq 4$. Write $\chi(1+p^2) = \mathbf{e}(b/p^{n-2})$ with $p \nmid b$. Define*
$$A(p) = -\frac{p^2}{\log_p(1+p^2)} \quad \text{and} \quad B(p) = A(p)(1 - \log_p(A(p)))\ ,$$
except when $p^n = 2^4$, in which case we define $B(p) = 13$. Then
$$\mathfrak{g}(\chi) = p^{n/2}\mathbf{e}\left(\frac{bB(p)}{p^n}\right)\chi(b)\cdot\begin{cases} \mathbf{e}\left(\frac{b}{8}\right) & \text{if } n \geq 4 \text{ is even,} \\ \mathbf{e}\left(\frac{(b^2-1)/2+b}{8}\right) & \text{if } n \geq 5 \text{ is odd.}\end{cases}$$

3. *If $p^n = 2^2$, or $p^n = 2^3$ and $\chi(-1) = 1$, we have $\mathfrak{g}(\chi) = p^{n/2}$, and if $p^n = 2^3$ and $\chi(-1) = -1$ we have $\mathfrak{g}(\chi) = p^{n/2}i$.*

Thanks to this theorem, we see that the computation of Gauss sums in the context of Dirichlet characters can be reduced to the computation of Gauss sums modulo p for prime p. This is, of course, the same as the computation of a Gauss sum for a character of $\mathbb{F}_p^*$.

We recall the available methods for computing a single Gauss sum of this type:

1. The naïve method, time $\widetilde{O}(p)$ (applicable in general, time $\widetilde{O}(N)$).
2. Using the Gross–Koblitz formula, also time $\widetilde{O}(p)$, but the implicit constant is much smaller, and also computations can be done modulo p or p^2, for instance, if desired (applicable only to $N = p$, or in the context of finite fields).
3. Using theta functions, time $\widetilde{O}(p^{1/2})$ (applicable in general, time $\widetilde{O}(N^{1/2})$).

6.3 General Complete Exponential Sums over $\mathbb{Z}/N\mathbb{Z}$

We have just seen the (perhaps surprising) fact that Gauss sums modulo p^a for $a \geq 2$ can be "explicitly computed". This is in fact a completely general fact. Let χ be a Dirichlet character modulo N, and let $F \in \mathbb{Q}[X]$ be integer valued. Consider the following *complete exponential sum*:

$$S(F, N) = \sum_{x \bmod N} \chi(x) e^{2\pi i F(x)/N} .$$

For this to make sense, we must, of course, assume that $x \equiv y \pmod{N}$ implies $F(x) \equiv F(y) \pmod{N}$, which is, for instance, the case if $F \in \mathbb{Z}[X]$. As we did for Gauss sums, using Chinese remaindering we can reduce the computation to the case where $N = p^a$ is a prime power. But the essential point is that if $a \geq 2$, $S(F, p^a)$ can be "explicitly computed", see [5] for the detailed statement and proof, so we are again reduced to the computation of $S(F, p)$.

A simplified version and incomplete version of the result when χ is the trivial character is as follows:

Theorem 6.11 *Let* $S = \sum_{x \bmod p^a} e^{2\pi i F(x)/p^a}$, *and assume that* $a \geq 2$ *and* $p > 2$. *Then under suitable assumptions on* F *we have the following:*

1. *If there does not exist* y *such that* $F'(y) \equiv 0 \pmod{p}$ *then* $S = 0$.
2. *Otherwise, there exists* $u \in \mathbb{Z}_p$ *such that* $F'(u) = 0$ *and* $v_p(F''(u)) = 0$, u *is unique, and we have*

$$S = p^{a/2} e^{2\pi i F(u)/p^a} g(u, p, a) ,$$

where $g(u, p, a) = 1$ *if* a *is even and otherwise*

$$g(u, p, a) = \left(\frac{F''(u)}{p}\right) i^{p(p-1)/2} .$$

Exercise 6.12 Let $F(x) = cx^3 + dx$ with c and d integers, and let p be a prime number such that $p \nmid 6cd$. The assumptions of the theorem will then be satisfied. Compute explicitly $\sum_{x \bmod p^a} e^{2\pi i F(x)/p^a}$ for $a \geq 2$. You will need to introduce a square root of $-3cd$ modulo p^a.

For instance, using a variant of the above theorem, it is immediate to prove the following result due to Salié:

Proposition 6.13 *The* Kloosterman sum $K(m, n, N)$ *is defined by*

$$K(m, n, N) = \sum_{x \in (\mathbb{Z}/N\mathbb{Z})^*} e^{2\pi i (mx + nx^{-1})/N} ,$$

where x *runs over the invertible elements of* $\mathbb{Z}/N\mathbb{Z}$. *If* $p > 2$ *is a prime such that* $p \nmid n$ *and* $a \geq 2$ *we have*

$$K(n,n,p^a)=\begin{cases}2p^{a/2}\cos(4\pi n/p^a) & \text{if } 2\mid a,\\ 2p^{a/2}\left(\dfrac{n}{p}\right)\cos(4\pi n/p^a) & \text{if } 2\nmid a \text{ and } p\equiv 1 \pmod 4,\\ -2p^{a/2}\left(\dfrac{n}{p}\right)\sin(4\pi n/p^a) & \text{if } 2\nmid a \text{ and } p\equiv 3 \pmod 4.\end{cases}$$

Note that it is immediate to reduce general $K(m, n, N)$ to the case $m = n$ and $N = p^a$, and to give formulas also for the case $p = 2$. As usual the case $N = p$ is *not* explicit, and, contrary to the case of Gauss sums where it is easy to show that $|\mathfrak{g}(\chi)| = \sqrt{p}$ for a primitive character χ, the bound $|K(m, n, p)| \le 2\sqrt{p}$ for $p \nmid nm$ due to Weil is much more difficult to prove, and in fact follows from his proof of the Riemann hypothesis for curves.

7 Numerical Computation of L-Functions

7.1 *Computational Issues*

Let $L(s)$ be a general L-function as defined in Sect. 1, and let N be its conductor. There are several computational problems that we want to solve. The first, but not necessarily the most important, is the numerical computation of $L(s)$ for given complex values of s. This problem is of very varying difficulty depending on the size of N and of the imaginary part of s (note that if the *real part* of s is quite large, the defining series for $L(s)$ converges quite well, if not exponentially fast, so there is no problem in that range, and by the functional equation the same is true if the real part of $1 - s$ is quite large).

The problems for $\Im(s)$ large are quite specific, and are already crucial in the case of the Riemann zeta function $\zeta(s)$. It is by an efficient management of this problem (for instance, by using the so-called *Riemann–Siegel formula*) that one is able to compute billions of nontrivial zeros of $\zeta(s)$. We will not consider these problems here, but concentrate on reasonable ranges of s.

The second problem is specific to general L-functions as opposed to L-functions attached to Dirichlet characters. For instance, in the general situation, we are given an L-function by an Euler product known outside of a finite and small number of "bad primes". Using recipes dating to the late 1960s and well explained in a beautiful paper of Serre [15], one can give the "gamma factor" $\gamma(s)$, and some (but not all) the information about the "conductor", which is the exponential factor, at least in the case of L-functions of varieties, or more generally of motives.

We will ignore these problems and assume that we know all the bad primes, gamma factor, conductor, and root number. Note that if we know the gamma factor and the bad primes, using the formulas that we will give below for different values of the argument it is easy to recover the conductor and the root number. What is most difficult to obtain are the Euler factors at the bad primes, and this is the object of current work.

7.2 Dirichlet L-Functions

Let χ be a Dirichlet character modulo N. We define the L-function attached to χ as the complex function

$$L(\chi, s) = \sum_{n\geq 1} \frac{\chi(n)}{n^s} .$$

Since $|\chi(n)| \leq 1$, it is clear that $L(\chi, s)$ converges absolutely for $\Re(s) > 1$. Furthermore, since χ is multiplicative, as for the Riemann zeta function we have an *Euler product*

$$L(\chi, s) = \prod_p \frac{1}{1 - \chi(p)/p^s} .$$

The denominator of this product being generically of degree 1, this is also called an L-function of degree 1, and conversely, with a suitable definition of the notion of L-function, one can show that these are the only L-functions of degree 1.

If f is the conductor of χ and χ_f is the character modulo f equivalent to χ, it is clear that

$$L(\chi, s) = \prod_{p\mid N,\, p\nmid f} (1 - \chi_f(p)p^{-s})L(\chi_f, s) ,$$

so if desired we can always reduce to primitive characters, and this is what we will do from now on.

Dirichlet L-series have important analytic and arithmetic properties, some of them conjectural (such as the Riemann Hypothesis), which should (again conjecturally) be shared by all global L-functions, see the discussion in the introduction. We first give the following:

Theorem 7.1 *Let χ be a* primitive *character modulo N, and let $e = 0$ or 1 be such that $\chi(-1) = (-1)^e$.*

1. *(Analytic continuation.) The function $L(\chi, s)$ can be analytically continued to the whole complex plane into a meromorphic function, which is in fact holomorphic except in the special case $N = 1$, $L(\chi, s) = \zeta(s)$, where it has a unique pole, at $s = 1$, which is simple with residue 1.*
2. *(Functional equation.) There exists a* functional equation *of the following form: letting $\gamma_{\mathbb{R}}(s) = \pi^{-s/2}\Gamma(s/2)$, we set*

$$\Lambda(\chi, s) = N^{(s+e)/2}\gamma_{\mathbb{R}}(s + e)L(\chi, s) ,$$

where e is as above. Then

$$\Lambda(\chi, 1-s) = \omega(\chi)\Lambda(\overline{\chi}, s) ,$$

where $\omega(\chi)$, the so-called root number, *is a complex number of modulus* 1 *given by the formula $\omega(\chi) = \mathfrak{g}(\chi)/(i^e N^{1/2})$.*

3. *(Special values.) For each integer $k \geq 1$ we have the* special values

$$L(\chi, 1-k) = -\frac{B_k(\chi)}{k} - \delta_{N,1}\delta_{k,1} ,$$

where δ is the Kronecker symbol, and the generalized Bernoulli numbers $B_k(\chi)$ *are easily computable algebraic numbers. In particular, when $k \not\equiv e \pmod 2$ we have $L(\chi, 1-k) = 0$ (except when $k = N = 1$).*
By the functional equation this is equivalent to the formula for $k \equiv e \pmod 2$, $k \geq 1$:

$$L(\chi, k) = (-1)^{k-1+(k+e)/2}\omega(\chi)\frac{2^{k-1}\pi^k\overline{B_k(\chi)}}{m^{k-1/2}k!} .$$

To state the next theorem, which for the moment we state for Dirichlet L-functions, we need still another important special function:

Definition 7.2 For $x > 0$ we define the *incomplete gamma function* $\Gamma(s, x)$ by

$$\Gamma(s, x) = \int_x^\infty t^s e^{-t}\,\frac{dt}{t} .$$

Note that this integral converges for *all* $s \in \mathbb{C}$, and that it tends to 0 exponentially fast when $x \to \infty$, more precisely $\Gamma(s, x) \sim x^{s-1}e^{-x}$. In addition (but this would carry us too far here) there are many efficient methods to compute it; see, however, the section on inverse Mellin transforms below.

Theorem 7.3 *Let χ be a* primitive *character modulo N. For all $A > 0$ we have:*

$$\Gamma\left(\frac{s+e}{2}\right)L(\chi, s) = \delta_{N,1}\pi^{s/2}\left(\frac{A^{(s-1)/2}}{s-1} - \frac{A^{s/2}}{s}\right) + \sum_{n\geq 1}\frac{\chi(n)}{n^s}\Gamma\left(\frac{s+e}{2}, \frac{\pi n^2 A}{N}\right)$$
$$+ \omega(\chi)\left(\frac{\pi}{N}\right)^{s-1/2}\sum_{n\geq 1}\frac{\overline{\chi}(n)}{n^{1-s}}\Gamma\left(\frac{1-s+e}{2}, \frac{\pi n^2}{AN}\right) .$$

Remarks 7.4 1. Thanks to this theorem, we can compute numerical values of $L(\chi, s)$ (for s in a reasonable range) in time $\widetilde{O}(N^{1/2})$.

2. The optimal value of A is $A = 1$, but the theorem is stated in this form for several reasons, one of them being that by varying A (for instance, taking $A = 1.1$ and $A = 0.9$) one can check the correctness of the implementation, or even compute the root number $\omega(\chi)$ if it is not known.

3. To compute values of $L(\chi, s)$ when $\Im(s)$ is large, one does not use the theorem as stated, but variants, see [13].
4. The above theorem, called the *approximate functional equation*, evidently implies the functional equation itself, so it seems to be more precise; however, this is an illusion since one can show that under very mild assumptions functional equations in a large class imply corresponding approximate functional equations.

7.3 Approximate Functional Equations

In fact, let us make this last statement completely precise. For the sake of simplicity we will assume that the L-functions have no poles (this corresponds for Dirichlet L-functions to the requirement that χ not be the trivial character). We begin by the following (where we restrict to certain kinds of gamma products, but it is easy to generalize; incidentally recall the *duplication formula* for the gamma function $\Gamma(s/2)\Gamma((s+1)/2) = 2^{1-s}\pi^{1/2}\Gamma(s)$, which allows the reduction of factors of the type $\Gamma(s+a)$ to several of the type $\Gamma(s/2+a')$ and conversely).

Definition 7.5 Recall that we have defined $\Gamma_{\mathbb{R}}(s) = \pi^{-s/2}\Gamma(s/2)$, which is the gamma factor attached to L-functions of even characters, for instance to $\zeta(s)$. A *gamma product* is a function of the type

$$\gamma(s) = f^{s/2} \prod_{1 \le i \le d} \Gamma_{\mathbb{R}}(s + b_j) ,$$

where $f > 0$ is a real number. The number d of gamma factors is called the *degree* of $\gamma(s)$.

Note that the b_j may not be real numbers, but in the case of L-functions attached to motives, they will always be, and in fact be integers.

Proposition 7.6 *Let γ be a gamma product.*

1. *There exists a function $W(t)$ called the* inverse Mellin transform *of γ such that*

$$\gamma(s) = \int_0^\infty t^s W(t)\, dt/t$$

for $\Re(s)$ sufficiently large (greater than the real part of the rightmost pole of $\gamma(s)$ suffices).

2. *$W(t)$ is given by the following* Mellin inversion formula *for $t > 0$:*

$$W(t) = \mathcal{M}^{-1}(\gamma)(t) = \frac{1}{2\pi i} \int_{\sigma - i\infty}^{\sigma + i\infty} t^{-s} \gamma(s)\, ds ,$$

for any σ larger than the real part of the poles of $\gamma(s)$.

3. *$W(t)$ tends to 0 exponentially fast when $t \to +\infty$. More precisely, as $t \to \infty$ we have*

$$W(t) \sim C \cdot (t/f^{1/2})^B \exp(-\pi d (t/f^{1/2})^{2/d})$$

with $B = (1 - d + \sum_{1 \le j \le d} b_j)/d$ and $C = 2^{(d+1)/2}/d^{1/2}$.

Definition 7.7 Let $\gamma(s)$ be a gamma product and $W(t)$ its inverse Mellin transform. The *incomplete gamma product* $\gamma(s, x)$ is defined for $x > 0$ by

$$\gamma(s, x) = \int_x^\infty t^s W(t) \frac{dt}{t} .$$

Note that this integral always converges since $W(t)$ tends to 0 exponentially fast when $t \to \infty$. In addition, thanks to the above proposition it is immediate to show the following:

Corollary 7.8 *1. For any σ larger than the real part of the poles of $\gamma(s)$ we have*

$$\gamma(s, x) = \frac{x^s}{2\pi i} \int_{\sigma - i\infty}^{\sigma + i\infty} \frac{x^{-z}\gamma(z)}{z - s} dz .$$

2. *For s fixed, as $x \to \infty$ we have with the same constants B and C as above*

$$\gamma(s, x) \sim \frac{C}{2\pi} x^s (x/f^{1/2})^{B - 2/d} \exp(-\pi d (x/f^{1/2})^{2/d})$$

so has essentially the same exponential decay as $W(x)$.

The first theorem, essentially due to Lavrik, which is an exercise in complex integration is as follows (recall that a function f is of *finite order* $\alpha \ge 0$ if for all $\varepsilon > 0$ and sufficiently large $|z|$ we have $|f(z)| \le \exp(|z|^{\alpha + \varepsilon})$):

Theorem 7.9 *For $i = 1$ and $i = 2$, let $L_i(s) = \sum_{n \ge 1} a_i(n) n^{-s}$ be Dirichlet series converging in some right half-plane $\Re(s) \ge \sigma_0$. For $i = 1$ and $i = 2$, let $\gamma_i(s)$ be gamma products having the same degree d. Assume that the functions $\Lambda_i(s) = \gamma_i(s) L_i(s)$ extend analytically to $\mathbb{C}$ into holomorphic functions of* finite order, *and that we have the functional equation*

$$\Lambda_1(k - s) = w \cdot \Lambda_2(s)$$

for some constant $w \in \mathbb{C}^*$ and some real number k.

Then for all $A > 0$, we have

$$\Lambda_1(s) = \sum_{n \ge 1} \frac{a_1(n)}{n^s} \gamma_1(s, nA) + w \sum_{n \ge 1} \frac{a_2(n)}{n^{k-s}} \gamma_2\Big(k - s, \frac{n}{A}\Big)$$

and symmetrically

$$\Lambda_2(s) = \sum_{n\geq 1} \frac{a_2(n)}{n^s}\gamma_2\Big(s, \frac{n}{A}\Big) + w^{-1}\sum_{n\geq 1}\frac{a_1(n)}{n^{k-s}}\gamma_1(k-s, nA)\;,$$

where $\gamma_i(s,x)$ are the corresponding incomplete gamma products.

Note that, as already mentioned, it is immediate to modify this theorem to take into account possible poles of $L_i(s)$.

Since the incomplete gamma products $\gamma_i(s,x)$ tend to 0 exponentially fast when $x \to \infty$, the above formulas are rapidly convergent series. We can make this more precise: if we write as above $\gamma_i(s,x) \sim C_i x^{B'_i} \exp(-\pi d(x/f_i^{1/2})^{2/d})$, since the convergence of the series is dominated by the exponential term, choosing $A = 1$, to have the nth term of the series less than e^{-D}, say, we need (approximately) $\pi d(n/f^{1/2})^{2/d} > D$, in other words $n > (D/(\pi d))^{d/2} f^{1/2}$, with $f = \max(f_1, f_2)$. Thus, if the "conductor" f is large, we may have some trouble. But this stays reasonable for $f < 10^8$, say.

The above argument leads to the belief that, apart from special values which can be computed by other methods, the computation of values of L-functions of conductor f requires at least $C \cdot f^{1/2}$ operations. It has, however, been shown by Hiary (see [10]), that if f is far from squarefree (for instance, if $f = m^3$ for Dirichlet L-functions), the computation can be done faster (in $\widetilde{O}(m)$ in the case $f = m^3$), at least in the case of Dirichlet L-functions.

For practical applications, it is very useful to introduce an additional function as a parameter. We state the following version due to Rubinstein (see [13]), whose proof is essentially identical to that of the preceding version. To simplify the exposition, we again assume that the L function has no poles (it is easy to generalize), but also that $L_2 = \overline{L_1}$.

Theorem 7.10 *Let $L(s) = \sum_{n\geq 1} a(n)n^{-s}$ be an L-function as above with functional equation $\Lambda(k-s) = w\overline{\Lambda}(s)$ with $\Lambda(s) = \gamma(s)L(s)$. For simplicity of exposition, assume that $L(s)$ has no poles in $\mathbb{C}$. Let $g(s)$ be an entire function such that for fixed s we have $|\Lambda(z+s)g(z+s)/z| \to 0$ as $\Im(z) \to \infty$ in any bounded strip $|\Re(z)| \leq \alpha$. We have*

$$\Lambda(s)g(s) = \sum_{n\geq 1}\frac{a(n)}{n^s} f_1(s,n) + \omega \sum_{n\geq 1}\frac{\overline{a(n)}}{n^{k-s}} f_2(k-s, n)\;,$$

where

$$f_1(s,x) = \frac{x^s}{2\pi i}\int_{\sigma-i\infty}^{\sigma+i\infty}\frac{\gamma(z)g(z)x^{-z}}{z-s}\,dz \quad \text{and} \quad f_2(s,x) = \frac{x^s}{2\pi i}\int_{\sigma-i\infty}^{\sigma+i\infty}\frac{\gamma(z)\overline{g(k-\overline{z})}x^{-z}}{z-s}\,dz\;,$$

where σ is any real number greater than the real parts of all the poles of $\gamma(z)$ and than $\Re(s)$.

Several comments are in order concerning this theorem:

1. As already mentioned, the proof is a technical but elementary exercise in complex analysis. In particular, it is very easy to modify the formula to take into account possible poles of $L(s)$, see [13] once again.
2. As in the unsmoothed case, the functions $f_i(s,x)$ are exponentially decreasing as $x \to \infty$. Thus, this gives fast formulas for computing values of $L(s)$ for reasonable values of s. The very simplest case of this approximate functional equation, even simpler than the Riemann zeta function, is for the computation of the value at $s = 1$ of the L-function of an *elliptic curve* E: if the sign of its functional equation is equal to $+1$ (otherwise $L(E,1) = 0$), the (unsmoothed) formula reduces to

$$L(E,1) = 2\sum_{n\geq 1} \frac{a(n)}{n} e^{-2\pi n/N^{1/2}} ,$$

where N is the conductor of the curve.
3. It is not difficult to show that as $n \to \infty$ we have a similar behavior for the functions $f_i(s,n)$ as in the unsmoothed case (Corollary 7.8), i.e.,

$$f_i(s,n) \sim C_i \cdot n^{B'_i} e^{-\pi d (n/N^{1/2})^{2/d}}$$

for some explicit constants C_i and B'_i (in the preceding example $d = 2$).
4. The theorem can be used with $g(s) = 1$ to compute values of $L(s)$ for "reasonable" values of s. When s is unreasonable, for instance, when $s = 1/2 + iT$ with T large (to check the Riemann hypothesis for instance), one chooses other functions $g(s)$ adapted to the computation to be done, such as $g(s) = e^{is\theta}$ or $g(s) = e^{-a(s-s_0)^2}$; I refer to Rubinstein's paper for detailed examples.
5. By choosing two very simple functions $g(s)$ such as a^s for two different values of a close to 1, one can compute numerically the value of the root number ω if it is unknown. In a similar manner, if the $a(n)$ are known but not ω nor the conductor N, by choosing a few easy functions $g(s)$ one can find them. But much more surprisingly, if almost nothing is known apart from the gamma factors and N, say, by cleverly choosing a number of functions $g(s)$ and applying techniques from numerical analysis such as singular value decomposition and least squares methods, one can prove or disprove (numerically of course) the existence of an L-function having the given gamma factors and conductor, and find its first few Fourier coefficients if they exist. This method has been used extensively by D. Farmer in his search for $\mathrm{GL}_3(\mathbb{Z})$ and $\mathrm{GL}_4(\mathbb{Z})$ Maass forms, by Poor and Yuen in computations related to the paramodular conjecture of Brumer–Kramer and abelian surfaces, and by A. Mellit in the search of L-functions of degree 4 with integer coefficients and small conductor. Although a fascinating and active subject, it would carry us too far afield to give more detailed explanations.

7.4 Inverse Mellin Transforms

We thus see that it is necessary to compute inverse Mellin transforms of some common gamma factors. Note that the exponential factors (either involving the conductor and/or π) are easily taken into account: if $\gamma(s) = \mathcal{M}(W)(s) = \int_0^\infty W(t)t^s\,dt/t$ is the Mellin transform of $W(t)$, we have for $a > 0$, setting $u = at$:

$$\int_0^\infty W(at)t^s\,dt/t = \int_0^\infty W(u)u^s a^{-s}\,du/u = a^{-s}\gamma(s)\;,$$

so the inverse Mellin transform of $a^{-s}\gamma(s)$ is simply $W(at)$.

As we have seen, there exists an explicit formula for the inverse Mellin transform, which is immediate from the Fourier inversion formula. We will see that although this looks quite technical, it is in practice very useful for computing inverse Mellin transforms.

Let us look at the simplest examples (omitting the exponential factor $f^{s/2}$ thanks to the above remark):

1. $\mathcal{M}^{-1}(\Gamma_{\mathbb{R}}(s)) = 2e^{-\pi x^2}$ (this occurs for L-functions of even characters, and in particular for $\zeta(s)$).
2. $\mathcal{M}^{-1}(\Gamma_{\mathbb{R}}(s+1)) = 2xe^{-\pi x^2}$ (this occurs for L-functions of odd characters).
3. $\mathcal{M}^{-1}(\Gamma_{\mathbb{C}}(s)) = 2e^{-2\pi x}$ (this occurs for L-functions attached to modular forms and to elliptic curves).
4. $\mathcal{M}^{-1}(\Gamma_{\mathbb{R}}(s)^2) = 4K_0(2\pi x)$ (this occurs, for instance, for Dedekind zeta functions of real quadratic fields). Here, $K_0(z)$ is a well-known special function called a K-Bessel function. Of course this is just a name, but it can be computed quite efficiently and can be found in all computer algebra packages.
5. $\mathcal{M}^{-1}(\Gamma_{\mathbb{C}}(s)^2) = 8K_0(4\pi x^{1/2})$.
6. $\mathcal{M}^{-1}(\Gamma_{\mathbb{C}}(s)\Gamma_{\mathbb{C}}(s-1)) = 8K_1(4\pi x^{1/2})/x^{1/2}$, where $K_1(z)$ is another K-Bessel function which can be defined by $K_1(z) = -K_0'(z)$.

Exercise 7.11 Prove all these formulas.

It is clear, however, that when the gamma factor is more complicated, we cannot write such "explicit" formulas, for instance, what must be done for $\gamma(s) = \Gamma_{\mathbb{C}}(s)\Gamma_{\mathbb{R}}(s)$ or $\gamma(s) = \Gamma_{\mathbb{R}}(s)^3$? In fact all of the above formulas involving K-Bessel functions are "cheats" in the sense that we have simply given a *name* to these inverse Mellin transform, without explaining how to compute them.

However, the Mellin inversion formula does provide such a method. The main point to remember (apart, of course, from the crucial use of the Cauchy residue formula and contour integration), is that the gamma function *tends to zero exponentially fast* on vertical lines, uniformly in the real part (this may seem surprising if you have never seen it since the gamma function grows so fast on the real axis, see appendix). This exponential decrease implies that in the Mellin inversion formula we can *shift* the line of integration without changing the value of the integral, as long as we take into account the residues of the poles which are encountered along the way.

The line $\Re(s) = \sigma$ has been chosen so that σ is larger than the real part of any pole of $\gamma(s)$, so shifting to the right does not bring anything. On the other hand, shifting toward the left shows that for any $r < 0$ not a pole of $\gamma(s)$ we have

$$W(t) = \sum_{\substack{s_0 \text{ pole of } \gamma(s) \\ \Re(s_0) > r}} \mathrm{Res}_{s=s_0}(t^{-s}\gamma(s)) + \frac{1}{2\pi i}\int_{r-i\infty}^{r+i\infty} t^{-s}\gamma(s)\,ds\,.$$

Using the reflection formula for the gamma function $\Gamma(s)\Gamma(1-s) = \pi/\sin(s\pi)$, it is easy to show that if r stays say half-way between the real part of two consecutive poles of $\gamma(s)$ then $\gamma(s)$ will tend to 0 exponentially fast on $\Re(s) = r$ as $r \to -\infty$, in other words that the integral tends to 0 (exponentially fast). We thus have the *exact formula*

$$W(t) = \sum_{s_0 \text{ pole of } \gamma(s)} \mathrm{Res}_{s=s_0}(t^{-s}\gamma(s))\,.$$

Let us see the simplest examples of this, taken from those given above.

1. For $\gamma(s) = \Gamma_{\mathbb{C}}(s) = 2 \cdot (2\pi)^{-s}\Gamma(s)$ the poles of $\gamma(s)$ are for $s_0 = -n$, n a positive or zero integer, and since $\Gamma(s) = \Gamma(s+n+1)/((s+n)(s+n-1)\cdots s)$, the residue at $s_0 = -n$ is equal to

$$2 \cdot (2\pi t)^n \Gamma(1)/((-1)(-2)\cdots(-n)) = (-1)^n(2\pi t)^n/n!\,,$$

   so we obtain $W(t) = 2\sum_{n\geq 0}(-1)^n(2\pi t)^n/n! = 2 \cdot e^{-2\pi t}$. Of course, we knew that!
2. For $\gamma(s) = \Gamma_{\mathbb{C}}(s)^2 = 4(2\pi)^{-2s}\Gamma(s)^2$, the inverse Mellin transform is $8K_0(4\pi x^{1/2})$ whose expansion we do *not* yet know. The poles of $\gamma(s)$ are again for $s_0 = -n$, but here all the poles are double poles, so the computation is slightly more complicated. More precisely, we have

$$\Gamma(s)^2 = \Gamma(s+n+1)^2/((s+n)^2(s+n-1)^2\cdots s^2)\,,$$

   so setting $s = -n + \varepsilon$ with ε small this gives

$$\begin{aligned}\Gamma(-n+\varepsilon)^2 &= \frac{\Gamma(1+\varepsilon)^2}{\varepsilon^2}\frac{1}{(1-\varepsilon)^2\cdots(n-\varepsilon)^2}\\ &= \frac{1+2\Gamma'(1)\varepsilon+O(\varepsilon^2)}{n!^2\varepsilon^2}(1+2\varepsilon/1)(1+2\varepsilon/2)\cdots(1+2\varepsilon/n)\\ &= \frac{1+2\Gamma'(1)\varepsilon+O(\varepsilon^2)}{n!^2\varepsilon^2}(1+2H_n\varepsilon)\,,\end{aligned}$$

   where we recall that $H_n = \sum_{1\leq j\leq n} 1/j$ is the harmonic sum. Since $(4\pi^2 t)^{-(-n+\varepsilon)} = (4\pi^2 t)^{n-\varepsilon} = (4\pi^2 t)^n(1 - \varepsilon\log(4\pi^2 t) + O(\varepsilon^2))$, it follows that

$$(4\pi^2 t)^{-(-n+\varepsilon)}\Gamma(-n+\varepsilon)^2 = \frac{(4\pi^2 t)^n}{n!^2\varepsilon^2}(1+\varepsilon(2H_n+2\Gamma'(1)-\log(4\pi^2 t)))\;,$$

so that the residue of $\gamma(s)$ at $s=-n$ is equal to $4((4\pi^2 t)^n/n!^2)(2H_n+2\Gamma'(1)-\log(4\pi^2 t))$. We thus have $2K_0(4\pi t^{1/2}) = \sum_{n\ge 0}((4\pi^2 t)^n/n!^2)(2H_n+2\Gamma'(1)-\log(4\pi^2 t))$, hence using the easily proven fact that $\Gamma'(1)=-\gamma$, where

$$\gamma = \lim_{n\to\infty}(H_n - \log(n)) = 0.57721566490\ldots$$

is Euler's constant, this gives finally the expansion

$$K_0(t) = \sum_{n\ge 0}\frac{(t/2)^{2n}}{n!^2}(H_n - \gamma - \log(t/2))\;.$$

Exercise 7.12 In a similar manner, or directly from this formula, find the expansion of $K_1(t)$.

Exercise 7.13 Like all inverse Mellin transforms of gamma factors, the function $K_0(x)$ tends to 0 exponentially fast as $x\to\infty$ (more precisely $K_0(x)\sim(2x/\pi)^{-1/2}e^{-x}$). Note that this is absolutely not "visible" on the expansion given above. Use this remark and the above expansion to write an algorithm which computes Euler's constant γ *very efficiently* to a given accuracy.

It must be remarked that even though the series defining the inverse Mellin transform converge for *all* $x>0$, one need a large number of terms before the terms become very small when x is large. For instance, we have seen that for $\gamma(s)=\Gamma(s)$ we have $W(t)=\mathcal{M}^{-1}(\gamma)(t)=\sum_{n\ge 0}(-1)^n t^n/n! = e^{-t}$, but this series is not very good for computing e^{-t}.

Exercise 7.14 Show that for $t>0$, to compute e^{-t} to any reasonable accuracy (even to 1 decimal) we must take at least $n>3.6\cdot t$ ($e=2.718...$), and work to accuracy at most e^{-2t} in an evident sense.

The reason that this is not a good way is that there is catastrophic cancelation in the series. One way to circumvent this problem is to compute e^{-t} as

$$e^{-t} = 1/e^t = 1/\sum_{n\ge 0}t^n/n!\;,$$

and the cancelation problem disappears. However, this is very special to the exponential function, and is not applicable, for instance, to the K-Bessel function.

Nonetheless, an important result is that for any inverse Mellin transform as above, or more importantly for the corresponding incomplete gamma product, there exist *asymptotic expansions* as $x\to\infty$, in other words nonconvergent series which, however, give a good approximation if limited to a few terms.

Let us take the simplest example of the incomplete gamma function $\Gamma(s,x) = \int_x^\infty t^s e^{-t}\,dt/t$. The *power series* expansion is easily seen to be (at least for s not a negative or zero integer, otherwise the formula must be slightly modified):

$$\Gamma(s,x) = \Gamma(s) - \sum_{n\geq 0}(-1)^n \frac{x^{n+s}}{n!(s+n)}\;,$$

which has the same type of (bad when x is large) convergence behavior as e^{-x}. On the other hand, it is immediate to prove by integration by parts that

$$\Gamma(s,x) = e^{-x}x^{s-1}\left(1 + \frac{s-1}{x} + \frac{(s-1)(s-2)}{x^2} + \cdots \right.$$
$$\left. + \frac{(s-1)(s-2)\cdots(s-n)}{x^n} + R_n(s,x)\right)\;,$$

and one can show that in reasonable ranges of s and x the modulus of $R_n(s,x)$ is smaller than the first "neglected term" in an evident sense. This is, therefore, quite a practical method for computing these functions when x is rather large.

Exercise 7.15 Explain why the asymptotic series above terminates when s is a strictly positive integer.

7.5 *Hadamard Products and Explicit Formulas*

This could be the subject of a course in itself, so we will be quite brief. I refer to Mestre's paper [11] for a precise and general statement (note that there are quite a number of evident misprints in the paper).

In Theorem 7.9, we assume that the L-series that we consider satisfy a functional equation, together with some mild growth conditions, in particular that they are of finite order. According to a well-known theorem of complex analysis, this implies that they have a so-called *Hadamard product*, see appendix. For instance, in the case of the Riemann zeta function, which is of order 1, we have

$$\zeta(s) = \frac{e^{bs}}{s(s-1)\Gamma(s/2)} \prod_\rho \left(1 - \frac{s}{\rho}\right) e^{s/\rho}\;,$$

where the product is over all nontrivial zeros of $\zeta(s)$ (i.e., such that $0 \leq \Re(\rho) \leq 1$), and $b = \log(2\pi) - 1 - \gamma$. In fact, this can be written in a much nicer way as follows: recall that $\Lambda(s) = \pi^{-s/2}\Gamma(s/2)\zeta(s)$ satisfies $\Lambda(1-s) = \Lambda(s)$. Then

$$s(s-1)\Lambda(s) = \prod_\rho \left(1 - \frac{s}{\rho}\right)\;,$$

where it is now understood that the product is taken as the limit as $T \to \infty$ of $\prod_{|\Im(\rho)| \le T}(1 - s/\rho)$.

However, almost all L-functions that are used in number theory not only have the above properties, but have also *Euler products*. Taking again the example of $\zeta(s)$, we have for $\Re(s) > 1$ the Euler product $\zeta(s) = \prod_p (1 - 1/p^s)^{-1}$. It follows that (in a suitable range of s) we have equality between two products, hence taking logarithms, equality between two *sums*. In our case the Hadamard product gives

$$\log(\Lambda(s)) = -\log(s(s-1)) + \sum_\rho \log(1 - s/\rho) ,$$

while the Euler product gives

$$\begin{aligned}\log(\Lambda(s)) &= -(s/2)\log(\pi) + \log(\Gamma(s/2)) - \sum_p \log(1 - 1/p^s) \\ &= -(s/2)\log(\pi) + \log(\Gamma(s/2)) + \sum_{p,k \ge 1} 1/(kp^{ks}) ,\end{aligned}$$

Equating the two sides gives a relation between on the one hand a sum over the nontrivial zeros of $\zeta(s)$, and on the other hand a sum over prime powers.

In itself, this is not very useful. The crucial idea is to introduce a test function F which we will choose to the best of our interests, and obtain a formula depending on F and some transforms of it.

This is in fact quite easy to do, and even though not very useful in this case, let us perform the computation for Dirichlet L-function of even primitive characters.

Theorem 7.16 *Let χ be an even primitive Dirichlet character of conductor N, and let F be a real function satisfying a number of easy technical conditions (see [11]). We have the* explicit formula*:*

$$\begin{aligned}&\sum_\rho \Phi(\rho) - 2\delta_{N,1} \int_{-\infty}^{\infty} F(x)\cosh(x/2)\,dx \\ &\quad = -\sum_{p,k \ge 1} \frac{\log(p)}{p^{k/2}}\left(\chi^k(p)F(k\log(p)) + \overline{\chi^k(p)}F(-k\log(p))\right) \\ &\quad + F(0)\log(N/\pi) \\ &\quad + \int_0^\infty \left(\frac{e^{-x}}{x}F(0) - \frac{e^{-x/4}}{1-e^{-x}}\frac{F(x/2)+F(-x/2)}{2}\right)dx ,\end{aligned}$$

where we set

$$\Phi(s) = \int_{-\infty}^{\infty} F(x)e^{(s-1/2)x}\,dx ,$$

and as above the sum on ρ is a sum over all the nontrivial zeros of $L(\chi, s)$ taken symmetrically ($\sum_\rho = \lim_{T\to\infty} \sum_{|\Im(\rho)|\le T}$).

Remarks 7.17 1. Write $\rho = 1/2 + i\gamma$ (if the GRH is true all γ are real, but even without GRH we can always write this). Then

$$\Phi(\rho) = \int_{-\infty}^{\infty} F(x)e^{i\gamma x}\,dx = \widehat{F}(\gamma)$$

is simply the value at γ of the *Fourier transform* $\widehat{F}$ of F.
2. It is immediate to generalize to odd χ or more general L-functions:

Exercise 7.18 After studying the proof, generalize to an arbitrary pair of L-functions as in Theorem 7.9.

Proof The proof is not difficult, but involves a number of integral transform computations. We will omit some detailed justifications which are in fact easy but boring.

As in the theorem, we set

$$\Phi(s) = \int_{-\infty}^{\infty} F(x)e^{(s-1/2)x}\,dx\,,$$

and we first prove some lemmas.

Lemma 7.19 *We have the inversion formulas valid for any $c > 1$:*

$$F(x) = e^{x/2}\int_{c-i\infty}^{c+i\infty} \Phi(s)e^{-sx}\,ds\,.$$

$$F(-x) = e^{x/2}\int_{c-i\infty}^{c+i\infty} \Phi(1-s)e^{-sx}\,ds\,.$$

Proof This is in fact a hidden version of the Mellin inversion formula: setting $t = e^x$ in the definition of $\Phi(s)$, we deduce that $\Phi(s) = \int_0^\infty F(\log(t))t^{s-1/2}\,dt/t$, so that $\Phi(s+1/2)$ is the Mellin transform of $F(\log(t))$. By Mellin inversion, we thus have for sufficiently large σ:

$$F(\log(t)) = \frac{1}{2\pi i}\int_{\sigma-i\infty}^{\sigma+i\infty} \Phi(s+1/2)t^{-s}\,ds\,,$$

so changing s into $s-1/2$ and t into e^x gives the first formula for $c = \sigma + 1/2$ sufficiently large, and the assumptions on F (which we have not given) imply that we can shift the line of integration to any $c > 1$ without changing the integral.

For the second formula, we simply note that

$$\Phi(1-s) = \int_{-\infty}^{\infty} F(x)e^{-(s-1/2)x}\,dx = \int_{-\infty}^{\infty} F(-x)e^{(s-1/2)x}\,dx\;,$$

so we simply apply the first formula to $F(-x)$. □

Corollary 7.20 *For any $c > 1$ and any $p \geq 1$ we have*

$$\int_{c-i\infty}^{c+i\infty} \Phi(s)p^{-ks}\,ds = F(k\log(p))p^{-k/2} \quad \text{and}$$
$$\int_{c-i\infty}^{c+i\infty} \Phi(1-s)p^{-ks}\,ds = F(-k\log(p))p^{-k/2}\;.$$

Proof Simply apply the lemma to $x = k\log(p)$. □

Note that we will also use this corollary for $p = 1$.

Lemma 7.21 *Denote as usual by $\psi(s)$ the logarithmic derivative $\Gamma'(s)/\Gamma(s)$ of the gamma function. We have*

$$\int_{c-i\infty}^{c+i\infty} \Phi(s)\psi(s/2) = \int_0^{\infty}\left(\frac{e^{-x}}{x}F(0) - \frac{e^{-x/4}}{1-e^{-x}}F(x/2)\right)dx \quad \text{and}$$
$$\int_{c-i\infty}^{c+i\infty} \Phi(1-s)\psi(s/2) = \int_0^{\infty}\left(\frac{e^{-x}}{x}F(0) - \frac{e^{-x/4}}{1-e^{-x}}F(-x/2)\right)dx\;.$$

Proof We use one of the most common integral representations of ψ, see Proposition 9.6.43 of [4]: we have

$$\psi(s) = \int_0^{\infty}\left(\frac{e^{-x}}{x} - \frac{e^{-sx}}{1-e^{-x}}\right)dx\;.$$

Thus, assuming that we can interchange integrals (which is easy to justify), we have, using the preceding lemma:

$$\begin{aligned}\int_{c-i\infty}^{c+i\infty} \Phi(s)\psi(s/2)\,ds &= \int_0^{\infty}\left(\frac{e^{-x}}{x}\int_{c-i\infty}^{c+i\infty}\Phi(s)\,ds\right.\\ &\qquad\left.-\frac{1}{1-e^{-x}}\int_{c-i\infty}^{c+i\infty}\Phi(s)e^{-(s/2)x}\,ds\right)dx\\ &= \int_0^{\infty}\left(\frac{e^{-x}}{x}F(0) - \frac{e^{-x/4}}{1-e^{-x}}F(x/2)\right)dx\;,\end{aligned}$$

proving the first formula, and the second follows by changing $F(x)$ into $F(-x)$. □

Proof of the theorem. Recall from above that if we set $\Lambda(s) = N^{s/2}\pi^{-s/2}\Gamma(s/2) L(\chi, s)$ we have the functional equation $\Lambda(1-s) = \omega(\chi)\Lambda(\overline{\chi}, s)$ for some $\omega(\chi)$ of modulus 1.

For $c > 1$, consider the following integral

$$J = \frac{1}{2i\pi}\int_{c-i\infty}^{c+i\infty} \Phi(s)\frac{\Lambda'(s)}{\Lambda(s)}\,ds\;,$$

which by our assumptions does not depend on $c > 1$. We shift the line of integration to the left (it is easily seen that this is allowed) to the line $\Re(s) = 1 - c$, so by the residue theorem we obtain

$$J = S + \frac{1}{2i\pi}\int_{1-c-i\infty}^{1-c+i\infty} \Phi(s)\frac{\Lambda'(s)}{\Lambda(s)}\,ds\;,$$

where S is the sum of the residues in the rectangle $[1-c, c] \times \mathbb{R}$. We first have possible poles at $s = 0$ and $s = 1$, which occur only for $N = 1$, and they contribute to S

$$-\delta_{N,1}(\Phi(0) + \Phi(1)) = -2\delta_{N,1}\int_{-\infty}^{\infty} F(x)\cosh(x/2)\,dx\;,$$

and of course second we have the contributions from the nontrivial zeros ρ, which contribute $\sum_\rho \Phi(\rho)$, where it is understood that zeros are counted with multiplicity, so that

$$S = -2\delta_{N,1}\int_{-\infty}^{\infty} F(x)\cosh(x/2)\,dx + \sum_\rho \Phi(\rho)\;.$$

On the other hand, by the functional equation we have $\Lambda'(1-s)/\Lambda(1-s) = -\overline{\Lambda}'(s)/\overline{\Lambda}(s)$ (note that this does not involve $\omega(\chi)$), where we write $\overline{\Lambda}(s)$ for $\Lambda(\overline{\chi}, s)$, so that

$$\begin{aligned}\int_{1-c-i\infty}^{1-c+i\infty} \Phi(s)\frac{\Lambda'(s)}{\Lambda(s)}\,ds &= \int_{c-i\infty}^{c+i\infty} \Phi(1-s)\frac{\Lambda'(1-s)}{\Lambda(1-s)}\,ds\\ &= -\int_{c-i\infty}^{c+i\infty} \Phi(1-s)\frac{\overline{\Lambda}'(s)}{\overline{\Lambda}(s)}\,ds\;.\end{aligned}$$

Thus,

$$\begin{aligned}S &= J - \frac{1}{2i\pi}\int_{1-c-i\infty}^{1-c+i\infty} \Phi(s)\frac{\Lambda'(s)}{\Lambda(s)}\,ds\\ &= \frac{1}{2i\pi}\int_{c-i\infty}^{c+i\infty}\left(\Phi(s)\frac{\Lambda'(s)}{\Lambda(s)} + \Phi(1-s)\frac{\overline{\Lambda}'(s)}{\overline{\Lambda}(s)}\right)ds\;.\end{aligned}$$

Now by definition, we have as above

$$\log(\Lambda(s)) = \frac{s}{2}\log(N/\pi) + \log\left(\Gamma\left(\frac{s}{2}\right)\right) + \sum_{p,k\geq 1} \frac{\chi^k(p)}{kp^{ks}}$$

(where the double sum is over primes and integers $k \geq 1$), so

$$\frac{\Lambda'(s)}{\Lambda(s)} = \frac{1}{2}\log(N/\pi) + \frac{1}{2}\psi(s/2) - \sum_{p,k\geq 1} \chi^k(p)\log(p)p^{-ks}\;,$$

and similarly for $\overline{\Lambda}'(s)/\overline{\Lambda}(s)$. Thus, by the above lemmas and corollaries, we have

$$S = \log(N/\pi)F(0) + J_1 - \sum_{p,k\geq 1} \frac{\log(p)}{p^{k/2}}(\chi^k(p)F(k\log(p)) + \overline{\chi^k(p)}F(-k\log(p)))\;,$$

where

$$J_1 = \int_0^\infty \left(\frac{e^{-x}}{x}F(0) - \frac{e^{-x/4}}{1-e^{-x}}\frac{F(x/2)+F(-x/2)}{2}\right)dx\;,$$

proving the theorem. □

This theorem can be used in several different directions, and has been an extremely valuable tool in analytic number theory. Just to mention a few:

1. Since the conductor N occurs, we can obtain *bounds* on N, assuming certain conjectures such as the generalized Riemann hypothesis. For instance, this is how Stark–Odlyzko–Poitou–Serre find *lower bounds for discriminants* of number fields. This is also how Mestre finds lower bounds for conductors of abelian varieties, and so on.
2. When the L-function has a zero at its central point (here of course it usually does not, but for more general L-functions it is important), this can give good upper bounds for the order of the zero.
3. More generally, suitable choices of the test functions can give information on the nontrivial zeros ρ of small imaginary part.

8 Some Useful Analytic Computational Tools

We finish this course by giving a number of little-known numerical methods which are not always directly related to the computation of L-functions, but which are often very useful.

8.1 The Euler–Maclaurin Summation Formula

This numerical method is *very* well known (there is in fact even a whole chapter in Bourbaki devoted to it!), and is as old as Taylor's formula, but deserves to be mentioned since it is very useful. We will be vague on purpose, and refer to [1] or Sect. 9.2 of [4] for details. Recall that the *Bernoulli numbers* are defined by the formal power series

$$\frac{T}{e^T-1}=\sum_{n\geq 0}\frac{B_n}{n!}T^n\,.$$

We have $B_0=0$, $B_1=-1/2$, $B_2=1/6$, $B_3=0$, $B_4=-1/30$, and $B_{2k+1}=0$ for $k\geq 1$.

Let f be a C^∞ function defined on $\mathbb{R}>0$. The basic statement of the Euler–MacLaurin formula is that there exists a constant $z=z(f)$ such that

$$\sum_{n=1}^{N}f(n)=\int_1^N f(t)\,dt+z(f)+\frac{f(N)}{2}+\sum_{1\leq k\leq p}\frac{B_{2k}}{(2k)!}f^{(2k-1)}(N)+R_p(N)\,,$$

where $R_p(N)$ is "small", in general smaller than the first neglected term, as in most asymptotic series.

The above formula can be slightly modified at will, first by changing the lower bound of summation and/or of integration (which simply changes the constant $z(f)$), and second by writing $\int_1^N f(t)\,dt+z(f)=z'(f)-\int_N^\infty f(t)\,dt$ (when f tends to 0 sufficiently fast for the integral to converge), where $z'(f)=z(f)+\int_1^\infty f(t)\,dt$.

The Euler–MacLaurin summation formula can be used in many contexts, but we mention the two most important ones.

• First, to have some idea of the size of $\sum_{n=1}^N f(n)$. Let us take an example. Consider $S_2(N)=\sum_{n=1}^N n^2\log(n)$. Note incidentally that

$$\exp(S_2(N))=\prod_{n=1}^{N}n^{n^2}=1^{1^2}2^{2^2}\cdots N^{N^2}\,.$$

What is the size of this generalized kind of factorial? Euler–MacLaurin tells us that there exists a constant z such that

$$S_2(N)=\int_1^N t^2\log(t)\,dt+z+\frac{N^2\log(N)}{2}+\frac{B_2}{2!}(N^2\log(N))'+\frac{B_4}{4!}(N^2\log(N))'''+\cdots\,.$$

We have $\int_1^N t^2\log(t)\,dt=(N^3/3)\log(N)-(N^3-1)/9$, $(N^2\log(N))'=2N\log(N)+N$, $(N^2\log(N))''=2\log(N)+3$, and $(N^2\log(N))'''=2/N$, so using $B_2=$

$1/6$ we obtain for some other constant z':

$$S_2(N) = \frac{N^3\log(N)}{3} - \frac{N^3}{9} + \frac{N^2\log(N)}{2} + \frac{N\log(N)}{6} + \frac{N}{12} + z' + O\left(\frac{1}{N}\right) ,$$

which essentially answers our question, up to the determination of the constant z'. Thus we obtain a generalized Stirling's formula:

$$\exp(S_2(N)) = N^{N^3/3+N^2/2+N/6}e^{-(N^3/9-N/12)}C ,$$

where $C = \exp(z')$ is an a priori unknown constant. In the case of the usual Stirling's formula we have $C = (2\pi)^{1/2}$, so we can ask for a similar formula here. And indeed, such a formula exists: we have

$$C = \exp(\zeta(3)/(4\pi^2)) .$$

Exercise 8.1 Do a similar (but simpler) computation for $S_1(N) = \sum_{1\le n\le N} n\log(n)$. The corresponding constant is explicit but more difficult (it involves $\zeta'(-1)$; more generally the constant in $S_r(N)$ involves $\zeta'(-r)$).

• The second use of the Euler–MacLaurin formula is to increase considerably the speed of convergence of slowly convergent series. For instance, if you want to compute $\zeta(3)$ directly using the series $\zeta(3) = \sum_{n\ge1} 1/n^3$, since the remainder term after N terms is asymptotic to $1/(2N^2)$ you will never get more than 15 or 20 decimals of accuracy. On the other hand, it is immediate to use Euler–MacLaurin:

Exercise 8.2 Write a computer program implementing the computation of $\zeta(3)$ (and more generally of $\zeta(s)$ for reasonable s) using Euler–MacLaurin, and compute it to 100 decimals.

A variant of the method is to compute limits: a typical example is the computation of Euler's constant

$$\gamma = \lim_{N\to\infty}\left(\sum_{n=1}^{N}\frac{1}{n} - \log(N)\right) .$$

Using Euler–MacLaurin, it is immediate to find the *asymptotic expansion*

$$\sum_{n=1}^{N}\frac{1}{n} = \log(N) + \gamma + \frac{1}{2N} - \sum_{k\ge1}\frac{B_{2k}}{2kN^{2k}}$$

(note that this is not a misprint, the last denominator is $2kN^{2k}$, not $(2k)!N^{2k}$).

Exercise 8.3 Implement the above, and compute γ to 100 decimal digits.

Note that this is *not* the fastest way to compute Euler's constant, the method using Bessel functions given in Exercise 7.13 is better.

8.2 Variant: Discrete Euler–MacLaurin

One problem with the Euler–MacLaurin method is that we need to compute the derivatives $f^{(2k-1)}(N)$. When k is tiny, say $k = 2$ or $k = 3$ this can be done explicitly. When $f(x)$ has a special form, such as $f(x) = 1/x^\alpha$, it is very easy to compute all derivatives. In fact, this is more generally the case when the expansion of $f(1/x)$ around $x = 0$ is known explicitly. But in general none of this is available.

One way around this is to use finite differences instead of derivatives: we can easily compute

$$\Delta_\delta(f)(x) = (f(x+\delta) - f(x-\delta))/(2\delta)$$

and iterates of this, where δ is some fixed and nonzero number. The choice of δ is essential: it should not be too large, otherwise $\Delta_\delta(f)$ would be too far away from the true derivative (which will be reflected in the speed of convergence of the asymptotic formula), and it should not be too small, otherwise catastrophic cancelation errors will occur. After numerous trials, the value $\delta = 1/4$ seems reasonable.

One last thing must be done: find the analogue of the Bernoulli numbers. This is a very instructive exercise which we leave to the reader.

8.3 Zagier's Extrapolation Method

The following nice trick is due to D. Zagier. Assume that you have a sequence u_n that you suspect of converging to some limit a_0 when $n \to \infty$ in a regular manner. How do you give a reasonable numerical estimate of a_0?

Assume, for instance, that as $n \to \infty$ we have $u_n = \sum_{0 \le i \le p} a_i/n^i + O(n^{-p-1})$ for any p. One idea would be to choosing for n suitable values and solve a linear system. This would in general be quite unstable and inaccurate. Zagier's trick is instead to proceed as follows: choose some reasonable integer k, say $k = 10$, set $u'_n = n^k u_n$, and compute the kth *forward difference* $\Delta^k(u'_n)$ of this sequence (the forward difference of a sequence w_n is the sequence $\Delta(w)_n = w_{n+1} - w_n$). Note that

$$u'_n = a_0 n^k + \sum_{1 \le i \le k} a_i n^{k-i} + O(1/n) .$$

The two crucial points are the following:

- The kth forward difference of a polynomial of degree less than or equal to $k - 1$ vanishes, and that of n^k is equal to $k!$.
- Assuming reasonable regularity conditions, the kth forward difference of an asymptotic expansion beginning at $1/n$ will begin at $1/n^{k+1}$.

Thus, under reasonable assumptions we have

$$a_0 = \Delta^k(v)_n/k! + O(1/n^{k+1}) ,$$

so choosing n large enough can give a good estimate for a_0.

A number of remarks concerning this basic method:

Remarks 8.4 1. It is usually preferable to apply this not to the sequence u_n itself, but for instance to the sequence u_{n+100}, if it is not too expensive to compute, since the first terms of u_n are usually far from the asymptotic expansion.

2. It is immediate to modify the method to compute further coefficients a_1, a_2, etc.
3. If the asymptotic expansion of u_n is (for instance) in powers of $1/n^{1/2}$, it is not difficult to modify this method, see below.

Example. Let us compute numerically the constant occurring in the first example of the use of Euler–MacLaurin that we have given. We set

$$u_N = \sum_{1\le n\le N} n^2\log(n) - (N^3/3 + N^2/2 + N/6)\log(N) + N^3/9 - N/12 .$$

We compute, for instance, that $u_{1000} = 0.0304456\cdots$, which has only 4 correct decimal digits. On the other hand, if we apply the above trick with $k = 12$ and $N = 100$, we find

$$a_0 = \lim_{N\to\infty} u_N = 0.0304484570583932707802515304696767\cdots$$

with 28 correct decimal digits: recall that the exact value is

$$\zeta(3)/(4\pi^2) = 0.0304484570583932707802515304711547 7\cdots .$$

Assume now that u_n has an asymptotic expansion in integral powers of $1/n^{1/2}$, i.e., $u_n = \sum_{0\le i\le p} a_i/n^{i/2} + O(n^{-(p+1)/2})$ for any p. We can modify the above method as follows. First write $u_n = v_n + w_n/n^{1/2}$, where $v_n = \sum_{0\le i\le q} a_{2i}/n^i + O(n^{-q-1})$ and $w_n = \sum_{0\le i\le q} a_{2i+1}/n^i + O(n^{-q-1})$ are two sequences as above. Once again we choose some reasonable integer k such as $k = 10$, and we now multiply the sequence u_n by $n^{k-1/2}$, so we set $u'_n = n^{k-1/2}u_n = n^{k-1/2}v_n + n^{k-1}w_n$. Thus, when we compute the kth forward difference we will have

$$\Delta^k(n^{k-1/2}v_n) = \frac{(k-1/2)(k-3/2)\cdots 1/2}{n^{1/2}}\left(a_0 + \sum_{0\le i\le q+k} b_{k,i}/n^i\right)$$

for certain coefficients $b_{k,i}$, while as above since $n^{k-1}w_n = P_{k-1}(n) + O(1/n)$ for some polynomial $P_{k-1}(n)$ of degree $k-1$, we have $\Delta^k(n^{k-1}w_n) = O(1/n^k)$. Thus we have essentially eliminated the sequence w_n, so we now apply the usual method to $v'_n = n^{1/2}\Delta^k(n^{k-1/2}v_n)$, which has an expansion in integral powers of $1/n$: we will thus have

$$\Delta^k(v'_n)/k! = ((k-1/2)(k-3/2)\cdots(1/2))a(0) + O(1/n^k)$$

(in fact we do not even have to take the same k for this last step).

This method can immediately be generalized to sequences u_n having an asymptotic expansion in integral powers of $n^{1/q}$ for small integers q.

8.4 Computation of Euler Sums and Euler Products

Assume that we want to compute numerically

$$S_1 = \prod_p \left(1 + \frac{1}{p^2}\right) ,$$

where here and elsewhere, the expression $\prod_p$ always means the product over all prime numbers. Trying to compute it using a large table of prime numbers will not give much accuracy: if we use primes up to X, we will make an error of the order of $1/X$, so it will be next to impossible to have more than 8 or 9 decimal digits.

On the other hand, if we simply notice that $1 + 1/p^2 = (1 - 1/p^4)/(1 - 1/p^2)$, by definition of the Euler product for the Riemann zeta function this implies that

$$S_1 = \frac{\zeta(2)}{\zeta(4)} = \frac{\pi^2/6}{\pi^4/90} = \frac{15}{\pi^2} = 1.519817754635066571658\cdots$$

Unfortunately, this is based on a special identity. What if we wanted instead to compute $S_2 = \prod_p (1 + 2/p^2)$? There is no special identity to help us here.

The way around this problem is to approximate the function of which we want to take the product (here $1 + 2/p^2$) by *infinite products* of values of the Riemann zeta function. Let us do it step by step before giving the general formula.

When p is large, $1 + 2/p^2$ is close to $1/(1 - 1/p^2)^2$, which is the Euler factor for $\zeta(2)^2$. More precisely, $(1 + 2/p^2)(1 - 1/p^2)^2 = 1 - 3/p^4 + 2/p^6$, so we deduce that

$$S_2 = \zeta(2)^2 \prod_p (1 - 3/p^4 + 2/p^6) = (\pi^4/36) \prod_p (1 - 3/p^4 + 2/p^6) .$$

Even though this looks more complicated, what we have gained is that the new Euler product converges *much* faster. Once again, if we compute it for p up to 10^8, say, instead of having 8 decimal digits we now have approximately 24 decimal digits (convergence in $1/X^3$ instead of $1/X$). But there is no reason to stop there: we have $(1 - 3/p^4 + 2/p^6)/(1 - 1/p^4)^3 = 1 + O(1/p^6)$ with evident notation and explicit formulas if desired, so we get an even better approximation by writing $S_2 = \zeta(2)^2/\zeta(4)^3 \prod_p (1 + O(1/p^6))$, with convergence in $1/X^5$. More generally, it is easy to compute by induction exponents $a_n \in \mathbb{Z}$ such that $S_2 = \prod_{2\le n\le N} \zeta(n)^{a_n} \prod_p(1 + O(1/p^{N+1}))$ (in our case $a_n = 0$ for n odd but this will not be true in general). It can be shown in essentially all examples that one can pass to the limit, and, for instance, here write $S_2 = \prod_{n\ge 2} \zeta(n)^{a_n}$.

Exercise 8.5 1. Compute explicitly the recursion for the a_n in the example of S_2.

2. More generally, if $S = \prod_p f(p)$, where $f(p)$ has a convergent series expansion in $1/p$ starting with $f(p) = 1 + 1/p^b + o(1/p^b)$ with $b > 1$ (not necessarily integral), express S as a product of zeta values raised to suitable exponents, and find the recursion for these exponents.

An important remark needs to be made here: even though the product $\prod_{n\ge 2} \zeta(n)^{a_n}$ may be convergent, it may converge rather slowly: remember that when n is large we have $\zeta(n) - 1 \sim 1/2^n$, so that in fact if the a_n grow like 3^n the product will not even converge. The way around this, which must be used even when the product converges, is as follows: choose a reasonable integer N, for instance $N = 50$, and compute $\prod_{p\le 50} f(p)$, which is of course very fast. Then the tail $\prod_{p>50} f(p)$ of the Euler product will be equal to $\prod_{n\ge 2} \zeta_{>50}(n)^{a_n}$, where $\zeta_{>N}(n)$ is the zeta function without its Euler factors up to N, in other words $\zeta_{>N}(n) = \zeta(n) \prod_{p\le N}(1 - 1/p^n)$. (I am assuming here that we have zeta values at integers as in the S_2 example above, but it is immediate to generalize.) Since $\zeta_{>N}(n) - 1 \sim 1/(N + 1)^n$, the convergence of our zeta product will of course be considerably faster.

Note that by using the power series expansion of the logarithm together with *Möbius inversion*, it is immediate to do the same for Euler *sums*, for instance, to compute $\sum_p 1/p^2$ and the like, see Sect. 10.3.6 of [4] for details. Using *derivatives* of the zeta function we can compute Euler sums of the type $\sum_p \log(p)/p^2$, and using antiderivatives we can compute sums of the type $\sum_p 1/(p^2 \log(p))$. We can even compute sums of the form $\sum_p \log(\log(p))/p^2$, but this is slightly more subtle: it involves taking derivatives with respect to the order of *fractional derivation*.

We can also compute products and sums over primes which involve Dirichlet characters, as long as their conductor is small, as well as such products and sums where the primes are restricted to certain congruence classes:

Exercise 8.6 Compute to 100 decimal digits

$$\prod_{p\equiv 1 \pmod 4} (1 - 1/p^2) \quad \text{and} \quad \prod_{p\equiv 1 \pmod 4} (1 + 1/p^2)$$

by using products of $\zeta(ns)$ and of $L(\chi_{-4}, ns)$ as above, where as usual χ_{-4} is the character $\left(\frac{-4}{n}\right)$.

8.5 Summation of Alternating Series

This is due to F. Rodriguez–Villegas, D. Zagier, and the author [7].

We have seen above the use of the Euler–Maclaurin summation formula to sum quite general types of series. If the series is *alternating* (the terms alternate in sign), the method cannot be used as is, but it is trivial to modify it: simply write

$$\sum_{n\geq 1}(-1)^n f(n) = \sum_{n\geq 1} f(2n) - \sum_{n\geq 1} f(2n-1)$$

and apply Euler–Maclaurin to each sum. One can even do better and avoid this double computation, but this is not what I want to mention here.

A completely different method which is much simpler since it avoids completely the computation of derivatives and Bernoulli numbers, due to the above authors, is as follows. The idea is to express (if possible) $f(n)$ as a *moment*

$$f(n) = \int_0^1 x^n w(x)\, dx$$

for some *weight function* $w(x)$. Then it is clear that

$$S = \sum_{n\geq 0}(-1)^n f(n) = \int_0^1 \frac{1}{1+x} w(x)\, dx\ .$$

Assume that $P_n(X)$ is a polynomial of degree n such that $P_n(-1) \neq 0$. Evidently

$$\frac{P_n(X) - P_n(-1)}{X+1} = \sum_{k=0}^{n-1} c_{n,k} X^k$$

is still a polynomial (of degree $n-1$), and we note the trivial fact that

$$\begin{aligned}
S &= \frac{1}{P_n(-1)} \int_0^1 \frac{P_n(-1)}{1+x} w(x)\, dx \\
&= \frac{1}{P_n(-1)} \left(\int_0^1 \frac{P_n(-1) - P_n(x)}{1+x} w(x)\, dx + \int_0^1 \frac{P_n(x)}{1+x} w(x)\, dx \right) \\
&= \frac{1}{P_n(-1)} \sum_{k=0}^{n-1} c_{n,k} f(k) + R_n\ ,
\end{aligned}$$

with

$$|R_n| \leq \frac{M_n}{|P_n(-1)|} \int_0^1 \frac{1}{1+x} w(x)\, dx = \frac{M_n}{|P_n(-1)|} S\ ,$$

and where $M_n = \sup_{x\in[0,1]} |P_n(x)|$. Thus, if we can manage to have $M_n/|P_n(-1)|$ small, we obtain a good approximation to S.

It is a classical result that the best choice for P_n are the shifted Chebychev polynomials defined by $P_n(\sin^2(t)) = \cos(2nt)$, but in any case we can use these polynomials and ignore that they are the best.

This leads to an incredibly simple algorithm which we write explicitly:

$d \leftarrow (3+\sqrt{8})^n$; $d \leftarrow (d+1/d)/2$; $b \leftarrow -1$; $c \leftarrow -d$; $s \leftarrow 0$; For $k = 0, \dots, n-1$ do:

$c \leftarrow b - c$; $s \leftarrow s + c \cdot f(k)$; $b \leftarrow (k+n)(k-n)b/((k+1/2)(k+1))$;

The result is s/d.

The convergence is in 5.83^{-n}.

It is interesting to note that, even though this algorithm is designed to work with functions f of the form $f(n) = \int_0^1 x^n w(x)\,dx$ with w continuous and positive, it is in fact valid outside its proven region of validity. For example:

Exercise 8.7 It is well known that the Riemann zeta function $\zeta(s)$ can be extended analytically to the whole complex plane, and that we have, for instance, $\zeta(-1) = -1/12$ and $\zeta(-2) = 0$. Apply the above algorithm to the *alternating* zeta function

$$\beta(s) = \sum_{n\geq 1}(-1)^{n-1}\frac{1}{n^s} = \left(1 - \frac{1}{2^{s-1}}\right)\zeta(s)$$

(incidentally, prove this identity), and by using the above algorithm, show the non-convergent "identities"

$$1 - 2 + 3 - 4 + \cdots = 1/4 \quad \text{and} \quad 1 - 2^2 + 3^2 - 4^2 + \cdots = 0\,.$$

Exercise 8.8 (*B. Allombert*) Let χ be a periodic arithmetic function of period m, say, and assume that $\sum_{0\leq j<m}\chi(j) = 0$ (for instance, $\chi(j) = (-1)^j$ with $m = 2$).

1. Using the same polynomials P_n as above, write a similar algorithm for computing $\sum_{n\geq 0}\chi(n)f(n)$, and estimate its rate of convergence.
2. Using this, compute to 100 decimals $L(\chi_{-3}, k) = 1 - 1/2^k + 1/4^k - 1/5^k + \cdots$ for $k = 1, 2$, and 3, and recognize the exact value for $k = 1$ and $k = 3$.

8.6 Numerical Differentiation

The problem is as follows: given a function f, say defined and C^∞ on a real interval, compute $f'(x_0)$ for a given value of x_0. To be able to analyze the problem, we will assume that $f'(x_0)$ is not too close to 0, and that we want to compute it to a given *relative accuracy*, which is what is usually required in numerical analysis.

The naïve, although reasonable, approach, is to choose a small $h > 0$ and compute $(f(x_0+h) - f(x_0))/h$. However, it is clear that (using the same number of function evaluations) the formula $(f(x_0+h) - f(x_0-h))/(2h)$ will be better. Let us analyze this in detail. For simplicity, we will assume that all the derivatives of f around x_0 that we consider are neither too small nor too large in absolute value. It is easy to modify the analysis to treat the general case.

Assume f computed to a relative accuracy of ε, in other words that we know values $\tilde{f}(x)$ such that $\tilde{f}(x)(1-\varepsilon) < f(x) < \tilde{f}(x)(1+\varepsilon)$ (the inequalities being reversed if $f(x) < 0$). The absolute error in computing $(f(x_0+h) - f(x_0-h))/(2h)$ is thus essentially equal to $\varepsilon|f(x_0)|/h$. On the other hand, by Taylor's theorem we have $(f(x_0+h) - f(x_0-h))/(2h) = f'(x_0) + (h^2/6)f'''(x)$ for some x close to x_0, so the absolute error made in computing $f'(x_0)$ as $(f(x_0+h) - f(x_0-h))/(2h)$ is close to $\varepsilon|f(x_0)|/h + (h^2/6)|f'''(x_0)|$. For a given value of ε (i.e., the accuracy to which we compute f) the optimal value of h is $(3\varepsilon|f(x_0)/f'''(x_0)|)^{1/3}$ for an absolute error of $(1/2)(3\varepsilon|f(x_0)f'''(x_0)|)^{2/3}$ hence a relative error of $(3\varepsilon|f(x_0)f'''(x_0)|)^{2/3}/(2|f'(x_0)|)$.

Since we have assumed that the derivatives have reasonable size, the relative error is roughly $C\varepsilon^{2/3}$, so if we want this error to be less than η, say, we need ε of the order of $\eta^{3/2}$, and h will be of the order of $\eta^{1/2}$.

Note that this result is not completely intuitive. For instance, assume that we want to compute derivatives to 38 decimal digits. With our assumptions, we choose h around 10^{-19}, and perform the computations with 57 decimals of relative accuracy. If for some reason or other we are limited to 38 decimals in the computation of f, the "intuitive" way would be also to choose $h = 10^{-19}$, and the above analysis shows that we would obtain only approximately 19 decimals. On the other hand, if we chose $h = 10^{-13}$, for instance, close to $10^{-38/3}$, we would obtain 25 decimals.

There are, of course, many other formulas for computing $f'(x_0)$, or for computing higher derivatives, which can all easily be analyzed as above. For instance (exercise), one can look for approximations to $f'(x_0)$ of the form $S = (\sum_{1\le i\le 3} \lambda_i f(x_0 + h/a_i))/h$, for any nonzero and pairwise distinct a_i, and we find that this is possible as soon as $\sum_{1\le i\le 3} a_i = 0$ (for instance, if $(a_1, a_2, a_3) = (-3, 1, 2)$ we have $(\lambda_1, \lambda_2, \lambda_3) = (-27, -5, 32)/20$), and the absolute error is then of the form $C_1/h + C2h^3$, so the same analysis shows that we should work with accuracy $\varepsilon^{4/3}$ instead of $\varepsilon^{3/2}$. Even though we have $3/2$ times more evaluations of f, we require less accuracy: for instance, if f requires time $O(D^a)$ to be computed to D decimals, as soon as $(3/2)\cdot((4/3)D)^a < ((3/2)D)^a$, i.e., $3/2 < (9/8)^a$, hence $a \ge 3.45$, this new method will be faster.

Perhaps, the best known method with more function evaluations is the approximation

$$f'(x_0) \approx (f(x-2h) - 8f(x-h) + 8f(x+h) - f(x+2h))/(12h) ,$$

which requires accuracy $\varepsilon^{5/4}$, and since this requires 4 evaluations of f, this is faster than the first method as soon as $2\cdot(5/4)^a < (3/2)^a$, in other words $a > 3.81$, and faster than the second method as soon as $(4/3)\cdot(5/4)^a < (4/3)^a$, in other words $a > 4.46$. To summarize, use the first method if $a < 3.45$, the second method if

$3.45 \leq a < 4.46$, and the third if $a > 4.46$. Of course this game can be continued at will, but there is not much point in doing so. In practice the first method is sufficient.

8.7 *Double Exponential Numerical Integration*

A remarkable although little-known technique invented around 1970 deals with *numerical integration* (the numerical computation of a definite integral $\int_a^b f(t)\,dt$, where a and b are allowed to be $\pm\infty$). In usual numerical analysis courses, one teaches very elementary techniques such as the trapezoidal rule, Simpson's rule, or more sophisticated methods such as Romberg or Gaussian integration. These methods apply to very general classes of functions $f(t)$, but are unable to compute more than a few decimal digits of the result, except for Gaussian integration which we will mention below.

However, in most mathematical (as opposed, for instance, to physical) contexts, the function $f(t)$ is *extremely regular*, typically holomorphic or meromorphic, at least in some domain of the complex plane. It was observed in the late 1960s by Takahashi and Mori [14] that this property can be used to obtain a *very simple* and *incredibly accurate* method to compute definite integrals of such functions. It is now instantaneous to compute 100 decimal digits, and takes only a few seconds to compute 500 decimal digits, say.

In view of its importance, it is essential to have some knowledge of this method. It can, of course, be applied in a wide variety of contexts, but note also that in his thesis [12], P. Molin has applied it specifically to the *rigorous* and *practical* computation of values of L-functions, which brings us back to our main theme.

There are two basic ideas behind this method. The first is in fact a theorem, which I state in a vague form: If F is a holomorphic function which tends to 0 "sufficiently fast" when $x \to \pm\infty$, x real, then the most efficient method to compute $\int_{\mathbb{R}} F(t)\,dt$ is indeed the trapezoidal rule. Note that this is a *theorem*, not so difficult but a little surprising nonetheless. The definition of "sufficiently fast" can be made precise. In practice, it means at least like e^{-ax^2} ($e^{-a|x|}$ is not fast enough), but it can be shown that the best results are obtained with functions tending to 0 *doubly exponentially fast* such as $\exp(-\exp(a|x|))$. Note that it would be (very slightly) worse to choose functions tending to 0 even faster.

To be more precise, we have an estimate coming, for instance, from the *Euler–Maclaurin summation formula*:

$$\int_{-\infty}^{\infty} F(t)\,dt = h \sum_{n=-N}^{N} F(nh) + R_N(h)\ ,$$

and under suitable holomorphy conditions on F, if we choose $h = a\log(N)/N$ for some constant a close to 1, the remainder term $R_N(h)$ will satisfy $R_n(h) =$

$O(e^{-bN/\log(N)})$ for some other (reasonable) constant b, showing exponential convergence of the method.

The second and, of course, crucial idea of the method is as follows: evidently not all functions are doubly exponentially tending to 0 at $\pm\infty$, and definite integrals are not all from $-\infty$ to $+\infty$. But it is possible to reduce to this case by using clever *changes of variable* (the essential condition of holomorphy must, of course, be preserved).

Let us consider the simplest example, but others that we give below are variations on the same idea. Assume that we want to compute

$$I = \int_{-1}^{1} f(x)\,dx\,.$$

We make the "magical" change of variable $x = \phi(t) = \tanh(\sinh(t))$, so that if we set $F(t) = f(\phi(t))$ we have

$$I = \int_{-\infty}^{\infty} F(t)\phi'(t)\,dt\,.$$

Because of the elementary properties of the hyperbolic sine and tangent, we have gained two things at once: first the integral from -1 to 1 is now from $-\infty$ to ∞, but most importantly the function $\phi'(t)$ is easily seen to tend to 0 doubly exponentially. We thus obtain an *exponentially good approximation*

$$\int_{-1}^{1} f(x)\,dx = h\sum_{n=-N}^{N} f(\phi(nh))\phi'(nh) + R_N(h)\,.$$

To give an idea of the method, if one takes $h = 1/200$ and $N = 500$, hence only 1000 evaluations of the function f, one can compute I to several hundred decimal places!

Before continuing, I would like to comment that in this theory many results are not completely rigorous: the method works very well, but the proof that it does is sometimes missing. Thus I cannot resist giving a *proven and precise* theorem due to P. Molin (which is, of course, just an example). We keep the above notation $\phi(t) = \tanh(\sinh(t))$, and note that $\phi'(t) = \cosh(t)/\cosh^2(\sinh(t))$.

Theorem 8.9 (P. Molin) *Let f be holomorphic on the disc $D = D(0,2)$ centered at the origin and of radius 2. Then for all $N \geq 1$, if we choose $h = \log(5N)/N$ we have*

$$\int_{-1}^{1} f(x)\,dx = h\sum_{n=-N}^{N} f(\phi(nh))\phi'(nh) + R_N\,,$$

where

$$|R_N| \le \left(e^4 \sup_D |f|\right) \exp(-5N/\log(5N)) .$$

Coming back to the general situation, I briefly comment on the computation of general definite integrals $\int_a^b f(t)\,dt$.

1. If a and b are finite, we can reduce to $[-1, 1]$ by affine changes of variable.
2. If a (or b) is finite and the function has an algebraic singularity at a (or b), we remove the singularity by a polynomial change of variable.
3. If $a = 0$ (say) and $b = \infty$, then if f does *not* tend to 0 exponentially fast (for instance, $f(x) \sim 1/x^k$), we use $x = \phi(t) = \exp(\sinh(t))$.
4. If $a = 0$ (say) and $b = \infty$ and if f does tend to 0 exponentially fast (for instance, $f(x) \sim e^{-ax}$ or $f(x) \sim e^{-ax^2}$), we use $x = \phi(t) = \exp(t - \exp(-t))$.
5. If $a = -\infty$ and $b = \infty$, use $x = \phi(t) = \sinh(\sinh(t))$ if f does not tend to 0 exponentially fast, and $x = \phi(t) = \sinh(t)$ otherwise.

The problem of *oscillating* integrals such as $\int_0^\infty f(x)\sin(x)\,dx$ is more subtle, but there does exist similar methods when, as here, the oscillations are completely under control.

Remark 8.10 The theorems are valid when the function is holomorphic in a sufficiently large region containing the path of integration. If the function is only *meromorphic*, with known poles, the direct application of the formulas may give totally wrong answers. However, if we take into account the poles, we can recover perfect agreement. Example of bad behavior: $f(t) = 1/(1 + t^2)$ (poles $\pm i$). Integrating on the intervals $[0, \infty]$, $[0, 1000]$, or even $[-\infty, \infty]$, which involve different changes of variables, give perfect results (the latter being somewhat surprising). On the other hand, integrating on $[-1000, 1000]$ gives a totally wrong answer because the poles are "too close", but it is easy to take them into account if desired.

Apart from the above pathological behavior, let us give a couple of examples where we must slightly modify the direct use of doubly exponential integration techniques.

• Assume, for instance, that we want to compute

$$J = \int_1^\infty \left(\frac{1 + e^{-x}}{x}\right)^2 dx ,$$

and that we use the built-in function `intnum` of `Pari/GP` for doing so. The function tends to 0 slowly at infinity, so we should compute it using the `GP` syntax `oo` to represent ∞, so we write `f(x)=((1+exp(-x))/x)^2;`, then `intnum(x=1,oo,f(x))`. This will give some sort of error, because the software will try to evaluate $\exp(-x)$ for large values of x, which it cannot do since there is exponent underflow. To compute the result, we need to split it into its slow part and fast part: when a function tends exponentially fast to 0 like $\exp(-ax)$, ∞ is represented as `[oo,a]`, so we write $J = J_1 + J_2$, with J_1 and J_2 computed by

`J1=intnum(x=1,[oo,1],(exp(-2*x)+2*exp(-x))/x^2);` and `J2=intnum(x=1,oo,1/x^2);` (which, of course, is equal to 1), giving

$$J = 1.3345252753723345485962398139190637\cdots .$$

Note that we could have tried to "cheat" and written directly `intnum(x=1,[oo,1],f(x))`, but the answer would be wrong, because the software would have assumed that $f(x)$ tends to 0 exponentially fast, which is not the case.

• A second situation where we must be careful is when we have "apparent singularities", which are not real singularities. Consider the function $f(x) = (\exp(x) - 1 - x)/x^2$. It has an apparent singularity at $x = 0$ but in fact it is completely regular. If you ask `J=intnum(x=0,1,f(x))`, you will get a result which is reasonably correct, but never more than 19 decimals, say. The reason is *not* due to a defect in the numerical integration routine, but more in the computation of $f(x)$: if you simply write `f(x)=(exp(x)-1-x)/x^2;`, the results will be bad for x close to 0.

Assuming that you want 38 decimals, say, the solution is to write

```
f(x)=if(x<10^(-10),1/2+x/6+x^2/24+x^3/120,(exp(x)-1-x)/x^2);
```

and now we obtain the value of our integral as

$$J = 0.59962032299535865949972137289656934022\cdots$$

8.8 The Use of Abel–Plana for Definite Summation

We finish this course by describing an identity, which is first quite amusing and second can be used efficiently for definite summation. Consider, for instance, the following theorem:

Theorem 8.11 *Define by convention* $\sin(n/10)/n$ *as equal to its limit* $1/10$ *when* $n = 0$, *and define* $\sum'_{n\geq 0} f(n)$ *as* $f(0)/2 + \sum_{n\geq 1} f(n)$. *We have*

$$\sum_{n\geq 0}{}' \left(\frac{\sin(n/10)}{n}\right)^k = \int_0^\infty \left(\frac{\sin(x/10)}{x}\right)^k$$

for $1 \leq k \leq 62$, *but not for* $k \geq 63$.

If you do not like all these conventions, replace the left-hand side by

$$\frac{1}{2\cdot 10^k} + \sum_{n\geq 1}\left(\frac{\sin(n/10)}{n}\right)^k .$$

It is clear that something is going on: it is the Abel–Plana formula. There are several forms of this formula, here is one of them:

Theorem 8.12 (Abel–Plana) *Assume that f is an entire function and that $f(z) = o(\exp(2\pi|\Im(z)|))$ as $|\Im(z)| \to \infty$ uniformly in vertical strips of bounded width, and a number of less important additional conditions which we omit. Then*

$$\sum_{m\geq 1} f(m) = \int_0^\infty f(t)\,dt - \frac{f(0)}{2} + i\int_0^\infty \frac{f(it)-f(-it)}{e^{2\pi t}-1}\,dt$$
$$= \int_{1/2}^\infty f(t)\,dt - i\int_0^\infty \frac{f(1/2+it)-f(1/2-it)}{e^{2\pi t}+1}\,dt\;.$$

In particular, if the function f is even, *we have*

$$\frac{f(0)}{2} + \sum_{m\geq 1} f(m) = \int_0^\infty f(t)\,dt\;.$$

Since we have seen above that using doubly exponential techniques it is easy to compute numerically a definite *integral*, the Abel–Plana formula can be used to compute numerically a *sum*. Note that in the first version of the formula there is an apparent singularity (but which is not a singularity) at $t = 0$, and the second version avoids this problem.

In practice, this summation method is very competitive with other methods if we use the doubly exponential method to compute $\int_0^\infty f(t)\,dt$, but most importantly if we use a variant of *Gaussian integration* to compute the complex integrals, since the nodes and weights for the function $t/(e^{2\pi t} - 1)$ can be computed once and for all by using continued fractions, see Sect. 9.4.

9 The Use of Continued Fractions

9.1 Introduction

The last idea that I would like to mention and that is applicable in quite different situations is the use of continued fractions. Recall that a continued fraction is an expression of the form

$$a_0 + \cfrac{b_0}{a_1 + \cfrac{b_1}{a_2 + \cfrac{b_2}{a_3 + \ddots}}}\;.$$

The problem of *convergence* of such expressions (when they are unlimited) is difficult and will not be considered here. We refer to any good textbook on the elementary properties of continued fractions. In particular, recall that if we denote by p_n/q_n the nth *partial quotient* (obtained by stopping at b_{n-1}/a_n) then both p_n and q_n satisfy the same recursion $u_n = a_n u_{n-1} + b_{n-1} u_{n-2}$.

We will mainly consider continued fractions representing *functions* as opposed to simply numbers. Whatever the context, the interest of continued fractions (in addition to the fact that they are easy to evaluate) is that they give essentially the *best possible* approximations, both for real numbers (this is the standard theory of *regular* continued fractions, where $b_n = 1$ and $a_n \in \mathbb{Z}_{\geq 1}$ for $n \geq 1$), and for functions (this is the theory of *Padé approximants*).

9.2 The Two Basic Algorithms

The first algorithm that we need is the following: assume that we want to expand a (formal) power series $S(z)$ (without loss of generality such that $S(0) = 1$) into a continued fraction:

$$S(z) = 1 + c(1)z + c(2)z^2 + \cdots = 1 + \cfrac{b(0)z}{1 + \cfrac{b(1)z}{1 + \cfrac{b(2)z}{1 + \ddots}}}\ .$$

The following method, called the *quotient-difference* (QD) algorithm does what is required:

We define two arrays $e(j,k)$ for $j \geq 0$ and $q(j,k)$ for $j \geq 1$ by $e(0,k) = 0$, $q(1,k) = c(k+2)/c(k+1)$ for $k \geq 0$, and by induction for $j \geq 1$ and $k \geq 0$:

$$\begin{aligned} e(j,k) &= e(j-1,k+1) + q(j,k+1) - q(j,k)\ , \\ q(j+1,k) &= q(j,k+1)e(j,k+1)/e(j,k)\ . \end{aligned}$$

Then $b(0) = c(1)$ and $b(2n-1) = -q(n,0)$ and $b(2n) = -e(n,0)$ for $n \geq 1$.

Three essential implementation remarks: first keeping the whole arrays is costly, it is sufficient to keep the latest vectors of e and q. Second, even if the $c(n)$ are rational numbers it is essential to do the computation with floating point approximations to avoid coefficient explosion. The algorithm can become unstable, but this is corrected by increasing the working accuracy. Third, it is, of course, possible that some division by 0 occurs, and this is in fact quite frequent. There are several ways to overcome this, probably the simplest being to multiply or divide the power series by something like $1 - z/\pi$.

The second algorithm is needed to *evaluate* the continued fraction for a given value of z. It is well known that this can be done from bottom to top (start at $b(n)z/1$, then $b(n-1)/(1+b(n)z/1)$, etc.), or from top to bottom (start at $(p(-1), q(-1)) = (1, 0)$, $(p(0), q(0)) = (1, 1)$, and use the recursion). It is in general better to evaluate from bottom to top, but before doing this we can considerably improve on the speed by using an identity due to Euler:

$$1 + \cfrac{b(0)z}{1 + \cfrac{b(1)z}{1 + \cfrac{b(2)z}{1 + \ddots}}} = 1 + \cfrac{B(0)Z}{Z + A(1) + \cfrac{B(1)}{Z + A(2) + \cfrac{B(2)}{Z + A(3) + \ddots}}},$$

where $Z = 1/z$, $A(1) = b(1)$, $A(n) = b(2n-2) + b(2n-1)$ for $n \geq 2$, $B(0) = b(0)$, $B(n) = -b(2n)b(2n-1)$ for $n \geq 1$. The reason for which this is much faster is that we replace n multiplications $(b(j) * z)$ plus n divisions by 1 multiplication plus approximately $1 + n/2$ divisions, counting as usual additions as negligible.

This is still not the end of the story since we can "compress" any continued fraction by taking, for instance, two steps at once instead of one, which reduces the cost . In any case this leads to a very efficient method for evaluating continued fractions.

9.3 Using Continued Fractions for Inverse Mellin Transforms

We have mentioned above that one can use asymptotic expansions to compute the incomplete gamma function $\Gamma(s, x)$ when x is large. But this method cannot give us great accuracy since we must stop the asymptotic expansion at its smallest term. We can, of course, always use the power series expansion, which has infinite radius of convergence, but when x is large this is not very efficient (remember the example of computing e^{-x}).

In the case of $\Gamma(s, x)$, continued fractions save the day: indeed, one can prove that

$$\Gamma(s, x) = \cfrac{x^s e^{-x}}{x + 1 - s - \cfrac{1(1-s)}{x + 3 - s - \cfrac{2(2-s)}{x + 5 - s - \ddots}}},$$

with precisely known speed of convergence. This formula is the best method for computing $\Gamma(s, x)$ when x is large (say $x > 50$), and can give arbitrary accuracy.

However, here we were in luck: we had an "explicit" continued fraction representing the function that we wanted to compute. Evidently, in general this will not be the case.

It is a remarkable idea of Dokchitser [9] that it does not really matter if the continued fraction is not explicit, at least in the context of computing L-functions, for instance, for inverse Mellin transforms. Simply do the following:

1. First compute sufficiently many terms of the asymptotic expansion of the function to be computed. This is very easy because our functions all satisfy a *linear differential equation* with polynomial coefficients, which gives a *recursion* on the coefficients of the asymptotic expansion.
2. Using the quotient-difference algorithm seen above, compute the corresponding continued fraction, and write it in the form due to Euler to evaluate it as efficiently as possible.
3. Compute the value of the function at all desired arguments by evaluating the Euler continued fraction.

The first two steps are completely automatic and rigorous. The whole problem lies in the third step, the evaluation of the continued fraction. In the case of the incomplete gamma function, we had a theorem giving us the speed of convergence. In the case of inverse Mellin transforms, not only do we not have such a theorem, but we do not even know how to prove that the continued fraction converges! However, experimentation shows that not only does the continued fraction converge, but rather fast, in fact at a similar speed to that of the incomplete gamma function.

Even though this step is completely heuristic, since its introduction by T. Dokchitser it is used in all packages computing L-functions since it is so useful. It would, of course, be nice to have a *proof* of its validity, but for now this seems completely out of reach, except for the simplest examples where there are at most two gamma factors (for instance, the problem is completely open for the inverse Mellin transform of $\Gamma(s)^3$).

9.4 Using Continued Fractions for Gaussian Integration and Summation

We have seen above the doubly exponential method for numerical integration, which is robust and quite generally applicable. However, an extremely classical method is *Gaussian integration*: it is orders of magnitude faster, but note the crucial fact that it is much less robust, in that it works much less frequently.

The setting of Gaussian integration is the following: we have a measure $d\mu$ on a (compact or infinite) interval $[a, b]$; you can, of course, think of $d\mu$ as $K(x)dx$ for some fixed function $K(x)$. We want to compute $\int_a^b f(x)d\mu$ by means of *nodes* and *weights*, i.e., for a given n compute x_i and w_i for $1 \le i \le n$ such that $\sum_{1\le i\le n} w_i f(x_i)$ approximates as closely as possible the exact value of the integral.

Note that *classical* Gaussian integration such as Gauss–Legendre integration (integration of a continuous function on a compact interval) is easy to perform because one can easily compute explicitly the necessary nodes and weights using standard *orthogonal polynomials*. What I want to stress here is that *general* Gaussian integration can be performed very simply using continued fractions, as follows.

In general, the measure $d\mu$ is (or can be) given through its *moments* $M_k = \int_a^b x^k d\mu$. The remarkably simple algorithm to compute the x_i and w_i using continued fractions is as follows:

1. Set $\Phi(z) = \sum_{k\geq 0} M_k z^{k+1}$, and using the quotient-difference algorithm compute $c(m)$ such that $\Phi(z) = c(0)z/(1 + c(1)z/(1 + c(2)z/(1 + \cdots)))$ (see the remark made above in case the algorithm has a division by 0; it may also happen that the odd or even moments vanish, so that the continued fraction is only in powers of z^2, but this is also easily dealt with).
2. For any m, denote as usual by $p_m(z)/q_m(z)$ the mth convergent obtained by stopping the continued fraction at $c(m)z/1$, and denote by $N_n(z)$ the reciprocal polynomial of $p_{2n-1}(z)/z$ (which has degree $n-1$) and by $D_n(z)$ the reciprocal polynomial of q_{2n-1} (which has degree n).
3. The x_i are the n roots of D_n (which are all simple and in the interval $]a, b[$), and the w_i are given by the formula $w_i = N_n(x_i)/D'_n(x_i)$.

By construction, this Gaussian integration method will work when the function $f(x)$ to be integrated is well approximated by polynomials, but otherwise will fail miserably, and this is why we say that the method is much less "robust" than doubly exponential integration.

The fact that Gaussian "integration" can also be used very efficiently for numerical *summation* was discovered quite recently by H. Monien. We explain the simplest case. Consider the measure on $]0, 1]$ given by $d\mu = \sum_{n\geq 1} \delta_{1/n}/n^2$, where δ_x is the Dirac measure centered at x. Thus, by definition $\int_0^1 f(x)d\mu = \sum_{n\geq 1} f(1/n)/n^2$. Let us apply the recipe given above: the kth moment M_k is given by $M_k = \sum_{n\geq 1}(1/n)^k/n^2 = \zeta(k+2)$, so that $\Phi(z) = \sum_{k\geq 1} \zeta(k+1)z^k$. Note that this is closely related to the digamma function $\psi(z)$, but we do not need this. Applying the quotient-difference algorithm, we write $\Phi(z) = c(0)z/(1 + c(1)z/(1 + \cdots))$, and compute the x_i and w_i as explained above. We will then have that $\sum_i w_i f(x_i)$ is a very good approximation to $\sum_{n\geq 1} f(1/n)/n^2$, or equivalently (changing the definition of f) that $\sum_i w_i f(y_i)$ is a very good approximation to $\sum_{n\geq 1} f(n)$, with $y_i = 1/x_i$.

To take essentially the simplest example, stopping the continued fraction after two terms we find that $y_1 = 1.0228086266\cdots$, $w_1 = 1.15343168\cdots$, $y_2 = 4.37108 2834\cdots$, and $w_2 = 10.3627543\cdots$, and (by definition) we have $\sum_{1\leq i\leq 2} w_i f(y_i) = \sum_{n\geq 1} f(n)$ for $f(n) = 1/n^k$ with $k = 2, 3, 4$, and 5.

10 Pari/GP Commands

In this section, we give some of the Pari/GP commands related to the subjects studied in this course, together with examples. Unless mentioned otherwise, the commands assume that the current default accuracy is the default, i.e., 38 decimal digits.

zeta(s): Riemann zeta function at s.

```
? zeta(3)
% = 1.2020569031595942853997381615114499908
? zeta(1/2+14*I)
% = 0.022241142609993589246213199203968626387
   - 0.10325812326645005790236309555257383451*I
```

lfuncreate(obj): create L-function attached to mathematical object obj.

lfun(pol,s): Dedekind zeta function of the number field K defined by pol at s. Identical to L=lfuncreate(pol); lfun(L,s).

```
? L = lfuncreate(x^3-x-1); lfunan(L,10)
% = [1, 0, 0, 0, 1, 0, 1, 1, 0, 0]
? lfun(L,1)
% = 0.36840932071582682111186846662888526986*x^-1 + O(x^0)
? lfun(L,2)
% = 1.1100010060250153929372222560595385375
```

lfunlambda(pol,s): same, but for the completed function $\Lambda_K(s)$, identical to lfunlambda(L,s) where L is as above.

```
? lfunlambda(L,2)
% = 0.41169121016707136240079852448689476625
```

lfun(D,s): L-function of quadratic character $(D/.)$ at s.
Identical to L=lfuncreate(D); lfun(L,s).

```
? lfun(-23,-2)
% = -48.000000000000000000000000000000000000
? lfun(5,-1)
% = -0.40000000000000000000000000000000000000
```

L1=lfuncreate(pol); L2=lfuncreate(1); L=lfundiv(L1,L2): L function attached to $\zeta_K(s)/\zeta(s)$.

```
? L1 = lfuncreate(x^3-x-1); L2 = lfuncreate(1);
? L = lfundiv(L1,L2); lfunan(L,14)
```

```
% = [1, -1, -1, 0, 0, 1, 0, 1, 0, 0, 0, 0, -1, 0]
```

`lfunetaquo`($[m_1, r_1; m_2, r_2]$): L-function of eta product $\eta(m_1\tau)^{r_1}\eta(m_2\tau)^{r_2}$, for instance with `[1,1;23,1]` or `[1,2;11,2]`.

```
? L1 = lfunetaquo([1,1;23,1]); lfunan(L1,14)
% = [1, -1, -1, 0, 0, 1, 0, 1, 0, 0, 0, 0, -1, 0]
? L2 = lfunetaquo([1,2;11,2]); lfunan(L2,14)
% = [1, -2, -1, 2, 1, 2, -2, 0, -2, -2, 1, -2, 4, 4]
```

`lfuncreate(ellinit(e))`: L-function of elliptic curve e, for instance with $e = [0, -1, 1, -10, -20]$.

```
? e = ellinit([0,-1,1,-10,-20]);
? L = lfuncreate(e); lfunan(L,14)
% = [1, -2, -1, 2, 1, 2, -2, 0, -2, -2, 1, -2, 4, 4]
```

`ellap(e,p)`: compute $a(p)$ for an elliptic curve e.

```
? ellap(e,nextprime(10^42))
% = -1294088699019102994696
```

`eta(q+O(q^B))^m`: compute the mth power of η to B terms.

```
? eta(q+O(q^5))^26
% = 1 - 26*q + 299*q^2 - 1950*q^3 + 7475*q^4 + O(q^5)
```

`D=mfDelta(); mfcoefs(D,B)`: compute $B+1$ terms of the Fourier expansion of Δ.

```
? D = mfDelta(); mfcoefs(D,7)
% = [0, 1, -24, 252, -1472, 4830, -6048, -16744]
```

`ramanujantau(n)`: compute Ramanujan's tau function $\tau(n)$ using the trace formula.

```
? ramanujantau(nextprime(10^7))
% = 110949191154874445294730241687634133420
```

`qfbhclassno(n)`: Hurwitz class number $H(n)$.

```
? vector(13,n,qfbhclassno(n-1))
% = [-1/12, 0, 0, 1/3, 1/2, 0, 0, 1, 1, 0, 0, 1, 4/3]
```

`qfbsolve(Q,n)`: solve $Q(x, y) = n$ for a binary quadratic form Q (contains in particular Cornacchia's algorithm).

```
? Q = Qfb(1,0,1); p = 10^16+61; qfbsolve(Q,p)
% = [86561206, 50071525]
```

`gamma(s)`: gamma function at s.

```
? gamma(1/4)*gamma(3/4)-Pi*sqrt(2)
% = 2.350988701644575016 E-38
```

`incgam(x,s)`: incomplete gamma function $\Gamma(s, x)$.

```
? incgam(1,5/2)
% = 0.082084998623898795169528674467159807838
```

`G=gammamellininvinit(A)`: initialize data for computing inverse Mellin transforms of $\prod_{1\le i\le d} \Gamma_{\mathbb{R}}(s + a_i)$, with $A = [a_1, \ldots, a_d]$.

`gammamellininv(G,t)`: inverse Mellin transform at t of A, with G initialized as above.

```
? G = gammamellininvinit([0,0]); gammamellininv(G,2)
% = 4.8848219774465217355974384319702281090 E-6
```

`K(nu,x)`: $K_\nu(x)$, K-Bessel function of (complex) index ν at x.

```
? 4*besselk(0,4*Pi)
% = 4.8848219774465217355974384319702281090 E-6
```

`sumnum(n=a,f(n))`: numerical summation of $\sum_{n\ge a} f(n)$ using discrete Euler–MacLaurin.

```
? sumnum(n=1,1/(n^2+n^(4/3)))
% = 0.95586324768586066988568837766973815238
```

`sumnumap(n=a,f(n))`: numerical summation of $\sum_{n\ge a} f(n)$ using Abel–Plana.

`sumnummonien(n=a,f(n))`: numerical summation using Monien's Gaussian summation method,

(there also exists `sumnumlagrange`, which can also be very useful).

`limitnum(n->f(n))`: limit of $f(n)$ as $n \to \infty$ using a variant of Zagier's method, assuming asymptotic expansion in integral powers of $1/n$ (also `asympnum` to obtain more coefficients).

```
? limitnum(n->(1+1/n)^n)
% = 2.7182818284590452353602874713526624978
? asympnum(n->(1+1/n)^n*exp(-1))
% = [1, -1/2, 11/24, -7/16, 2447/5760, -959/2304,...]
```

`sumeulerrat(f(x))`: $\sum_{p\geq 2} f(p)$, p ranging over primes (more general variant exists form $\sum_{p\geq a} f(p^s)$).

```
? sumeulerrat(1/(x^2+x))
% = 0.33022992626420324101509458808674476056
```

`prodeulerrat(f(x))`: $\prod_{p\geq 2} f(p)$, p ranging over primes, with same variants.

```
? prodeulerrat((1-1/x)^2*(1+2/x))
% = 0.28674742843447873410789271278983844644
```

`sumalt(n=a,(-1)^n*f(n))`: $\sum_{n\geq a}(-1)^n f(n)$, assuming f positive.

```
? sumalt(n=1,(-1)^n/(n^2+n))
% = -0.38629436111989061883446424291635313615
```

`f'(x)` (or `deriv(f)(x)`): numerical derivative of f at x.

```
? -zeta'(-2)
% = 0.030448457058393270780251530471154776647
? zeta(3)/(4*Pi^2)
% = 0.030448457058393270780251530471154776647
```

`intnum(x=a,b,f(x))`: numerical computation of $\int_a^b f(x)\,dx$ using general doubly exponential integration.

`intnumgauss(x=a,b,f(x))`: numerical integration using Gaussian integration.

```
? intnum(t=0,1,lngamma(t+1))
% = -0.081061466795327258219670263594382360139
```

For instance, for 500 decimal digits, after the initial computation of nodes and weights in both cases (`intnuminit(0,1)` and `intnumgaussinit()`) this examples requires 2.5 s by doubly exponential integration but only 0.25 s by Gaussian integration.

11 Three Pari/GP Scripts

11.1 *The Birch–Swinnerton-Dyer Example*

Here is a list of commands which implements the explicit BSD example given in Sect. 2.7, again assuming the default accuracy of 38 decimal digits.

```
? E = ellinit([1,-1,0,-79,289]); /* initialize */
? N = ellglobalred(E)[1] /* compute conductor */
% = 234446
? /* define the integral $f(x)$ */
? f(x) = intnum(t=1,[oo,x],exp(-x*t)*log(t)^2);
? /* check that f(100) is small enough for 38D */
? f(100)
% = 7.2... E-50
? A = ellan(E,8000); /* compute 8000 coefficients */
? /* Note that $2\pi 8000/sqrt(N) > 100$ */
? S = sum(n=1,8000,A[n]*f(2*Pi*n/sqrt(N)))
% = 9.02... E-35 /* almost 0 */
? /* compute APPARENT order of vanishing of L(E,s) */
? ellanalyticrank(E)[1]
% = 4
```

Note that for illustrative purposes we use the `intnum` command to compute $f(x)$, corresponding to the use of doubly exponential integration, but in the present case there are methods which are orders of magnitude faster. The last command, which is almost immediate, implements these methods.

11.2 *The Beilinson–Bloch Example*

The code for the explicit Beilinson–Bloch example seen in Sect. 2.8 is simpler (I have used the integral representation of $g(u)$, but of course I could have used the series expansion instead.):

```
? e(u) =
{
  my(E = ellinit([0,u^2+1,0,u^2,0]));
  lfun(E,2)*ellglobalred(E)[1];
}
? g(u) =
{
  my(S);
```

```
    S = 2*Pi*intnum(t=0,1,asin(t)/(t*sqrt(1-(t/u)^2)));
    S+Pi^2*acosh(u);
  }
? e(5)/g(5)
% = 8.000000000000000000000000000000000000
? /* we obtain perfect accuracy */
? /* for example: */
? for(u = 2,18,print1(bestappr(e(u)/g(u),10^6)," "))
% = 1 2 4/11 8 32 8 4/3 8 32 64 8 96 256 48 16 16 192
```

11.3 The Mahler Measure Example

```
? L=lfunetaquo([2,1;4,1;6,1;12,1]);
\\ Equivalently L=lfuncreate(ellinit([0,-1,0,-4,4]));
? lfun(L,3)
% = 0.95050371329356644983179739940014855951
? (Pi^2/36)*(Catalan*Pi+intnum(t=0,1,asin(t)*asin(1-t)/t))
% = 0.95050371329356644983179739940014855950
```

12 Appendix: Selected Results

12.1 The Gamma Function

The Gamma function, denoted by $\Gamma(s)$, can be defined in several different ways. My favorite is the one I give in Sect. 9.6.2 of [4], but for simplicity I will recall the classical definition. For $s \in \mathbb{C}$ we define

$$\Gamma(s) = \int_0^\infty e^{-t} t^s \frac{dt}{t} .$$

It is immediate to see that this converges if and only if $\Re(s) > 0$ (there is no problem at $t = \infty$, the only problem is at $t = 0$), and integration by parts shows that $\Gamma(s+1) = s\Gamma(s)$, so that if $s = n$ is a positive integer, we have $\Gamma(n) = (n-1)!$. We can now *define* $\Gamma(s)$ for all complex s by using this recursion backwards, i.e., setting $\Gamma(s) = \Gamma(s+1)/s$. It is then immediate to check that $\Gamma(s)$ is a meromorphic function on $\mathbb{C}$ having poles at $s = -n$ for $n = 0, 1, 2, \ldots$, which are simple with residue $(-1)^n/n!$.

The gamma function has numerous additional properties, the most important being recalled below:

1. (Stirling's formula for large $\Re(s)$): as $s \to \infty$, $s \in \mathbb{R}$ (say, there is a more general formulation) we have $\Gamma(s) \sim s^{s-1/2} e^{-s} (2\pi)^{1/2}$.

2. (Stirling's formula for large $\Im(s)$): as $|T| \to \infty$, $\sigma \in \mathbb{R}$ being fixed (say, once again there is a more general formulation), we have $|\Gamma(\sigma + iT)| \sim |T|^{\sigma-1/2} e^{-\pi|T|/2} (2\pi)^{1/2}$. In particular, it tends to 0 exponentially fast on vertical strips.
3. (Reflection formula): we have $\Gamma(s)\Gamma(1-s) = \pi/\sin(\pi s)$.
4. (Duplication formula): we have $\Gamma(s)\Gamma(s+1/2) = 2^{1-2s}\pi^{1/2}\Gamma(2s)$ (there is also a more general distribution formula giving $\prod_{0\le j<N} \Gamma(s+j/N)$ which we do not need). Equivalently, if we set $\Gamma_{\mathbb{R}}(s) = \pi^{-s/2}\Gamma(s/2)$ and $\Gamma_{\mathbb{C}}(s) = 2 \cdot (2\pi)^{-s}\Gamma(s)$, we have $\Gamma_{\mathbb{R}}(s)\Gamma_{\mathbb{R}}(s+1) = \Gamma_{\mathbb{C}}(s)$.
5. (Link with the beta function): let a and b in $\mathbb{C}$ with $\Re(a) > 0$ and $\Re(b) > 0$. We have

$$B(a,b) := \int_0^1 t^{a-1}(1-t)^{b-1}\,dt = \frac{\Gamma(a)\Gamma(b)}{\Gamma(a+b)}\ .$$

12.2 Order of a Function: Hadamard Factorization

Let F be a holomorphic function in the whole of $\mathbb{C}$ (it is immediate to generalize to the case of meromorphic functions, but for simplicity we stick to the holomorphic case). We say that F has *finite order* if there exists $\alpha \ge 0$ such that as $|s| \to \infty$ we have $|F(s)| \le e^{|s|^\alpha}$. The infimum of such α is called the order of F. It is an immediate consequence of Liouville's theorem that functions of order 0 are polynomials. Most functions occurring in number theory, and in particular all L-functions occurring in this course, have order 1. The Selberg zeta function, which we do not consider, is also an interesting function and has order 2.

The Weierstrass–Hadamard factorization theorem is the following:

Theorem 12.1 *Let F be a holomorphic function of order ρ, set $p = \lfloor \rho \rfloor$, let $(a_n)_{n\ge1}$ be the nonzero zeros of F repeated with multiplicity, and let m be the order of the zero at $z = 0$. There exists a polynomial P of degree at most p such that for all $z \in \mathbb{C}$ we have*

$$F(z) = z^m e^{P(z)} \prod_{n\ge1} \left(1 - \frac{z}{a_n}\right) \exp\left(\frac{z/a_n}{1} + \frac{(z/a_n)^2}{2} + \cdots + \frac{(z/a_n)^p}{p}\right)\ .$$

In the case of order 1 which is of interest to us, this reads

$$F(z) = B \cdot z^m e^{Az} \prod_{n\ge1} \left(1 - \frac{z}{a_n}\right) e^{z/a_n}\ .$$

For example, we have

$$\sin(\pi z) = \pi z \prod_{n\ge1} \left(1 - \frac{z^2}{n^2}\right) \quad \text{and} \quad \frac{1}{\Gamma(z+1)} = e^{\gamma z} \prod_{n\ge1} \left(1 + \frac{z}{n}\right) e^{-z/n}\ ,$$

where as usual $\gamma = 0.57721\cdots$ is Euler's constant.

Exercise 12.2 1. Using these expansions, prove the reflection formula and the duplication formula for the gamma function, and find the distribution formula giving $\prod_{0\le j<N}\Gamma(s+j/N)$.

2. Show that the above expansion for the sine function is equivalent to the formula expressing $\zeta(2k)$ in terms of Bernoulli numbers.
3. Show that the above expansion for the gamma function is equivalent to the Taylor expansion

$$\log(\Gamma(z+1)) = -\gamma z + \sum_{n\ge 2}(-1)^n\frac{\zeta(n)}{n}z^n\;,$$

and prove the validity of this Taylor expansion for $|z|<1$, hence of the above Hadamard product.

12.3 Elliptic Curves

We will not need the abstract definition of an elliptic curve. For us, an elliptic curve E defined over a field K will be a nonsingular projective curve defined by the (affine) generalized Weierstrass equation with coefficients in K:

$$y^2 + a_1xy + a_3y = x^3 + a_2x^2 + a_4x + a_6\;.$$

This curve has a *discriminant* (obtained essentially by completing the square and computing the discriminant of the resulting cubic), and the essential property of being nonsingular is equivalent to the discriminant being nonzero.

This curve has a unique point $\mathcal{O}$ at infinity, with projective coordinates $(0:1:0)$. Using chord and tangents, one can define an addition law on this curve, and the first essential (but rather easy) result is that it is an *abelian group law* with neutral element $\mathcal{O}$, making E into an algebraic group.

In the case where $K=\mathbb{Q}$ (or more generally a number field), a deeper theorem due to Mordell states that the group $E(\mathbb{Q})$ of rational points of E is a *finitely generated abelian group*, i.e., is isomorphic to $\mathbb{Z}^r\oplus E(\mathbb{Q})_{\text{tors}}$, where $E(\mathbb{Q})_{\text{tors}}$ (the torsion subgroup) is a finite group, and the integer r is called the (algebraic) *rank* of the curve.

Still in the case $K=\mathbb{Q}$, for all prime numbers p except a finite number, we can *reduce* the equation modulo p, thus obtaining an elliptic curve over the finite field $\mathbb{F}_p$. Using an algorithm due to J. Tate, we can find first a *minimal Weierstrass equation* for E, second the behavior of E reduced at the "bad" primes in terms of so-called *Kodaira symbols*, and third the algebraic *conductor* N of E, product of the bad primes raised to suitable exponents (and other important quantities).

The deep theorem of Wiles et al. tells us that the L-function of E (as defined in the main text) is equal to the L-function of a rational Hecke eigenform in the modular form space $M_2(\Gamma_0(N))$, where N is the conductor of E.

A weak form of the Birch and Swinnerton-Dyer conjecture says that the algebraic rank r is equal to the analytic rank defined as the order of vanishing of the L-function of E at $s = 1$.

References

1. N. Bourbaki, *Développement tayloriens généralisés. Formule sommatoire d'Euler–MacLaurin*, Fonctions d'une variable réelle, Chap. 6.
2. H. Cohen, *A Course in Computational Algebraic Number Theory (fourth corrected printing)*, Graduate Texts in Math. **138**, Springer, 2000.
3. H. Cohen, *Number Theory I, Tools and Diophantine Equations*, Graduate Texts in Math. **239**, Springer, 2007.
4. H. Cohen, *Number Theory II, Analytic and Modern Tools*, Graduate Texts in Math. **240**, Springer, 2007.
5. H. Cohen, *A p-adic stationary phase theorem and applications*, preprint.
6. H. Cohen and F. Strömberg, *Modular Forms: A Classical Approach*, Graduate Studies in Math. **179**, American Math. Soc., (2017).
7. H. Cohen, F. Rodriguez-Villegas, and D. Zagier, *Convergence acceleration of alternating series*, Exp. Math. **9** (2000), 3–12.
8. H. Cohen and D. Zagier, *Vanishing and nonvanishing theta values*, Ann. Sci. Math. Quebec **37** (2013), pp 45–61.
9. T. Dokchitser, *Computing special values of motivic L-functions*, Exp. Math. **13** (2004), 137–149.
10. G. Hiary, *Computing Dirichlet character sums to a power-full modulus*, ArXiv preprint arXiv:1205.4687v2.
11. J.-F. Mestre, *Formules explicites et minorations de conducteurs de variétés algébriques*, Compositio Math. **58** (1986). pp. 209–232.
12. P. Molin, *Intégration numérique et calculs de fonctions L*, Thèse, Université Bordeaux I (2010).
13. M. Rubinstein, *Computational methods and experiments in analytic number theory*, In: Recent Perspectives in Random Matrix Theory and Number Theory, F. Mezzadri and N. Snaith, eds (2005), pp. 407–483.
14. H. Takashi and M. Mori, *Double exponential formulas for numerical integration*, Publications of RIMS, Kyoto University (1974), 9:721–741.
15. J.-P. Serre, *Facteurs locaux des fonctions zêta des variétés algébriques (définitions et conjectures)*, Séminaire Delange–Pisot–Poitou **11** (1969–1970), exp. 19, pp. 1–15.

Exponential Diophantine Equations

Florian Luca

Abstract This paper is a very gentle introduction to solving exponential Diophantine equations using the technology of linear forms in logarithms of algebraic numbers.

1 Introduction

This is a very gentle introduction on how to solve certain exponential Diophantine equations. The paper contains two main parts, one on binary recurrent sequences (Sects. 2–10) and one on linear forms in logarithms (Sects. 11–14), as well as a short part explaining how to use LLL in order to reduce the initial bounds coming from applying linear forms in logarithms. In Sects. 2–10, we take a brief tour of the theory of linear recurrent sequences of order 2. We define the Lucas sequences in Sect. 5 and observe that the sequence of Fibonacci numbers is an example of such a sequence. The most important result here is the Theorem of Primitive Divisors, which is formulated in its most general form in Sect. 7, but is proved only for Lucas sequences with integer roots in Sect. 6 using results on Cyclotomic Polynomials briefly outlined in Sect. 4 and results on the indices of appearance of primes in Lucas sequences in Sect. 5. We present some applications in Sect. 8, in particular related to the largest prime factor of $x^2 - 1$ when x tends to infinity in the set of natural numbers. Sections 9 and 10 contain some exercises and notes. In Sect. 11, we give various specific statements which belong to what is known as the theory of lower bounds for linear forms in logarithms of algebraic numbers. To illustrate the importance of this machinery, we give three examples in Sect. 12. We shall see, for example, that the largest Fibonacci number having only one repeated digit in its decimal expansion is 55 and give two

F. Luca (✉)
School of Mathematics, University of the Witwatersrand, Private Bag X3,
Wits 2050, Johannesburg, South Africa
e-mail: florian.luca@wits.ac.za

F. Luca
Faculty of Sciences, Department of Mathematics, University of Ostrava,
30 Dubna 22, 701 03 Ostrava 1, Czech Republic

I. Inam and E. Büyükasik (eds.), *Notes from the International Autumn School on Computational Number Theory*, Tutorials, Schools, and Workshops in the Mathematical Sciences, https://doi.org/10.1007/978-3-030-12558-5_4

proofs of that, one using linear forms in complex logarithms and one using linear forms in p-adic logarithms. Sections 13 and 14 contain some problems/exercises and notes. In the last section, we give an example on how to use the LLL algorithm to reduce the large bounds from Baker's method to a range, which can be checked with a computer. The proposed exercises and problems can be attempted with the methods explained in the previous sections. There are certain little computations along the way which can be done with either Maple or Mathematica. Versions of these notes exist on the Internet in .pdf form from short courses give by the author in Mérida, Venezuela (2007), Debrecen, Hungary (2009), Dangbo, Benin (2014), and Krakow, Poland (2017).

2 Binary Recurrent Sequences

Definition 2.1 Let $k \geq 1$ be an integer. A sequence $(u_n)_{n\geq 0} \subset \mathbb{C}$ is called linearly recurrent of order k if the recurrence

$$u_{n+k} = a_1 u_{n+k-1} + a_2 u_{n+k-2} + \cdots + a_k u_n \tag{1}$$

holds for all $n \geq 0$ with some fixed coefficients $a_1, \ldots, a_k \in \mathbb{C}$.

Let us suppose that $a_k \neq 0$ (for if not, the sequence $(u_n)_{n\geq 0}$ satisfies a linear recurrence of order smaller than k). If $a_1, \ldots, a_k \in \mathbb{Z}$ and $u_0, \ldots, u_{k-1} \in \mathbb{Z}$, then, by induction on n, we get that u_n is an integer for all $n \geq 0$. The polynomial

$$f(X) = X^k - a_1 X^{k-1} - \cdots - a_k \in \mathbb{C}[X]$$

is called the *characteristic polynomial* of $(u_n)_{n\geq 0}$. Suppose that

$$f(X) = \prod_{i=1}^{s} (X - \alpha_i)^{\sigma_i},$$

where $\alpha_1, \ldots, \alpha_s$ are the distinct roots of $f(X)$ with multiplicities $\sigma_1, \ldots, \sigma_s$, respectively.

Proposition 2.2 *Suppose that $f(X) \in \mathbb{Z}[X]$ has k distinct roots. Then there exist constants $c_1, \ldots, c_k \in \mathbb{K} = \mathbb{Q}(\alpha_1, \ldots, \alpha_k)$ such that the formula*

$$u_n = \sum_{i=1}^{k} c_i \alpha_i^n \qquad \textit{holds for all } n \geq 0. \tag{2}$$

Proof Let

$$u(z) = \sum_{n\geq 0} u_n z^n.$$

We observe that

$$\begin{aligned} u(z)(1 - a_1 z - \cdots - a_k z^k) &= u_0 + (u_1 - u_0 a_1)z + \cdots \\ &\quad + \sum_{m \geq k} (u_m - a_1 u_{m-1} - \cdots - a_k u_{m-k}) z^m \\ &:= P(z), \end{aligned}$$

where $P(z) = \sum_{m=0}^{k-1} (u_m - a_1 u_{m-1} - \cdots - a_m u_0) z^m \in \mathbb{C}[z]$. Thus,

$$\begin{aligned} u(z) &= \frac{P(z)}{1 - a_1 z - \cdots - a_k z^k} = \frac{P(z)}{z^k f(1/z)} = \frac{P(z)}{z^k \prod_{i=1}^k (1/z - \alpha_i)} \\ &= \frac{P(z)}{\prod_{i=1}^k (1 - z\alpha_i)} = \sum_{i=1}^k \frac{c_i}{1 - z\alpha_i} \end{aligned}$$

for some coefficients $c_i \in \mathbb{K}$. For the last step, we have used the theory of the partial fractions together with the fact that the roots $\alpha_1, \ldots, \alpha_k$ are distinct and the degree of $P(z)$ is smaller than k. If

$$|z| < \rho := \min\{|\alpha_i|^{-1} : i = 1, \ldots, k\},$$

then we can write

$$\frac{1}{1 - z\alpha_i} = \sum_{n \geq 0} (z\alpha_i)^n = \sum_{n \geq 0} \alpha_i^n z^n \quad \text{for all } n \geq 0.$$

Thus, for $|z| < \rho$, we get that

$$\sum_{n \geq 0} u_n z^n = u(z) = \sum_{i=1}^k c_i \sum_{n \geq 0} \alpha_i^n z^n = \sum_{n \geq 0} \left(\sum_{i=1}^k c_i \alpha_i^n \right) z^n.$$

Identifying coefficients, we get the relation (2).

If $k = 2$, the sequence $(u_n)_{n \geq 0}$ is called *binary recurrent*. In this case, its characteristic polynomial is of the form

$$f(X) = X^2 - a_1 X - a_2 = (X - \alpha_1)(X - \alpha_2).$$

Suppose that $\alpha_1 \neq \alpha_2$. Proposition 2.2 tells us that

$$u_n = c_1 \alpha_1^n + c_2 \alpha_2^n \quad \text{for all } n \geq 0. \tag{3}$$

Definition 2.3 A binary recurrent sequence $(u_n)_{n \geq 0}$ whose general term is given by formula (3) is called nondegenerate if $c_1 c_2 \alpha_1 \alpha_2 \neq 0$ and α_1 / α_2 is not a root of 1.

3 Examples of Binary Recurrent Sequences

3.1 The Fibonacci and Lucas Sequences

The Fibonacci sequence $(F_n)_{n\geq 0}$ is given by $F_0 = 0,\ F_1 = 1$, and $F_{n+2} = F_{n+1} + F_n$ for all $n \geq 0$. Its characteristic equation is

$$f(X) = X^2 - X - 1 = (X - \alpha)(X - \beta),$$

where $\alpha = (1 + \sqrt{5})/2$ and $\beta = (1 - \sqrt{5})/2$. In order to find c_1 and c_2 starting with formula (3), we give to n the values 0 and 1 and obtain the system

$$c_1 + c_2 = F_0 = 0, \qquad c_1\alpha + d_1\beta = F_1 = 1.$$

Solving it, we get $c_1 = 1/\sqrt{5},\ c_2 = -1/\sqrt{5}$. Since $\sqrt{5} = (\alpha - \beta)$, we can write

$$F_n = \frac{\alpha^n - \beta^n}{\alpha - \beta} \qquad \text{for all } n \geq 0. \tag{4}$$

A sequence related to the Fibonacci sequence is the Lucas sequence $(L_n)_{n\geq 0}$ given by $L_0 = 2,\ L_1 = 1$, and $L_{n+2} = L_{n+1} + L_n$ for all $n \geq 0$. It has the same characteristic equation as the Fibonacci sequence, therefore, there exist two constants d_1 and d_2 such that

$$L_n = d_1\alpha^n + d_2\beta^n \qquad \text{for all } n \geq 0.$$

Giving to n the values 0 and 1, we get that

$$d_1 + d_2 = L_0 = 2, \qquad d_1\alpha + d_2\beta = L_1 = 1.$$

Solving the above system of linear equations, we see that $d_1 = d_2 = 1$, and so

$$L_n = \alpha^n + \beta^n \qquad \text{for all } n \geq 0. \tag{5}$$

Using the formulas (4) and (5), one can easily prove various formulas which involve F_n and L_n.

Example 3.1 The formula

$$L_n^2 - 5F_n^2 = 4(-1)^n$$

holds for all $n \geq 0$. In fact,

$$\begin{aligned} L_n^2 - 5F_n^2 &= (\alpha^n + \beta^n)^2 - 5\left(\frac{\alpha^n - \beta^n}{\alpha - \beta}\right)^2 \\ &= (\alpha^n + \beta^n)^2 - (\alpha^n - \beta^n)^2 = 4(\alpha\beta)^n = 4(-1)^n, \end{aligned}$$

where we used the fact that $(\alpha - \beta)^2 = 5$.

3.2 Binary Recurrences Associated to Pell Equations with $N = \pm 1$

Let $d > 1$ be an integer which is not a perfect square and let (x_1, y_1) be the minimal solution in positive integers of the equation

$$x^2 - dy^2 = \pm 1. \tag{6}$$

We put

$$\zeta = x_1 + \sqrt{d}y_1 \quad \text{and} \quad \eta = x_1 - \sqrt{d}y_1.$$

There is a theorem from the theory of Pell equations, which tells us that (x_1, y_1) always exists and that all positive integer solutions (x, y) of the Eq. (6) are of the form $(x, y) = (x_\ell, y_\ell)$ for some positive integer ℓ, where

$$x_\ell + \sqrt{d}y_\ell = (x_1 + \sqrt{d}y_1)^\ell = \zeta^\ell.$$

Conjugating the above relation (in $\mathbb{Q}[\sqrt{d}]$), we get

$$x_\ell - \sqrt{d}y_\ell = (x_1 - \sqrt{d}y_1)^\ell = \eta^\ell.$$

From here, we deduce that

$$x_\ell = \frac{\zeta^\ell + \eta^\ell}{2} \quad \text{and} \quad y_\ell = \frac{\zeta^\ell - \eta^\ell}{2\sqrt{d}} \quad \text{for all } \ell \geq 1. \tag{7}$$

It turns out to be useful to put $(x_0, y_0) = (1, 0)$ so that formula (7) holds also with $\ell = 0$.

It is easy to see that $(x_\ell)_{\ell \geq 0}$ and $(y_\ell)_{\ell \geq 0}$ are binary recurrent sequences of characteristic equation

$$\begin{aligned} f(X) &= X^2 - a_1 X - a_2 = (X - \zeta)(X - \eta) \\ &= X^2 - (\zeta + \eta)X + \zeta\eta = X^2 - 2x_1 X \pm 1. \end{aligned}$$

3.3 Binary Recurrences Associated to Pell Equations with $N \neq \pm 1$

Let $d > 1$ be an integer which is not a perfect square and let N be a nonzero integer. Let (u, v) be a positive integer solution of the equation

$$u^2 - dv^2 = N. \tag{8}$$

Theorem 3.2 *Let $d > 1$ be an integer which is not a perfect square and let $N \neq 0$. Then all the positive integer solutions (u, v) of the Eq. (8) belong to a finite set of binary recurrent sequences. These sequences have the same characteristic equation $X^2 - 2x_1X + 1 = 0$, where (x_1, y_1) is the minimal solution in positive integers of the equation $x^2 - dy^2 = 1$.*

Proof Let us observe first that if there is a solution (u_0, v_0) of the Eq. (8), then there are infinitely many of them. In fact, let (x_1, y_1) be the minimal solution of $x^2 - dy^2 = 1$, and put $\zeta = x_1 + \sqrt{d}y_1$, $\eta = x_1 - \sqrt{d}y_1$. Putting

$$u_\ell + \sqrt{d}v_\ell = (u_0 + \sqrt{d}v_0)\zeta^\ell$$

and conjugating (in $\mathbb{Q}[\sqrt{d}]$), we get

$$u_\ell - \sqrt{d}v_\ell = (u_0 - \sqrt{d}v_0)\eta^\ell.$$

Multiplying the two relations, above we get

$$u_\ell^2 - dv_\ell^2 = (u_0^2 - dv_0^2)(\zeta\eta)^\ell = N,$$

from where we read that (u_ℓ, v_ℓ) is also a solution of the Pell equation (8). Observe that

$$u_\ell = \frac{c_1\zeta^\ell + c_2\eta^\ell}{2} \quad \text{and} \quad v_\ell = \frac{c_1\zeta^\ell - c_2\eta^\ell}{2\sqrt{d}},$$

where $c_1 = u_0 + \sqrt{d}v_0$ and $c_2 = u_0 - \sqrt{d}v_0$. The sequences $(u_\ell)_{\ell\geq 0}$ and $(v_\ell)_{\ell\geq 0}$ are both binary recurrent of characteristic equation

$$f(X) = (X - \zeta)(X - \eta) = X^2 - (\zeta + \eta)X + \zeta\eta = X^2 - 2x_1X + 1.$$

Finally, let us prove that all the positive integer solutions (u, v) of the Eq. (8) are obtained the way we described above from some "small" positive integer solution (u_0, v_0). Let (u, v) be a positive integer solution of the Eq. (8). Assuming that $u + \sqrt{d}v \leq |N|\zeta$, there are only finitely many possibilities for the pair (u, v). Suppose now that $u + \sqrt{d}v > |N|\zeta$, and let $\ell \geq 1$ be the minimal positive integer ℓ such that $(u + \sqrt{d}v)\zeta^{-\ell} \leq N\zeta$. We write

$$\begin{aligned} u_0 + \sqrt{d}v_0 &= (u + \sqrt{d}v)\zeta^{-\ell} = (u + \sqrt{d}v)(x_\ell - \sqrt{d}y_\ell) \\ &= (ux_\ell - dvy_\ell) + \sqrt{d}(-uy_\ell + vx_\ell). \end{aligned}$$

Since u and v are positive, by the definition of ℓ, we have that $u_0 + v_0\sqrt{d} > |N|$, for if not then we would have $0 < u_0 + \sqrt{d}v_0 < |N|$, therefore

$$(u + \sqrt{d}v)\zeta^{-(\ell-1)} < (u_0 + \sqrt{d}v_0)\zeta < |N|\zeta,$$

contradicting the way we chose ℓ. Let us now prove that u_0 and v_0 are positive. It is clear that at least one of them is positive. Since $u_0^2 - dy_0^2 = N$, we have that $|u_0 - \sqrt{d}v_0| = |N|/(u_0 + \sqrt{d}v_0) < 1$. If u_0 and v_0 would have opposite signs, then $1 > |u_0 - \sqrt{d}v_0| = |u_0| + \sqrt{d}|v_0|$, which is not possible. The above argument shows, therefore, that for every positive integer solution (u, v) of the Pell equation (8) there is some nonnegative integer ℓ and some positive integer solution (u_0, v_0) of the Eq. (8) with $u_0 + \sqrt{d}v_0 < |N|\zeta$, such that

$$u + \sqrt{d}v = (u_0 + \sqrt{d}v_0)\zeta^{\ell},$$

which, via the argument from the beginning of this section, confirms that (u, v) belongs to a finite union of binary recurrent sequences.

Up to now, we looked at some algebraic properties of linear recurrent sequences of order 2. Next, we delve into their number theoretic properties. In particular, we look at a very powerful theorem on the divisibility of elements of certain linear recurrences of order by primes. But before, we take a detour and talk about cyclotomic polynomials.

4 Cyclotomic Polynomials

The family of polynomials of the form $X^n - 1$ has the very important property that their roots are precisely the nth roots of unity. However, many of these polynomials have common factors: for example, for any positive integers m and n, we have that $(X^m - 1, X^n - 1) = X^{(m,n)} - 1$. These common factors show that there are mth roots of unity that are also nth roots of unity. Thus, to remove all common factors among these polynomials, we can define a polynomial $\Phi_n(X)$ whose roots are all nth roots of unity, yet are not mth roots of unity for any $m < n$.

Definition 4.1 The polynomial

$$\Phi_n(X) = \prod_{\substack{d:(d,n)=1 \\ d<n}} (X - e^{2i\pi d/n})$$

is called the *nth cyclotomic polynomial*. Note that its degree is $\phi(n)$, the Euler function of n, which counts the number of positive integers $k \leq n$ with $(k, n) = 1$.

Example 4.2 The relation

$$X^n - 1 = \prod_{d|n} \Phi_d(X) \tag{9}$$

is very important. To prove it, look at the list

$$\frac{1}{n}, \frac{2}{n}, \ldots, \frac{n}{n}$$

and write in this list every fraction k/n as a reduced fraction ℓ/d with $(\ell, d) = 1$ and some divisor d of n. It is easy to see that all divisors d of n appear in this way, and for each fixed d, all ℓs which have $1 \le \ell \le d$ and are coprime to d appear as well. Since the root $e^{2\pi i\ell/n}$ as a root of $X^n - 1$ is in fact a root of $\Phi_d(X)$, we get (9). To see how it works, note that

$$(1/6, 2/6, 3/6, 4/6, 5/6, 6/6) = (1/6, 1/3, 1/2, 2/3, 5/6, 1).$$

The last number 1 leads to the only root of $\Phi_1(X) = X - 1$. The number $1/2$ corresponds to the only root of unity of order 2, which is -1, so $\Phi_2(X) = X + 1$. The numbers $1/3$ and $2/3$ correspond to cubic roots of unity so they participate in $\Phi_3(X) = X^2 + X + 1$. Finally, $1/6$ and $5/6$ lead to the two roots of unity of order 6 and $\Phi_6(X) = X^2 - X + 1$.

Proposition 4.3 *For all $n \in \mathbb{N}$, $\Phi_n(X) \in \mathbb{Z}[X]$.*

Proof We proceed by strong induction on n. Clearly $\Phi_1(X) = X - 1 \in \mathbb{Z}[X]$. Now assume, for all $n \le k$, that $\Phi_n(X) \in \mathbb{Z}[X]$. Let

$$L_n(X) = \prod_{\substack{d|n \\ d<n}} \Phi_d(X).$$

We must then have that $\Phi_n(X) \cdot L_n(X) = X^n - 1$. Since $L_n(X) \in \mathbb{Z}[X]$, from polynomial division, we have that $\Phi_n(X) \in \mathbb{Q}[X]$. We also have that $\Phi_n(X) \in \overline{\mathbb{Z}}[X]$ by definition (here, $\overline{\mathbb{Z}}$ is the ring of algebraic integers), so $\Phi_n(X) \in \mathbb{Q}[X] \cap \overline{\mathbb{Z}}[X] = \mathbb{Z}[X]$.

5 Lucas Sequences

Definition 5.1 A Lucas sequence $\{u_n\}_{n\ge0}$ is a linearly recurrent sequence of order 2

$$u_{n+2} = ru_{n+1} + su_n, \qquad n \ge 0$$

with nonzero integers r, s such that $(r, s) = 1$, $u_0 = 0$, $u_1 = 1$, and the ratio α/β of the roots α, β of the quadratic $x^2 - rx - s = 0$ is not a root of unity.

Note that

$$(\alpha, \beta) = \left(\frac{r + \sqrt{r^2 + 4s}}{2}, \frac{r - \sqrt{r^2 + 4s}}{2}\right).$$

Also, $r = \alpha + \beta$ and $s = -\alpha\beta$. The initial conditions lead to

$$u_n = \frac{\alpha^n - \beta^n}{\alpha - \beta}, \qquad n = 0, 1, \ldots.$$

Example 5.2 The Fibonacci sequence $\{F_n\}_{n\geq 0}$ is a Lucas sequence since

$$F_n = \frac{\alpha^n - \beta^n}{\alpha - \beta}, \qquad (\alpha, \beta) = \left(\frac{1+\sqrt{5}}{2}, \frac{1-\sqrt{5}}{2}\right).$$

Its companion Lucas sequence $\{L_n\}_{n\geq 0}$ has $L_n = \alpha^n + \beta^n = F_{2n}/F_n$.

Example 5.3 Let $d > 0$ be an integer not a square and (x_n, y_n) be all positive integer solutions to the Pell equation $x^2 - dy^2 = \pm 1$. Let (x_1, y_1) be the smallest such solution. Then

$$x_n + y_n\sqrt{d} = (x_1 + y_1\sqrt{d})^n.$$

Put $\alpha = x_1 + y_1\sqrt{d}$ and $\beta = x_1 - y_1\sqrt{d}$. Then $\alpha + \beta = 2x_1$ and $\alpha\beta = \pm 1$. Furthermore,

$$x_n = \frac{\alpha^n + \beta^n}{2} \quad \text{and} \quad y_n = \frac{\alpha^n - \beta^n}{2\sqrt{d}},$$

so in fact $\{y_n/y_1\}_{n\geq 1}$ is the Lucas sequence of roots (α, β) and characteristic polynomial $x^2 - rx - s$, where $r = 2x_1$ and $s = \pm 1$.

Let $\{u_n\}_{n\geq 0}$ be a Lucas sequence.

Proposition 5.4 *If $p \mid s$ is a prime, then p does not divide u_n for any $n \geq 1$.*

Proof Since $u_1 = 1$, $u_2 = r$ and $(r, s) = 1$, it follows that if $p \mid s$ and $p \mid u_n$ for some $n \geq 1$, then $n \geq 3$. Let n_0 be the minimal n with $p \mid u_n$. Since $u_{n_0} = ru_{n_0-1} + su_{n_0-2}$ and p divides both u_{n_0} and s, we get that $p \mid ru_{n_0-1}$, and since $(r, s) = 1$, we get $p \mid u_{n_0-1}$, in contradiction with the definition of n_0. This contradiction proves the proposition.

Proposition 5.5 *Let $\{u_n\}_{n\geq 1}$ be a Lucas sequence. Then $(u_m, u_n) = u_{(m,n)}$.*

Proof Note that $u_{(m,n)} \mid (u_m, u_n)$, because if $a \mid b$ then $u_a \mid u_b$. To see this, write $b = a\ell$ and then

$$\frac{u_b}{u_a} = \frac{(\alpha^a)^\ell - (\beta^a)^\ell}{\alpha^a - \beta^a} = (\alpha^a)^{\ell-1} + \cdots + (\beta^a)^{\ell-1}.$$

The expression in the right-hand side is an integer (either because it is symmetric in α, β, therefore a polynomial with integer coefficients in $\alpha + \beta = r$ and $\alpha\beta = -s$, or because it is both a rational number and an algebraic integer). Hence, $u_a \mid u_b$.

Now, we prove that $(u_m, u_n) \mid u_{(m,n)}$. Let $D = (u_m, u_n)$ and assume that $m > n$. Then

$$\alpha^m - \beta^m = \alpha^{m-n}(\alpha^n - \beta^n) + \beta^n(\alpha^{m-n} - \beta^{m-n}),$$

leading to $u_m = \alpha^{m-n}u_n + \beta^n u_{m-n}$. Since D divides u_m and u_n, we get that D divides $\beta^n u_{m-n}$. Here, we say that an algebraic integer η divides an algebraic integer λ if their ratio is an algebraic integer. Thus, D divides $(\alpha\beta)^n u_{m-n} = (-s)^n u_{m-n}$, and since by the previous proposition D and s are coprime, we get that $D \mid u_{m-n}$. Continuing in this way, we use the Euclidean algorithm to arrive at the conclusion that D divides $u_{(m,n)}$. Thus, $(u_m, u_n) \mid u_{(m,n)}$. Hence, $(u_m, u_n) = u_{(m,n)}$.

We put $\Delta = (\alpha - \beta)^2 = r^2 + 4s$. For an odd prime p and an integer a, we use $\left(\frac{a}{p}\right)$ for the Legendre symbol of a with respect to p, which is

$$\left(\frac{a}{p}\right) = \begin{cases} 0 & \text{if} \quad p \mid a; \\ 1 & \text{if} \quad a \equiv x^2 \pmod{p} \text{ and } p \nmid a; \\ -1 & \text{otherwise.} \end{cases}$$

Proposition 5.6 *Assume $p \nmid s$ is odd. The following hold:*

(i) If $p \mid \Delta$, then $p \mid u_p$.
(ii) If $p \nmid \Delta$, and $\alpha \in \mathbb{Q}$, then $p \mid u_{p-1}$.
(iii) If $p \nmid \Delta$ and $\alpha \notin \mathbb{Q}$, then

$$p \mid \begin{cases} u_{p-1} & \textit{if } \left(\frac{\Delta}{p}\right) = 1, \\ u_{p+1} & \textit{if } \left(\frac{\Delta}{p}\right) = -1. \end{cases}$$

Proof (i) Since $p \mid (\alpha - \beta)^2$, we know that $\alpha \equiv \beta \pmod{p}$. Here and in what follows we say that two algebraic integers γ, δ are congruent modulo a nonzero integer m if $(\gamma - \delta)/m$ is an algebraic integer. Hence,

$$u_p = \frac{\alpha^p - \beta^p}{\alpha - \beta} = \alpha^{p-1} + \alpha^{p-2}\beta + \cdots + \beta^{p-1} \equiv p \cdot \alpha^{p-1} \equiv 0 \pmod{p}.$$

(ii) Here, $p \nmid \alpha$, $p \nmid \beta$, and $p \nmid (\alpha - \beta)$. We use Fermat's Little Theorem to get that $\alpha^{p-1} \equiv \beta^{p-1} \equiv 1 \pmod{p}$, so

$$u_{p-1} = \frac{\alpha^{p-1} - \beta^{p-1}}{\alpha - \beta} \equiv 0 \pmod{p}.$$

(iii) Expand

$$2^p \alpha^p = r^p + \binom{p}{1} r^{p-1}\sqrt{\Delta} + \cdots + \sqrt{\Delta} \cdot \Delta^{(p-1)/2} \equiv r + \sqrt{\Delta} \cdot \Delta^{(p-1)/2} \pmod{p}. \tag{10}$$

If $\left(\frac{\Delta}{p}\right) = 1$, then $\Delta^{(p-1)/2} \equiv 1 \pmod{p}$. The above leads to $2^p \alpha^p \equiv 2\alpha \pmod{p}$. Since $2^p \equiv 2 \pmod{p}$ and p is odd, we get that p divides $\alpha(\alpha^{p-1} - 1)$. Thus, $p \mid (\alpha\beta)(\alpha^p - 1)$ and since $\alpha\beta = s$ is coprime to p, we get $\alpha^{p-1} \equiv 1 \pmod{p}$. The same argument works with α replaced by β, so we get that $\alpha^{p-1} \equiv \beta^{p-1} \pmod{p}$.

Hence, p divides $(\alpha - \beta)u_{p-1}$, therefore it divides also Δu_{p-1}, and since p is coprime to Δ, we get that $p \mid u_{p-1}$.

This was when $\left(\frac{\Delta}{p}\right) = 1$. Assume now that $\left(\frac{\Delta}{p}\right) = -1$. Equation (10) together with the fact that $\Delta^{(p-1)/2} \equiv -1 \pmod{p}$ now shows that $\alpha^p \equiv \beta \pmod{p}$, therefore $\alpha^{p+1} \equiv -s \pmod{p}$. The same argument works with α replaced by β so we get $\beta^{p+1} \equiv -s \pmod{p}$. Thus, $\alpha^{p+1} \equiv \beta^{p+1} \pmod{p}$, which leads, via an argument similar to the one above, to the conclusion that $p \mid u_{p+1}$.

Lemma 5.7 *Assume that $p \mid u_m$, $m \mid n$ and $p \mid u_n/u_m$. Then $p \mid n/m$.*

Proof Note that $u_n/u_m = ((\alpha^m)^{n/m} - (\beta^m)^{n/m})/(\alpha^m - \beta^m)$ is the n/mth term of the Lucas sequence $\{U_k\}_{k\geq 1}$ of roots (α^m, β^m) and discriminant $(\alpha^m - \beta^m)^2 = \Delta u_m^2$, a multiple of p. By (i) of the previous lemma, $p \mid U_p$ and by the hypothesis, $p \mid U_{n/m}$. Hence, $p \mid (U_p, U_{n/m}) = U_{(p,n/m)}$, and the conclusion follows since if $p \nmid n/m$, then $(p, n/m) = 1$, so $p \mid U_1 = 1$, a contradiction.

Lemma 5.8 *Assume that m is a positive integer and p is an odd prime such that $p \mid u_m$. Then $p \mid u_{mp}/u_m$, but $p^2 \nmid u_{mp}/u_m$.*

Proof Note that

$$\frac{u_{mp}}{u_m} = \frac{(\alpha^m)^p - (\beta^m)^p}{\alpha^m - \beta^m}.$$

We are given that $p \mid u_m$, so $\alpha^m - \beta^m = p\lambda$ for some algebraic integer λ. Therefore

$$\frac{u_{mp}}{u_m} = \sum_{i=0}^{p-1} (\alpha^m)^i (\beta^m)^{p-i-1} = \sum_{i=0}^{p-1} (\beta^m + p\lambda)^i (\beta^m)^{p-i-1}.$$

We expand this sum and take it modulo p^2:

$$\begin{aligned}\frac{u_{mp}}{u_m} &\equiv p(\beta^m)^{p-1} + \sum_{i=0}^{p-1} ip\lambda\beta^{p-2} \equiv p(\beta^m)^{p-1} + p\lambda(\beta^m)^{p-2}\frac{p(p-1)}{2} \\ &\equiv p(\beta^m)^{p-1} \pmod{p^2}.\end{aligned}$$

This means that p^2 does not divide the fraction from the left-hand side above.

Definition 5.9 For a prime p, let $z(p)$ be the order of appearance of p in $\{u_n\}_{n\geq 0}$; i.e., the minimal positive integer k such that $p \mid u_k$.

If such k does not exist (for example, if $p \mid s$), we put $z(p) = \infty$.

Exercise 5.10 If $p \mid u_n$, then $z(p) \mid n$.

Proof Note that $p \mid u_n$ and $p \mid u_{z(p)}$, so $p \mid (u_n, u_{z(p)}) = u_{(n,z(p))}$. But we have that $(n, z(p)) \geq z(p)$ since $z(p)$ is minimal, but $(n, z(p)) \leq z(p)$ from the definition of the greatest common divisor; so, $(n, z(p)) = z(p)$, therefore $z(p) \mid n$. □

Exercise 5.11 Let $\psi(p)$ be the period of the Fibonacci sequence modulo p. Then $\psi(p) \in \{z(p), 2z(p), 4z(p)\}$.

Theorem 5.12 *If $p \nmid s\Delta$, then $p \equiv \pm 1 \pmod{z(p)}$.*

Proof Since $p \nmid \Delta$, we have that p divides either u_{p-1} or u_{p+1}, while $p \mid u_{z(p)}$. Hence, $p \mid u_{(z(p),p\pm 1)} = u_{z(p)}$. Thus, $z(p) \mid p \pm 1$.

6 Zsigmondy's Theorem

We now have the tools we need to prove Zsigmondy's Theorem.

Definition 6.1 $p \mid u_n$ is a *primitive divisor* of u_n if $p \nmid \Delta$ and $z(p) = n$.

Theorem 6.2 (Zsigmondy) *If the roots of the characteristic polynomial of the Lucas sequence $\{u_n\}_{n\geq 0}$ are integers, and if $n > 6$, then u_n has a primitive divisor.*

Proof Assume that $a > b > 0$. Let

$$\Phi_n(X, Y) = \prod_{\substack{1\leq k\leq n\\(k,n)=1}} (X - e^{2\pi ik/n}Y) \in \mathbb{Z}[X, Y]$$

be the homogenization of the cyclotomic polynomial for n. Recall that the homogenization of a polynomial

$$f(X) = a_d X^d + a_a X^{d-1} + \cdots + a_0, \qquad a_d \neq 0,$$

is simply the homogeneous polynomial

$$f(X, Y) = a_d X^d + a_1 X^{d-1} Y + \cdots + a_0 Y^d.$$

The formula

$$X^n - 1 = \prod_{d\mid n} \Phi_d(X)$$

homogenizes to

$$X^n - Y^n = \prod_{d\mid n} \Phi_d(X, Y). \tag{11}$$

We write $\Phi_n(a, b) = AB$, where each prime $p \mid B$ is primitive and the primes $q \mid A$ are not primitive. Let $p \mid A$. Then $p \mid u_n$ and $p \mid u_d$ for some $d < n$. We may assume that $d \mid n$ (otherwise, we replace d by (d, n)). Since $p \mid A \mid \Phi_n(A, B) \mid u_n/u_d$, where the last divisibility relation follows from Eq. (11) with $(X, Y) = (a, b)$, we get that $p \mid (u_d, u_n/u_d)$, so, by Lemma 5.7, we get that $p \mid n/d$. Thus, $d \mid n/p$, therefore

$p \mid u_d \mid u_{n/p}$. But $A \mid \Phi_n(a, b) \mid u_n/u_{n/p}$. By Lemma 5.8(iii) with $d = n/p$, we get that $p \| A$ if p is odd. When $p = 2$, then u_n is odd when a and b have different parities or when n is odd. If $2\|n$, then $A \mid u_n/u_2 = ((a^2)^{n/2} - (b^2)^{n/2})/(a^2 - b^2)$ is again odd, and if $4 \mid n$, then $A \mid u_n/u_{n/2} = a^{n/2} + b^{n/2} = (a^{n/4})^2 + (b^{n/4})^2$, and since a and b are odd, we have that the above sum of two odd squares is congruent to 2 (mod 4). Thus, the exponent of 2 in the factorization of A is at most 1 also. This shows that A is square free and $A \mid n$. But note that, by the Principle of Inclusion and Exclusion,

$$\Phi_n(a, b) = \frac{a^n - b^n}{\prod_{p|n}(a^{n/p} - b^{n/p})} \cdot \frac{\prod_{\substack{p<q \\ pq|n}}(a^{n/pq} - b^{n/pq})}{\prod_{\substack{p<q<r \\ pqr|n}}(a^{n/pqr} - b^{n/pqr})} \cdots. \tag{12}$$

Since

$$a^d > a^d - b^d = (a - b)(a^{d-1} + \ldots + b^{d-1}) \geq a^{d-1}, \tag{13}$$

it follows that a lower bound on $\Phi_n(a, b)$ is obtained by replacing each factor $a^d - b^d$ from the top of (12) by a^{d-1} and each factor from bottom of (9) by a^d thus getting

$$\Phi_n(a, b) \geq a^{n - \sum_{p\mid n} \frac{n}{p} + \sum_{pq \mid n} \frac{n}{pq} - \cdots - 2^{\omega(n)-1}} \geq a^{\phi(n) - 2^{\omega(n)-1}},$$

where $\omega(n)$ is the number of distinct prime factors of n. Since $A \leq n$, we get that

$$B \geq \frac{a^{\phi(n) - 2^{\omega(n)-1}}}{n}. \tag{14}$$

Thus, if $B = 1$, (which is equivalent to saying that u_n has no primitive prime factors), we then get that

$$\phi(n) - 2^{\omega(n)-1} \leq \log_2 n$$

where $\log_2 n$ is the logarithm to base 2. Note that if $p_1, \ldots, p_k$ are all the prime factors of n, then $2^k \leq p_1 \cdots p_k \leq n$, so $k \leq \log_2 n$. Thus,

$$\phi(n) = n \prod_{i=1}^{k} \left(1 - \frac{1}{p_i}\right) \geq n \prod_{i=2}^{\lfloor \log_2 n + 1 \rfloor} \left(1 - \frac{1}{i}\right) = \frac{n}{\lfloor \log_2 n + 1 \rfloor}.$$

Also, since $n \geq 2 \cdot 3 \cdot 5^{k-2}$, we get that $4^{k-2} < n/6$, so $2^{k-1} < 2\sqrt{n/6} < \sqrt{n}$. Thus,

$$\log_2 n \geq \phi(n) - 2^{k-1} \geq \frac{n}{\log_2(2n)} - \sqrt{n}, \tag{15}$$

and Mathematica shows that $n \leq 200$. Thus, $k \leq 4$, and the inequality (15) becomes

$$7 \geq \phi(n) - 8$$

giving $\phi(n) \le 15$ (so $\phi(n) \le 14$ because $\phi(n)$ is even for $n > 2$), so we conclude that $14 \ge \phi(n) \ge n(1-1/2)(1-1/3)(1-1/5)(1-1/7)$ giving $n \le 122$, so $k \le 3$. Hence,

$$6 \ge \phi(n) - 4,$$

giving $10 \ge \phi(n) \ge n(1-1/2)(1-1/3)(1-1/5)$, yielding $n \le 37$. This gives

$$5 \ge \phi(n) - 4,$$

so $\phi(n) \le 9$ (therefore $\phi(n) \le 8$), giving $8 \ge \phi(n) \ge (1-1/2)(1-1/3)(1-1/5)$, so $n \le 30$. For $n \in [7, 30]$, $\phi(n) - 2^{k-1} > 0$, so for each such positive integer n the inequality $n > a^{\phi(n)-2^{k-1}}$ gives a bound on a. Now one checks all cases to conclude that the theorem holds. The theorem fails for $n = 6$ not only because $\phi(n) - 2^{k-1} = 0$ for $n = 6$ but also because when $u_n = 2^n - 1$, then $u_6 = 2^6 - 1 = 63 = 3^2 \cdot 7 = u_2^2 \cdot u_3$.

If the characteristic polynomial of $(u_n)_{n \ge 0}$ has real roots (not necessarily integers), then exceptions to Primitive Divisors appear up to the index $n = 12$. This theorem was proved by Carmichael [9] in 1913.

Example 6.3 The first 20 terms of the Fibonacci sequence are

1, 1, 2, 3, 5, 8, 13, 21, 34, 55, 89, 144, 233, 377, 610, 987, 1597, 2584, 4181, 6765.

Observe that $F_1 = F_2 = 1$, $F_5 = 5$ (and $5 = \Delta$), $F_6 = 2^3$ (and $2 \mid F_3$), $F_{12} = 144 = 2^4\, 3^2$ (and $2 \mid F_3,\ 3 \mid F_4$), and all the other terms in the above list have primitive divisors.

7 The Primitive Divisor Theorem

The Primitive Divisor Theorem is an extension of Zsigmondy's theorem and Carmichael's theorem and says that if $n \notin \{1, 2, 3, 4, 6\}$, then u_n has a primitive divisor except in finitely many instances all of which are known.

Theorem 7.1 *If $n \notin \{1, 2, 3, 4, 6\}$, then u_n has a primitive divisor except when $((\alpha_1, \alpha_2), n)$ is of the form*

$$\left(\pm((a_1 + \sqrt{\Delta})/2, (a_1 - \sqrt{\Delta})/2)), n\right),$$

where (a_1, Δ, n) is one of the following triples:

n	(a_1, Δ)
5	$(1, 5), (1, -7), (2, -40), (1, -11), (1, -15), (12, -76), (12, -1364)$
7	$(1, -7), (1, -19)$
8	$(2, -24), (1, -7)$
10	$(2, -8), (5, -3), (5, -47)$
12	$(1, 5), (1, -7), (1, -11), (2, -56), (1, -15), (1, -19)$
13	$(1, -7)$
18	$(1, -7)$
30	$(1, -7)$

The preceding theorem is due to Zsigmondy [47] and rediscovered, independently, by Birkhoff and Vandiver [5] more than 10 years later for the case in which the roots (α_1, α_2) are integers. Later, Carmichael [9] proved it for the case when the roots are real. Bilu, Hanrot, and Voutier [4], building upon prior work of Schinzel [38] and Stewart [41], finished off the case when the roots are complex nonreal.

8 Applications

Proposition 8.1 *The largest solution of the equation*

$$F_n = m_1!m_2! \cdots m_k!$$

with $2 \le m_1 \le m_2 \le \cdots \le m_k$ *is* $F_{12} = 3!\,4! = 2!^2 3!^2$.

Proof If $n > 12$, then, by Theorem 7.1, F_n has a primitive prime factor p. Since $p \equiv \pm 1 \pmod{n}$, we get that $m_k \ge p \ge n - 1$. By the inequality $m_k! \ge (m_k/e)^{m_k}$, we get that $m_k! \ge ((n-1)/e)^{n-1}$. Since $n > 12$, we have that $(n-1)/e > \alpha$, so $m_k! > \alpha^{n-1}$. We have obtained

$$F_n = m_1! \cdots m_k! \ge m_k! > \alpha^{n-1},$$

which is false because $F_n \le \alpha^{n-1}$ for all n. This last inequality can be proved by checking it for $n = 1,\ 2$ and by using induction for $n \ge 3$ via the recurrence

$$F_n = F_{n-1} + F_{n-2} \le \alpha^{n-2} + \alpha^{n-3} = \alpha^{n-1}.$$

In 1844, E. Catalan [10] conjectured that the only solution in positive integers of the equation

$$x^m - y^n = 1 \qquad m \ge 2,\ n \ge 2 \tag{16}$$

is $3^2 - 2^3 = 1$. This conjecture was proved by Mihăilescu [32] in 2002. It is clear that we may assume that m and n are primes, because if (x, y, m, n) is a solution and $p \mid m$ and $q \mid n$ are primes, then $(x^{m/p}, y^{n/q}, p, q)$ is also a solution. Furthermore, m and n are distinct, because if $m = n$, then

$$1 = x^m - y^m = (x - y)(x^{m-1} + \cdots + y^{m-1}) > x - y \ge 1,$$

which is impossible.

In what follows, we present a result of Lebesque of 1850 [25] concerning this equation using the language of primitive divisors of terms of Lucas sequences.

Proposition 8.2 *Equation* (16) *has no solution with* $n = 2$.

Proof Assume that there is a solution of the equation

$$x^m = y^2 + 1.$$

It is clear that y is even, because if y were odd, then $y^2 + 1 \equiv 2 \pmod 4$ cannot be a perfect power. Factoring the right-hand side in $\mathbb{Z}[i]$, we get $(y + i)(y - i) = x^m$. The two numbers $y + i$ and $y - i$ are coprime in $\mathbb{Z}[i]$. In fact, if q is some prime of $\mathbb{Z}[i]$ dividing both $y + i$ and $y - i$, then $q \mid (y + i) - (y - i) = 2i \mid 2$, and $q \mid x^m$ where x is odd (because y is even), which is a contradiction. Thus, since $y + i$ and $y - i$ are coprime and their product is x^m, we infer that there exists $\alpha_1 = a + bi \in \mathbb{Z}[i]$ such that $y + i = \zeta\alpha_1^m$, where ζ is a unit. Furthermore, $x = a^2 + b^2$. The only units of $\mathbb{Z}[i]$ are ± 1, $\pm i$ of finite orders dividing 4, and since m is odd, we can replace α_1 by one of its associates (for example, by $\zeta^m\alpha_1$) and conclude that $\zeta\alpha_1^m = \zeta^{m^2}\alpha_1^m = (\zeta^m\alpha_1)^m$. Thus, we can take $\zeta = 1$ to infer that

$$y + i = \alpha_1^m.$$

Conjugating the above relation and eliminating y between the two equations, we get

$$2i = \alpha_1^m - \alpha_2^m, \qquad \text{where} \qquad \alpha_2 = \overline{\alpha_1}.$$

Since $\alpha_1 - \alpha_2 = 2bi$ divides $2i$, we get that $b = \pm 1$ and

$$\pm 1 = \frac{\alpha_1^m - \alpha_2^m}{\alpha_1 - \alpha_2}. \tag{17}$$

Interchanging α_1 and α_2, we may assume that $b = 1$. The right-hand side of the above equation is the mth term of a Lucas sequence with $a_1 = \alpha_1 + \alpha_2 = 2a$ and $a_2 = -(\alpha_1\alpha_2) = -(a^2 + 1) = -x$, which is odd. Thus, a_1 are a_2 coprime. We need that α_1/α_2 is not a root of 1. If it were, since it is also in $\mathbb{Q}[i]$, then it would be ± 1 or $\pm i$. If $\alpha_1/\alpha_2 = \pm 1$, we get either that $a + i = a - i$, which is false, or that $a + i = -a + i$, leading to $a = 0$, so $x = 1$, which is also false. If $\alpha_1/\alpha_2 = \pm i$, we then get $a + i = ai + 1$, so either $a = 1$, or $a + i = -ai - 1$, so $a = -1$. In both cases, we get $x = 2$, contradicting the fact that x is odd. Thus, the right-hand side of (17) is the mth term of a Lucas sequence, and clearly it has no primitive divisors. Now either $m \in \{2, 3, 4, 6\}$, or the triple (α_1, α_2, m) appears in the table from Theorem 7.1. A rapid inspection of the table shows that none of the pairs of roots (α_1, α_2) appearing in the table has components in $\mathbb{Q}[i]$, therefore it must be the case that $m = 3$. In this way, we get

$$\pm 1 = \frac{(a+i)^3 - (a-i)^3}{2i} = (a+i)^2 + (a+i)(a-i) + (a-i)^2 = 3a^2 - 1,$$

which gives us $3a^2 \in \{0, 2\}$, and none of these possibilities leads to a solution of the initial equation.

Let $P(m)$ be the maximal prime factor of m.

Proposition 8.3 *The largest solution of* $P(x^2 - 1) \leq 29$ *is*

$$36171409^2 - 1 = 2^5 \cdot 3 \cdot 5 \cdot 7^3 \cdot 11 \cdot 13^3 \cdot 17 \cdot 23 \cdot 29^2.$$

Proof We write $x^2 - 1 = dy^2$, where d is squarefree. The only primes that can divide d are $\{2, 3, 5, 7, 11, 13, 17, 19, 23, 29\}$. Thus, there are only 1023 possible values for d. For each one of such d, let $(x_1(d), y_1(d))$ be the minimal solution of the Pell equation $x^2 - dy^2 = 1$, and let $(x_\ell(d), y_\ell(d))$ be its ℓth solution. Let $u_\ell = y_\ell(d)/y_1(d)$, which, from what we have seen is a Lucas sequence (see Example 5.3). It follows, by Theorem 7.1, that if $\ell \geq 31$, then $y_\ell(d)$ has a prime factor $\geq \ell - 1 \geq 30$. Thus, for each of the 1023 values of d it suffices to generate the first 30 terms of the sequence $(x_\ell(d))_{\ell \geq 1}$ and observe that all the positive integers x such that $P(x^2 - 1) \leq 29$ are obtained in this way. A quick computation with Mathematica shows that out of all these $1023 \cdot 30 = 30690$ candidates, $x = 36171409$ is the largest solution to our problem. Other interesting examples are

$$16537599^2 - 1 = 2^{12} \cdot 5^2 \cdot 7^2 \cdot 11 \cdot 17 \cdot 19 \cdot 23^2 \cdot 29;$$
$$12901780^2 - 1 = 3^2 \cdot 11 \cdot 19^4 \cdot 23^2 \cdot 29^3.$$

9 Problems

1. Formulate and prove an analogue of Proposition 2.2 for the case when the roots $\alpha_1, \ldots, \alpha_s$ of $f(X)$ are not necessarily simple.

2. Prove that $(1 + \sqrt{5})/2 = [\overline{1}]$, and that the kth convergent of $(1 + \sqrt{5})/2$ is F_{k+1}/F_k. The notation $[\overline{1}]$ stands for the periodic continued fraction $[1, 1, \ldots]$ (see also Sect. 11.5).

3. Find the simple continued fraction of $(F_{10n+1}/F_{10n})^5$.

4. Let $n \geq 2$ be an integer. Prove that n does not divide $2^n - 1$.

5. Let $k \geq 2$ and $n_1, n_2, \ldots, n_k$ positive integers such that

$$n_2 \mid 2^{n_1} - 1, \quad n_3 \mid 2^{n_2} - 1, \quad \ldots \quad n_k \mid 2^{n_{k-1}} - 1, \quad n_k \mid 2^{n_1} - 1.$$

Prove that $n_1 = \cdots = n_k = 1$.

6. Determine if there exists an integer n with precisely 2000 distinct prime factors such that $n \mid 2^n + 1$.

7. Let m and n be positive integers such that $A = ((m+3)^n + 1)/(3m)$ is an integer. Prove that A is odd.

8. Prove that

$$\sum_{k=0}^{n} \binom{2n+1}{2k+1} 2^{3k}$$

is not a multiple of 5 for any $n \geq 0$.

9. Let $p_1, \ldots, p_n$ be distinct primes. Prove that $2^{p_1 \cdots p_n} + 1$ has at least 4^n divisors.

10. Let $b,\ m,\ n$ be positive integers with $b > 1$ and $m \neq n$ such that $b^m - 1$ and $b^n - 1$ have the same prime factors. Prove that $b + 1$ is a power of 2.

11. (i) Suppose that $2^n + 1$ is prime. Prove that n is a power of 2.
(ii) Suppose that $4^n + 2^n + 1$ is prime. Prove that n is a power of 3.

12. Determine all positive integers n such that $(2^n + 1)/n^2$ is an integer.

13. Determine all pairs (n, p) of positive integers with p prime, $n < 2p$, and such that $p^{n-1} + 1$ is divisible by n^{p-1}.

14. Determine all triples (a, m, n) of positive integers such that $a^m + 1 \mid (a+1)^n$.

15. (i) Find all positive integers n such that $3^n - 1$ is divisible by 2^n.
(ii) Find all positive integers n such that $9^n - 1$ is divisible by 7^n.

16. Prove that the Diophantine equation $(n-1)! + 1 = n^k$ has no solutions with $n > 5$.

17. Prove that

$$F_{2n-1}^2 + F_{2n+1}^2 + 1 = 3F_{2n-1}F_{2n+1}$$

holds for all $n \geq 1$.

18. The sequence $(a_n)_{n \geq 1}$ is given by

$$a_1 = 1, \quad a_2 = 12, \quad a_3 = 20, \quad a_{n+3} = 2a_{n+2} + 2a_{n+1} - a_n.$$

Prove that $1 + 4a_n a_{n+1}$ is a perfect square for all $n \geq 1$.

19. Prove that every positive integer can be represented as a sum of Fibonacci numbers in such a way that there are no two consecutive ones.

20. Prove that the largest solution of the equation

$$F_{n_1} F_{n_2} \cdots F_{n_k} = m_1! m_2! \cdots m_\ell!$$

with $1 \le n_1 < n_2 < \cdots < n_k$ and $2 \le m_1 \le m_2 \le \cdots \le m_\ell$ is

$$F_1 F_2 F_3 F_4 F_5 F_6 F_8 F_{10} F_{12} = 11!.$$

21. Let (x_ℓ, y_ℓ) be the ℓth solution of the Pell equation $x^2 - dy^2 = 1$. For a positive integer m, let $P(m)$ be the largest prime factor of m. Prove that if $P(x_\ell) \le 5$, then $\ell = 1$.

22. If $p \ge 5$ is prime, prove that the Diophantine equation

$$x^2 + 2^a \cdot 3^b = y^p$$

has no solutions $x \ge 1,\ a \ge 0,\ b \ge 0$ and $\gcd(x, y) = 1$.

23. Find all Fibonacci numbers of the form $\dfrac{1}{n+1}\dbinom{2n}{n}$.

24. Prove that the only solution of the equation

$$\frac{x^n - 1}{x - 1} = y^2$$

with n odd is $(x, n, y) = (3, 5, 11)$.

25. Let $\phi(n)$ be the Euler function of n. Prove that if $\phi(n)^2 \mid n^2 - 1$, then $n \in \{1, 2, 3\}$.

26. Let $\phi(n)$ and $\sigma(n)$ be the Euler function and the sum of divisors of n, respectively.

(i) Prove $P(\phi(n)\sigma(n))$ tends to infinity with n.
(ii) Prove that $P(\phi(n)\sigma(n)) \ge (1 + o(1)) \log\log n$ as $n \to \infty$.

27. Find all positive integers n such that $n! \mid \sigma(n!)$.

28. Suppose that $m > n \ge 0$ and that $2^m - 2^n$ divides $3^m - 3^n$. Prove that $2^m - 2^n$ divides $x^m - x^n$ for all positive integers x.

29. Find all positive integers n such that $\binom{2n}{n} \mid \sigma(\sigma(\binom{2n}{n}))$.

10 Notes

Problem 1 is well known in the theory of linearly recurrent sequences (see [20], or [40] for example). For Problem 3, see [12]. Problems 4–19 are from the book [44]. Some of these problems have appeared in mathematical competitions at national levels in various countries, or at the International Mathematical Olympiad. While their solutions are considered, in general, a collection of isolated tricks, we leave it to the solver to note that most of these problems are pretty easy consequences of either Theorem 7.1, or of the arguments that appear in the lemmas that we used to prove this beautiful result. For Problem 22, see [28]. Problem 25 is a result of Křížek and Luca [23]. Problem 27 is due to Pomerance, see [36]. For Problem 28, see [37]. For Problem 29, see [30].

11 Linear Forms in Logarithms

11.1 Statements

In 1966, A. Baker [1] gave an effective lower bound on the absolute value of a nonzero linear form in logarithms of algebraic numbers; that is, for a nonzero expression of the form

$$\sum_{i=1}^{n} b_i \log \alpha_i,$$

where $\alpha_1, \ldots, \alpha_n$ are algebraic numbers and $b_1, \ldots, b_n$ are integers. His result marked the dawn of the era of effective resolution of the Diophantine equations of certain types, namely the ones that can be reduced to exponential ones; i.e., where the unknown variables are in the exponents. Many of the computer programs available today which are used to solve Diophantine equations (PARI, MAGMA, KASH, etc.) use some version of Baker's inequality. For our purpose, we shall give some of the Baker-type inequalities available today which are easy to apply. We start with some preliminaries about algebraic numbers. Let α be an algebraic number of degree d. Let

$$f(x) = \sum_{i=0}^{d} a_i x^{d-i} \in \mathbb{Z}[x]$$

be the minimal polynomial of α with $a_0 > 0$ and $(a_0, \ldots, a_d) = 1$. We then put $H(\alpha) := \max\{|a_i| : i = 0, \ldots, d\}$ and call it the *height* of α. Now write

$$f(X) = a_0 \prod_{i=1}^{d} (X - \alpha^{(i)}),$$

where $\alpha = \alpha^{(1)}$. The numbers $\alpha^{(i)}$ are called the *conjugates* of α. The *logarithmic height* of α is

$$h(\alpha) = \frac{1}{d}\left(\log |a_0| + \sum_{i=1}^{d} \log \max\{|\alpha^{(i)}|, 1\}\right).$$

Example 11.1 If $\alpha = p/q$ is a rational number, where p and $q > 0$ are coprime integers, then $H(\alpha) = \max\{|p|, q\}$ and $h(\alpha) = \log \max\{|p|, q\}$.

11.2 The Complex Case

The following result is due to Matveev [31]. Let $\mathbb{L}$ be a number field of degree D; that is, a finite extension of degree D of $\mathbb{Q}$. Let $\alpha_1, \ldots, \alpha_n$ be nonzero elements of $\mathbb{L}$ and let $b_1, \ldots, b_n$ be integers. Put

$$B = \max\{|b_1|, \ldots, |b_n|\},$$

and

$$\Lambda = \prod_{i=1}^{n} \alpha_i^{b_i} - 1.$$

Let $A_1, \ldots, A_n$ be positive real numbers such that

$$A_j \geq h'(\alpha_j) := \max\{Dh(\alpha_j), |\log \alpha_j|, 0.16\} \qquad \text{for all } j = 1, \ldots, n. \tag{18}$$

Here and everywhere else log is a principal determination of the natural logarithm. With these notations, Matveev proved the following theorem (see also Theorem 9.4 in [8]).

Theorem 11.2 *If $\Lambda \neq 0$, then*

$$\log |\Lambda| > -3 \cdot 30^{n+4}(n+1)^{5.5} D^2(1 + \log D)(1 + \log nB)A_1 A_2 \cdots A_n. \tag{19}$$

If furthermore $\mathbb{L}$ is real, then

$$\log \Lambda > -1.4 \cdot 30^{n+3} n^{4.5} D^2(1 + \log D)(1 + \log B)A_1 A_2 \cdots A_n. \tag{20}$$

11.3 The p-Adic Case

In this section, we shall present a p-adic version of a lower bound for linear forms in logarithms of algebraic numbers due to Kunrui Yu [45]. Let π be a prime ideal in $\mathcal{O}_{\mathbb{L}}$. Let e_π and f_π be its indices of *ramification* and *inertia*, respectively. It is known that if $p \in \mathbb{Z}$ is the unique prime number such that $\pi \mid p$, then

$$p\mathcal{O}_{\mathbb{L}} = \prod_{i=1}^{k} \pi_i^{e_i},$$

where $\pi_1, \ldots, \pi_k$ are prime ideals of $\mathcal{O}_{\mathbb{L}}$. The prime π is one of the primes π_i, say π_1, and its e_π equals e_1. The number f_π is the dimension of π over its prime field $\mathbb{Z}/p\mathbb{Z}$, or, to say it differently, can be computed via the formula $\#\mathcal{O}_{\mathbb{L}}/\pi = p^{f_\pi}$. Observe that $e_\pi f_\pi \le D$. For an algebraic number $\alpha \in \mathbb{L}$, we write $\mathrm{ord}_\pi(\alpha)$ for the exponent of π in the factorization of the fractional ideal $\alpha\mathcal{O}_{\mathbb{L}}$ generated by α inside $\mathbb{L}$. Let

$$H_j \ge \max\{h(\alpha_j), \log p\} \qquad \text{for all } j = 1, \ldots, n.$$

In case we are in $\mathbb{Q}$, π is a prime, $D = 1$, $e_\pi = f_\pi = 1$. With the above definitions and notations, Yu proved the following result.

Theorem 11.3 *If $\Lambda \neq 0$, then*

$$\mathrm{ord}_\pi(\Lambda) \le 19(20\sqrt{n+1}D)^{2(n+1)} e_\pi^{n-1} \frac{p^{f_\pi}}{(f_\pi \log p)^2} \log(e^5 nD) H_1 \cdots H_n \log B. \tag{21}$$

11.4 Linear Forms in Two Logarithms

The modern French school of transcendence theory developed some lower bounds for linear forms in two logarithms of algebraic numbers, which have a slightly worse dependence in the parameter $\log B$ than the Baker–Matveev–Yu bounds, but have the property that the multiplicative constants involved are much smaller. Consequently, in applications they yield better results. Let us now see two of their results.

Let α_1 and α_2 be algebraic numbers and put $\mathbb{L} = \mathbb{Q}[\alpha_1, \alpha_2]$. Next, let $D_1 = D$, if $\mathbb{L}$ is real, and $D_1 = D/2$, otherwise. Suppose that

$$A_j \ge \max\{D_1 h(\alpha_j), |\log \alpha_j|, 1\} \qquad \text{for both } j = 1,\ 2.$$

Let

$$\Gamma = b_2 \log \alpha_2 - b_1 \log \alpha_1, \tag{22}$$

and

$$b' = \frac{b_1}{A_2} + \frac{b_2}{A_1}.$$

With these notations, Laurent, Mignotte, and Nesterenko [24] proved the following theorem. Recall that two nonzero complex numbers α, β are multiplicatively independent if the only solution of the equation $\alpha^x \beta^y = 1$ in integers x, y is $x = y = 0$ (see also Exercise 42).

Theorem 11.4 *(i) If α_1 and α_2 are multiplicatively independent, then*

$$\log|\Gamma| \ge -30.9 \left(\max\{D_1 \log b', 21, D_1/2\}\right)^2 A_1 A_2 \tag{23}$$

(ii) If furthermore α_1 and α_2 are real and positive, then

$$\log|\Gamma| \geq -23.34\left(\max\{D_1\log b' + 0.14D_1, 21, D_1/2\}\right)^2 A_1A_2 \qquad (24)$$

Suppose next that π is a prime ideal in $\mathcal{O}_{\mathbb{L}}$ which does not divide $\alpha_1\alpha_2$. Furthermore, let $D_2 = D/f_\pi$ and let g be the minimal positive integer such that both $\alpha_1^g - 1$ and $\alpha_2^g - 1$ belong to π. Suppose further that

$$H_j \geq \max\{D_2 h(\alpha_j), \log p\} \qquad \text{holds for both } j = 1, 2.$$

We put, as before,

$$\Lambda = \alpha_1^{b_1}\alpha_2^{b_2} - 1,$$

and

$$b'' = \frac{b_1}{H_2} + \frac{b_2}{H_1}.$$

With these notations, Bugeaud and Laurent [7] proved the following theorem.

Theorem 11.5 *Suppose that α_1 and α_2 are multiplicatively independent. Then,*

$$\operatorname{ord}_\pi(\Lambda) \leq \frac{24pgH_1H_2D_2^2}{(p-1)(\log p)^4}\left(\max\{\log b'' + \log\log p + 0.4, 10(\log p)/D_2, 10\}\right)^2.$$

11.5 Reducing the Bounds

The upper bounds provided by the lower bounds for the linear forms in logarithms are in general too large to allow any meaningful computation, so they need to be reduced. There is an entire theory concerning reducing such bounds, an algorithm called LLL, from the name of its inventors (Lenstra–Lenstra–Lovász), which we will describe in the next section with an application. For now, we resume ourselves to the case of two logarithms, a case in which the following reduction lemma based on continued fractions expansions of the involved numbers is surprisingly effective. Let us recall what is a continued fraction expansion.

Given $\alpha = \alpha_0 \in \mathbb{R}\backslash\mathbb{Q}$, let $(a_n)_{n\geq 0}$ be defined as

$$a_k = \lfloor \alpha_k \rfloor, \qquad \alpha_{k+1} = \frac{1}{\alpha_k - a_k} \qquad \text{for all } k \geq 0.$$

For each $n \geq 0$, put

$$\frac{p_n}{q_n} = a_0 + \cfrac{1}{a_1 + \cfrac{1}{a_2 + \cdots + \cfrac{1}{a_{n-1} + \cfrac{1}{a_n}}}}.$$

The rational number p_n/q_n is called the nth convergent of α and is denoted $[a_0, \ldots, a_n]$. The sequences $\{p_n\}_{n\geq 0}$ and $\{q_n\}_{n\geq 0}$ satisfy

$$\begin{array}{ll} p_0 = a_0, & q_0 = 1; \\ p_1 = a_0 a_1 + 1, & q_1 = a_1; \\ p_k = a_k p_{k-1} + p_{k-2}, & q_k = a_k q_{k-1} + q_{k-2}; \end{array}$$

Further,

$$\left|\alpha - \frac{p_n}{q_n}\right| < \frac{1}{q_n q_{n+1}} \quad \text{for all} \quad n \geq 0.$$

Hence, p_n/q_n converges to α and we write $\alpha = [a_0, a_1, \ldots]$. The last expression is referred to as the continued fraction of α.

The following result is a variation of a lemma of Baker and Davenport [2] and is due to Dujella and Pethő [17]. For a real number x, we use $\|x\| = \min\{|x - n| : n \in \mathbb{Z}\}$ for the distance from x to the nearest integer.

Lemma 11.6 *Let M be a positive integer and p/q be a convergent of the continued fraction of the irrational γ such that $q > 6M$ and let μ be some real number. Let $\varepsilon = \|\mu q\| - M\|\gamma q\|$. If $\varepsilon > 0$, then there is no solution to the inequality*

$$0 < |m\gamma - n + \mu| < AB^{-m}$$

in positive integers m and n with

$$\frac{\log(Aq/\varepsilon)}{\log B} \leq m \leq M.$$

12 Applications

In this section, we give two applications of the results mentioned in the previous section.

12.1 Matt's Equation

At the Awesome Math Camp at UT Dallas (June 8–25, 2014) while this lecturer was teaching the Advanced Number Theory Course (NT3), Matt Babbitt observed that

$$2^x - 3y^2 = 13$$

has the solutions $(x, y) = (4, 1),\ (8, 9)$, and wondered if there are others or how would one go to prove that these are the only ones. We prove this here.

Theorem 12.1 *The only positive integer solutions* (x, y) *to*

$$2^x - 3y^2 = 13$$

are $(x, y) = (4, 1),\ (8, 9)$.

Proof Modulo 3 we see that x is even. Write $x = 2z$, we get

$$(2^z)^2 - 3y^2 = 13.$$

With $d = 3$ and $N = 13$, this is a norm form equation. The fundamental solution (u, v) to $u^2 - 3v^2 = 1$ is $(2, 1)$, so that $\zeta = 2 + \sqrt{3}$. We search for solutions to

$$X^2 - 3Y^2 = 13,$$

with $X + \sqrt{3}Y \in [\sqrt{13}, \sqrt{13}(2 + \sqrt{3})] = [3.6055\ldots, 13.4561\ldots]$ getting the possibilities $4 + \sqrt{3} = 5.73205\ldots$ and $5 + 2\sqrt{3} = 8.4641\ldots$. However, since in fact $5 + 2\sqrt{3} = (4 - \sqrt{3})(2 + \sqrt{3})$, it follows that

$$X = u_n \qquad \text{for some} \qquad n \in \mathbb{Z},$$

where

$$2^z = u_n = \frac{(4 + \sqrt{3})\zeta^n + (4 - \sqrt{3})\zeta^{-n}}{2}. \tag{25}$$

Assume first $n \geq 0$. Then

$$\left|2^{z+1}(4 + \sqrt{3})^{-1}\zeta^{-n} - 1\right| = \frac{4 - \sqrt{3}}{(4 + \sqrt{3})\zeta^{2n}} < \frac{1}{\zeta^{2|n|}}. \tag{26}$$

In case $n < 0$, we get

$$\left|2^{z+1}(4 - \sqrt{3})^{-1}\zeta^{2n} - 1\right| = \frac{4 + \sqrt{3}}{(4 - \sqrt{3})\zeta^{-2n}} < \frac{1}{\zeta^{2|n|-1}}. \tag{27}$$

We apply Matveev for the left-hand side. Here, $n = 3, \alpha_1 = 2,\ \alpha_2 = 4 \pm \sqrt{3}, \alpha_3 = \zeta$, $b_1 = z + 1,\ b_2 = -1,\ b_3 = -|n|$. Equation (25) shows that

$$2^z > \zeta^{|n|} > 2^{|n|},$$

so $B = z + 1$. Here $\mathbb{L} = \mathbb{Q}[\sqrt{3}]$, so $D = 2$ and $\mathbb{L}$ is real. The left-hand sides in (26) and (27) are nonzero, since one of them being zero means

$$(4 \pm \sqrt{3})\zeta^{|n|} = 2^{z+1}$$

and taking norms in $\mathbb{L}$ we get

$$13 = N((4 \pm \sqrt{3})\zeta^{|n|}) = N(2^{z+1}) = 2^{2z+2},$$

a contradiction. The minimal polynomials of $\alpha_1, \alpha_2, \alpha_3$ are

$$\begin{aligned} &X - 2, \\ &X^2 - 8X + 13 = (X - (4 + \sqrt{3}))(X - (4 - \sqrt{3}), \\ &X^2 - 4X + 1 = (X - (2 + \sqrt{3}))(X - (2 - \sqrt{3})), \end{aligned}$$

so that

$$\begin{aligned} h(\alpha_1) &= \log 2, \\ h(\alpha_2) &= \frac{1}{2}\left(\log(4 + \sqrt{3}) + \log(4 - \sqrt{3})\right) = \frac{\log 13}{2}, \\ h(\alpha_3) &= \frac{1}{2}\log(2 + \sqrt{3}). \end{aligned}$$

So, we choose the bounds $A_2 = 2\log 2,\ A_2 = \log 13,\ A_3 = \log(2 + \sqrt{3})$. Denoting by Λ the left-hand side of either (26) or (27), we get

$$\log \Lambda > -1.4 \times 30^6 \times 3^{4.5} \times 2^2(1 + \log 2)(1 + \log(z + 1))(2\log 2)(\log 13)(\log(2 + \sqrt{3})).$$

The left–hand side above is $> -5 \times 10^{12}(1 + \log((z + 1)))$. Putting this into (26), (27), we get

$$-5 \times 10^{12} \log(1 + \log((z + 1))) < -(2|n| - 1)\log \zeta,$$

or

$$(2|n| - 1)\log \zeta < 5 \times 10^{12}(1 + \log((z + 1))). \tag{28}$$

Since from (25) we also have

$$2^z < (4 + \sqrt{3})\zeta^{|n|} < 8\zeta^{|n|}, \tag{29}$$

we get $|n| \log \zeta > (z-3) \log 2$. Putting this into (28), we get

$$(z-3)2\log 2 - \log \zeta < 5 \times 10^{12}(1 + \log((z+1))).$$

Mathematica tells us that $z < 1.3 \times 10^{14}$. Now we have to lower the bound. For this, we go back to (26) and (27) and with

$$\Gamma = (z+1)\log 2 - |n| \log \zeta - \log(4 \pm \sqrt{3}).$$

We note that

$$0 < e^{\Gamma} - 1 = \frac{4 \mp \sqrt{3}}{(4 \pm \sqrt{3})\zeta^{2|n|}},$$

so $\Gamma > 0$. Since then $\Gamma < e^{\Gamma} - 1$, we get from (26) and (27) that

$$0 < \Gamma < \frac{4 \mp \sqrt{3}}{(4 \pm \sqrt{3})\zeta^{2|n|}}.$$

Dividing across by $\log \zeta$, we get that

$$0 < m\gamma - |n| + \mu < \frac{4 \mp \sqrt{3}}{(4 \pm \sqrt{3})(\log \zeta)\zeta^{2|n|}} < \frac{2}{\zeta^{2|n|}} < \frac{2 \times 8^2 \times 2^2}{2^{2(z+1)}} = \frac{A}{B^m},$$

where

$$m = z+1, \quad \gamma = \frac{\log 2}{\log \zeta}, \quad \mu = \frac{\log(4 \pm \sqrt{3})}{\log \zeta}, \quad A = 512, \quad B = 4.$$

Now we have set ourselves up to apply Lemma 11.6. We take $M = 1.3 \times 10^{14}$. The first 32 terms of the continued fraction for η are

[0, 1, 1, 8, 1, 317, 1, 5, 2, 4, 2, 2, 2, 3, 1, 1, 1, 1, 30, 1, 1, 3, 1, 1, 11, 1, 16, 3, 1, 3, 4, 2].

Further,

$$\frac{p_{31}}{q_{31}} = \frac{749871422424301}{1424732176854786}.$$

Thus, $q_{31} = 1.42 \cdots \times 10^{15} > 10^{15} > 6M$. Computing $\varepsilon = \| \|q\mu\| - M\|q\gamma\|$ for the value $q = q_{31}$ and $M = 1.3 \times 10^{14}$, we get $\varepsilon > 0.15$. Now Lemma 11.6 tells us that

$$z + 1 < \frac{\log(Aq/\varepsilon)}{\log B} \le \frac{\log(512 q_{31}/0.15)}{\log 4} = 31.0383\ldots,$$

so that $z \leq 30$. Hence, $|n| < z \log 2/\log \zeta \leq 30 \log 2 \log \zeta = 15.7897\ldots$, so we conclude that $|n| \leq 15$. We now generate u_n for $|n| \leq 15$, and get the following numbers:

4, 5, 11, 16, 40, 59, 149, 220, 556, 821, 2075, 3064, 7744, 11435, 28901, 42676, 107860, 159269, 402539, 594400, 1502296, 2218331, 5606645, 8278924, 20924284, 30897365, 78090491, 115310536, 291437680, 430344779, 1087660229.

The only powers of 2 are 4 and 16.

12.2 Rep-Digit Fibonacci Numbers

Recall that a *rep-digit* is a positive integer having only one distinct digit when written in base 10. The concept can be generalized to every base $b > 1$ the resulting numbers being called *base b rep-digits*. In this section, we look at those Fibonacci numbers F_n which are rep-digits. Putting d for the repeated digit and assuming that F_n has m digits, the problem reduces to finding all the solutions of the Diophantine equation

$$F_n = \overline{dd\cdots d}_{(10)} = d \cdot 10^{m-1} + d \cdot 10^{m-2} + \cdots + d = d\left(\frac{10^m - 1}{10 - 1}\right), \quad d \in \{1, \ldots, 9\}. \tag{30}$$

The result is the following.

Proposition 12.2 *The largest solution of Eq.* (30) *is* $F_{10} = 55$.

The above result was first proved by Luca in 2000 [26]. That proof was elementary in that only congruences and Quadratic Reciprocity was used. Here, we give two nonelementary proofs using linear forms in logarithms.

Proof Suppose that $n > 1000$. We start by proving something a bit weaker.

12.2.1 Obtaining Some Bound on n

With $\alpha = (1+\sqrt{5})/2$ and $\beta = (1-\sqrt{5})/2$, Eq. (30) can be rewritten as

$$\frac{\alpha^n - \beta^n}{\sqrt{5}} = d\left(\frac{10^m - 1}{9}\right).$$

We rewrite the above equation by separating on the one side the *large terms* and in the other side the *small terms*. That is, we rewrite the equation under the form

$$|\alpha^n - (d\sqrt{5}/9)10^m| = |\beta^n - d\sqrt{5}/9| \leq \alpha^{-1000} + \sqrt{5} < 2.5. \tag{31}$$

We need some estimates for m in terms of n. By induction over n, it is easy to prove that

$$\alpha^{n-2} < F_n < \alpha^{n-1} \qquad \text{for all } n \geq 3.$$

Thus,

$$\alpha^{n-2} < F_n < 10^m \qquad \text{or} \qquad n < (\log 10/\log\alpha)m + 2,$$

and

$$10^{m-1} < F_n < \alpha^{n-1}.$$

On the other hand,

$$\begin{aligned} n &> (\log 10/\log\alpha)(m-1) + 1 = (\log 10/\log\alpha)m - (\log 10/\log\alpha - 1) \\ &> (\log 10/\log\alpha)m - 4. \end{aligned}$$

We then deduce that

$$n \in [c_1 m - 4, c_1 m + 2] \quad \text{with} \quad c_1 = \log 10/\log\alpha = 4.78497\ldots \tag{32}$$

Since $c_1 > 4$, we check easily that for $n > 1000$ we have $n = \max\{m, n\}$. We now rewrite inequality (31) as

$$\Lambda = |(d\sqrt{5}/9)\alpha^{-n}10^m - 1| < \frac{2.5}{\alpha^n} < \frac{1}{\alpha^{n-2}},$$

which leads to

$$\log\Lambda < -(n-2)\log\alpha. \tag{33}$$

We compare this upper bound with the lower bound on the quantity Λ given by Theorem 11.2. Observe first that Λ is not zero, for if it were, then $\sqrt{5}$ would be of the form $q\alpha^n$ with some $q \in \mathbb{Q}$. In particular, since on the one hand its square is 5 and on the other hand it is of the form $q^2\alpha^{-2n}$, we get that $\alpha^{2n} \in \mathbb{Q}$, which is false for any $n > 0$. With the notations of that theorem, we take

$$\alpha_1 = d\sqrt{5}/9, \ \alpha_2 = \alpha, \ \alpha_3 = 10; \qquad b_1 = 1, \ b_2 = -n, \ b_3 = m.$$

Observe that $\mathbb{L} = \mathbb{Q}[\alpha_1, \alpha_2, \alpha_3] = \mathbb{Q}[\sqrt{5}]$, so $D = 2$. The above comments show that $B = n$. We note also that the conjugates of α_1, α_2, and α_3 are

$$\alpha_1' = -d\sqrt{5}/9, \ \alpha_2' = \beta, \ \alpha_3' = 10.$$

Furthermore, α_2 and α_3 are algebraic integers, while the minimal polynomial of α_1 over $\mathbb{Q}$ is

$$(X - \alpha_1)(X - \alpha_1') = X^2 - 5d^2/81.$$

Thus, the minimal polynomial of α_1 over the integers is a divisor of $81X^2 - 5d^2$. Hence,

$$h(\alpha_1) < \frac{1}{2}\left(\log 81 + 2\log\sqrt{5}\right) = \frac{1}{2}\log(405) < 3.01.$$

Clearly,

$$h(\alpha_2) = \frac{1}{2}(\log\alpha + 1) < 0.75, \qquad h(\alpha_3) = \log 10 < 2.31.$$

Hence, we can take $A_1 = 6.02,\ A_2 = 1.5,\ A_3 = 4.62$ and then the inequalities (18) hold. Theorem 11.2 tells us that

$$\log\Lambda > -1.4 \cdot 30^6 \cdot 3^{4.5} \cdot 4 \cdot (1 + \log 4) \cdot 6.02 \cdot 1.5 \cdot 4.62(1 + \log n).$$

Comparing this last inequality with (33), leads to

$$(n-2)\log\alpha < 1.4 \cdot 30^6 \cdot 3^{4.5} \cdot 4 \cdot (1 + \log 4) \cdot 6.02 \cdot 1.5 \cdot 4.62(1 + \log n),$$

giving

$$n - 2 < 1.2 \cdot 10^{14}(1 + \log n).$$

Mathematica tells us that $n < 4.5 \cdot 10^{15}$.

12.2.2 Reducing the Bound

Observe that in the equality

$$1 - (d\sqrt{5}/9)\alpha^{-n}10^m = \frac{1}{\alpha^n}\left(\beta^n - \frac{d\sqrt{5}}{9}\right)$$

the right-hand side is negative. Thus, writing

$$z = \log\alpha_1 - n\log\alpha_2 + m\log\alpha_3,$$

we get that

$$-\frac{2.5}{\alpha^n} < 1 - e^z < 0.$$

In particular, $z > 0$. Furthermore, since $n > 1000$, the right- hand side exceeds $-1/2$, therefore $e^z < 1.5$. We thus have that

$$0 < e^z - 1 < \frac{2.5e^z}{\alpha^n} < \frac{4}{\alpha^n}.$$

Since $e^z - 1 > z$, we get

$$0 < m \log \alpha_3 - n \log \alpha_2 + \log \alpha_1 < \frac{4}{\alpha^n},$$

which can be rewritten as

$$0 < m \left(\frac{\log \alpha_3}{\log \alpha_2} \right) - n + \left(\frac{\log \alpha_1}{\log \alpha_2} \right) < \frac{4}{\alpha^n \log \alpha_2} < \frac{9}{\alpha^n}.$$

Since

$$|1 - (d\sqrt{5}10^m/\alpha^n)| < 1,$$

we have that

$$\frac{d\sqrt{5}10^m}{\alpha^n} < 2,$$

therefore

$$\alpha^n > \frac{d\sqrt{5}10^m}{2} > 10^m.$$

We have obtained

$$0 < m \left(\frac{\log \alpha_3}{\log \alpha_2} \right) - n + \left(\frac{\log \alpha_1}{\log \alpha_2} \right) < \frac{9}{10^m}. \tag{34}$$

Since $n < 4.5 \cdot 10^{15}$, inequality (32) shows that $m < 9.5 \cdot 10^{14}$. With

$$\gamma = \frac{\log \alpha_3}{\log \alpha_2}, \quad \mu = \frac{\log \alpha_1}{\log \alpha_2}, \quad A = 9, \quad B = 10,$$

we get

$$0 < m\gamma - n + \mu < \frac{A}{B^m},$$

where $m < M := 10^{15}$. The conditions to apply Lemma 11.6 are fulfilled. Observe that

$$\frac{p_{35}}{q_{35}} = C_{35} = \frac{970939497358931987}{202914354378543655} \quad \text{for } \gamma$$

and $q_{35} > 202914354378543655 > 2 \cdot 10^{17} > 6M$. We compute

$$M\|q_{35}\gamma\| = 0.00216711\ldots < 0.01.$$

For each one of the values of $d \in \{1, \ldots, 9\}$, we compute $\|q_{35}\mu\|$. The minimal value of this expression is obtained when $d = 5$ and is

$$0.029\ldots > 0.02.$$

Thus, we can take $\varepsilon = 0.01 < 0.02 - 0.01 < \|q_{35}\mu\| - M\|q_{35}\gamma\|$. Since

$$\frac{\log(Aq_{35}/\varepsilon)}{\log B} = 21.2313\ldots,$$

Lemma 11.6 tells us that there is no solution in the range $m \in [22, 10^{15}]$. Thus, $m \leq 21$ and now inequality (32) tells us that $n \leq 102$. However, we have assumed that $n > 1000$. To finish, we use Mathematica to print the values of all the Fibonacci numbers modulo 10000 (that is, their last four digits) and convince ourselves that there are no Fibonacci numbers F_n which are rep-digits in the range $11 \leq n \leq 1000$.

12.2.3 Using Linear Forms in Two p-Adic Logarithms

Given that the multiplicative constant that appears in Theorem 11.2 is very large, it is better to use the Theorems from Sect. 11.4 whenever this is possible. In what follows, we illustrate this phenomenon with our problem. We rewrite our equation

$$\frac{\alpha^n - \varepsilon\alpha^{-n}}{\sqrt{5}} = \frac{d(10^m - 1)}{9}, \qquad \text{with } \varepsilon = (-1)^n \in \{\pm 1\},$$

as

$$\alpha^n + \frac{d\sqrt{5}}{9} - \varepsilon\alpha^{-n} = \frac{d2^m\sqrt{5}^{2m+1}}{9},$$

or, equivalently, as

$$\alpha^{-n}(\alpha^n - z_1)(\alpha^n - z_2) = \frac{d2^m\sqrt{5}^{2m+1}}{9}, \tag{35}$$

where

$$z_{1,2} = \frac{-d\sqrt{5} \pm \sqrt{5d^2 + 324\varepsilon}}{18}$$

are the solutions to the quadratic equation

$$z^2 + \frac{d\sqrt{5}}{9}z - \varepsilon = 0.$$

The left-hand side of Eq. (35) is not zero, an observation which is necessary in order to apply the machinery of lower bounds for linear forms in logarithms of algebraic numbers. Assume first that $d \neq 9$. We then claim that z_1 and α are multiplicatively independent and that the same is true with z_2 instead of z_1. To see this, observe that if $\varepsilon = -1$, then $5d^2 + 324\varepsilon < 0$, while if $\varepsilon = 1$, then $5d^2 + 324\varepsilon$ is coprime to 5 and it is not a perfect square for any $d \in \{1, \ldots, 8\}$. Thus, $z_{1,2}$ are of the form

$x_1\sqrt{5} \pm x_2\sqrt{e}$ with $x_1,\ x_2 \in \mathbb{Q}^*$ and some squarefree integer $e \neq 0,\ 1,\ 5$. Hence, no power of $z_{1,2}$ can be in $\mathbb{Q}[\sqrt{5}]$, which contains the powers of α. Let $\mathbb{L} = \mathbb{Q}[z_1, z_2]$. Then $\mathbb{Q}[\sqrt{5}] \subset \mathbb{L}$ and $D = 4$. Let π be a prime ideal in $\mathbb{L}$ dividing $\sqrt{5}$. Equation (35) gives us

$$2m + 1 \leq \mathrm{ord}_\pi(\alpha^n - z_1) + \mathrm{ord}_\pi(\alpha^n - z_2).$$

To bound the two orders from the right-hand side above, we use Theorem 11.5. Observe that all the conjugates of $z_{1,2}$ are of the form

$$\frac{\pm d\sqrt{5} \pm \sqrt{5d^2 + 324\varepsilon}}{18},$$

and their absolute values are $\leq (8\sqrt{5} + \sqrt{5 \cdot 64 + 324})/18 < 2.41$. Moreover, the minimal polynomial of $z_{1,2}$ over the integers divides

$$9^2\left(X^2 - \frac{d\sqrt{5}}{9}X - \varepsilon\right)\left(X^2 + \frac{d\sqrt{5}}{9}X - \varepsilon\right) = 81(X^2 - \varepsilon)^2 - 5d^2X^2 \in \mathbb{Z}[X],$$

therefore

$$h(z_{1,2}) < \frac{\log 81 + 4 \cdot \log 2.41}{4} < 2.$$

Observe next that 5 is a perfect square in $\mathbb{Q}[\sqrt{5}] \subseteq \mathbb{L}$. Furthermore, 5 splits into two distinct prime ideals in the quadratic field $\mathbb{Q}[\sqrt{5d^2 + 324\varepsilon}] \subset \mathbb{L}$ because

$$\left(\frac{5d^2 + 324\varepsilon}{5}\right) = \left(\frac{324\varepsilon}{5}\right) = \left(\frac{18^2}{5}\right)\left(\frac{\pm 1}{5}\right) = 1.$$

Hence, we have that $e_\pi = 2$ and $f_\pi = 1$ giving $D_2 = 4$. From the preceding calculations, we know that $h(\alpha) < 0.75$. Thus, taking $\alpha_1 = \alpha,\ \alpha_2 = z_{1,2},\ b_1 = n,\ b_2 = 1$, it follows that we can take

$$H_1 = 3 > \max\{4h(\alpha), \log 5\},$$

and

$$H_2 = 8 > \max\{4h(z_{1,2}), \log 5\}.$$

To find g, observe that

$$z_1^2 \equiv -\frac{d\sqrt{5}}{9}z_1 + \varepsilon \pmod{\pi} \equiv \varepsilon \pmod{\pi},$$

so that $z_1^4 \equiv 1 \pmod{\pi}$. The same happens when we replace z_1 by z_2. Finally,

$$(2\alpha)^4 = (1+\sqrt{5})^4 \equiv 1 \pmod{\pi}$$

and since $2^4 = 16 \equiv 1 \pmod{5}$, we have that $\alpha^4 \equiv 1 \pmod{\pi}$. Thus, we can take $g = 4$. Clearly,

$$b'' \le \frac{n}{8} + \frac{1}{3} < \frac{n+3}{8}.$$

Since $n < 4.8m + 2$ (see inequality (32)), we have that

$$b'' < 0.6m + 3.8.$$

Finally, let us see that π cannot divide both $\alpha^n - z_1$ and $\alpha^n - z_2$, since if it were, then it would divide their difference

$$z_1 - z_2 = \frac{\sqrt{5d^2 + 324\varepsilon}}{9},$$

which is an algebraic number whose norm is a rational number having both the numerator and the denominator coprime to 5. Thus, applying Theorem 11.5 for $\text{ord}_\pi(\alpha^n - z_i)$ for one of the indices $i = 1$ or 2 and the fact that $\text{ord}_\pi(\alpha^n - z_j) = 0$ for the other index j such that $\{i, j\} = \{1, 2\}$, and using also the fact that $2.5 \log 5 < 10$ gives us

$$2m + 1 < \frac{24 \cdot 5 \cdot 3 \cdot 8 \cdot 16}{4(\log 5)^4} (\max\{\log(0.6m + 3.8) + \log\log 5 + 0.4, 10\})^2,$$

which implies that

$$2m + 1 < 1717 \, (\max\{\log(0.6m + 3.8) + 1, 10\})^2.$$

If the maximum is 10, then $2m + 1 < 1717 \cdot 10^2$, so $m < 858500$. In the other case,

$$2m + 1 < 1717(\log(0.6m + 3.8) + 1)^2,$$

giving $m < 104808$. Thus, $m < 104808$, leading to $n < 501506$. This is much better than the bound we have obtained by applying Theorem 11.2. In order to finish, it suffices to use Mathematica to generate $F_n \pmod{10^7}$ for all $1 \le n \le 501506$ (that is, generate the last 6 digits of F_n) and verify that there are no solutions to our problem for $n > 10$. It is interesting to note that

$$\begin{aligned}
F_{142266} &\equiv 1888888 \pmod{10^7}\\
F_{238103} &\equiv 5777777 \pmod{10^7}\\
F_{242740} &\equiv 9555555 \pmod{10^7}\\
F_{252314} &\equiv 8777777 \pmod{10^7}\\
F_{490387} &\equiv 9333333 \pmod{10^7}
\end{aligned}$$

but the above positive integers n are the only ones in the interval $[1, 501506]$ such that the last five digits of F_n are all equal to each other and nonzero.

Assume now that $d = 9$. We then get that $F_n = 10^m - 1$. Thus, $10^m = F_n + 1$. On the other hand, for all positive integers $k \geq 0$, the formulas

$$F_{4k} + 1 = F_{2k-1}L_{2k+1}, \qquad F_{4k+1} + 1 = F_{2k+1}L_{2k}, \tag{36}$$

$$F_{4k+2} + 1 = F_{2k+2}L_{2k}, \qquad F_{4k+3} + 1 = F_{2k+1}L_{2k+2} \tag{37}$$

hold. In particular, $10^m = F_{(n-\delta)/2}L_{(n+\delta)/2}$, where $\delta \in \{\pm 1, \pm 2\}$ and $n \equiv \delta \pmod 2$. However, if $n > 26$, then $(n - \delta)/2 > 12$, and by Theorem 7.1, $F_{(n-\delta)/2}$ has a primitive prime factor $p > 12$ that does not divide 10^m. Thus, our equation has no solutions with $d = 9$ and $n > 26$. A calculation by hand finishes the proof.

13 Problems

30. Let $\mathscr{P} = \{p_1, \cdots, p_k\}$ be a finite set of primes and put

$$\mathscr{S} = \{\pm p_1^{\alpha_1} \cdots p_k^{\alpha_k} : \alpha_i \in \mathbb{Z},\ i = 1, \ldots, k\}$$

for the set of all rational numbers which when written in reduced form have both the numerator and denominator divisible only by primes in $\mathscr{P}$. Prove, using the theorems from the Sects. 11.2 and 11.3, that the equation

$$x + y = 1, \qquad x,\ y \in \mathscr{S}$$

has only finitely many solutions (x, y) which, in practice (namely, given $\mathscr{P}$) can be computed explicitly.

31. Prove that given integers $a > b > 1$ the Pillai equation

$$a^x - b^y = a^{x_1} - b^{y_1}, \qquad (x, y) \neq (x_1, y_1) \tag{38}$$

has only finitely many positive integer solutions (x, y, x_1, y_1).

32. Compute all the solutions of the equation

$$2^x + 3^y = 5^z$$

in nonnegative integers $x,\ y,\ z$.

33. Compute all solutions of the equation

$$5^u 7^v - 5^w = 7^x + 1$$

in nonnegative integers $u,\ v,\ w,\ x$.

34. Prove that if $k \geq 2$ and $d > 0$ are integers such that $(k-1)d+1,\ (k+1)d+1$, and $4kd+1$ are all three perfect squares, then $d = 16k^3 - 4k$.

35. Prove that if $k \geq 2$ and $d > 0$ are integers such that $F_{2(k-1)}d+1,\ F_{2k}d+1$ and $F_{2(k+1)}d+1$ are all three perfect squares, then $d = 4F_{2k-1}F_{2k}F_{2k+1}$.

36. Let p_k be the kth prime number. Find all the solutions of the Diophantine equation

$$n! + 1 = p_k^a p_{k+1}^b \qquad \text{with } p_{k-1} \leq n < p_k$$

in integer unknowns $n \geq 1,\ k \geq 2,\ a \geq 0,\ b \geq 0$.

37. (i) Prove that if $F_n = x^k$ with $n \geq 3,\ x \geq 2,\ k \geq 2$ integers, then k is bounded (that is, there are no perfect powers of exponent arbitrarily large in the Fibonacci sequence).

(ii) Prove that the conclusion of (i) holds also when we replace F_n by the nth term of a nondegenerate binary recurrent sequence (see Definition 2.3) whose roots are real.

38. Using Theorem 11.2, prove that there are only finitely many triples (α, β, n) with $n \notin \{1, 2, 3, 4, 6\}$ and coprime integers $\alpha + \beta = r$ and $\alpha\beta = s$ and such that furthermore α/β is not a root of 1, such that the nth term of the Lucas sequence of roots α and β does not admit a primitive divisor. This statement if a weak version of Theorem 7.1 whose proof was given only under the hypothesis that α and β are integers.

39. Recall that a palindrome in base $b \geq 2$ is a positive integer such that the sequence of its base b reads the same from left to right as from right to left. Prove that the number $99 = \overline{1100011}_{(2)}$ is the largest positive integer of the form $10^n \pm 1$ which furthermore is a binary palindrome.

40. Prove that $F_{10} = 55$ is the largest Fibonacci number which in base 10 is the concatenation of two other Fibonacci numbers (i.e., is the largest solution of the equation $F_n = \overline{F_\kappa F_\ell}_{(10)}$).

41. Let $(u_n)_{n\geq 0}$ be the ternary recurrent sequence such that $u_0 = u_1 = 0,\ u_2 = 1$ and $u_{n+3} = u_{n+1} + u_n$. Find all solutions of the equation $u_n = F_m$ in nonnegative integers n and m.

42. Let $(u_n)_{n\geq 1}$ and $(v_m)_{m\geq 1}$ be linearly recurrent sequences such that

$$u_n = c_1\alpha_1^n + \sum_{i=2}^{s} c_i\alpha_i^n; \qquad v_m = d_1\beta_1^m + \sum_{j=2}^{t} d_j\beta_j^m,$$

where $c_1 d_1 \neq 0$, $|\alpha_1| > \max\{|\alpha_i| : i = 2, \ldots, s\}$ and $|\beta_1| > \max\{|\beta_j| : j = 2, \ldots, t\}$. Suppose furthermore that α_1 and β_1 are multiplicatively independent; that is, the only solution to the equation

$$\alpha_1^x \beta_1^y = 1$$

in integers x and y is $x = y = 0$. Prove that the Diophantine equation $u_n = v_m$ has only finitely many, effectively computable, positive integer solutions (m, n).

43. Let $P(m)$ be the largest prime factor of m. Prove that

$$P(2^p - 1) > c_1 p \log p$$

holds with some absolute positive constant c_1.

44. Let $(u_n)_{n\geq 0}$ be a binary recurrent sequence which is nondegenerate. Prove that $P(|u_n|) > c_2 n^{1/(d+1)}$ holds with a positive constant c_2 depending on $\{u_n\}_{n\geq 0}$, where $d = [\mathbb{K} : \mathbb{Q}]$ and $\mathbb{K} = \mathbb{Q}[\alpha]$ is the splitting field of the characteristic equation of the sequence $(u_n)_{n\geq 0}$.

45. Prove that there are only finitely many positive integer solutions (p, a, k) with $p \geq 3$ prime such that

$$a^{p-1} + (p-1)! = p^k.$$

Can you compute them all?

46. (i) Prove that if $k \geq 1$ is fixed, then there are only finitely many positive integer solutions of the equation

$$F_n = m_1! + m_2! + \cdots + m_k!, \qquad \text{with } 1 \leq m_1 \leq \cdots \leq m_k.$$

(ii) Prove that when $k = 2$ the largest solution of the equation appearing at (i) is $F_{12} = 3! + 4!$.

47. (i) Prove that

$$P(2^n + 3^n + 5^n) \to \infty \qquad \text{as} \qquad n \to \infty.$$

(ii) Deduce that there are only finitely many n such that $P(2^n + 3^n + 5^n) < 23$. Can you find them all? For example,

$$2^1 + 3^1 + 5^1 = 2 \cdot 5;$$

$$2^2 + 3^2 + 5^2 = 2 \cdot 19;$$

$$2^3 + 3^3 + 5^3 = 2^5 \cdot 5;$$

$$2^4 + 3^4 + 5^4 = 2 \cdot 19^2;$$

$$2^5 + 3^5 + 5^5 = 2^3 \cdot 5^2 \cdot 17.$$

Are there other values of n?

14 Notes

An elementary proof of Proposition 12.2 appears in [26]. The conclusion of Problem 30 is known as the theorem of the finiteness of the number of solutions of an equation in two $\mathscr{S}$-units. Problem 31 has been studied by Pillai (see [35]). See the first two chapters of the book [40]. See [33] for a problem which is slightly more general, and [6] for an elementary treatment of Problems 32 and 33. The results of Problems 34 and 35 are due to Dujella [14] and [15]. A set $\{a_1, \ldots, a_m\}$ of positive integers such that $a_i a_j + 1$ is a perfect square is called a *Diophantine m-tuple*. It is conjectured that $m \leq 4$. Dujella [16], proved that $m \leq 5$ and that there can be at most finitely many Diophantine quintuples which are, in practice, effectively computable. Recently, He, Togbé, and Ziegler announced a proof that there are no Diophantine quintuples (see [22]). All the solutions to the equation appearing in Problem 36 were computed by Luca in [27]. The largest one is $5! + 1 = 11^2$. Concerning Problem 37, it is known that the largest perfect power in the Fibonacci sequence is $F_{12} = 144$ (see [8]). The result of (ii) was obtained independently by Pethő [34] and by Shorey and Stewart [39]. For Problem 38, see Stewart's paper [41], or the more modern paper [4]. Problem 39 is due to Luca and Togbé [29]. Problem 40 is a result of Banks and Luca [3]. For Problem 41, see De Weger's paper [13]. The result of Problem 43 was obtained independently by Erdős and Shorey in [19] and Stewart in [43]. The result of Problem 44 is due to Stewart [42]. Computing all triples (a, p, k) of Problem 45 was a problem proposed by Erdős and Graham in [18] and solved by Yu and Liu in [46]. The only solutions are $(a, p, k) = (1, 3, 1), (1, 5, 2), (5, 3, 3)$. The result of Problem 46 is due to Grossman and Luca [21].

15 An Application of the LLL Algorithm

Using theorems from the previous sections, one can find all solutions of the equation

$$x + y = z$$

where $(x, y) = 1$ and $x,\ y,\ z$ are integers having all their prime factors in some finite set $\mathscr{P}$ of primes. All such solutions can be effectively bounded. In general, the bounds arising from linear forms in logs are very large.

To reduce them, one can use the LLL. We illustrate this by a concrete example.

Proposition 1 *The equation*

$$|2^a \cdot 3^b \cdot 5^c - 7^d \cdot 11^e \cdot 13^f| < 10$$

has 69 *nonnegative integer solutions* (a, b, c, d, e, f). *The largest is*

$$2^5 \cdot 3 \cdot 5^2 - 7^4 = -1.$$

Proof Let $B = \max\{a, b, c, d, e, f\}$. A search in the box $B \le 23$ gives the 69 solutions mentioned in the statement.

From now on, we assume that $B \ge 24$. If $B \in \{a, b, c\}$, then by Yu's theorem, we have

$$\begin{aligned} B &= \operatorname{ord}_p(7^d \cdot 11^e \cdot 13^f - (-k)) \\ &\le 19(20\sqrt{5})^{2.5} \frac{p}{(\log p)^2} \log(e^5 4) \log |k| \log 7 \log 11 \log 13 \log B, \end{aligned}$$

where $k = 2^a \cdot 3^b \cdot 5^c - 7^d \cdot 11^e \cdot 13^f$, and $p \in \{2, 3, 5\}$. Since $p/(\log p)^2 \le 4.2$, we get

$$B \le 19 \cdot 20^{10} \cdot 5^5 \cdot 4.2 \cdot \log 7 \cdot \log 11 \cdot \log 13 \cdot \log 9 \cdot \log(e^5 \cdot 4) \log B,$$

which gives $B < 2.8 \cdot 10^{22}$.

If B is one of $\{c, d, e\}$, then by the same theorem we have

$$\begin{aligned} B &= \operatorname{ord}_p(2^a \cdot 3^b \cdot 5^c - k) \\ &\le 19(20 \cdot \sqrt{5})^{2.5} \frac{p}{(\log p)^2} \log(e^5 \cdot 4)(\log p)^4 \log B, \end{aligned}$$

where $p \in \{7, 11, 13\}$. This gives

$$B \le 19 \cdot 20^{10} \cdot 5^5 \cdot 17(\log 17)^2 \cdot \log(e^5 \cdot 4) \log B,$$

or $B < 3 \cdot 10^{23}$. Thus, $B < 3 \cdot 10^{23}$. We now take

$$\Lambda = -a \log 2 - b \log 3 - c \log 5 + d \log 7 + e \log 11 + f \log 13.$$

If B is one of $\{a, b, c\}$, then

$$|e^\Lambda - 1| \le \frac{10}{2^B} < \frac{1}{2^{B-4}}.$$

In the other case

$$|1 - e^{-\Lambda}| \le \frac{10}{7^B} \le \frac{1}{7^{B-2}} < \frac{1}{2^{B-2}}.$$

Since $B \ge 24$, the above inequalities give

$$|\Lambda| < \frac{1}{2^{B-5}}. \tag{39}$$

The next step consists in finding a lower bound on $|\Lambda|$ when

$$\max\{|a|, |b|, |c|, |d|, |e|, |f|\} = B < 3 \cdot 10^{22}.$$

For this we describe the LLL algorithm. We use the version from Henri Cohen's book (see Chap. 2.6 in [11]).

The following algorithm finds a lower bound for

$$\mathbf{m} = \min\{|\sum_{i=1}^{n} x_i \log \alpha_i| : |x_i| \le X, \ (x_1, \ldots, x_n) \ne 0\},$$

where $\log \alpha_1, \ldots, \log \alpha_n$ are logarithms of algebraic numbers, real and independent over $\mathbb{Q}$.

Step 1. *Let $C > (nX)^n$. We form the $n \times n$ matrix given by*

$$\begin{pmatrix} 1 & 0 & 0 & \cdots & 0 & 0 \\ 0 & 1 & 0 & \cdots & 0 & 0 \\ 0 & 0 & 1 & \cdots & 0 & 0 \\ \cdots & \cdots & \cdots & \cdots & \cdots & \cdots \\ 0 & 0 & 0 & \cdots & 1 & 0 \\ \lfloor C \log \alpha_1 \rfloor & \lfloor C \log \alpha_2 \rfloor & \lfloor C \log \alpha_3 \rfloor & \cdots & \lfloor C \alpha_{n-1} \rfloor & \lfloor C \log \alpha_n \rfloor \end{pmatrix}.$$

Step 2. *For the lattice Γ generated by the columns of the above matrix we compute a reduced basis (see below). Let this basis be $\mathbf{b}_1, \ldots, \mathbf{b}_n$.*

Step 3. *We compute the Gram–Schmidt basis $\mathbf{b}_1^*, \ldots, \mathbf{b}_n^*$,* of $\mathbf{b}_1, \ldots, \mathbf{b}_n$. *That is,*

$$\mathbf{b}_i^* = \mathbf{b}_i - \sum_{1 \le j < i} \mu_{i,j} \mathbf{b}_j^* \qquad \text{where} \qquad \mu_{i,j} = \frac{\mathbf{b}_i \cdot \mathbf{b}_j^*}{\mathbf{b}_j^* \cdot \mathbf{b}_j^*}.$$

Step 4. *We compute*

$$c_1 = \max_{1 \le i \le n} \left\{ \frac{\|\mathbf{b}_1\|}{\|\mathbf{b}_i^*\|} \right\},$$

and

$$d = \frac{\|\mathbf{b}_1\|}{c_1}.$$

Step 5. *If $d^2 > nX^2$ and $\sqrt{d^2 - nX^2} > (1 + nX)/2$, then*

$$\mathbf{m} > \frac{\sqrt{d^2 - nX^2} - (1 + nX)/2}{C}.$$

In the above algorithm, the reduced basis is a basis satisfying the following technical condition for $\{\mathbf{b}_1, \ldots, \mathbf{b}_n\}$:

(i) $|\mu_{i,j}| \le 1/2$ for all $i = 1, \ldots, n$ and $j < i$;
(ii) for $i \ge 2$ we have

$$\|\mathbf{b}_i^* + \mu_{i,i-1} \mathbf{b}_{i-1}^*\|^2 \ge \frac{3}{4} \|\mathbf{b}_{i-1}\|^2.$$

Given the basis there is an algorithm which produces a reduced basis in polynomial time. Various software like Mathematica, MAPLE, etc., have this algorithm as part of their libraries, so we do not explain how such a basis is found, we just limit ourself to what we need.

Going back to our example, we take $X = 10^{23}$ and $C > (6 \cdot 10^{23})^6$. We can take for example $C > 10^{150}$. We compute the reduced basis $\mathbf{b}_1, \ldots, \mathbf{b}_6$. For example, the first vector is

$$\mathbf{b}_1 = \begin{pmatrix} 4546080841331590178058124 \\ 4374795057043709667281085 \\ -1869249439796078878453047 \\ 2212132469572082716281439 \\ 5664246846772059856891579 \\ 1345434977173738964138866 \end{pmatrix}$$

We compute $\mathbf{b}_1^*, \ldots, \mathbf{b}_6^*$, and afterwards c_1. The numbers $\|\mathbf{b}_1\|/\|\mathbf{b}_i^*\|$ for $1 \le i \le 6$ are

$$1,\ 0.884943,\ 0.905395,\ 0.849188,\ 0.712015,\ 0.480536,$$

so $c_1 = 1$. Then $d = \|\mathbf{b}_1\| > 9 \cdot 10^{24}$. Thus,

$$d^2 > 81 \cdot 10^{46} > nX^2 = 6(10^{23})^2 = 6 \cdot 10^{46}.$$

Therefore

$$\frac{\sqrt{d^2 - nX^2} - (1 + nX)/2}{C} > 10^{-126}.$$

Thus, together with (39), we got

$$\begin{aligned} 10^{-126} &< |-a\log 2 - b\log 3 - c\log 5 + d\log 7 + e\log 11 + f\log 13| \\ &< \frac{1}{2^{B-5}}, \end{aligned}$$

which gives

$$(B-5)\log 2 < 126\log 10,$$

so $B \le 423$.

Thus, in one step we got from 10^{23} down to 423. We now apply the algorithm again. We take $C = 10^{21} > (6 \cdot 423)^6$. After reducing the basis of the lattice we can take $c_1 = 1.43$ and $d = \|\mathbf{b}_1\|/c_1 > 3283$. Thus, the new bound is

$$|\Lambda| \ge \frac{(3283^2 - 6 \cdot 423^2)^{1/2} - (1 + 6 \cdot 423)/2}{10^{21}} \ge \frac{1.8}{10^{18}},$$

and then

$$(B-5)\log 2 \leq 18\log 10 - \log(1.8),$$

so $B \leq 23$, which is a contradiction.

Acknowledgements We thank the referee for a careful reading of the manuscript and for comments which improved the quality of this paper.

References

1. A. Baker, "Linear forms in logarithms of algebraic numbers. I, II, III", *Mathematika* **13** (1966); 204–216, ibid. **14** (1967), 102–107; ibid. **14** (1967), 220–228.
2. A. Baker and H. Davenport, "The equations $3x^2 - 2 = y^2$ and $8x^2 - 7 = z^2$", *Quart. J. Math. Oxford Ser. (2)* **20** (1969), 129–137.
3. W. D. Banks and F. Luca, "Concatenations with binary recurrent sequences", *J. Integer Seq.* **8** (2005), Article 05.1.3, 19pg.
4. Yu. Bilu, G. Hanrot and P. M. Voutier, "Existence of primitive divisors of Lucas and Lehmer numbers. With an appendix by M. Mignotte", *J. Reine Angew. Math.* **539** (2001), 75–122.
5. G. D. Birkhoff and H. S. Vandiver, "On the integral divisors of $a^n \pm b^n$", *Ann. Math. (2)* **5** (1904), 173–180.
6. J. L. Brenner y L. L. Foster, "Exponential Diophantine equations", *Pacific J. Math.* **101** (1982), 263–301.
7. Y. Bugeaud and M. Laurent, "Minoration effective de la distance p-adique entre puissances de nombres algébriques", *J. Number Theory* **61** (1996), 311–342.
8. Y. Bugeaud, M. Mignotte and S. Siksek, "Classical and modular approaches to exponential Diophantine equations. I. Fibonacci and Lucas perfect powers", *Ann. of Math. (2)* **163** (2006), 969–1018.
9. R. D. Carmichael,"On the numerical factors of the arithmetic forms $\alpha^n \pm \beta^n$", *Ann. of Math. (2)* **15** (1913), 30–70.
10. E. Catalan, "Note extraite d'une lettre adressée à l'editeur", *J. Reine Angew. Math.* **27** (1844), 192.
11. H. Cohen, *A course in computational algebraic number theory*, Springer 2000.
12. H. Cohn and B. M. M. de Weger, "Solution to **10457**", *Amer. Math. Monthly* **104** (1997), 875.
13. B. M. M. de Weger, "Padua and Pisa are exponentially far apart", *Publ. Mat.* **41** (1997), 631–651.
14. A. Dujella, "The problem of the extension of a parametric family of Diophantine triples", *Publ. Math. Debrecen* **51** (1997), 311–322.
15. A. Dujella, "A proof of the Hoggatt-Bergum conjecture", *Proc. Amer. Math. Soc.* **127** (1999), 1999–2005.
16. A. Dujella, "There are only finitely many Diophantine quintuples", *J. Reine Angew. Math.* **566** (2004), 183–214.
17. A. Dujella and A. Pethő, "A generalization of a theorem of Baker and Davenport", *Quart. J. Math. Oxford Ser. (2)* **49** (1998), 291–306.
18. P. Erdős and R. L. Graham, *Old and new problems and results in combinatorial number theory*, Monographies de L'Enseignement Mathématique **28** (1980) Université de Genève, 128 pp.
19. P. Erdős and T. N. Shorey, "On the greatest prime factor of $2^p - 1$ for a prime p and other expressions", *Acta Arith.* **30** (1976), 257–265.
20. G. Everest, A. van der Poorten, Alf, I. Shparlinski and T. Ward, *Recurrence sequences*, Mathematical Surveys and Monographs, **104** American Mathematical Society, Providence, RI, 2003.

21. G. Grossman and F. Luca, "Sums of factorials in binary recurrence sequences", *J. Number Theory* **93** (2002), 87–107.
22. B, He, A. Togbé and V. Ziegler, "There are no Diophantine quintuples", arXiv:org/abs/1610.04020 to appear in *Transactions of the Amer. Math. Soc.*.
23. M. Křížek and F. Luca, "On the solutions of the congruence $n^2 \equiv 1 \pmod{\phi^2(n)}$", *Proc. Amer. Math. Soc.* **129** (2001), 2191–2196.
24. M. Laurent, M. Mignotte and Yu. Nesterenko,"Formes linéaires en deux logarithmes et déterminants d'interpolation", *J. Number Theory* **55** (1995), 285–321.
25. V. A. Lebesgue, "Sur l'impossibilité en nombres entiers de l'uation $x^m = y^2 + 1$", *Nouv. Ann. Math.* **9** (1850), 178–181.
26. F. Luca, "Fibonacci and Lucas numbers with only one distinct digit", *Portugal. Math.* **57** (2000), 243–254.
27. F. Luca, "On a conjecture of Erdős and Stewart", *Math. Comp.* **70** (2001), 893–896.
28. F. Luca, "On the equation $x^2 + 2^a 3^b = y^n$", *Int. J. Math. Math. Sci.* **29** (2002), 239–244.
29. F. Luca and A. Togbé, "When is $10^n \pm 1$ a binary palindrome?", *C.R. Acad. Sci. Paris* **346** (2008), 487–489.
30. F. Luca and J. L. Varona, "Multiperfect numbers on lines of the Pascal triangle", *J. Number Theory* **129** (2009), 1136–1148.
31. E. M. Matveev, "An explicit lower bound for a homogeneous rational linear form in logarithms of algebraic numbers. II", *Izv. Ross. Akad. Nauk Ser. Mat.* **64** (2000), 125–180; English translation in *Izv. Math.* **64** (2000), 1217–1269.
32. P. Mihăilescu, "Primary cyclotomic units and a proof of Catalan's conjecture", *J. Reine Angew. Math.* **572** (2004), 167–195.
33. D. Z. Mo and R. Tijdeman, "Exponential Diophantine equations with four terms", *Indag. Math. (N.S.)* **3** (1992), 47–57.
34. A. Pethő, "Perfect powers in second order linear recurrences", *J. Number Theory* **15** (1982), 5–13.
35. S. S. Pillai, "On $a^x - b^y = c$", *J. Indian Math. Soc. (N.S.)* **2** (1936), 119–122.
36. C. Pomerance and U. Everling, "Solution to **10331**", *Amer. Math. Monthly* **103** (1996), 701–702.
37. H. Ruderman and C. Pomerance, "Solution to **E2468**", *Amer. Math. Monthly* **84** (1977), 59–60.
38. A. Schinzel, "Primitive divisors of the expression $A^n - B^n$ in algebraic number fields", *J. Reine Angew. Math.* **268/269** (1974), 27–33.
39. T. N. Shorey and C. L. Stewart, "On the Diophantine equation $ax^{2t} + bx^t y + cy^2 = d$ and pure powers in recurrence sequences", *Math. Scand.* **52** (1983), 24–36.
40. T. N. Shorey and R. Tijdeman, *Exponential Diophantine equations*, Cambridge Tracts in Mathematics, **87** Cambridge University Press, Cambridge, 1986.
41. C. L. Stewart, "Primitive divisors of Lucas and Lehmer numbers", in *Transcendence theory: advances and applications (Proc. Conf., Univ. Cambridge, Cambridge, 1976)*, 79–92. Academic Press, London, 1977.
42. C. L. Stewart, "On divisors of terms of linear recurrence sequences", *J. Reine Angew. Math.* **333** (1982), 12–31.
43. C. L. Stewart, "On divisors of Fermat, Fibonacci, Lucas and Lehmer numbers. III", *J. London Math. Soc. (2)* **28** (1983), 211–217.
44. P. Vandendriessche and H. J. Lee, *Problems in Elementary Number Theory*, https://www.math.muni.cz/bulik/vyuka/pen-20070711.pdf
45. K. Yu: "p-adic logarithmic forms and group varieties II", *Acta Arith.* **89** (1999), 337–378.
46. K. Yu and D. Liu, "A complete resolution of a problem of Erdős and Graham", in *Symposium on Diophantine Problems (Boulder, CO, 1994), Rocky Mountain J. Math.* **26** (1996), 1235–1244.
47. K. Zsigmondy, "Zur Theorie der Potenzreste", *Monatsh. Math.* **3** (1892), 265–284.

Part II
Research Contributions

Nullstellensatz via Nonstandard Methods

Haydar Göral

Abstract In this short note, we survey some degree and height bound results for arithmetic Nullstellensatz from the literature. We also introduce the notion of height functions, ultraproducts and nonstandard extensions. As our main remark, we find height bounds for polynomial rings over integral domains via nonstandard methods.

1 Introduction

The arithmetic version of the Nullstellensatz states that if $f_1, \ldots, f_s$ belong to the polynomial ring $\mathbb{Z}[X_1, \ldots, X_n]$ without a common zero in $\mathbb{C}$, then there exist a in $\mathbb{Z} \setminus \{0\}$ and $g_1, \ldots, g_s$ in $\mathbb{Z}[X_1, \ldots, X_n]$ such that

$$a = f_1 g_1 + \cdots + f_s g_s. \tag{1}$$

By $\deg f$, we mean the total degree of the polynomial f in several variables. Finding degree bounds for the ideal membership problem for fields, and finding height bounds for a and $g_1, \ldots, g_s$ above have received extensive attention using computational methods, and also via nonstandard methods. To explain this, let K be a field. If $f_0, f_1, \ldots, f_s$ in $K[X_1, \ldots, X_n]$ all have degree less than D and f_0 is in the ideal $\langle f_1, \ldots, f_s \rangle$, then

$$f_0 = \sum_{i=1}^{s} f_i h_i \tag{2}$$

for certain h_i whose degrees are bounded by a constant $C = C(n, D)$ depending only on n and D. Hermann [9] was the first one who proved the existence of the bound $C(n, D)$ above using computational methods. Seidenberg [17] reconsidered Hermann's result. Furthermore, Seidenberg showed that in fact one may

H. Göral (✉)
Department of Mathematics, Faculty of Sciences, Dokuz Eylül University, Tinaztepe Yerleşkesi, 35390 Buca/İzmir, Turkey
e-mail: hgoral@gmail.com

I. Inam and E. Büyükasik (eds.), *Notes from the International Autumn School on Computational Number Theory*, Tutorials, Schools, and Workshops in the Mathematical Sciences, https://doi.org/10.1007/978-3-030-12558-5_5

take $C(n, D) = (2D)^{2^n}$. In other words, one has the degree bound effectively to test the ideal membership problem. The doubly exponential bound $(2D)^{2^n}$ can be improved immensely, if we take $f_0 = 1$ and deal with Nullstellensatz. By the fundamental work of Kollár [11], if $f_0 = 1$ then one can choose $C(n, D) = D^n$ for $D \geq 3$. When $f_0 = 1$, the case $D = 2$ was investigated by Sombra [18], and we can take $C(n, 2) = 2^{n+1}$. Nonstandard analysis also gives noneffective criteria for the ideal membership problem. The result of [9] was also proved by van den Dries and Schmidt [3] using nonstandard methods, however the bound $C(n, D)$ is ineffective with this method. We also refer the reader to [15, 16]. Nonstandard analysis yields existence results ineffectively, but it prevents us from doing many computations. Once we have an existence result, then one can pursue an effective result. In this note, we follow a similar approach as of [3], and we apply nonstandard methods in order to prove the existence of bounds for the complexity of the coefficients of h_i in (2) by taking $f_0 = 1$. We also define an abstract height function (see Sect. 2) on integral domains which generalizes the absolute value function and the height function on the field of algebraic numbers. A height function measures the arithmetic complexity of the coefficients of polynomials over $R[X_1, \ldots, X_n]$, where R is an integral domain. The notion of a height function is a fundamental concept in number theory. The Mordell–Weil theorem states that elliptic curves over number fields are finitely generated abelian groups and the canonical height function (the Néron-Tate height) attached to them plays a significant role in the proof. Basically, this finiteness theorem is based on another finiteness result which is Northcott's theorem, and it states that there are only finitely many algebraic numbers of bounded degree and bounded height. In particular, in a number field we have only finitely many elements of bounded height. The ideal membership problem over $\mathbb{Z}$ and for some other rings were also considered. A precise degree bound in the ideal membership problem over $\mathbb{Z}$ was deduced in [1]. Effective height estimates for a and $g_1, \ldots, g_s$ in arithmetic Nullstellensatz (in Eq. (1)) were studied in [12, 14], however these bounds may also depend on s. Besides, finding effective bounds in Nullstellensatz is also related to theoretical computer science, see [4].

In this note, we assume that all rings are commutative with unity. Moreover, throughout this note R stands for an integral domain and K for its field of fractions. The symbol h denotes a height function on R which will be defined in the next section. Our goal in this note is to show how nonstandard analysis can be useful to obtain bounds in commutative algebra. We prove the existence of the constant c_2 in the following result using nonstandard methods (see also Remark 1). We note that our constant c_2 for the height function does not depend on R or s, but it is ineffective. Precisely, our main remark in this note is the following:

Main Remark *Given $n \geq 1$, $D \geq 1$, $H \geq 1$ and a function $\theta : \mathbb{N} \to \mathbb{N}$, there are two constants $c_1(n, D)$ and $c_2(n, D, H, \theta)$ such that for all rings R with a height function* h *of θ-type, if $f_1, \ldots, f_s$ in $R[X_1, \ldots, X_n]$ have no common zero in K^{alg} with* $\deg(f_i) \leq D$ *and* $\mathrm{h}(f_i) \leq H$, *then there exist nonzero a in R and $h_1, \ldots, h_s$ in $R[X_1, \ldots, X_n]$ such that*

(i) $a = f_1h_1 + \cdots + f_sh_s$
(ii) $\deg(h_i) \leq c_1$
(iii) $\mathrm{h}(a), \mathrm{h}(h_i) \leq c_2$.

Remark 1 There is also a standard proof of the Main Remark as follows very briefly: Using the degree bound $c_1 = (2D)^{2^n}$ of [17] for the polynomials $g_1, \ldots, g_s$ in the Bézout expression $1 = f_1g_1 + \cdots + g_sf_s$, one translates the problem into solving a system of linear equations over K with precise number of unknowns equations. As the height function satisfies some additive and multiplicative properties (see Sect. 2), we can derive a height bound c_2 using the Gauss–Jordan method or Corollary 1 as used in our approach. However, this computational method can be complicated since the bounds for the height function depend on θ which is implicitly given. Our approach for the Main Remark will be nonstandard analysis (see the next section for the definition) as in [3], and it is a typical application of compactness theorem from model theory. For a similar result on the field of algebraic numbers, see [7, Theorem 2].

Remark 2 Both of the constants c_1 and c_2 do not depend on s because the K-vector space $V(n, D) = \{f \in K[X_1, \ldots, X_n] : \deg(f) \leq D\}$ is finite dimensional over K. In fact, the dimension is $q(n, D) = \binom{n+D}{n}$. Thus, given $1 = f_1g_1 + \cdots + f_sg_s$, we may always assume $s \leq q = q(n, D)$.

Short Outline: In the next section, we introduce the notion of height functions and we give a quick introduction to ultraproducts and nonstandard analysis. Then in the same section, we give some known results from commutative algebra. In Sect. 3, we prove our Main Remark.

2 Preliminaries

2.1 Height Functions

In this subsection, we define generalized height functions (see [6, 7]) ultraproducts and nonstandard extensions.

Definition 1 Let $\theta : \mathbb{N} \to \mathbb{N}$ be a function. The function

$$\mathrm{h} : R \to [0, \infty)$$

is said to be a height function of θ-type if for any x and y in R with $\mathrm{h}(x) \leq n$ and $\mathrm{h}(y) \leq n$, then both $\mathrm{h}(x + y) \leq \theta(n)$ and $\mathrm{h}(xy) \leq \theta(n)$ hold. We say that h is a height function on R if h is a height function of θ-type for some $\theta : \mathbb{N} \to \mathbb{N}$.

One can extend the height function h to the polynomial ring $R[X_1, \ldots, X_n]$ as

$$\mathrm{h}\Big(\sum_\alpha a_\alpha X^\alpha\Big) = \max_\alpha \mathrm{h}(a_\alpha),$$

where $X^\alpha = X_1^{\alpha_1} \cdots X_n^{\alpha_n}$ is a monomial in $R[X_1, \ldots, X_n]$. Observe that this extension does not have to be a height function on the polynomial ring. Now we give some examples of height functions that generalize absolute values and the height function on the field of algebraic numbers.

Example 1

- ***Absolute values***: If $(R, |\cdot|)$ is an absolute valued ring then $\mathrm{h}(x) = |x|$ is a height function of θ-type where $\theta(n) = n^2 + 1$.
- ***The degree function***: The degree function on $R[X_1, \ldots, X_n]$ is a height function of θ-type where $\theta(n) = 2n$.
- ***The height function*** H ***on the field of algebraic numbers***: By convention we define $\mathrm{H}(0) = 1$. Let α be a non-zero algebraic number in $\overline{\mathbb{Q}}$ with irreducible polynomial $f(X) \in \mathbb{Z}[X]$ of degree d. Write

$$f(X) = a_d(X - \alpha_1) \cdots (X - \alpha_d),$$

where $\alpha_1, \ldots, \alpha_d \in \overline{\mathbb{Q}}$ are all conjugates of α and $a_d \in \mathbb{Z}$. Then the height of α is defined as

$$\mathrm{H}(\alpha) = \Bigg(|a_d| \prod_{|\alpha_j| \geq 1} |\alpha_j|\Bigg)^{1/d}.$$

The height function measures the arithmetic complexity of an algebraic number and it behaves well under arithmetic operations:

- For a non-zero rational number a/b where a and b are coprime integers,

$$\mathrm{H}(a/b) = \max\{|a|, |b|\}.$$

- For all α in $\overline{\mathbb{Q}}$ and $n \in \mathbb{N}$, we have $\mathrm{H}(\alpha^n) = \mathrm{H}(\alpha)^n$.
- For all α and β in $\overline{\mathbb{Q}}$, we have $\mathrm{H}(\alpha + \beta) \leq 2\mathrm{H}(\alpha)\mathrm{H}(\beta)$.
- For all α and β in $\overline{\mathbb{Q}}$, we have $\mathrm{H}(\alpha\beta) \leq \mathrm{H}(\alpha)\mathrm{H}(\beta)$.
- For all non-zero α in $\overline{\mathbb{Q}}$, we have $\mathrm{H}(1/\alpha) = \mathrm{H}(\alpha)$.

Observe that H is a height function of θ-type where $\theta(n) = 2n^2$. As mentioned in the introduction, Northcott's theorem yields that if F is a number field and B is a real number, then the set

$$\{\alpha \in F : \mathrm{H}(\alpha) \leq B\}$$

is finite. More details and properties of the height function on $\overline{\mathbb{Q}}$ can be found in [2, 10].

2.2 Ultraproducts, Nonstandard Extensions and Height Functions

Definition 2 *(Language)* A language $\mathscr{L}$ is given by specifying the following data:

(i) A set of function symbols $\mathscr{F}$ and positive integers n_f for each f in $\mathscr{F}$,
(ii) A set of relation symbols $\mathscr{R}$ and positive integers n_E for each E in $\mathscr{R}$,
(iii) A set of constant symbols $\mathscr{C}$.

The number n_f indicates that f is a function of n_f variables and the number n_E indicates that E is an n_E-ary relation. For instance, the language $\{+, -, \cdot, 0, 1\}$ is called the language of rings where $+, -, \cdot$ are binary function symbols and 0,1 are constants. Now, we define structures with respect to a language $\mathscr{L}$.

Definition 3 *(Structure)* An $\mathscr{L}$-structure $\mathbb{M}$ is given by the following data:

(i) A nonempty set M called the domain of $\mathbb{M}$,
(ii) A function $f^{\mathbb{M}} : M^{n_f} \to M$ for each $f \in \mathscr{F}$,
(iii) A set $E^{\mathbb{M}} \subseteq M^{n_E}$ for each $E \in \mathscr{R}$,
(iv) An element $c^{\mathbb{M}} \in M$ for each $c \in \mathscr{C}$.

We regard $f^{\mathbb{M}}$, $E^{\mathbb{M}}$ and $c^{\mathbb{M}}$ as the interpretations of f, E and c, and we write the structure as $\mathbb{M} = (M, f^{\mathbb{M}}, E^{\mathbb{M}}, c^{\mathbb{M}} : f \in \mathscr{F}, E \in \mathscr{R}, c \in \mathscr{C})$. For example, a ring is an $\mathscr{L}$-structure $(R, +, -, \cdot, 0, 1)$ in the language of rings.

Next, we define ultraproducts and nonstandard extensions which we will need in our approach for the Main Remark.

Definition 4 *(Filter)* A *filter* D on $\mathbb{N}$ is a set of subsets of $\mathbb{N}$ such that

(i) If A is in D and $A \subseteq B$, then B is also in D.
(ii) If A and B are in D, then so is $A \cap B$.
(iii) $\mathbb{N} \in D, \emptyset \notin D$.

A filter D on $\mathbb{N}$ is called an *ultrafilter* (maximal) if for each $X \subseteq \mathbb{N}$, exactly one of X and $\mathbb{N} \setminus X$ belongs to D. By Zorn's lemma, one can show that every filter is contained in an ultrafilter. An ultrafilter is said to be *non-principal* if it contains all cofinite sets, in other words it does not contain any finite set. Now we fix an ultrafilter D and let $(\mathbb{M}_n)_n$ be a sequence of $\mathscr{L}$-structures. Two elements (two infinite tuples) $(x_n)_n$ and $(y_n)_n$ from $\prod_{n\in\mathbb{N}} \mathbb{M}_n$ are said to be *D-equivalent* and denoted by $(x_n)_n \equiv_D (y_n)_n$, if $\{n : x_n = y_n\} \in D$. The D-equivalence class $(x_n)_n/D$ of $(x_n)_n$ is defined as

$$\{(y_n)_n \in \prod_{n\in\mathbb{N}} \mathbb{M}_n : (y_n)_n \equiv_D (x_n)_n\}.$$

The *ultraproduct*

$$\mathcal{U} = \prod_{n \in \mathbb{N}} \mathbb{M}_n / D$$

of the structures $\mathbb{M}_n$ is defined by the D-equivalence classes, and it is an $\mathcal{L}$-structure as follows:

- For a function symbol $f \in \mathcal{F}$ of arity k,

$$f^{\mathcal{U}} : \left((a_n^1)_n/D, \ldots, (a_n^k)_n/D\right) \mapsto (f^{\mathbb{M}_n}(a_n^1, \ldots, a_n^k))_n/D.$$

- For a relation symbol $E \in \mathcal{R}$ of arity k,

$$E^{\mathcal{U}} = \left\{\left((a_n^1)_n/D, \ldots, (a_n^k)_n/D\right) \in \mathcal{U}^k : \{n \in \mathbb{N} : (a_n^1, \ldots, a_n^k) \in E^{\mathbb{M}_n}\} \in D\right\}.$$

- For a constant symbol $c \in \mathcal{C}$,

$$c^{\mathcal{U}} = (c^{\mathbb{M}_n})/D.$$

One can see that this definition is well-defined. If all the structures $\mathbb{M}_n$ are the same structure $\mathbb{M}$, then the ultraproduct is called the *ultrapower*. The notion of ultraproducts and their properties apply to many-sorted structures.

Definition 5 (*Nonstandard Extension of a Structure*) Let $\mathbb{M}$ be a nonempty structure. A *nonstandard extension* $^*\mathbb{M}$ of $\mathbb{M}$ is an ultrapower of $\mathbb{M}$ with respect to a non-principal ultrafilter on $\mathbb{N}$.

Now let $\mathbb{M}$ be a nonempty structure and $^*\mathbb{M}$ be a nonstandard extension of $\mathbb{M}$ with respect to a non-principal ultrafilter D on $\mathbb{N}$. We can regard each element x of $\mathbb{M}$ as the class of the constant sequence $(x)_n/D$ of $^*\mathbb{M}$. Identifying $\mathbb{M}$ as a subset of $^*\mathbb{M}$, the structure $\mathbb{M}$ becomes an elementary substructure of $^*\mathbb{M}$, this means that they satisfy exactly the same first-order sentences. For a subset A of $\mathbb{M}$, the set $^*A$ is defined to be the set

$$\{(a_n)_n/D : \{n : a_n \in A\} \in D\}.$$

Note that $^*A$ contains A. Every function on a subset A of $\mathbb{M}$ extends to $^*A$ coordinatewise and this is well-defined. This means that if f is a function from A to B, then f extends to a function $^*f$ from $^*A$ to $^*B$. More precisely, let $a = (a_n)_n/D$ be an element of $^*A$. Then $^*f(a) = (f(a_n)_n)/D$ which is an element of $^*B$. The totally ordered field $^*\mathbb{R}$ is called hyperreals and the order is defined by as follows: $(x_n)_n/D \leq (y_n)_n/D$ if $\{n : x_n \leq y_n\} \in D$. The elements $^*\mathbb{R} \setminus \mathbb{R}$ are called *nonstandard real numbers*. Let

$$\mathbb{R}_{\text{fin}} = \{x \in {}^*\mathbb{R} : |x| < n \text{ for some } n \in \mathbb{N}\}$$

be the set of all finite nonstandard real numbers. The elements in ${}^*\mathbb{R} \setminus \mathbb{R}_{\text{fin}}$ are called infinite. Note that $\mathbb{R}_{\text{fin}}$ is a subring of ${}^*\mathbb{R}$ containing $\mathbb{R}$. Let $\varepsilon = (1/n)_n/D$. Observe that ε is a nonstandard positive number and $\varepsilon < 1/n$ for all $n \geq 1$, in other words ε is an infinitesimal.

Let R_m be a ring with a height function ht_m of θ-type, where $m \in \mathbb{N}$. Let K_m be the field of fractions of R_m. Let (S, h_S) be the ultraproduct of the structures $(R_m, \text{ht}_\text{m})$ with respect to a non-principal ultrafilter D on $\mathbb{N}$. Then S becomes a domain with coordinatewise addition and multiplication. As discussed before, similarly the function $\theta : \mathbb{N} \to \mathbb{N}$ extends to a function ${}^*\theta : {}^*\mathbb{N} \to {}^*\mathbb{N}$, where ${}^*\mathbb{N}$ is a nonstandard extension of $\mathbb{N}$ with respect to D. Let $a = (a_m)_m/D$ be an element of S, where $a_m \in R_m$. Then, $\text{h}_\text{S}(a)$ is defined as $(\text{ht}_\text{m}(a_m))_m/D$ and this is again well-defined. Note that h_S is a function from S to ${}^*\mathbb{R}$, where ${}^*\mathbb{R}$ is a nonstandard extension of $\mathbb{R}$ with respect to D. In fact, if x, y in S with $\text{h}_\text{S}(x) \leq n$ and $\text{h}_\text{S}(y) \leq n$, where $n \in {}^*\mathbb{N}$, then we have both $\text{h}_\text{S}(x + y) \leq {}^*\theta(n)$ and $\text{h}_\text{S}(xy) \leq {}^*\theta(n)$. Define

$$S_{\text{fin}} = \{x \in S : \text{h}_\text{S}(x) \in \mathbb{R}_{\text{fin}}\}.$$

By the properties of the height function h_S, we see that S_{fin} is a subring of S. Put $L = \text{Frac}(S_{\text{fin}})$ which is a subfield of $F = \text{Frac}(S)$. Note that the ultraproduct of the fields K_m with respect to D is the field of fractions of S. For more detailed information about ultraproducts, nonstandard analysis and model theory, the reader might consult [5, 8, 13].

2.3 *Proper Ideals and Degree Bounds*

In this subsection, we give some results from commutative algebra that we need for the Main Remark.

Lemma 1 *Let F be a field and $f_1, \ldots, f_s \in F[X_1, \ldots, X_n]$. The following are equivalent:*

(1) $1 \in \langle f_1, \ldots, f_s \rangle$,

(2) $f_1, \ldots, f_s$ *have no common zeros in* F^{alg},

(3) $f_1, \ldots, f_s$ *have no common zeros in any field extension* F_1 *of* F.

Proof Cleary if $1 \in \langle f_1, \ldots, f_s \rangle$, then $f_1, \ldots, f_s$ have no common zeros in any field extension F_1 of F, thus the first condition gives the third condition. The third condition immediately implies the second condition. Now assume that the second condition holds and we will prove the first condition. By Hilbert's Nullstellensatz, there are $g_1, \ldots, g_s \in F^{\text{alg}}[X_1, \ldots, X_n]$ such that

$$1 = f_1 g_1 + \cdots + f_s g_s.$$

This is a system of linear equations when we consider the coefficients of all the polynomials. Therefore the linear system $1 = f_1 Y_1 + \cdots + f_s Y_s$ has a solution in

F^{alg}. Now by the Gauss–Jordan method, one can see that this linear system has a solution in F. So there are $h_1, \ldots, h_s \in F[X_1, \ldots, X_n]$ such that $1 = f_1 h_1 + \cdots + f_s h_s$.

Corollary 1 *Let $F \subseteq F_1$ be a field extension and $I \subset F[X_1, \ldots, X_n]$ be a proper ideal. Then the ideal $I F_1[X_1, \ldots, X_n]$ is also proper in $F_1[X_1, \ldots, X_n]$.*

Proof Let $I \subset F[X_1, \ldots, X_n]$ be a proper ideal. Then since I is finitely generated, $I = \langle f_1, \ldots, f_s \rangle$ for some $f_1, \ldots, f_s \in F[X_1, \ldots, X_n]$. By Lemma 1, we see that $f_1, \ldots, f_s$ have a common zero in a field extension F_2 of F. Moreover, we can assume that F_2 also contains F_1. So by Lemma 1 again, $I F_1[X_1, \ldots, X_n] \neq F_1[X_1, \ldots, X_n]$.

Recall the result of [9, 17] from the introduction (see also [3, Theorem 1.11]):

Theorem 1 *[9, 17] If $f_0, f_1, \ldots, f_s$ in $K[X_1, \ldots, X_n]$ all have degree less than D and f_0 is in $\langle f_1, \ldots, f_s \rangle$, then $f_0 = \sum_{i=1}^{s} f_i h_i$ for certain h_i whose degrees are bounded by a constant $c_1 = c_1(n, D)$ depending only on n and D.*

From the nonstandard point of view, the proof of Theorem 1 follows from the fact that ${}^*(K[X_1, \ldots, X_n])$ is a faithfully flat ${}^*K[X_1, \ldots, X_n]$-module, see [3, Theorem 1.8]. Note that ${}^*K[X_1, \ldots, X_n] \subsetneq {}^*(K[X_1, \ldots, X_n])$, as the degree function is unbounded.

3 Proof of the Main Remark

By Theorem 1, the constant c_1 exists and it only depends on n and D. Now we prove the existence of the constant c_2. Assume that n, D, H and θ are given. To prove the Main Remark, we will first show that there is a constant $c_3(n, D, H, \theta)$ such that for all rings R with a height function h of θ-type, if $f_1, \ldots, f_s$ in $R[X_1, \ldots, X_n]$ have no common zero in K^{alg} with $\deg(f_i) \leq D$ and $\mathrm{h}(f_i) \leq H$, then there exist $h_1, \ldots, h_s$ in $K[X_1, \ldots, X_n]$ such that $1 = f_1 h_1 + \cdots + f_s h_s$, $\deg(h_i) \leq c_1$ and $\mathrm{h}(e) \leq c_3$ where e is an element that occurs as a numerator or denominator of some coefficient of some h_i. In fact, one can see that the existence of c_2 and the existence of c_3 are equivalent.

Proof of the existence of c_3: To obtain a contradiction, suppose that c_3 does not exist. Therefore, for every $m \geq 1$ there exist an integral domain R_m with a height function $\mathrm{ht_m}$ of θ-type and $f_1, \ldots, f_s$ in $R_m[X_1, \ldots, X_n]$ with $\deg f_i \leq D$ and $\mathrm{ht_m}(f_i) \leq H$ witnessing to this. This means that in the field of fractions K_m of R_m, there exist $g_1, \ldots, g_s$ in $K_m[X_1, \ldots, X_n]$ with $\deg g_i \leq c_1$ and

$$1 = f_1 g_1 + \cdots + f_s g_s,$$

but for all $h_1, \ldots, h_s \in K_m[X_1, \ldots, X_n]$ with $\deg h_i \leq c_1$,

$$1 = f_1h_1 + \cdots + f_sh_s$$

implies $\max_j \mathrm{ht_m}(a_j) > m$ where $a_j \in R_m$ is an element that occurs as a numerator or denominator of some coefficient of some h_i. As the constant c_1 exists, the counterexamples above are coming from formulas φ_m of uniformly bounded length and we are quantifying over these tuples coming from R. Now consider the ultraproduct of structures $(R_m, \mathrm{ht_m})$ with respect to a non-principal ultrafilter D on $\mathbb{N}$, and call it $(S, \mathrm{h_S})$, as defined in Sect. 2.2. Let F be the field of fractions of S. Recall that $\mathrm{h_S}$ is a function from S to $^*\mathbb{R}$ of θ-type. By the counterexamples above, there are polynomials $f_1, \ldots, f_s$ in $S_{\mathrm{fin}}[X_1, \ldots, X_n]$ of degrees less than D and heights less than H such that the linear system

$$f_1Y_1 + \cdots + f_sY_s = 1$$

has a solution in $F[X_1, \ldots, X_n]$, but not in $L[X_1, \ldots, X_n]$, where L is field of fractions of S_{fin} and $F = \mathrm{Frac}(S)$. This contradicts Corollary 1, since the ideal $\langle f_1, \ldots, f_s\rangle$ is proper in $L[X_1, \ldots, X_n]$, but it is not proper in $F[X_1, \ldots, X_n]$. $\square$

Hence for any integral domain R with a height function h of θ-type, we know that given $f_1, \ldots, f_s \in R[X_1, .., X_n]$ with no common zeros in K^{alg} with $\deg(f_i) \leq D$ and $\mathrm{h}(f_i) \leq H$, there are $h_1, \ldots, h_s$ in $K[X_1, \ldots, X_n]$ such that $1 = f_1h_1 + \cdots + f_sh_s$ and $\deg(h_i) \leq c_1(n, D)$. Moreover $s \leq q(n, D)$ and $\mathrm{h}(e) \leq c_3(n, D, H, \theta)$ where $e \in R$ is an element which occurs as a numerator or denominator for some coefficient of some h_i, and $q(n, D) = \binom{n+D}{n}$ as defined in Remark 2. Let $b_1, \ldots, b_m$ be all the elements in R that occur as a denominator for some coefficient of some h_i. Note that $m = m(n, D) \leq q^2$ depends only on n and D. Also we know that $\mathrm{h}(b_i) \leq c_3$. Put $a = b_1 \cdots b_m$. By the multiplicative properties of the height function, we get $\mathrm{h}(a) \leq c_4(n, D, H, \theta)$ for some c_4. Now we see that

$$a = \sum_{i=1}^{s} f_i \cdot (ah_i),$$

f_i and ah_i are in $R[X_1, .., X_n]$ and $\deg(ah_i) = \deg(h_i) \leq c_1$. Moreover, again by the multiplicative properties of the height function, we have $\mathrm{h}(ah_i) \leq c_5(n, D, H, \theta)$. Now take $c_2 = \max(c_4, c_5)$. Therefore, we obtain (i), (ii) and (iii) from the Main Remark for the constant c_2 as defined above. $\square$

References

1. M. Aschenbrenner, *Ideal membership in polynomial rings over the integers*, J. Amer. Math. Soc. **17** (2004), 407–441.
2. E. Bombieri, W. Gubler, Heights in Diophantine Geometry, Cambridge University Press; 1 edition (September 24, 2007).

3. L. van den Dries and K. Schmidt, *Bounds in the theory of polynomial rings over fields. A nonstandard approach*, Inventiones Math. **76** (1984), 77–91.
4. M. Giusti, J. Heintz, J. Sabia, *On the efficiency of effective Nullstellensätze*, Comput. Complexity 3 (1993), 56–95.
5. R. Goldblatt, Lectures on the Hyperreals, A Introduction to Nonstandard Analysis Springer-Verlag, New York, 1998.
6. H. Göral, Model Theory of Fields and Heights, Ph.D thesis, Lyon, 2015.
7. H. Göral, *Height Bounds, Nullstellensatz and Primality*, to appear in Comm. in Algebra, 46 (2018), no. 10, 4463–4472.
8. C. W. Henson, Foundation of Nonstandard Analysis, A Gentle Introduction to Nonstandard Extensions, Lecture Notes.
9. G. Hermann *Die Frage der endlich vielen Schritte in der Theorie der Polynomideale*, Math. Ann. **95** (1926), 736–788.
10. M. Hindry, J.H. Silverman, Diophantine Geometry, An Introduction, Springer-Verlag, 2000.
11. J. Kollár, *Sharp Effective Nullstellensatz*, J. Amer. Math. Soc. Volume 1, Number 4, 1988.
12. T. Krick, L.M. Pardo, M. Sombra *Sharp estimates for the arithmetic Nullstellensatz*, Duke Math. J. 109 (2001), no. 3, 521–598.
13. D. Marker, Model Theory: An Introduction, Springer-Verlag, New York, 2002.
14. P. Philippon, *Dénominateurs dans le théorème des zeros de Hilbert*, Acta Arith. 58 (1990), 1–25.
15. A. Robinson, *Théorie métamathématiques des idéaux*, collection de logique mathématiques, Dactyl-offset. Gauthier-Villars, E. Nauwelaerts, Ser A, Paris-Louvain, 1955.
16. A. Robinson, *Some problems of definability in the lower predicate calculus*, J. Symbolic Logic, Volume 25, Issue 2 (1960), 171.
17. A. Seidenberg, *Constructions in algebra*, Trans. AMS 197, 273–313 (1974).
18. M. Sombra, *Sparse Effective Nullstellensatz*, Advances in Applied Mathematics, Volume 22, Issue 2, February 1999, Pages 271–295.

On the $(1 + u^2 + u^3)$-Constacyclic and Cyclic Codes Over the Finite Ring $F_2 + uF_2 + u^2F_2 + u^3F_2 + vF_2$

G. Gözde Güzel, Abdullah Dertli and Yasemin Çengellenmiş

Abstract In this paper a new finite ring is introduced along with its algebraic properties. In addition, a new Gray map is defined on the ring. The Gray images of both the cyclic and the $(1 + u^2 + u^3)$-constacyclic codes over the finite ring are found to be permutation equivalent to binary quasicyclic codes.

1 Introduction

Studies on error-correcting codes over finite rings were raised by Blake in the1970s [3, 4]. In order to obtain codes on finite fields using finite rings, mapping between the ring and the field has to be defined. This mapping, called a Gray map, is a linear distance-preserving map. Using mapping, the first Gray images of codes on ring Z_4 were defined, which were binary non-linear codes with good parameters [9]. This turning point in algebraic coding theory increased the volume of work on different types of rings, with different properties, such as finite chain rings, Frobenious rings, and finite non-chain rings [1, 2, 5, 6, 8, 10–15]. Thus, in recent studies, a large number of optimal codes have been obtained over finite fields using Gray images of cyclic and constacyclic codes for many various finite rings [2, 7, 8, 10, 11, 13].

The paper is organized as follows. In Sect. 2, a new finite ring is introduced and some knowledge about it is provided. In Sect. 3, the relationship between the cyclic

G. G. Güzel (✉)
Ipsala Vocational College, Trakya University, Edirne, Turkey
e-mail: ggodeguzel@trakya.edu.tr

A. Dertli
Department of Mathematics, Faculty of Science and Arts, Ondokuz Mayis University, Samsun, Turkey
e-mail: abdullah.dertli@gmail.com

Y. Çengellenmiş
Department of Mathematics, Faculty of Science, Trakya University, Edirne, Turkey
e-mail: ycengellenmis@gmail.com

I. Inam and E. Büyükasik (eds.), *Notes from the International Autumn School on Computational Number Theory*, Tutorials, Schools, and Workshops in the Mathematical Sciences, https://doi.org/10.1007/978-3-030-12558-5_6

and $(1+u^2+u^3)$-constacyclic codes and the finite ring $R = F_2 + uF_2 + u^2F_2 + u^3F_2 + vF_2$ is given. In Sect. 4, a new Gray map on R is defined. Gray images of both cyclic codes with odd length and $\lambda-$ constacyclic codes over the finite ring.

2 Preliminaries

Consider the quotient ring $F_2[u, v]/\langle u^4, v^2 - v, uv = vu\rangle$. For conditions $u^4 = 0$, $v^2 = v$, and $uv = vu = 0$, ring $F_2 + uF_2 + u^2F_2 + u^3F_2 + vF_2$ is isomorphic to the quotient ring $F_2[u, v]/\langle u^4, v^2 - v, uv = vu\rangle$. Let R be the ring $F_2 + uF_2 + u^2F_2 + u^3F_2 + vF_2$ with properties $u^4 = 0$, $v^2 = v$, and $uv = vu = 0$. R is a finite and commutative ring and its characteristic is 2 because of the finite field F_2 with characteristic 2.

Elements of ring R, given with obvious addition and multiplication operations, are written as $0, u^2+v+1, u^3+v, u^2, u^2+u, u^3+u, u^3+u^2+v+1, u^2+u+1, u^3+u^2+u+1, u^3, u^2+u+v, u^3+u^2, u^3+u^2+u+v+1, u^3+u^2+u+v, 1, u^3+v+1, u+v+1, u^3+u+v+1, u^3+u+1, v+1, u^3+u^2+1, v, u, u+1, u^2+1, u^3+u+v, u^3+1, u^3+u^2+v, u^2+u+v+1, u^3+u^2+u, u+v, u^2+v$.

Each unit element of ring R takes the form $1+bu+cu^2+du^3$. Namely, units are $1, 1+u, 1+u^2, 1+u+u^2, 1+u^3, 1+u+u^3, 1+u+u^2+u^3, 1+u^2+u^3$.

The ideals of R are $<0>, <v>, <u^3>, <u^2>, <u+u^2>, <1+u^2+v>, <v+u^3>, <u^2+v>, <u+v+u^2>, <1>$. Ring R has two maximal ideals: $<1+v>$ and $<u+v>$. Therefore, R is a semilocal ring. Since all ideals of ring R are principally generated, R is a principal ideal ring.

Definition 1 A linear code C over R (or F_q) of length n is an R-submodule (F_q subspace) of R^n (F_q^n). The elements of a linear code are called codewords.

Definition 2 If D is a linear code over F_2, of length n, then the Hamming weight $w_H(d)$ of a codeword $d = (d_0, ..., d_{n-1})$ in D is the number of non-zero coordinates.

Definition 3 The minimum Hamming distance of D is defined as $d_H(D)$=min $d_H(d, d')$, where for any $d' \in D$, $d \neq d'$ and $d_H(d, d')$ is the Hamming distance between two codewords with $d_H(d, d') = w_H(d - d')$.

Definition 4 Let λ be a unit in R. Let σ and ν be cyclic and λ-constacyclic shift, respectively. A code of length n is cyclic if and only if $\sigma(C) = C$; a code of length n is λ-constacyclic if and only if $\nu(C) = C$.

Definition 5 Let $a \in F_2^{6n}$ with $a = (a_0, a_1, ..., a_{6n-1}) = \left(a^{(0)}|a^{(1)}|a^{(2)}|a^{(3)}|a^{(4)}|a^{(5)}\right)$, $a^{(i)} \in F_2^n$ for $i = 0, 1, 2, 3, 4, 5$. Let $\sigma^{\otimes 6}$ be a map from F_2^{6n} to F_2^{6n} given by $\sigma^{\otimes 6}(a) = \left(\sigma\left(a^{(0)}\right)|\sigma\left(a^{(1)}\right)|\sigma\left(a^{(2)}\right)|\sigma\left(a^{(3)}\right)|\sigma\left(a^{(4)}\right)|\sigma\left(a^{(5)}\right)\right)$, where σ is a cyclic shift from F_2^n to F_2^n given by $\sigma\left(a^{(i)}\right) = ((a^{(i,n-1)}), (a^{(i,0)}), (a^{(i,1)}), ..., (a^{(i,n-2)}))$ for every $a^{(i)} = (a^{(i,0)}, ..., a^{(i,n-1)})$, where $a^{(i,j)} \in F_2$, $j = 0, 1, ..., n-1$. A code of length $6n$ over F_2 is said to be a binary quasicyclic code of index 6 and length $6n$, if $\sigma^{\otimes 6}(C) = C$.

Proposition 1 *A subset C of R^n is a cyclic code of length n if and only if its polynomial representation is an ideal of $R[x]/\langle x^n - 1\rangle$.*

Proposition 2 *A subset C of R^n is a λ-constacyclic code of length n if and only if its polynomial representation is an ideal of $R[x]/\langle x^n - \lambda\rangle$.*

3 λ-Constacyclic Codes of Odd Length

In order to define constacyclic codes over any ring, unit elements not equal to 1 must be chosen. In this paper, to make the Gray image more appropriate and the code length an odd number, a unit element is chosen as follows: $\lambda = (1 + u^2 + u^3)$.

Remark 1 For $\lambda = (1 + u^2 + u^3)$, note that $\lambda^n = \lambda$ if n is odd and $\lambda^n = 1$ if n is even. Throughout this section, we consider length n as odd in order to use this property.

Proposition 3 *Let n be odd. Then μ is a map of $R[x]/\langle x^n - 1\rangle$ to $R[x]/\langle x^n - \lambda\rangle$, defined by $\mu(f(x)) = f(\lambda x)$, where $\lambda = 1 + u^2 + u^3$ and μ is a ring isomorphisms.*

Proof Let $f(x) \equiv g(x)\, mod(x^n - 1)$, there are some polynomials like $h(x) \in R[x]$ such that $f(x) - g(x) = h(x)(x^n - 1)$. When the equality is rewritten by replacing x with λx, we obtain the equality $f(\lambda x) - g(\lambda x) = h(\lambda x)(\lambda^n x^n - 1)$. Since n is odd, the equalities $\lambda^n = \lambda$ and $\lambda^2 = 1$ hold. By using these truths, if the last equality is rearranged then we get $\mu(f(x)) - \mu(g(x)) = h(\lambda x)\lambda(x^n - \lambda)$. So, $\mu(f(x)) \equiv \mu(g(x))\, mod(x^n - \lambda)$. Therefore, μ is a well-defined map. Since the same process can be reversed, μ is injective.

For any $f(x)$ and $g(x)$ in $R[x]/\langle x^n - 1\rangle$, using the definition of addition in the functions and the definition of μ, $\mu(f(x) + g(x)) = \mu((f + g)(x)) = (f + g)(\lambda x) = f(\lambda x) + g(\lambda x) = \mu(f(x)) + \mu(g(x))$. μ is a homomorphism, since it can also be proved in a similar manner for multiplication. Hence μ is a ring isomorphism.

The immediate corollary of the above proposition is as follows:

Corollary 1 *If n is odd, then I is an ideal of $R[x]/\langle x^n - 1\rangle$ if and only if $\mu(I)$ is an ideal of $R[x]/\langle x^n - \lambda\rangle$, where $\lambda = 1 + u^2 + u^3$.*

Corollary 2 *Let n be odd and C be a subset of R^n. Let $\overline{\mu}$ be the permutation of R^n, defined by $\overline{\mu}(c_0, c_1, ..., c_{n-1}) = \left(c_0, \lambda c_1, ..., \lambda^i c_i, ..., \lambda^{n-1} c_{n-1}\right)$. Then, C is a cyclic code if and only if $\overline{\mu}(C)$ is a λ-constacyclic code, where $\lambda = 1 + u^2 + u^3$.*

Proof Let C be a cyclic code. Consider its polynomial representation as again C. Then C is an ideal in $R[x]/\langle x^n - 1\rangle$. Since the above corollary is satisfied, $\mu(C)$ is an ideal in $R[x]/\langle x^n - \lambda\rangle$. Hence, $\overline{\mu}(C)$ is a linear λ-constacyclic code, where $\lambda = 1 + u^2 + u^3$.

Table 1 Gray weights of elements of the finite ring R

$w_G(x)$	x
0	0
1	u^2+v
2	$u^2, u^2+u, u^3+u, u^2+u+1, u^3+u^2, u, u+1, u^2+1, u^3+1, u^3+u^2+u$
3	$u^3+v, u^3+u^2+v+1, u^2+u+v, u^3+u^2+u+v,$ $u+v+1, v+1, v, u^3+u+v, u^2+u+v+1, u+v$
4	$u^3+u^2+u+1, u^3, 1, u^3+u+1, u^3+u^2+1$
5	$u^2+v+1, u^3+u^2+u+v+1, u^3+v+1, u^3+u+v+1, u^3+u^2+v$

4 A New Gray Map and Gray Images of λ-Constacyclic and Cyclic Codes

In this section, we will define a Gray map ϕ, from R to ${F_2}^6$, written as $\phi(a+bu+cu^2+du^3+ev)=(a+b+c+d+e, a+c+d+e, a, e, a+b+d, d)$.

The Gray map ϕ can be extended to R^n as follows:

$\phi : R^n \rightarrow {F_2}^{6n}$

$(z_0, z_1, ..., z_{n-1}) \mapsto \phi(z_0, z_1, ..., z_{n-1}) = (a_0+b_0+c_0+d_0+e_0, ..., a_{n-1}+b_{n-1}+c_{n-1}+d_{n-1}+e_{n-1}, a_0+c_0+d_0+e_0, ..., a_{n-1}+c_{n-1}+d_{n-1}+e_{n-1}, a_0, ..., a_{n-1}, e_0, ..., e_{n-1}, a_0+b_0+d_0, ..., a_{n-1}+b_{n-1}+d_{n-1}, d_0, ..., d_{n-1})$ where $z_i = a_i + b_i u + c_i u^2 + d_i u^3 + e_i v$ with $a_i, b_i, c_i, d_i, e_i \in F_2$ for $0 \leq i \leq n-1$.

The Gray weight of element x of R is defined as the Hamming weight of the Gray image of x (provided in Table 1).

Definition 6 Let C be a linear code over R of length n. For any codeword $c=(c_0, ..., c_{n-1})$, the Gray weight of c is defined as $w_G(c)=\sum_{i=0}^{n-1} w_G(c_i)$ and the minimum Gray distance of C is defined as $d_G(C)$= min $d_G(c, c^{'})$, where for any $c^{'} \in C, c \neq c^{'}$ and $d_G(c, c^{'})$ is the Gray distance between two codewords with $d_G(c, c^{'}) = w_G(c-c^{'})$.

Proposition 4 *The Gray map ϕ is linear and a distance-preserving map from R^n (Gray distance) to F_2^{6n} (Hamming distance).*

Proof For any x, y in R, where $x=a_1+b_1u+c_1u^2+d_1u^3+e_1v$ and $y=a_2+b_2u+c_2u^2+d_2u^3+e_2v$, it holds that $\phi(x+y)=\phi((a_1+a_2)+(b_1+b_2)u+(c_1+c_2)u^2+(d_1+d_2)u^3+(e_1+e_2)v)$. By using the definition of ϕ, $\phi(x+y)=(a_1+a_2+b_1+b_2+c_1+c_2+d_1+d_2+e_1+e_2, a_1+a_2+c_1+c_2+d_1+d_2+e_1+e_2, a_1+a_2, e_1+e_2, a_1+a_2+b_1+b_2+d_1+d_2, d_1+d_2)=(a_1+b_1+c_1+d_1+e_1, a_1+c_1+d_1+e_1, a_1, e_1, a_1+b_1+d_1, d_1)+(a_2+b_2+c_2+d_2+e_2, a_2+c_2+d_2+e_2, a_2, e_2, a_2+b_2+d_2, d_2)=\phi(x)+\phi(y)$. For $k \in F_2$, $\phi(kx)=(k(a_1+b_1+c_1+d_1+e_1), k(a_1+c_1+d_1+e_1), ka_1, ke_1, k(a_1+b_1+d_1), kd_1) =$

$k(a_1+b_1+c_1+d_1+e_1, a_1+c_1+d_1+e_1, a_1, e_1, a_1+b_1+d_1, d_1) = k\phi(x)$. So, ϕ is a linear map.

For any x, y in R^n, $d_G(x, y) = w_G(x-y) = w_H(\phi(x-y)) = d_H(\phi(x), \phi(y))$. So, ϕ is isometric.

Theorem 1 *If C is (n, M, d_G) linear code over R, then $\phi(C)$ is $(6n, M, d_H)$ linear code over F_2, where $d_G = d_H$.*

4.1 Gray Images of λ-Constacyclic Codes over the Finite Ring R

Proposition 5 *Let ν and σ be as provided above. Then, $\phi\nu = \pi\sigma^{\otimes 6}\phi$, where π is a permutation of $\{0, 1, ..., 6n-1\}$, defined by $\pi = (0, n)(4n, 5n)$.*

Proof Let $z = (z_0, z_1, ..., z_{n-1})$ be in R^n. Let $a_i, b_i, c_i, d_i, e_i \in F_2$ for $0 \le i \le n-1$, such that $z_i = a_i + b_i u + c_i u^2 + d_i u^3 + e_i v$. Then, $\nu(z) = ((1+u^2+u^3) z_{n-1}, z_0, z_1, ..., z_{n-2})$. Note that $(1+u^2+u^3)z_{n-1} = a_{n-1} + b_{n-1}u + (a_{n-1} + c_{n-1})u^2 + (a_{n-1} + b_{n-1} + d_{n-1})u^3 + e_{n-1}v$. From definitions of the Gray map and ν cyclic shift, we get $\phi(\nu(z)) = (a_{n-1} + c_{n-1} + d_{n-1} + e_{n-1}, a_0 + b_0 + c_0 + d_0 + e_0, ..., a_{n-2} + b_{n-2} + c_{n-2} + d_{n-2} + e_{n-2}, a_{n-1} + b_{n-1} + c_{n-1} + d_{n-1} + e_{n-1}, a_0 + c_0 + d_0 + e_0, ..., a_{n-2} + c_{n-2} + d_{n-2} + e_{n-2}, a_{n-1}, a_0, ..., a_{n-2}, e_{n-1}, e_0, ..., e_{n-2}, d_{n-1}, a_0 + b_0 + d_0, ..., a_{n-2} + b_{n-2} + d_{n-2}, a_{n-1} + b_{n-1} + d_{n-1}, d_0, ..., d_{n-2})$.

On the other hand, by applying $\sigma^{\otimes 6}$ to $\phi(z) = (a_0 + b_0 + c_0 + d_0 + e_0, ..., a_{n-1} + b_{n-1} + c_{n-1} + d_{n-1} + e_{n-1}, a_0 + c_0 + d_0 + e_0, ..., a_{n-1} + c_{n-1} + d_{n-1} + e_{n-1}, a_0, ..., a_{n-1}, e_0, ..., e_{n-1}, a_0 + b_0 + d_0, ..., a_{n-1} + b_{n-1} + d_{n-1}, d_0, ..., d_{n-1})$, we have

$\sigma^{\otimes 6}(\phi(z)) = (a_{n-1} + b_{n-1} + c_{n-1} + d_{n-1} + e_{n-1}, a_0 + b_0 + c_0 + d_0 + e_0, ..., a_{n-2} + b_{n-2} + c_{n-2} + d_{n-2} + e_{n-2}, a_{n-1} + c_{n-1} + d_{n-1} + e_{n-1}, a_0 + c_0 + d_0 + e_0, ..., a_{n-2} + c_{n-2} + d_{n-2} + e_{n-2}, a_{n-1}, a_0, ..., a_{n-2}, e_{n-1}, e_0, ..., e_{n-2}, a_{n-1} + b_{n-1} + d_{n-1}, a_0 + b_0 + d_0, ..., a_{n-2} + b_{n-2} + d_{n-2}, d_{n-1}, d_0, ..., d_{n-2})$.

If we apply the permutation π to $\sigma^{\otimes 6}(\phi(z))$, then we get $\pi(\sigma^{\otimes 6}(\phi(z))) = (a_{n-1} + c_{n-1} + d_{n-1} + e_{n-1}, a_0 + b_0 + c_0 + d_0 + e_0, ..., a_{n-2} + b_{n-2} + c_{n-2} + d_{n-2} + e_{n-2}, a_{n-1} + b_{n-1} + c_{n-1} + d_{n-1} + e_{n-1}, a_0 + c_0 + d_0 + e_0, ..., a_{n-2} + c_{n-2} + d_{n-2} + e_{n-2}, a_{n-1}, a_0, ..., a_{n-2}, e_{n-1}, e_0, ..., e_{n-2}, d_{n-1}, a_0 + b_0 + d_0, ..., a_{n-2} + b_{n-2} + d_{n-2}, a_{n-1} + b_{n-1} + d_{n-1}, d_0, ..., d_{n-2})$. Hence, $\phi(\nu(z)) = \pi(\sigma^{\otimes 6}(\phi(z)))$.

Theorem 2 *The Gray image of a $(1+u^2+u^3)$-constacyclic code of length n over R is permutation equivalent to a binary quasicyclic code of index 6 and length $6n$.*

Proof Let C be a $(1+u^2+u^3)$-constacyclic code over R. So, $\nu(C) = C$. By applying ϕ, $\phi(\nu(C)) = \phi(C)$ is held. By using Proposition 5, we obtain $\pi(\sigma^{\otimes 6}(\phi(C))) = \phi(C)$. Therefore, this means that $\phi(C)$ is permutation equivalent to a binary quasicyclic code of index 6 and length $6n$.

4.2 Gray Images of Cyclic Codes over the Finite Ring R

Proposition 6 *Let $\overline{\mu}$ be as provided above. If n is odd then $\phi\overline{\mu} = \gamma'\phi$, where $\gamma'(z_0, z_1, ..., z_{6n-1}) = (z_{\gamma(0)}, z_{\gamma(1)}, ..., z_{\gamma(6n-1)})$, such that $\gamma = \xi_1\xi_2$, which is a permutation of $\{0, 1, ..., 6n-1\}$, where $\xi_1 = (1, n+1)(3, n+3)...(n-2, 2n-2)$ and $\xi_2 = (4n+1, 5n+1)(4n+3, 5n+3)...(4n+(n-2), 5n+(n-2))$.*

Proof It is obvious that this proof can be made similar to the proof of Proposition 5.

Theorem 3 *If n is odd and Δ is the Gray image of a cyclic code over R, then $\gamma'(\Delta)$ is permutation equivalent to a binary quasicyclic code of index 6 and length 6n.*

Proof Let $\Delta = \phi(C)$ be a Gray image of a cyclic code C over R. From Proposition 6, $\phi(\overline{\mu}(C)) = \gamma'(\phi(C)) = \gamma'(\Delta)$. Recall that Corollary 2 states that $\overline{\mu}(C)$ is a $(1+u^2+u^3)$-constacyclic code D over the ring R. So, we get $\phi(\overline{\mu}(C)) = \phi(D)$. According to Theorem 2, $\phi(D)$ is permutation equivalent to a binary quasicyclic code of index 6 and length $6n$. The expected result is satisfied.

Corollary 3 *The Gray image of a cyclic code with odd length n over ring R is permutation equivalent to a binary quasicyclic code of index 6 and length 6n.*

5 Conclusion and Future Work

In this paper, a new finite ring $R := F_2 + uF_2 + u_2^{2F} + u_2^{3F} + vF_2$ is introduced, where $u^4 = 0$, $v^2 = v$, $uv = vu = 0$. The isomorphism between $\lambda-$ constacyclic codes and cyclic codes of odd length over the ring R is proved, where $\lambda = (1 + u^2 + u^3)$. A new Gray map on finite ring R is defined. It is obtained that the Gray images of $(1+u^2+u^3)$-constacyclic codes and cyclic codes of odd length over the finite ring R are permutation equivalent to binary quasicyclic codes of index 6.

The ring can be extended for conditions $u^k = 0$ and $v^m = v$ and similar properties can be studied under these conditions. Similar results can be examined for the remaining unit elements in the same ring. Moreover, the structure of cyclic codes can be investigated finding optimal codes over these rings in this manner.

References

1. Amarra M.C.V., Nemenzo F.R., 2008, *On $(1-u)-$ cyclic codes over $F_{p^k} + uF_{p^k}$*
2. Aydin N., Cengellenmis Y., Dertli A., 2017, *On some constacyclic codes over $Z_4[u]/(u^2-1)$, their Z_4 images, and new codes*, Des. Codes Cryptogr., https://doi.org/10.1007/s10623-017-0392-y.
3. Blake I.F., 1972, *Codes over certain rings*, Inform. Control, 20: 396–404.
4. Blake I.F., 1975, *Codes over integer residue rings*, Inform. Control, 29: 295–300.

5. Cengellenmis Y., 2009, *On* $(1 - u^m)-$ *cyclic codes over* $F_2 + uF_2 + u^2F_2 + ... + u^mF_2$, International Journal of Contemporary Math. Sci., 4: 987–992.
6. Dertli A., Cengellenmis Y., 2016, *On* $(1 + u)-$ *cyclic and cyclic codes over* $F_2 + uF_2 + vF_2$, European J. of Pure and Applied Math., 9: 305–313.
7. Dougherty S.T., Salturk E. , *Constacyclic codes over local rings of order* 16, to be submitted.
8. Gao J.,2015, *Linear codes and* $(1 + uv)-$ *constacyclic codes over* $R[v]/(v^2 + v)$, IEICE Transactions on Fundamentals, E98-A: 1044–1048.
9. Hammons Jr. A.R., Kumar P.V., .Calderbank A.R, Sloane N.J.A., Solé P., 1994, *The* Z_4- *linearity of Kerdock, Preparata, Goethal, and related codes*, IEEE Trans. Inform. Theory, 40: 301–319.
10. Kai X., Zhu S., Wang L., 2012, *A family of constacyclic codes over* $F_2 + uF_2 + vF_2 + uvF_2$, J Sysst Sci Complex, 25: 1032–1040.
11. Karadeniz S., Yildiz B., 2011, $(1 + v)-$ *constacyclic codes over* $F_2 + uF_2 + vF_2 + uvF_2$, Journal of Franklin Ins., 348: 2625–2632.
12. Liao D., Tang Y., 2012, *A class of constacyclic codes over* $R + vR$ *and its Gray image*, Int. J. Communications, Network and System Sciences, 5: 222–227.
13. Qian J.F., Zang L.N., Zhu S.X., 2006, $(1 + u)-$ *constacyclic and cyclic codes over* $F_2 + uF_2$, Appl. Math. Lett., 19: 820–823.
14. Qian J.F., Zang L.N., Zhu S.X., 2006, *Constacyclic and cyclic codes over* $F_2 + uF_2 + u^2F_2$, IEICE Transactions on Fundamentals of Electronics Communications and Computer Sciences, 2011, E89-A(6): 1863–1865.
15. Zhu S., Wang L., *A class of constacyclic codes over* $F_p + vF_p$ *and its Gray image*, Discrete Mathematics, 311: 2677–2682.

On Higher Congruences Between Cusp Forms and Eisenstein Series. II.

Bartosz Naskręcki

Abstract We study congruences between cuspidal modular forms and Eisenstein series at levels which are square-free integers and for equal even weights. This generalizes our previous results from (Naskręcki, Computations with modular forms, Contrib. Math. Comput. Sci., vol. 6, pp. 257–277. Springer, Cham (2014) [17]) for prime levels and provides further evidence for the sharp bounds obtained under restrictive ramification conditions. We prove an upper bound on the exponent in the general square-free situation and also discuss the existence of the congruences when the coefficients belong to the rational numbers and weight equals 2.

1 Introduction

Let N be a square-free positive integer and f be a cuspidal newform of level $\Gamma_0(N)$ and even weight k. We consider in this paper congruences between Fourier coefficients $a_n(f)$ of the newform f and coefficients $a_n(E)$ of a suitably normalized Eisenstein series E of weight k modulo prime powers of an ideal λ lying in the coefficient field K_f of f and a positive integer r

$$a_n(f) \equiv a_n(E) \pmod{\lambda^r} \tag{1}$$

for all $n \geq 0$. In the classical setting of N prime and weight $k = 2$, the existence of such congruences for $E = E_2$ was established by Mazur in [16].

In this paper, we focus on the extension of the results and computations of [17]. We present a summary of a large volume of computations performed with MAGMA and a construction of the algorithm that is used to find congruences of type (1). We also present a partial classification of congruences when $\mathcal{O}_f = \mathbb{Q}$. The existence of the higher weight congruences for prime levels was discussed in [10]. Sufficient con-

B. Naskręcki (✉)
Faculty of Mathematics and Computer Science, Adam Mickiewicz University, Umultowska 87, 61-614 Poznań, Poland
e-mail: nasqret@gmail.com

I. Inam and E. Büyükasik (eds.), *Notes from the International Autumn School on Computational Number Theory*, Tutorials, Schools, and Workshops in the Mathematical Sciences, https://doi.org/10.1007/978-3-030-12558-5_7

ditions for existence of congruences for composite levels in weight 2 were obtained by Yoo in [21, 22]. The lower bounds on the congruence exponent in weight 2 was discussed by Hsu [12] and Berger–Klosin–Kramer [2]. Congruences on the level of Galois representations were studied recently by Billerey and Menares in [3, 4]. Lecouturier [14] computed the rank g_p of the completion of the Hecke algebra acting on cuspidal modular forms of weight 2 and level $\Gamma_0(N)$ at the p-maximal Eisenstein ideal using the knowledge of the exponent of the Eisenstein congruence.

In [15], Martin proved the existence of congruences between cuspidal and Eisenstein modular forms for levels which are not square-free. It would be interesting to find more computational examples of such congruences, which seem to be very sparsely distributed.

With the notation of Sect. 2, we present the main theorems of this paper.

Theorem 1.1 *Let $p_1, \ldots, p_t$ be different prime factors of N a square-free integer and let $k > 2$. Suppose that $f \in \mathcal{S}_k(N)^{\mathrm{new}}$ is a newform which is congruent to the Eisenstein eigenform $E = [p_1]^+ \circ \ldots \circ [p_t]^+ E_k \in \mathcal{E}_k(N)$ modulo a power $r > 0$ of a maximal ideal $\lambda \subset \mathcal{O}_f$. If ℓ is the residual characteristic of λ, we obtain the bound*

$$r \leq ord_{\lambda}(\ell) \cdot v_{\ell}\left(-\frac{B_k}{2k}\prod_{i=1}^{t}(1-p_i)\right).$$

The upper bound predicted by the theorem above is discussed numerically in Sect. 6. It turns out that, in general, it is optimal but in the case when the ideal λ is ramified above ℓ, the ramification degree $ord_{\lambda}(\ell)$ seems to be the right upper bound in most cases.

When we turn to the eigenforms with weight 2 and rational coefficients, the existence of the congruences between cuspidal newforms and Eisenstein series is limited to only finitely many prime powers. In this case, we investigated modular forms of levels with two prime factors.

Theorem 1.2 *Let p,q be two different primes. Suppose that $f \in \mathcal{S}_2(pq)^{\mathrm{new}}$ is a newform with rational coefficients and let E be an eigenform in $\mathcal{E}_2(pq)$. Let ℓ be a prime number and $r > 0$ an integer such that the congruence (16) holds for all $n \geq 0$. Then, one of two conditions holds:*

(1) $\ell^r \in \{2, 3, 4, 5\}$ or
(2) $\ell^r = 7$ and $E = [13]^-[2]^+E_2$.

Our numerical results discussed in Sect. 7 lead to certain further speculations about the upper bound for the congruences of type (1).

Question 1.1 *Suppose that the congruence* (1) *holds for a prime ideal λ above a rational prime ℓ and an exponent $r > 0$. Let e denote the order $ord_{\lambda}(\ell)$ and suppose that $e > 1$. For $k = 2$ and level N prime, we checked that $r \leq e$ for every prime $\ell > 3$ for $N \leq 13009$. In a similar fashion, for any square-free integer N and a weight k described in Table 3, the same conclusion holds except for two counterexamples* *found only in weight 2, described in Table 8. Based on those observations, we ask the following two questions:*

- *Assume that $e > 1$, the level N is prime and weight $k = 2$. Is it true that for every prime number $\ell > 3$ a congruence of type* (1) *satisfies the condition $r \leq e$?*
- *Assume that $e > 1$, the level N is square-free and weight $k \geq 2$ is even. Is it true that for every fixed prime number ℓ there are only finitely many congruences of type* (1) *which satisfy the condition $r > e$?*

The questions above are discussed in detail in Sect. 7.

Summary of the Paper

In Sect. 2, we describe a standard basis for the Eisenstein subspace of modular forms of square-free level that consists of the eigenforms. The material in this section is rather classical but we did not find a convenient reference which contained all the necessary results. In Sect. 3, we prove Theorem 3.2 using the results of Atkin–Lehner from [1] as our main tool. In Sect. 4, we study the congruence between cuspidal eigenforms of weight 2 and with rational coefficients and Eisenstein eigenforms. A result of Katz [13] and the theorem of Mazur [16] allow us to conclude that there are only finitely many prime powers for which the congruences exist. A refined statement is proved in Theorem 4.2. In Sects. 5, 6, and 7, we described an improved version of the algorithm that finds congruences of desired shape for fixed levels and weights (cf. [17]). We then discuss the numerical results contained in the attached tables and formulate some of them as corollaries from the computations. A complete database of congruences is available on request.

Notation

Let B_k denote the kth Bernoulli with $B_1 = -\frac{1}{2}$ and let $\sigma_{k-1}(n) = \sum_{m|n} m^{k-1}$ denote the divisor function for any integer $k \geq 2$. Let $E_k = -B_k/(2k) + \sum_{n=1}^{\infty} \sigma_{k-1}(n) q^n$ denote the Eisenstein series of weight k and level 1, where $q = \exp(2\pi i \tau)$ for τ in the upper half-plane $\mathscr{H}$. Let $\Gamma_0(N) = \left\{ \begin{pmatrix} a & b \\ c & d \end{pmatrix} \in \mathrm{SL}_2(\mathbb{Z}) : N \mid c \right\}$ denote the Hecke congruence subgroup of level N and $\mathscr{M}_k(N)$ the space of modular forms of weight k and level N with respect to the group $\Gamma_0(N)$. Let $\mathscr{S}_k(N)$, $\mathscr{E}_k(N)$, and $\mathscr{S}_k(N)^{\mathrm{new}}$ denote, respectively, the subspace of cusp forms, the subspace of Eisenstein series, and the subspace of newforms in $\mathscr{M}_k(N)$. On $\mathscr{M}_k(N)$, we have the action of the Hecke algebra $\mathbb{T}_N$, where T_p is the Hecke operator with index p, $p \nmid N$ and U_p if $p \mid N$. Let $a_n(f)$ denote the nth Fourier coefficient of f expanded at infinity.

2 Standard Basis of Eisenstein Eigenforms

In this section, we present a convenient basis of Eisenstein eigenforms in $\mathscr{E}_k(N)$ for all $k \geq 2$ with respect to the Hecke algebra $\mathbb{T}_N$. We believe that the presented material is not new; however, due to a lack of complete reference, we present full proofs here. Let us denote by A_d a linear map from $\mathscr{M}_k(N)$ to $\mathscr{M}_k(Nd)$ such that $A_d : f(\tau) \mapsto f(d\tau)$. The operator A_d is just a normalized slash operator $A_d(f) = d^{1-k} f \mid_k \gamma$, where

$$\gamma = \begin{pmatrix} d & 0 \\ 0 & 1 \end{pmatrix}.$$

We quote now a theorem of Atkin–Lehner which will be used at several places.

Lemma 2.1 ([1, Lemma 15]) *Let f be a modular form in $\mathscr{M}_k(N)$. We have the following relation between different Hecke operators acting on f*

$$(T_q \circ U_p)(f) = (U_p \circ T_q)(f) \ \textit{for}\ p \neq q, \tag{2}$$

$$(T_q \circ A_d)(f) = (A_d \circ T_q)(f) \ \textit{for}\ (q, d) = 1, \tag{3}$$

$$(U_q \circ A_d)(f) = (A_d \circ U_q)(f) \ \textit{for}\ (q, d) = 1. \tag{4}$$

For any $k > 2$, the series E_k is an eigenform in $\mathscr{M}_k(1)$ with respect to the Hecke algebra $\mathbb{T}_1$. In particular, for any T_n acting on $\mathscr{M}_k(1)$ for $k > 2$, we have

$$T_n(E_k) = a_n(E_k)E_k = \sigma_{k-1}(n)E_k. \tag{5}$$

We also record three simple identities related to σ_k functions. Let n be a positive integer and p a prime number such that $p \mid n$. For any $k \geq 2$, we have

$$\sigma_{k-1}(np) + p^{k-1}\sigma_{k-1}(n/p) = \sigma_{k-1}(p)\sigma_{k-1}(n), \tag{6}$$

$$\sigma_{k-1}(n) - p^{k-1}\sigma_{k-1}(n/p) = \sigma_{k-1}(np) - p^{k-1}\sigma_{k-1}(n), \tag{7}$$

$$\sigma_{k-1}(np) - \sigma_{k-1}(n) = p^{k-1}(\sigma_{k-1}(n) - \sigma_{k-1}(n/p)). \tag{8}$$

For a fixed positive integer d, we define two additional linear operators

$$[d]^+ := T_1 - d^{k-1}A_d : \mathscr{M}_k(N) \to \mathscr{M}_k(Nd),$$
$$[d]^- := T_1 - A_d : \mathscr{M}_k(N) \to \mathscr{M}_k(Nd),$$

where T_1 is the natural inclusion of $\mathscr{M}_k(N)$ into $\mathscr{M}_k(Nd)$.

Proposition 2.2 *If d, e are two positive integers and $\delta, \varepsilon \in \{+, -\}$, then operators $[d]^\delta$ and $[e]^\varepsilon$ commute.*

Proof By definition, the operators A_d and A_e commute, so the proposition follows. □

We compute the action of U_p and $[p]^{\pm}$ on E_k explicitly. We adopt the convention that $\sigma_{k-1}(r) = 0$ for any $r \in \mathbb{Q} \setminus \mathbb{Z}$.

Lemma 2.3 *Let $k > 2$ and p be a prime number. We have equalities*

$$U_p([p]^+ E_k) = [p]^+ E_k, \tag{9}$$

$$U_p([p]^- E_k) = p^{k-1}[p]^- E_k. \tag{10}$$

Proof Fix the integer $k > 2$ and a prime p. We denote by F, the form $[p]^+ E_k$, it lies in $\mathscr{M}_k(p)$. The nth Fourier coefficient of $U_p F$ is as follows:

$$a_n(U_p F) = a_{np}(F) = a_{np}(E_k - p^{k-1} A_p E_k) = a_{np}(E_k) - p^{k-1} a_n(E_k).$$

From the definition of the series E_k, we finally obtain

$$a_n(U_p F) = \sigma_{k-1}(np) - p^{k-1}\sigma_{k-1}(n).$$

On the other hand, the nth Fourier coefficient of F is equal to

$$\sigma_{k-1}(n) - p^{k-1}\sigma_{k-1}(n/p).$$

An application of the identity (7) shows that $U_p F = F$.

A similar reasoning combined with Eq. (8) proves the second statement of the lemma. □

For a square-free level N, we can now express the action of the Hecke algebra on a specific Eisenstein eigenform.

Lemma 2.4 *Let $k > 2$. Fix a positive integer t and a nonnegative integer $t \geq r$ and distinct prime numbers $p_1, \ldots, p_t$. Let N be a product of those primes. The form*

$$E = [p_1]^+ \circ \ldots \circ [p_r]^+ \circ [p_{r+1}]^- \circ \ldots \circ [p_t]^- E_k \in \mathscr{E}_k(\Gamma_0(N))$$

is an eigenform with respect to $\mathbb{T}_N$. Explicitly, the generators act as follows:

$$\begin{aligned} T_n E &= \sigma_{k-1}(n) E, && (n, N) = 1 \\ U_{p_i} E &= E, && 1 \leq i \leq r \\ U_{p_i} E &= p_i^{k-1} E, && r+1 \leq i \leq t \end{aligned}$$

Proof Let ℓ be a prime number not dividing N. Equality (3) and the definitions of $[p]^+$ and $[p]^-$ imply that operators T_ℓ and $[p_i]^{\pm}$ commute for any i in the range $\{1, \ldots, t\}$ and for any choice of the sign $\pm$. It follows that

$$T_\ell E = [p_1]^+ \circ \ldots \circ [p_r]^+ \circ [p_{r+1}]^- \circ \ldots \circ [p_t]^- (T_l E_k).$$

Equality (5) implies that $T_\ell E = \sigma_{k-1}(\ell)E$. The operator T_{ℓ^s} for a fixed $s > 1$ equals $P(T_\ell)$ for a specific choice of $P \in \mathbb{Z}[x]$, so $T_{\ell^s}E = P(\sigma_{k-1}(\ell))E$. The polynomial P is determined by the recurrence relation

$$T_{\ell^s} = T_\ell T_{\ell^{s-1}} - \ell^{s-1}T_{\ell^{s-2}}.$$

If we put $n = \ell^{s-1}$ in Eq. (6), the equation $P(\sigma_{k-1}(\ell)) = \sigma_{k-1}(\ell^s)$ follows, so $T_{\ell^s}E = \sigma_{k-1}(\ell^s)E$. For a given n coprime to N, the equation $T_n E = \sigma_{k-1}(n)E$ follows now from the definition of T_n and the fact that σ_{k-1} is a multiplicative function.

Let i be a fixed number in the set $\{1, \ldots, r\}$. Equation (4) implies that $U_{p_j} \circ [p_i]^+ = [p_i]^+ \circ U_{p_j}$ and $U_{p_j} \circ [p_i]^- = [p_i]^- \circ U_{p_j}$ for any $j \neq i$. Proposition 2.2 implies that the form E can be written as

$$E = [p_1]^+ \circ \ldots \circ [p_{i-1}]^+ \circ [p_{i+1}]^+ \circ \ldots \circ [p_r]^+ \circ [p_{r+1}]^- \circ \ldots \circ [p_t]^- \circ [p_i]^+ E_k.$$

and U_{p_i} acts on E in the following way:

$$U_{p_i} E = [p_1]^+ \circ \ldots \circ [p_{i-1}]^+ \circ [p_{i+1}]^+ \circ \ldots \circ [p_r]^+ \circ [p_{r+1}]^- \circ \ldots \circ [p_t]^- \circ U_{p_i}[p_i]^+ E_k.$$

Equation $U_{p_i}E = E$ is a direct consequence of Eq. (9). For $i > r$, we proceed in a similar way to show $U_{p_i}E = p_i^{k-1}E$. The Hecke algebra $\mathbb{T}_N$ is generated by operators T_n for $(n, N) = 1$ and U_{p_i} for $1 \leq i \leq t$, so the above argument shows that E is an eigenform with respect to T_N. □

We now construct a basis of eigenforms for $k > 2$ and N square-free. If $N = N^- N^+$ is a decomposition into two possibly trivial factors, we define

$$E^{(k)}_{N^-,N^+} = [q_1]^{\varepsilon_1} \circ \ldots [q_t]^{\varepsilon_t} E_k \tag{11}$$

where t is the number of prime factors of N and $q_1, \ldots, q_t$ are the prime factors of N. For i in $\{1, \ldots, t\}$ we define

$$\varepsilon_i = \begin{cases} +, & \text{if } q_i | N^+, \\ -, & \text{if } q_i | N^-. \end{cases}$$

For $N = 1$, we have only one form $E^{(k)}_{1,1} = E_k$. We will often drop the upper index in $E^{(k)}_{N^-,N^+}$ and write E_{N^-,N^+}, if it is clear from the context that the weight equals k.

Theorem 2.5 *Let $k > 2$ and let N be a square-free integer. The set*

$$B := \{E^{(k)}_{N^-,N^+} : N = N^- N^+\}$$

forms a $\mathbb{C}$-basis of the vector space $\mathscr{E}_k(N)$. Each element of this basis is an eigenform with respect to the Hecke algebra $\mathbb{T}_N$. The cardinality of the basis is 2^t where t is the number of prime factors of N.

Proof Forms from the set B are linearly independent because they have different sets of eigenvalues with respect to the Hecke algebra $\mathbb{T}_N$, cf. Lemma 2.4. Let $d(N)$ denote the number of divisors of N. We can choose N^- from $d(N)$ possible divisors of N, the factor N^+ is determined by this choice. Hence, the cardinality of B equals $d(N) = 2^t$. But from [8, Theorem 3.5.1] and [8, p.103], we know that the dimension of the space $\mathcal{E}_k(N)$ equals 2^t, so B is a basis of this space. □

Corollary 2.6 *Let $k > 2$ and let N be a square-free integer with prime factors $p_1, \ldots, p_t$. Choose a form $E_{N^-,N^+} \in \mathcal{E}_k(N)$ which is an eigenform. Let $a_0(E_{N^-,N^+})$ denote the initial coefficient of the q-expansion of E at infinity. Then*

$$a_0(E_{N^-,N^+}) = -\tfrac{B_k}{2k} \prod_{i=1}^{t} (1 - p_i^{k-1}), \quad \text{if } N^- = 1$$
$$a_0(E_{N^-,N^+}) = 0, \quad \text{if } N^- > 1$$

Proof Observe that for any form f and prime p, we have $a_0([p]^- f) = 0$. The operators $[\cdot]^-$ and $[\cdot]^+$ commute, so when $N^- > 1$, we can write E_{N^-,N^+} as $[p]^- f$, where p is prime and f is a form in $\mathcal{E}_k(N/p)$, and hence $a_0(E_{N^-,N^+}) = 0$. Now, for any form f and prime p, we obtain

$$a_0([p]^+ f) = a_0(f)(1 - p^{k-1}). \tag{12}$$

So if $N^- = 1$, we obtain

$$a_0(E_{N^-,N^+}) = -\frac{B_k}{2k} \prod_{i=1}^{t} (1 - p_i^{k-1})$$

if we apply successively equation (12) to each factor of N. Finally, we recall that $a_0(E_k) = -\frac{B_k}{2k}$. □

In weight $k = 2$, the series E_2 does not define a modular form in $\mathcal{M}_2(1)$, so in order to find the basis of eigenforms in $\mathcal{E}_2(N)$, we need to modify the argument above. It is well known that for a prime p, the form $[p]^+ E_2$ is a modular form in $\mathcal{E}_2(p)$.

Lemma 2.7 *Let p be a prime number. The form $[p]^+ E_2 \in \mathcal{E}_2(p)$ is an eigenform with respect to the Hecke algebra $\mathbb{T}_p$. The Fourier coefficient $a_1([p]^+ E_2)$ is 1 and for a prime $q \neq p$, the qth Fourier coefficient of $[p]^+ E_2$ is $q + 1$. The following identities hold:*

$$U_p([p]^+ E_2) = [p]^+ E_2,$$
$$T_n([p]^+ E_2) = a_n(E_2)[p]^+ E_2, \quad \text{for } (n, Np) = 1.$$

Proof Let $\ell \neq p$ be a prime number. For a fixed integer n, we obtain

$$a_n(T_\ell([p]^+ E_2)) = \sigma_1(n\ell) - p\sigma_1(n\ell/p) + \ell\sigma_1(n/\ell) - \ell p\sigma_1(n/(\ell p)).$$

On the other hand, we know that

$$(1+\ell)a_n([p^+]E_2) = (1+\ell)(\sigma_1(n) - p\sigma_1(n/p)).$$

We now apply Eq. (6) to obtain

$$a_n(T_\ell([p]^+E_2)) = (1+\ell)a_n([p^+]E_2).$$

It is easy to see that $a_n(U_p[p]^+E_2) = a_{np}([p]^+E_2)$ and

$$a_{np}([p]^+E_2) = \sigma_1(np) - p\sigma_1(n) = \sigma_1(n) - p\sigma_1(n/p) = a_n([p]^+E_2).$$

The third equation is a consequence of Eq. (7). Hence, the form $[p]^+E_2$ is an eigenform with respect to U_p and any T_ℓ for $\ell \neq p$, so it is an eigenform with respect to $\mathbb{T}_p$. The second equation from the statement of the lemma follows from the definition of T_n and from the multiplicativity of σ_1. From the definition, we also obtain that $a_1([p]^+E_2) = a_1(E_2) = \sigma_1(1) = 1$ and also $a_q([p]^+E_2) = a_q(E_2) = \sigma_1(q) = 1+q$ for any prime $q \neq p$. □

Lemma 2.8 *Let $N > 1$ be a square-free integer. Suppose $f \in \mathcal{E}_2(N)$ is an eigenform with respect to $\mathbb{T}_N$ such that $a_1(f) = 1$ and $a_q(f) = 1+q$ for any prime $q \nmid N$. For a fixed prime $p \nmid N$, the forms $[p]^+f$ and $[p]^-f \in \mathcal{M}_2(Np)$ are eigenforms with respect to $\mathbb{T}_{Np}$. The following identities hold:*

$$\begin{aligned}
U_p([p]^+f) &= [p]^+f,\\
U_p([p]^-f) &= p[p]^-f,\\
T_n([p]^+f) &= a_n(f)[p]^+f, \quad \textit{for } (n, Np) = 1,\\
T_n([p]^-f) &= a_n(f)[p]^-f, \quad \textit{for } (n, Np) = 1.
\end{aligned}$$

Moreover, $a_1([p]^\pm f) = 1$ and $a_q([p]^\pm f) = 1+q$ for any prime $q \nmid Np$.

Proof Let ℓ be a prime not dividing Np. Formula (3) implies that

$$(T_\ell \circ [p]^+)f = ([p]^+ \circ T_\ell)f.$$

The form f is normalized so $T_\ell f = a_\ell(f)f$ and it follows that

$$T_\ell([p]^+f) = a_\ell(f)[p]^+f.$$

In a similar way, we show that $T_\ell([p]^-f) = a_\ell(f)[p]^-f$. From the multiplicativity of σ_1, definition of E_2 and of T_n for $(n, Np) = 1$, we obtain the third and fourth equations from the statement of this lemma.

The equality $U_p([p]^-f) = p \cdot [p]^-f$ is equivalent to

$$a_{np}(f) - a_n(f) = p(a_n(f) - a_{n/p}(f)). \tag{13}$$

If f is a normalized eigenform, then $a_{np}(f) = a_n(f)a_p(f)$ for $p \nmid n$. Since $p \nmid N$, we obtain $a_p(f) = 1 + p$ and Eq. (13) holds. In the case, $n = n'p^\alpha$ for $\alpha > 0$ Eq. (13) is equivalent to

$$a_{p^{\alpha+1}}(f) = (p+1)a_{p^\alpha}(f) - pa_{p^{\alpha-1}}(f).$$

This clearly holds because f is an eigenform for $\mathbb{T}_N$ and we, therefore, have the recurrence relation $T_{p^{\alpha+1}} = T_p T_{p^\alpha} - pT_{p^{\alpha-1}}$ and $a_p(f) = 1 + p$. It is possible to show that $U_p([p]^+ f) = [p]^+ f$ in a similar fashion. Finally, the equalities $a_1([p]^\pm f) = 1$ and $a_q([p]^\pm f) = 1 + q$ follow from the assumptions made on f and from definitions of $[p]^\pm$. □

For $k = 2$, we can adopt the notation $E^{(k)}_{N^-,N^+}$ from Eq. (11) with one small exception: we require that $N^+ > 1$.

Theorem 2.9 *Let $N > 1$ be a square-free integer. The set*

$$B := \{E^{(2)}_{N^-,N^+} : N = N^- N^+, N^+ > 1\}$$

forms a $\mathbb{C}$-basis of the vector space $\mathscr{E}_2(N)$. Each element of this basis is an eigenform with respect to the Hecke algebra $\mathbb{T}_N$. The cardinality of the basis is $2^t - 1$ where t is the number of prime factors of N.

Proof We virtually repeat the proof of Theorem 2.5 replacing Lemma 2.4 with Lemma 2.8. The set B has one less element in this case and we compare it with [8, Theorem 3.5.1] to prove that B is a basis of the $\mathscr{E}_2(N)$. □

Remark 2.10 Theorem 2.9 is proved in [22, Sect. 2] in another way and the proof requires additional tools which are not necessary in our proof.

Corollary 2.11 *Let N be a square-free integer with prime factors $p_1, \ldots, p_t$. Choose a form $E = E_{N^-,N^+} \in \mathscr{E}_2(N)$ which is an eigenform. Then*

$$a_0(E_{N^-,N^+}) = -\tfrac{B_2}{4}\prod_{i=1}^{t}(1-p_i), \quad \text{if } N^- = 1$$

$$a_0(E_{N^-,N^+}) = 0, \quad \text{if } N^- > 1$$

Proof If $N^- > 1$, then E_{N^-,N^+} is of the form $[p]^- h$ for some $h \in \mathscr{E}_2(N/p)$ and a prime $p \mid N$, hence $a_0(E_{N^-,N^+}) = 0$. For the case $N^- = 1$, we simply use that $a_0([p]^+ h) = a_0(h)(1-p)$. □

3 Upper Bound of Congruences

In this section, we discuss a general upper bound for the exponent of congruences between cuspidal eigenforms and eigenforms in the Eisenstein subspace for square-free levels N and even weights $k \geq 2$. The theorems proved here generalize the results obtained previously in [17].

Lemma 3.1 ([1, Theorem 3]) *Let N be a square-free integer and let $k \geq 2$ be an even integer. If $f \in \mathscr{S}_k(N)^{new}$ is a newform, then for any $p \mid N$, we have*

$$a_p(f) = -\lambda_p p^{k/2-1},$$

where $\lambda_p \in \{\pm 1\}$.

Let K be a number field and $\mathscr{O}_K$ its ring of integers. For an element $\alpha \in \mathscr{O}_K$ and a maximal ideal $\lambda \subset \mathscr{O}_K$, let $\mathrm{ord}_\lambda(\alpha)$ denote the integer that satisfies the condition

$$n \leq \mathrm{ord}_\lambda(\alpha) \iff \lambda^n \mid \alpha\mathscr{O}_K.$$

We can naturally extend ord_λ to a function on $K^\times$. For a prime $\ell \in \mathbb{Z}$, let v_ℓ denote the standard ℓ-adic valuation on $\mathbb{Q}^\times$. For any $a \in \mathbb{Q}^\times$, we have $\mathrm{ord}_\lambda(a) = \mathrm{ord}_\lambda(\ell)v_\ell(a)$ where ℓ is the field characteristic of $\mathscr{O}_K/\lambda$.

Let K_f denote the field of coefficients of the newform $f \in \mathscr{S}_k(N)^{\mathrm{new}}$ and by $\mathscr{O}_f$ its ring of integers.

Let $f, g \in \mathscr{M}_k(N)$ be two eigenforms and K be a field that contains the composite of K_f and K_g. We say that f and g are congruent modulo a power λ^r of a maximal ideal $\lambda \in \mathscr{O}_K$ if and only if

$$a_n(f) \equiv a_n(E) \pmod{\lambda^r}$$

for all $n \geq 0$, where $\{a_n(f)\}$ and $\{a_n(g)\}$ are Fourier coefficient of the q-expansion at infinity of f and g, respectively.

Theorem 3.2 *Let $p_1, \ldots, p_t$ be different prime factors of N and let $k \geq 2$. Suppose that $f \in \mathscr{S}_k(N)^{\mathrm{new}}$ is a newform which is congruent to the Eisenstein eigenform $E = [p_1]^+ \circ \ldots \circ [p_t]^+ E_k \in \mathscr{E}_k(N)$ modulo a power $r > 0$ of a maximal ideal $\lambda \subset \mathscr{O}_f$. If ℓ is the residual characteristic of λ, we obtain the bound*

$$r \leq ord_\lambda(\ell) \cdot v_\ell\left(-\frac{B_k}{2k}\prod_{i=1}^{t}(1-p_i)\right).$$

Proof Let $p \mid N$ be a prime. From Lemma 3.1, we know that $a_p(f) = -\lambda_p p^{k/2-1}$. On the other hand, $a_p(E) = a_1(U_p E)$ and from Lemma 2.4, it follows that $a_p(E) = 1$. The congruence

$$a_p(f) \equiv a_p(E) \pmod{\lambda^r}$$

implies that $-\lambda_p p^{k/2-1} \equiv 1 \pmod{\lambda^r}$ and by squaring both sides, we obtain an equation

$$1 - p^{k-2} \equiv 0 \pmod{\lambda^r}. \tag{14}$$

Since f is a cusp form, $a_0(E) \equiv a_0(f) = 0 \pmod{\lambda^r}$ holds and by Corollary 2.6, we obtain

$$-\frac{B_k}{2k}\prod_{i=1}^{t}(1-p_i^{k-1}) \equiv 0 \pmod{\lambda^r}.$$

We observe that $1-p_i^{k-1} = (1-p_i^{k-2}) + p_i^{k-2}(1-p_i)$. Equation (14) holds for each p_i under the assumption $\ell \nmid N$. It follows that

$$-\frac{B_k}{2k}\prod_{i=1}^{t}(1-p_i) \equiv 0 \pmod{\lambda^r}.$$

For $k \geq 2$, we have the inequality $ord_\lambda(1-p_i^{k-1}) \geq ord_\lambda(1-p_i)$ for each i, and hence

$$r \leq ord_\lambda\left(-\frac{B_k}{2k}\prod_{i=1}^{t}(1-p_i)\right).$$

□

Corollary 3.3 *Let $p_1, \ldots, p_t$ be different prime factors of N and let $k \geq 2$. Suppose $f \in \mathscr{S}_k(N)^{\text{new}}$ is a newform which is congruent to the Eisenstein eigenform $E = [p_1]^{\varepsilon_1} \circ \ldots \circ [p_t]^{\varepsilon_t} E_k \in \mathscr{E}_k(N)$ modulo a power λ^r of a maximal ideal $\lambda \subset \mathscr{O}_f$. If we assume that $a_0(E) = 0$ and $p_i \notin \lambda$ for every $\varepsilon_i = -$, then we have the following bound for the congruence exponent:*

$$r \leq \min\{\min_{i,\varepsilon_i=+} ord_\lambda(1-p_i^{k-2}), \min_{i,\varepsilon_i=-} ord_\lambda(1-p_i^{k})\}.$$

Moreover, for every i such that $\varepsilon_i = +$, we have $p_i \notin \lambda$.

Proof We apply Lemma 3.1 to the congruence $a_{p_i}(f) \equiv a_{p_i}(E) \pmod{\lambda^r}$. After squaring both sides, we obtain the condition

$$p_i^{k-2} \equiv \begin{cases} 1, & \text{for } \varepsilon_i = +, \\ p_i^{2(k-1)}, & \text{for } \varepsilon_i = -. \end{cases} \tag{15}$$

The exponent r is less than or equal to $ord_\lambda(1-p_i^{k-2})$ when $\varepsilon_i = +$. Also, r is at most equal to $ord_\lambda(1-p_i^{k})$ when $\varepsilon_i = -$, because $p_i \notin \lambda$ by assumption. For each i such that $\varepsilon_i = +$, it follows from the congruence (15) that $1-p_i^{k-2} \in \lambda^r$. So, $1-p_i^{k-2} \in \lambda$ and then $p_i \notin \lambda$. □

4 Rational Congruences

We have proved in [17, Sect. 5.8] that for a prime N and a newform $f \in \mathscr{S}_2(\Gamma_0(N))^{\text{new}}$ with rational coefficients there exists a system of congruences

$$a_n(f) \equiv a_n(E) \pmod{\ell^r} \tag{16}$$

for all $n \geq 0$, $E = [N]^+E_2$ and a rational prime ℓ only for triples $(\ell, r, N) \in \{(3, 1, 19), (3, 1, 37), (5, 1, 11), (2, 1, 17)\}$ (only finitely many systems) and also for $(\ell, r, N) \in \{(2, 1, u^2 + 64) : 2 \nmid u\}$ (conjecturally infinitely many triples).

Lemma 4.1 *Let f be a newform $f \in \mathscr{S}_2(\Gamma_0(N))^{\text{new}}$ with rational coefficients and N a square-free number. Suppose we have an eigenform $E \in \mathscr{E}_2(N)$ and the congruence (16) holds for all $n \geq 0$, then*

$$(\ell, r) \in \{(2, 1), (2, 2), (2, 3), (3, 1), (3, 2), (5, 1), (7, 1)\}.$$

Proof We know that the Fourier coefficients of f at infinity are integers [8, Theorem 6.5.1] and for every prime $q \nmid N$

$$a_q(f) \equiv 1 + q \pmod{\ell^r}. \tag{17}$$

There exists an elliptic curve $\mathscr{E}$ over $\mathbb{Q}$ of conductor N such that for a prime q of good reduction for $\mathscr{E}$, $a_q(f) = q + 1 - |\mathscr{E}(\mathbb{F}_q)|$, [7, Chap. II, Sect. 2.6]. By the theorem of Katz, there exists a $\mathbb{Q}$-isogenous curve $\mathscr{E}'$ such that $\mathscr{E}'(\mathbb{Q})$ contains an ℓ^r-torsion point. By the theorem of Mazur [16], it follows that $\ell^r \in \{2, 3, 4, 5, 7, 8, 9\}$. □

Elliptic curves with conductor N a product of two primes were partially classified in [19]. This result allows us to discard the congruences with $\ell^r \in \{8, 9\}$.

Theorem 4.2 *Let p,q be two different primes. Suppose that $f \in \mathscr{S}_2(pq)^{\text{new}}$ is a newform with rational coefficients and let E be an eigenform in $\mathscr{E}_2(pq)$. Let ℓ be a prime number and $r > 0$ an integer such that the congruence (16) holds for all $n \geq 0$. Then, one of two conditions holds*

(1) $\ell^r \in \{2, 3, 4, 5\}$ or
(2) $\ell^r = 7$ and $E = [13]^-[2]^+E_2$.

Proof Let $N = pq$ be odd. Then, $a_2(f) \equiv 3 \pmod{\ell^r}$. From the Hasse–Weil bound, it follows that $|a_2(f)| \leq 2\sqrt{2} < 3$. Hence, $\ell^r \mid (3 - a_2(f)) < 6$ so we conclude (1). When $N = pq$ is even and $N = 6$, then the set of cusp forms is empty. So, we can assume that $p = 2$ and $q > 3$. Then, the inequality $|a_3(f)| \leq 2\sqrt{3} < 4$ and the congruence $a_3(f) \equiv 4 \pmod{\ell^r}$ holds, and hence $\ell^r \mid (4 - a_3(f)) < 8$. For $\ell^r = 7$ by [19, Theorem 3.6], it follows that $N = 26$. We compute that the space $\mathscr{S}_2(26)^{\text{new}}$ is of dimension 2 and spanned by the forms f_1, f_2 with the following Fourier expansions:

$$f_1 = q - q^2 + q^3 + q^4 - 3q^5 - q^6 - q^7 - q^8 - 2q^9 + 3q^{10} + 6q^{11} + \ldots$$
$$f_2 = q + q^2 - 3q^3 + q^4 - q^5 - 3q^6 + q^7 + q^8 + 6q^9 - q^{10} - 2q^{11} + \ldots$$

The space $\mathscr{E}_2(26)$ has a basis consisting of three eigenforms

$$[2]^-[13]^+E_2, \quad [13]^-[2]^+E_2, \quad [2]^+[13]^+E_2.$$

Table 1 $a_n(f_i) \equiv a_n(E_{N^-,N^+}) \pmod{2^3}$, $n \geq 0$, $f_i \in \mathscr{S}_2(\Gamma_0(N))^{\text{new}}$

	N	N^-	N^+	Form
1	714	17	42	f_9
2	1482	1	1482	f_{12}
3	1482	19	78	f_{12}
4	1554	1	1554	f_{14}
5	1554	37	42	f_{14}

Table 2 $a_n(f_i) \equiv a_n(E_{N^-,N^+}) \pmod{3^2}$, $n \geq 0$, $f_i \in \mathscr{S}_2(\Gamma_0(N))^{\text{new}}$

	N	N^-	N^+	Form
1	102	17	6	f_3
2	210	7	30	f_5
3	690	23	30	f_{11}
4	930	31	30	f_{15}
5	1974	329	6	f_9
6	4074	97	42	f_{12}
7	4074	1	4074	f_{12}
8	4290	1	4290	f_{29}

Lemma 2.8 implies that

$$a_2([2]^-[13]^+E_2) = 2,$$
$$a_2([13]^-[2]^+E_2) = 1,$$
$$a_2([2]^+[13]^+E_2) = 1.$$

The Sturm bound is 7 by Theorem 5.1, so we only have to compare 7 initial coefficients to verify the desired congruence. By a direct computation, we see that f_2 is congruent to $[13]^-[2]^+E_2$ modulo 7. The form f_1 is not congruent to any of the given Eisenstein eigenforms modulo 7. □

Remark 4.3 If N has more than two prime factors, we can find examples of congruences where $\ell^r \in \{8, 9\}$. In Tables 1 and 2, we present such examples. The index notation f_i of the modular forms is described in Sect. 6.1.

5 Algorithmic Search for Congruences

Our main goal in this section is to describe an effective algorithm that allows one to find congruences between cuspidal eigenforms and Eisenstein series for a large class of square-free conductors. Our approach follows [20] and an adaptation of Sturm's algorithm given in [6].

Theorem 5.1 *Let $p_1, \ldots, p_t$ be different prime numbers and $k \geq 2$. Let $N = p_1 \cdot \ldots \cdot p_t$ and f be a newform in $\mathscr{S}_k(N)^{\text{new}}$. We fix a natural number r and a maximal ideal λ in $\mathcal{O}_f$. Let E be an eigenform in $\mathscr{E}_k(N)$. If the congruence*

$$a_n(f) \equiv a_n(E) \bmod \lambda^r \tag{18}$$

holds for all $n \leq k(\prod_i (p_i + 1))/12$, then it holds for all $n \geq 0$.

Proof This is a simple adaptation of [6, Proposition 1]. □

In our algorithm, it will be sufficient to check the condition (18) for indices n that are prime numbers below the Sturm bound $B := k(\prod_i (p_i + 1))/12$.

Corollary 5.2 *With the assumptions as in Theorem 5.1 suppose that for primes $n \leq k(\prod_i (p_i + 1))/12$ the congruence (18) holds, then the congruence (18) holds for all natural numbers $n \geq 0$.*

Proof This follows immediately from Theorem 5.1 since f and E are normalized eigenforms. □

Lemma 5.3 *Let N be a square-free integer which is a product of prime numbers $p_1, \ldots, p_t$ and $k \geq 2$ be an integer. Let $\{\varepsilon_i\}_{i=1,\ldots,t}$ be a collection of symbols $\varepsilon_i \in \{+, -\}$. Let $f \in \mathscr{S}_k(N)^{\text{new}}$ be a newform and $E \in \mathscr{E}_k(N)$ an eigenform $E = [p_1]^{\varepsilon_1} \circ \ldots \circ [p_t]^{\varepsilon_t} E_k \in \mathscr{E}_k(N)$. Suppose that there exists a prime ideal λ in $\mathcal{O}_f$ and a positive integer r for which the congruence $a_n(f) \equiv a_n(E) \pmod{\lambda^r}$ holds for all integers n. Let ℓ denote the characteristic of the field $\mathcal{O}_f/\lambda$. One of the following conditions holds:*

(1) If $k \geq 2$ and $\varepsilon_1 = \ldots = \varepsilon_t = +$, then $\ell \mid -\frac{B_k}{2k} \prod_i (1 - p_i)$.
(2) If $k = 2$ and $\varepsilon_i = -$ for some i, then $\ell \mid GCD(\{1 - p_j^2 : \varepsilon_j = -\})$.
(3) If $k > 2$ and $\varepsilon_i = -$ for some i, then $\ell \mid GCD(\{1 - p_j^k : \varepsilon_j = -\} \cup \{1 - p_j^{k-2} : \varepsilon_j = +\})$.

Proof The lemma follows from the Theorem 3.2 and Corollary 3.3. □

5.1 Algorithm

Description of the algorithm: For a fixed integer $k \geq 2$, a square-free integer N, a prime number ℓ, and a fixed eigenform $E \in \mathscr{E}_k(N)$, the algorithm checks for which newforms $f \in \mathscr{S}_k(N)$ there is a congruence between f and E modulo λ^r, where the characteristic of the ideal λ is ℓ and $r > 0$ is the maximal possible.

Input: An even number $k \geq 2$, a square-free integer $N > 1$, a prime number ℓ, and an eigenform $E \in \mathscr{E}_k(N)$.

Steps of the Algorithm:

(1) Check whether $a_0(E)$ is 0. If yes, then proceed to Step 2. If no, then check if $\nu_\ell(a_0(E)) > 0$. If yes, then go to Step 2. If no, then terminate the algorithm.
(2) Compute subsets C_i of newforms in $\mathscr{S}_k(N)$ such that each two element in C_i are Galois conjugate
(3) For each set C_i pick one representative and create a set $F_{N,k}$ of those representatives for all i.
(4) Compute the Sturm bound $B = (k/12)[\mathrm{SL}_2(\mathbb{Z}) : \Gamma_0(N)]$.
(5) For each form $f \in F_{N,k}$ compute the coefficient field K_f.
(6) For each $f \in F_{N,k}$ create a set $S_{\ell,f}$ that is made of prime ideals that appear in the factorization of $\ell\mathscr{O}_f$.
(7) For each element $f \in F_{N,k}$ and $\lambda \in S_{\ell,f}$ compute the number

$$r_\lambda = \min \left\{ ord_\lambda \left(a_q(f) - a_q(E) \right) \mid q \leq B \right\}.$$

The minimum runs over prime numbers q. If $r_\lambda > 0$, then return a triple (f, λ, r_λ).

Output: Set of triples (f, λ, r) such that

$$a_n(f) \equiv a_n(E) \pmod{\lambda^r}$$

for all $n \geq 0$ and if for some $s > 0$, we have

$$a_n(f) \equiv a_n(E) \pmod{\lambda^s}$$

for all $n \geq 0$, then $s \leq r$. Remark: it might happen that the list will be empty.

Validity of the Algorithm: In Step 1, we check if the congruence (18) is possible. Step 2 amounts to a finite number of computational steps for a fixed level N and weight k by using, for instance, modular symbols. Moreover, we can represent each newform by a finite number of bits (e.g., by using the modular symbols representation). The number r_λ in Step 6 satisfies the output condition because of Corollary 5.2. Since N is square-free, the constant B is equal to the constant from Corollary 5.2.

6 Numerical Data

In this section, we present the computational data that was gathered while running Algorithm 5.1 under the restrictions of Lemma 5.3. We performed a check that includes weights k between 2 and 24 and square-free levels N up to 4559. More precise bounds are presented in Table 3. Our main computational resource was the cluster Gauss at the University of Luxembourg maintained by Prof. Gabor Wiese. This computer has 20 CPU units of type Inter(R) Xeon(R) CPU E7-4850 @ 2.00 GHz

Table 3 Weight k and corresponding maximal level N

k	2	4	6	8	10	12	14	16	18	20	22	24
$N \leq$	4559	922	302	202	193	102	94	94	94	94	94	94

and approximately 200 GB of RAM memory. We used the computer algebra package MAGMA [5] and the set of instructions MONTES [11] which greatly enhances the efficiency of computations performed on number fields with large discriminants. However, it took about 4 weeks of the computational time under full CPU load of the Gauss cluster (around 13440 CPU hours) to finish the calculations.

6.1 Description of Data in the Tables

Let f_i be as usual a newform in $\mathcal{S}_k(N)^{\mathrm{new}}$ where $k \geq 2$ and N is square-free. The index i is associated to the particular form by the algorithm presented in [7, Chap. IV], described in detail in MAGMA manual.[1] The number d will denote the degree of the extension K_f over $\mathbb{Q}$. Let $\lambda \subset \mathcal{O}_{f_i}$ be a prime ideal with residue characteristic ℓ. Let e denote the ramification degree $\mathrm{ord}_\lambda(\ell)$ and f the degree of the residue field extension $[\mathcal{O}_{f_i}/\lambda : \mathbb{F}_\ell]$. We consider the Eisenstein eigenform $E_{N^-,N^+} \in \mathcal{E}_k(N)$ with $N = N^-N^+$ such that

$$a_n(f_i) \equiv a_n(E_{N^-,N^+}) \pmod{\lambda^r} \tag{19}$$

for all $n \geq 0$. Assume that the positive integer r is maximal, i.e., there is no congruence of type (19) with ideal exponent r' greater than r. The number m will denote the maximum over s which satisfy simultaneous congruences

$$a_{p_j}(f_i) \equiv a_{p_j}(E_{N^-,N^+}) \pmod{\ell^s}, \quad 1 \leq j \leq t,$$

$$a_0(f_i) \equiv a_0(E_{N^-,N^+}) \pmod{\ell^s}.$$

Observe that m depends on the choice of N, N^+, N^-, f_i, and λ. An upper bound for the exponent r is the product $m \cdot e$. In general, the bound $m \cdot e$ might be smaller than the upper bound computed in Theorem 3.2 and Corollary 3.3. We also use specific labels to indicate different prime ideals λ that occur in the factorization of $\ell\mathcal{O}_{f_i}$. These labels are described in the MONTES package documentation.[2] Hence, in the column labeled by λ, we use the notation λ_j to denote a specific prime ideal with respect to the MONTES labeling. Similarly, in the column "form", we let f_i denote a specific newform that will appear.

[1] http://magma.maths.usyd.edu.au/magma/handbook/text/1545.

[2] http://www-ma4.upc.edu/~guardia/MontesAlgorithm.html.

Table 4 Typical row of data

N	N^-	N^+	k	ℓ	m	Form	λ	r	e	f	d
2651	1	2651	2	5	2	f_1	λ_1	2	2	1	35

Table 5 Congruences that satisfy $r > 2$ and $m > 1$, one for each pair (r, ℓ)

	N	N^-	N^+	k	ℓ	m	Form	λ	r	e	f	d
1	2	2	1	22	2	10	f_1	λ_1	8	1	1	1
2	2159	127	17	2	2	7	f_1	λ_1	7	1	1	56
3	78	78	1	8	2	3	f_1	λ_1	6	2	1	2
4	34	2	17	10	2	4	f_1	λ_1	5	2	1	2
5	1459	1	1459	2	3	5	f_1	λ_1	5	1	1	71
6	94	2	47	18	2	7	f_1	λ_2	4	1	1	18
7	146	2	73	6	3	2	f_1	λ_3	4	2	1	9
8	78	2	39	22	2	3	f_1	λ_4	3	1	1	5
9	163	1	163	10	3	4	f_1	λ_1	3	1	1	62
10	443	443	1	4	5	4	f_1	λ_1	3	1	1	60
11	1373	1	1373	2	7	3	f_1	λ_1	3	1	1	60
12	2663	1	2663	2	11	3	f_1	λ_2	3	1	1	132
13	239	239	1	4	13	4	f_1	λ_1	3	1	1	37

Example 6.1 In Table 4, we describe an example of a typical row of data in our congruence database. We read from it that a newform $f_1 \in \mathscr{S}_2(2651)^{\text{new}}$ is congruent to the Eisenstein series $E_{1,2651}$ modulo a power λ_1^2, where the ideal λ_1 is of residue characteristic 5 and its ramification degree e above $\ell = 5$ equals 2. Field degree $[K_{f_1} : \mathbb{Q}]$ is 35 and $\mathscr{O}_{f_1}/\lambda_1 = \mathbb{F}_5$. Theoretical upper bound for r is $m \cdot e = 4$ but our congruence appears only with the maximal exponent $r = 2$.

Example 6.2 In Table 5, we present for each pair (r, ℓ) one congruence for which r is maximal in the whole range described in Table 3. In case, there were more than one suitable pair (r, ℓ), we chose a specific pair and some k. Moreover, we sort the data by descending value of r.

Example 6.3 In Table 6, we describe some examples of congruences that satisfy the nontrivial bound $r \leq e$ with $m > 1$. We refer to Corollary 7.2, for a precise statement of our observation.

Table 6 Exemplary congruences that satisfy conditions: $e > 1, m > 1, \ell > 3$

	N	N^-	N^+	k	ℓ	m	Form	λ	r	e	f	d
1	31	31	1	10	5	2	f_1	λ_3	1	2	1	13
2	33	11	3	12	11	2	f_1	λ_4	1	2	1	6
3	33	11	3	12	11	2	f_1	λ_4	1	2	1	5
4	35	5	7	6	5	2	f_1	λ_1	1	2	1	2
5	35	5	7	6	5	2	f_1	λ_1	1	2	1	4
6	35	35	1	8	5	2	f_1	λ_2	1	2	1	5
7	35	35	1	12	5	2	f_1	λ_3	1	2	1	4
8	35	35	1	12	5	2	f_1	λ_3	1	2	1	6
9	35	5	7	14	5	2	f_1	λ_3	1	2	1	6
10	35	5	7	14	5	2	f_1	λ_3	1	2	1	8
11	35	35	1	16	5	2	f_1	λ_3	2	3	1	7
12	35	35	1	16	5	2	f_1	λ_4	1	2	1	9
13	35	35	1	16	5	2	f_1	λ_3	2	3	1	9
14	55	5	11	12	5	2	f_1	λ_2	1	2	1	11
15	55	5	11	12	5	2	f_1	λ_2	1	2	1	8
16	79	79	1	6	7	2	f_1	λ_1	1	2	1	19
17	79	79	1	12	7	2	f_1	λ_1	1	2	1	33
18	101	101	1	4	5	2	f_1	λ_1	1	3	1	9
19	101	101	1	8	5	2	f_1	λ_2	1	3	1	26
20	101	101	1	12	5	2	f_1	λ_2	1	3	1	42
21	107	107	1	4	5	2	f_1	λ_1	1	2	1	16
22	107	107	1	8	5	2	f_1	λ_1	1	2	1	28
23	133	7	19	8	7	3	f_1	λ_3	1	3	1	16
24	133	7	19	8	7	3	f_1	λ_3	1	3	1	16

7 Summary of Computational Results

In this paragraph, we summarize the large-scale numerical computations that established the existence of congruences for square-free levels N and weights k in the range predicted by Table 3. We will say that *there exists a congruence that satisfies* a condition $\mathscr{W} = \mathscr{W}(r, d, e, f, N^-, N^+, \ell, m)$, if we can find a weight k and level N such that there exists a newform $f \in \mathscr{S}_k(N)^{\text{new}}$ and an Eisenstein eigenform $E \in \mathscr{E}_k(N)$ that satisfy (19) for an ideal $\lambda \in \mathscr{O}_f$ and a positive integer r and such that the values of $r, d, e, f, N^-, N^+, \ell$ and m associated with this congruence satisfy the condition $\mathscr{W}$.

Corollary 7.1 *Let N be a square-free number depending on the weight as described in Table 3. In Table 7, we present the number of different congruences of type (19) that can be found in the presented range. In the column denoted by $r \geq 0$, we count the*

Table 7 Number of congruences of type (19) for fixed values of k

k	$r \geq 0$	$r > 0$	$m \cdot e = r > 0$	$m \cdot e > r > 0$
2	277447	62937	38805	24132
4	64232	13922	9208	4714
6	17300	3629	2475	1154
8	10755	2149	1517	632
10	9248	1483	1106	377
12	5738	1055	787	268
14	5276	1020	756	264
16	6010	1113	817	296
18	6995	1235	922	313
20	10735	1914	1428	486
22	8853	1425	1025	400
24	10359	1555	1153	402

Table 8 Congruences that satisfy $e > 1$, $m > 1$, $\ell > 3$, and $r > e$

	N	N^-	N^+	k	ℓ	m	Form	λ	r	e	f	d
1	2495	499	5	2	5	3	f_1	λ_1	3	2	1	55
2	3998	1999	2	2	5	3	f_1	λ_1	3	2	1	44

number of pairs (f, λ) returned by Algorithm 5.1. In the column "$r > 0$", we count the number of congruences, in the column "$m \cdot e = r > 0$", we count the number of congruences with maximal exponent $r = m \cdot e$ and the last column has a similar meaning.

Corollary 7.2 *For (N, k) from range in Table 3, there exists 96 congruences that satisfy $e > 1$, $m > 1$, and $\ell > 3$. Except for the cases described in Table 8, we have the bound $r \leq e$.*

Remark 7.3 Corollary 7.2 extends similar computations performed in [17] for prime levels N and weight $k = 2$. It was verified there that for primes $N \leq 13009$, the property $r \leq e$ holds for all $\ell > 3$ and $e > 1$. It is an open question, if there are infinitely many such congruences for all possible ranges of N and k.

Corollary 7.4 *Let $k = 2$. For $N \leq 4559$ square-free and for any $d \leq 222$, we found congruences (19) if $d \notin D$, where*

$$D = \{169, 175, 178, 192, 197, 204, 207, 208, 211, 214, 215, 216, 217, 218, 219, 220, 221\}.$$

Table 9 Selected congruences sorted by the degree d

	N	N^-	N^+	k	ℓ	m	Form	λ	r	e	f	d
1	131	1	131	2	13	1	f_1	λ_1	1	1	1	10
2	311	1	311	2	5	1	f_1	λ_1	1	1	1	22
3	479	1	479	2	239	1	f_1	λ_3	1	1	1	32
4	719	1	719	2	359	1	f_1	λ_2	1	1	1	45
5	839	1	839	2	419	1	f_1	λ_1	1	1	1	51
6	1031	1	1031	2	5	1	f_1	λ_1	1	1	1	60
7	1399	1	1399	2	233	1	f_1	λ_2	1	1	1	71
8	1487	1	1487	2	743	1	f_1	λ_1	1	1	1	0
9	1559	1	1559	2	19	1	f_1	λ_2	1	1	1	90
10	1931	1	1931	2	5	1	f_1	λ_1	1	1	1	101
11	2111	1	2111	2	5	1	f_1	λ_2	1	1	1	112
12	2351	1	2351	2	5	2	f_1	λ_1	1	1	1	123
13	2591	1	2591	2	5	1	f_1	λ_2	1	1	1	136
14	2879	1	2879	2	1439	1	f_1	λ_1	1	1	1	148
15	2903	1	2903	2	1451	1	f_1	λ_2	1	1	1	150
16	2999	1	2999	2	1499	1	f_1	λ_1	1	1	1	161
17	3359	1	3359	2	23	1	f_1	λ_1	1	1	1	174
18	3659	1	3659	2	31	1	f_1	λ_1	1	1	1	181
19	3671	1	3671	2	5	1	f_1	λ_1	1	5	1	193
20	3911	1	3911	2	5	1	f_1	λ_1	1	2	1	202
21	4079	1	4079	2	2039	1	f_1	λ_2	1	1	1	212
22	4391	1	4391	2	5	1	f_1	λ_4	1	1	1	222

Remark 7.5 In [9] the authors study, the existence of newforms f with large degree coefficient field K_f. The computations from Corollary 7.4 and Table 9 suggest that we can both find newforms that have large degree of K_f and that are congruent to an Eisenstein eigenform. In Figure 1, we show that the growth of d as a function of least N is roughly a linear function. The way we present data in Table 9 is as follows: we assume $N^- = 1$, in the ith row we present a congruence such that $d \geq 10i$ for the least possible N. All values of N that we found are prime numbers.

Corollary 7.6 *For $k = 2$ and level N less or equal to* 4559, *there exist a congruence for any level except $N = 13$ or 22, for which the space $\mathscr{S}_2(N)^{\text{new}}$ is zero.*

Corollary 7.7 *For $k = 2$ and $N^- > 1$, there exists* 54077 *congruences for levels $N \leq 4559$ and* 8860 *congruences for $N^- = 1$ and levels $N \leq 4559$.*

Remark 7.8 In [21, Theorem 4.1.2], it is assumed that either $N_- = 1$ and the number of prime divisors of N is odd and $\ell \mid \phi(N)$ or the number of prime divisors of N is even and N^- is a prime number such that $N^- \equiv -1 \pmod{\ell}$. In several cases

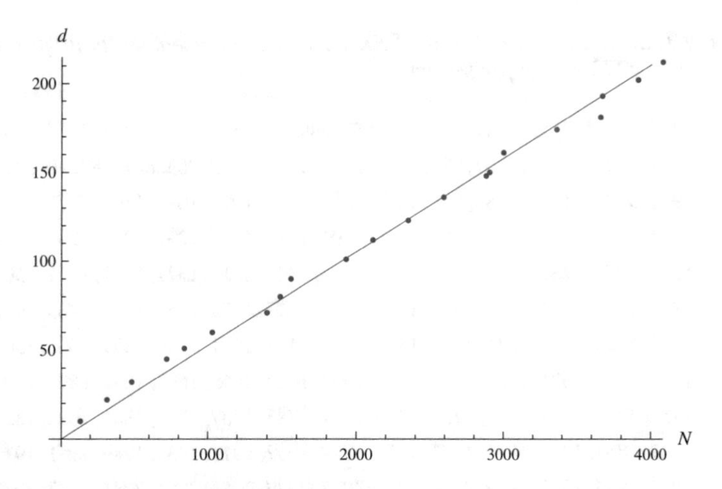

Fig. 1 Growth of degree d as a function of least level N for data from Table 9

Table 10 Weight k and number of congruences that satisfy $f > 2$

k	2	4	6	8	10	12	14	16	18	20	22	24
n.c.	993	177	20	4	0	0	0	2	4	2	0	0

Table 11 Weight k and the number of congruences such that $\ell \mid N$

k	2	4	6	8	10	12	14	16	18	20	22	24
n.c.	27771	4839	1366	1070	609	583	605	708	726	1323	990	1033

described in the above corollary, the assumptions of [21, Theorem 4.1.2] are satisfied. In those cases, we obtain a congruence for the coefficients a_p with $p \mid N$, which is not assumed in [21, Theorem 4.1.2]. Moreover, some examples of the previous corollary suggest that the assumptions of [21, Theorem 4.1.2] can be made weaker.

Corollary 7.9 *Let (N, k) be a pair of integers that fit into the range of Table 3. In Table 10, we present the number of corresponding congruences (abbreviated n.c. in the table) with $f > 2$.*

Corollary 7.10 *For $N \leq 4559$ and $k = 2$, there exist 30 congruences that satisfy $e = 17$ and $\ell = 2$. In that range, there is no congruence such that the ramification exponent e is larger than 17.*

Corollary 7.11 *Let (N, k) be the numbers from the range in Table 3. Then, in the described ranges, there is an appropriate number of congruence (n.c.) that satisfy the condition $\ell \mid N$. Values are presented in Table 11.*

Corollary 7.12 *Let $k = 2$ and $N \leq 4559$. There are congruences for all prime characteristic $\ell \leq 2273$ except for the set*

$$\{353, 389, 457, 463, 523, 541, 569, 571, 587, 599, 613, 617, 631, 643, 647, 677, 701, 733, 757, 769, 773, 787, 797, 821, 823, 827, 839, 857, 859, 863, 881, 887, 907, 929, 941, 947, 971, 977, 983, 991, 1021, 1051, 1061, 1091, 1097, 1109, 1117, 1151, 1153, 1163, 1171, 1181, 1187, 1193, 1201, 1213, 1217, 1231, 1237, 1249, 1259, 1277, 1279, 1283, 1291, 1297, 1301, 1303, 1307, 1319, 1321, 1327, 1361, 1367, 1373, 1381, 1399, 1423, 1427, 1429, 1433, 1447, 1453, 1459, 1471, 1483, 1487, 1489, 1493, 1523, 1531, 1543, 1549, 1553, 1567, 1571, 1579, 1597, 1607, 1609, 1613, 1619, 1621, 1627, 1637, 1657, 1663, 1667, 1669, 1693, 1697, 1699, 1709, 1721, 1723, 1741, 1747, 1753, 1759, 1777, 1783, 1787, 1789, 1801, 1823, 1831, 1847, 1861, 1867, 1871, 1873, 1877, 1879, 1907, 1913, 1933, 1949, 1951, 1979, 1987, 1993, 1997, 1999, 2011, 2017, 2027, 2029, 2053, 2081, 2083, 2087, 2089, 2099, 2111, 2113, 2131, 2137, 2143, 2153, 2161, 2179, 2203, 2207, 2213, 2221, 2237, 2239, 2243, 2251, 2267, 2269\}.$$

Acknowledgements The author would like to thank Wojciech Gajda for his excellent supervision of author's Ph.D. project. He also thanks Xavier Guitart, Kimball Martin, Hwajong Yoo, and Gabor Wiese for reading the earlier version of the manuscript and their valuable comments. Finally, he thanks both anonymous referees for their useful comments and corrections. The author was supported by the Polish National Science Centre research grant 2012/05/N/ST1/02871. This paper is partially based on the results obtained in the author's Ph.D. thesis [18].

References

1. Atkin, A.O.L., Lehner, J.: Hecke operators on $\Gamma_0(m)$. Math. Ann. **185**, 134–160 (1970). https://doi.org/10.1007/BF01359701
2. Berger, T., Klosin, K., Kramer, K.: On higher congruences between automorphic forms. Math. Res. Lett. **21**(1), 71–82 (2014). https://doi.org/10.4310/MRL.2014.v21.n1.a5
3. Billerey, N., Menares, R.: On the modularity of reducible mod l Galois representations. Math. Res. Lett. **23**(1), 15–41 (2016). https://doi.org/10.4310/MRL.2016.v23.n1.a2
4. Billerey, N., Menares, R.: Strong modularity of reducible Galois representations. Trans. Amer. Math. Soc. **370**(2), 967–986 (2018). https://doi.org/10.1090/tran/6979
5. Bosma, W., Cannon, J., Playoust, C.: The Magma algebra system. I. The user language. J. Symbolic Comput. **24**(3-4), 235–265 (1997). Computational algebra and number theory (London, 1993)
6. Chen, I., Kiming, I., Rasmussen, J.B.: On congruences mod p^m between eigenforms and their attached Galois representations. J. Number Theory **130**(3), 608–619 (2010). https://doi.org/10.1016/j.jnt.2009.09.012
7. Cremona, J.: Algorithms for modular elliptic curves, second edn. Cambridge University Press, Cambridge (1997)
8. Diamond, F., Shurman, J.: A first course in modular forms, *Graduate Texts in Mathematics*, vol. 228. Springer-Verlag, New York (2005)

9. Dieulefait, L.V.c., Jiménez Urroz, J., Ribet, K.A.: Modular forms with large coefficient fields via congruences. Res. Number Theory **1**, Art. 2, 14 (2015). https://doi.org/10.1007/s40993-015-0003-9
10. Dummigan, N., Fretwell, D.: Ramanujan-style congruences of local origin. J. Number Theory **143**, 248–261 (2014). https://doi.org/10.1016/j.jnt.2014.04.008
11. Guàrdia, J., Montes, J., Nart, E.: Higher Newton polygons and integral bases. J. Number Theory **147**, 549–589 (2015). https://doi.org/10.1016/j.jnt.2014.07.027
12. Hsu, C.: Higher congruences between newforms and Eisenstein series of squarefree level. ArXiv e-prints (2017). arXiv:1706.05589
13. Katz, N.: Galois properties of torsion points on abelian varieties. Invent. Math. **62**(3), 481–502 (1981)
14. Lecouturier, E.: Higher Eisenstein elements, higher Eichler formulas and rank of Hecke algebras. ArXiv e-prints (2017). arXiv: 1709.09114
15. Martin, K.: The basis problem revisited. ArXiv e-prints (2018). arXiv:1804.04234
16. Mazur, B.: Modular curves and the Eisenstein ideal. Inst. Hautes Études Sci. Publ. Math. (47), 33–186 (1978) (1977)
17. Naskręcki, B.: On higher congruences between cusp forms and Eisenstein series. In: Computations with modular forms, *Contrib. Math. Comput. Sci.*, vol. 6, pp. 257–277. Springer, Cham (2014). https://doi.org/10.1007/978-3-319-03847-6_10
18. Naskręcki, B.: Ranks in families of elliptic curves and modular forms. Adam Mickiewicz University (2014). Ph.D. thesis
19. Sadek, M.: On elliptic curves whose conductor is a product of two prime powers. Math. Comp. **83**(285), 447–460 (2014). https://doi.org/10.1090/S0025-5718-2013-02726-3
20. Sturm, J.: On the congruence of modular forms. In: Number theory (New York, 1984–1985), *Lecture Notes in Math.*, vol. 1240, pp. 275–280. Springer, Berlin (1987). https://doi.org/10.1007/BFb0072985
21. Yoo, H.: Modularity of residually reducible Galois representations and Eisenstein ideals. ProQuest LLC, Ann Arbor, MI (2013). Thesis (Ph.D.)–University of California, Berkeley
22. Yoo, H.: The index of an Eisenstein ideal and multiplicity one. Math. Z. **282**(3-4), 1097–1116 (2016). https://doi.org/10.1007/s00209-015-1579-4

Lucas Numbers Which are Products of Two Balancing Numbers

Zafer Şiar

Abstract In this paper, we find all Lucas numbers, which are products of two balancing numbers.

1 Introduction

Let P and Q be nonzero integers. Generalized Fibonacci and Lucas sequences are defined as follows:

$$U_0(P, Q) = 0, U_1(P, Q) = 1, \text{ and } U_{n+2}(P, Q) = PU_{n+1}(P, Q) + QU_n(P, Q)$$

and

$$V_0(P, Q) = 2, V_1(P, Q) = P, \text{ and } V_{n+2}(P, Q) = PV_{n+1}(P, Q) + QV_n(P, Q)$$

for $n \geq 0$, respectively. $U_n(P, Q)$ and $V_n(P, Q)$ are called n-th generalized Fibonacci numbers and n-th generalized Lucas numbers, respectively. It is well known that

$$U_n = U_n(P, Q) = \frac{a^n - b^n}{a - b} \text{ and } V_n = V_n(P, Q) = a^n + b^n, \tag{1}$$

where $a = \frac{P+\sqrt{P^2+4Q}}{2}$ and $b = \frac{P-\sqrt{P^2+4Q}}{2}$ are the roots of the characteristic equation $x^2 - Px - Q = 0$. If $P = Q = 1$, then we have the Fibonacci and Lucas sequences $(F_n)_{n\geq 0}$ and $(L_n)_{n\geq 0}$, respectively. The terms of the sequence $(U_n(2, 1))_{n\geq 0}$ give us Pell numbers denoted by P_n. If $P = 6$ and $Q = -1$, then the elements of the sequence $(U_n)_{n\geq 0}$ turn out to be balancing numbers and n-th balancing number is denoted by B_n. It can be seen from (1) that

Z. Şiar (✉)
Department of Mathematics, Bingöl University, Bingöl, Turkey
e-mail: zsiar@bingol.edu.tr

I. Inam and E. Büyükasik (eds.), *Notes from the International Autumn School on Computational Number Theory*, Tutorials, Schools, and Workshops in the Mathematical Sciences, https://doi.org/10.1007/978-3-030-12558-5_8

$$L_n = \alpha^n + \beta^n \tag{2}$$

and

$$B_n = \frac{\lambda^n - \delta^n}{4\sqrt{2}} \tag{3}$$

for every $n \geq 0$, where $\alpha = \frac{1+\sqrt{5}}{2}$, $\beta = \frac{1-\sqrt{5}}{2}$ and $\lambda = 3 + 2\sqrt{2}$, $\delta = 3 - 2\sqrt{2}$. Also, it is well known that the relation between n-th Lucas number L_n and α is given by

$$\alpha^{n-1} \leq L_n \leq 2\alpha^n \tag{4}$$

for $n \geq 0$. The inequality (4) can be proved by induction. It is clear that $5 < \lambda < 6$, $0 < \delta < 1$ and $\lambda\delta = 1$. Therefore

$$B_n < \frac{\lambda^n}{4\sqrt{2}}. \tag{5}$$

For more information concerning these sequences, one can consult [2, 7, 8, 11].

In [6], Farrokhi proved that if $r > 2,\ m > 2$, then there is no Fibonacci number F_n such that $F_n = F_r F_m$. In [9], the authors showed that if $r > 1,\ m > 1$ there is no Lucas number L_n such that $L_n = L_r L_m$. In [8], Keskin and Karaatlı have given a similar result for Balancing numbers. Besides, in [10], the authors generalized the results of papers in [6, 8, 9]. Here, they proved that when $P > 1$ and $Q = \pm 1$, there is no generalized Fibonacci number U_n such that $U_n = U_m U_r$ for $1 < r < m$, and there is no generalized Lucas number V_n such that $V_n = V_m V_r$ for $m > 1$ and $r > 1$. Moreover, the authors [6, 8–10] have used only divisibility properties of these sequences and congruences in order to solve these equations. In [4], Ddamulira et al. have found all Fibonacci numbers or Pell numbers which are products of two numbers from the other sequences. More clearly, they showed that all positive integer solutions (k, m, n) of the equations $F_k = P_m P_n$ and $P_k = F_m F_n$ have $k = 1, 2, 5, 12$ and $k = 1, 2, 3, 7$, respectively. Motivated by the studies of Ddamulira et al., in this paper, we determine all solutions of the Diophantine equation

$$L_k = B_n B_m \tag{6}$$

in nonnegative integers n, m, and k with $n \geq m \geq 1$.

This study can be viewed as a continuation of the previous works on this subject. We follow the approach and the method presented in [4]. Namely, we will use Matveev's result [12] like in [4]. This result uses Baker's theory which is based on lower bounds for a nonzero linear form in logarithms of algebraic numbers. In Sect. 2, we introduce necessary lemmas and theorems. Then, in Sect. 3, we prove our main theorem.

2 Auxiliary Results

Recently, many authors have used Baker's theory which is based on lower bounds for a nonzero linear form in logarithms of algebraic numbers to solve Diophantine equations such as Eq. (6). Since such bounds are of crucial importance in effectively solving of Diophantine equations, we start with recalling some basic notions from algebraic number theory.

Let η be an algebraic number of degree d with minimal polynomial

$$a_0x^d + a_1x^{d-1} + ... + a_d = a_0 \prod_{i=1}^{d} \left(X - \eta^{(i)}\right) \in \mathbf{Z}[x],$$

where the a_i's are relatively prime integers with $a_0 > 0$ and $\eta^{(i)}$'s are conjugates of η. Then

$$h(\eta) = \frac{1}{d}\left(\log a_0 + \sum_{i=1}^{d} \log\left(\max\left\{|\eta^{(i)}|, 1\right\}\right)\right) \tag{7}$$

is called the logarithmic height of η. In particular, if $\eta = a/b$ is a rational number with $\gcd(a, b) = 1$ and $b > 1$, then $h(\eta) = \log(\max\{|a|, b\})$.

The following properties of logarithmic height are found in many works stated in references:

$$h(\eta \pm \gamma) \leq h(\eta) + h(\gamma) + \log 2, \tag{8}$$

$$h(\eta\gamma^{\pm 1}) \leq h(\eta) + h(\gamma), \tag{9}$$

$$h(\eta^s) = |s|h(\eta). \tag{10}$$

The following theorem, deduced from Corollary 2.3 of Matveev [12], provides a large upper bound for the subscript n in Eq. (6) (also see Theorem 9.4 in [3]).

Theorem 1 *Assume that $\gamma_1, \gamma_2, ..., \gamma_t$ are positive real algebraic numbers in a real algebraic number field K of degree D, $b_1, b_2, ..., b_t$ are rational integers, and*

$$\Lambda := \gamma_1^{b_1} ... \gamma_t^{b_t} - 1$$

is not zero. Then

$$|\Lambda| > \exp\left(-1.4 \cdot 30^{t+3} \cdot t^{4.5} \cdot D^2(1 + \log D)(1 + \log B)A_1A_2...A_t\right),$$

where

$$B \geq \max\{|b_1|, ..., |b_t|\},$$

and $A_i \geq \max\{Dh(\gamma_i), |\log \gamma_i|, 0.16\}$ for all $i = 1, ..., t$.

The following lemma, proved by Dujella and Pethő [5], is a variation of a lemma of Baker and Davenport [1]. And this lemma will be used to reduce the upper bound for the subscript n in Eq. (6). Let the function $||\cdot||$ denote the distance from x to the nearest integer, that is, $||x|| = \min\{|x-n| : n \in Z\}$ for a real number x. Then, we have

Lemma 1 *Let M be a positive integer, let p/q be a convergent of the continued fraction of the irrational number γ such that $q > 6M$, and let A, B, μ be some real numbers with $A > 0$ and $B > 1$. Let $\varepsilon := ||\mu q|| - M||\gamma q||$. If $\varepsilon > 0$, then there exists no solution to the inequality*

$$0 < |u\gamma - v + \mu| < AB^{-w},$$

in positive integers u, v, and w with

$$u \leq M \text{ and } w \geq \frac{\log(Aq/\varepsilon)}{\log B}.$$

The following theorem and lemma are given in [3, 7], respectively.

Theorem 2 *The only perfect powers in the Fibonacci sequence are $F_0 = 0$, $F_1 = F_2 = 1$, $F_6 = 8$, and $F_{12} = 144$. The only perfect powers in the Lucas sequence are $L_1 = 1$ and $L_3 = 4$.*

Lemma 2 *For any integer $n \geq 3$, the inequalities*

$$\lambda^{n-0.99} < B_n < \lambda^{n-0.98}$$

hold.

3 Main Theorem

Theorem 3 *Let $1 \leq m \leq n$ and $k \geq 0$ be integers. If $L_k = B_n B_m$, then $(n, m, k) = (1, 1, 1)$.*

Assume that the equation $B_n B_m = L_k$ holds for $1 \leq m \leq n$. If $m = n$, we have $L_k = B_n^2$, which is possible only for $k = 1$ and $n = 1$ by Theorem 2. Therefore, we assume that $m < n$. Let $n \leq 30$. Here, we wrote a short program in *Mathematica.* Using it, we see that $k < 213$. In that case, with the help of *Mathematica* program, we obtain only the solution $(n, m, k) = (1, 1, 1)$ in the range $0 < m < n \leq 30$ for $k < 213$. From now on, assume that $n > 30$. Therefore, $k \geq 107$. Using the inequality (4) and Lemma 2, we get the inequality

$$\alpha^{k-1} \leq L_k = B_n B_m < \lambda^{n+m-1}.$$

From this, it follows that $k < (n+m-1)\,c+1$, where $c = \frac{\log\lambda}{\log\alpha} = 3.6631418....$ Since $m < n$, it is clear that $k < 8n$.

On the other hand, rearranging the equation $L_k = B_n B_m$ as

$$\alpha^k - \frac{\lambda^{n+m-1}}{32} = -\left(\beta^k + \frac{\lambda^n\delta^m + \lambda^m\delta^n - \delta^{n+m}}{32}\right)$$

and taking absolute values, we obtain

$$\begin{aligned}\left|\alpha^k - \frac{\lambda^{n+m}}{32}\right| &= \frac{32\cdot|\beta|^k + \lambda^n\delta^m + \lambda^m\delta^n + \delta^{n+m}}{32} \\ &= \frac{32\cdot|\beta|^k + \lambda^{n-m} + \delta^{n-m} + \delta^{n+m}}{32} \\ &< \frac{\lambda^{n-m}+1}{32} < \frac{\lambda^{n-m+1}}{32},\end{aligned}$$

where we have used the fact that $\lambda\delta = 1$. If we divide both sides of the above inequality by $\frac{\lambda^{n+m}}{32}$, we get

$$\left|32\alpha^k\lambda^{-(n+m)} - 1\right| < \frac{1}{\lambda^{2m-1}}. \tag{11}$$

Now, let us apply Theorem 1 with $\gamma_1 := 32,\ \gamma_2 := \alpha,\ \gamma_3 := \lambda$ and $b_1 := 1,\ b_2 := k,\ b_3 := -(n+m)$. Note that the numbers γ_i for $i = 1, 2, 3$ are positive real numbers and elements of the field $K = Q(\sqrt{2}, \sqrt{5})$. It can be easily seen that the degree of the field K is 4. So $D = 4$. It can be shown that the number $\Lambda_1 := 32\alpha^k\lambda^{-(n+m)} - 1$ is nonzero. Otherwise, if $\Lambda_1 = 0$, then, we get

$$\alpha^k\lambda^{-(n+m)} = \alpha^k\delta^{n+m} = 1/32.$$

Here, $1/32$ is not an algebraic integer in the field $K = Q(\sqrt{2}, \sqrt{5})$, whereas $\alpha^k\delta^{n+m}$ is an algebraic integer. This is a contradiction. Moreover, since $h(\gamma_1) = \log 32 = 3.46574...$, $h(\gamma_2) = \frac{\log\alpha}{2} = \frac{0.4812...}{2}$, and $h(\gamma_3) = \frac{\log\lambda}{2} = \frac{1.76275...}{2}$ by (7), we can take $A_1 := 14,\ A_2 := 1$, and $A_3 = 3.6$. Also, since $k < 8n$, we can take $B = 8n$. Thus, taking into account the inequality (11) and using Theorem 1, we obtain

$$\frac{1}{\lambda^{2m-1}} > |\Lambda_1| > \exp\left(-1.4\cdot 30^6\cdot 3^{4.5}\cdot 4^2(1+\log 4)(1+\log 8n)\cdot 14\cdot 3.6\right)$$

and so

$$(2m-1)\log\lambda < 1.4\cdot 30^6\cdot 3^{4.5}\cdot 4^2(1+\log 4)(1+\log 8n)\cdot 14\cdot 3.6.$$

By a simple computation, it follows that

$$2m\log\lambda < 2.77\cdot 10^{14}(1+\log 8n). \tag{12}$$

Now, rearranging the equation $L_k = B_n B_m$ as $\frac{\alpha^k}{B_m} - \frac{\lambda^n}{4\sqrt{2}} = -\left(\frac{\beta^k}{B_m} + \frac{\delta^n}{4\sqrt{2}}\right)$ and taking absolute values in here, we obtain

$$\left|\frac{\alpha^k}{B_m} - \frac{\lambda^n}{4\sqrt{2}}\right| \leq \frac{|\beta|^k}{B_m} + \frac{\delta^n}{4\sqrt{2}} < 1,$$

where we have used the fact that $k \geq 107$ and $n > 30$. Dividing both sides of the above inequality by $\frac{\lambda^n}{4\sqrt{2}}$, we get

$$\left|\frac{4\sqrt{2}\alpha^k\lambda^{-n}}{B_m} - 1\right| < \frac{4\sqrt{2}}{\lambda^n} < \frac{6}{\lambda^n}. \tag{13}$$

Thus, we again apply Theorem 1 to the inequality (13) with $\gamma_1 := \alpha,\ \gamma_2 := \lambda,\ \gamma_3 := \frac{B_m}{4\sqrt{2}}$ and $b_1 := k,\ b_2 := -n,\ b_3 := -1$. As one can see that the numbers γ_1, γ_2, and γ_3 are positive real numbers and elements of the field $K = Q(\sqrt{2}, \sqrt{5})$, so $D = 4$. Similarly to the argument used to prove that $\Lambda_1 \neq 0$, one can verify that $\Lambda_2 = \frac{4\sqrt{2}\alpha^k\lambda^{-n}}{B_m} - 1 \neq 0$. Also, since $h(\gamma_1) = \frac{\log\alpha}{2} = \frac{0.4812...}{2}$ and $h(\gamma_2) = \frac{\log\lambda}{2} = \frac{1.76275...}{2}$ by (7), we can take $A_1 := 1$ and $A_2 = 3.6$. Besides, using the properties (5) and (9), we get

$$h(\gamma_3) = h\left(B_m(4\sqrt{2})^{-1}\right) \leq h\left(B_m\right) + h\left(4\sqrt{2}\right) \leq \log B_m + \log 4\sqrt{2} \leq m\log\lambda.$$

So we can take $A_3 := 4m\log\lambda$. Also, since $k < 8n$, it follows that $B = 8n$. Thus, taking into account the inequality (13) and using Theorem 1, we obtain

$$\frac{6}{\lambda^n} > |\Lambda_2| > \exp\left(-C(1+\log 4)(1+\log 8n)\cdot 3.6\cdot 4m\log\lambda\right)$$

or

$$n\log\lambda - \log 6 < C(1+\log 4)(1+\log 8n)\cdot 3.6\cdot 4m\log\lambda, \tag{14}$$

where $C = 1.4 \cdot 30^6 \cdot 3^{4.5} \cdot 4^2$. Inserting the inequality (12) into the last inequality, a computer search with *Mathematica* gives us that $n < 3.6 \cdot 10^{31}$.

Now, let us try to reduce the upper bound on n applying Lemma 1. Let

$$z_1 := k\log\alpha - (n+m)\log\lambda + \log 32.$$

Then

$$\left|1 - e^{z_1}\right| < \frac{1}{\lambda^{2m-1}}$$

by (11). If $z_1 \geq 0$, then we have the inequality $z_1 \leq e^{z_1} - 1 = |1 - e^{z_1}| < \frac{1}{\lambda^{2m-1}}$. If $z_1 < 0$, then $|1 - e^{z_1}| = 1 - e^{z_1} < \frac{1}{\lambda^{2m-1}} < \frac{1}{2}$. From this, we get $e^{|z_1|} < 2$, and therefore

$$0 < |z_1| < e^{|z_1|} - 1 = e^{|z_1|} \left|1 - e^{z_1}\right| < \frac{2}{\lambda^{2m-1}}.$$

In both cases, the inequality

$$0 < |z_1| < \frac{2}{\lambda^{2m-1}}$$

is valid. That is,

$$0 < |k \log \alpha - (n+m) \log \lambda + \log 32| < \frac{2}{\lambda^{2m-1}}.$$

Dividing this inequality by $\log \lambda$, we get

$$0 < \left| k \left(\frac{\log \alpha}{\log \lambda} \right) - (n+m) + \left(\frac{\log 32}{\log \lambda} \right) \right| < 6.62 \cdot \lambda^{-2m}. \tag{15}$$

We now put

$$\gamma := \frac{\log \alpha}{\log \lambda}, \ \mu := \frac{\log 32}{\log \lambda}, \ A := 6.62, \ B := \lambda, \ \textit{and } w := 2m.$$

If we take $M := 2.88 \cdot 10^{32}$, then we get $q_{63} > 6M$, where q_{63} is the denominator of the 63th convergent of γ. In this case, a quick computation using *Mathematica* gives us the inequality

$$\varepsilon = ||\mu q_{63}|| - M||\gamma q_{63}|| \geq 0.449403.$$

Hence, we can apply Lemma 1 to the inequality (15). Using *Mathematica*, we can say that the inequality (15) has no solution for $2m \geq 44.9945$. So $m \leq 22$. Substituting this upper bound for m into (14), we obtain $n < 7.26 \cdot 10^{16}$.

Now, let

$$z_2 := k \log \alpha - n \log \lambda + \log \left(\frac{4\sqrt{2}}{B_m} \right).$$

In this case,

$$\left|1 - e^{z_2}\right| < \frac{6}{\lambda^n}$$

by (13). It is clear that $\frac{6}{\lambda^n} < \frac{1}{4}$ since $n > 30$. If $z_2 > 0$, then $0 < z_2 < e^{z_2} - 1 < \frac{6}{\lambda^n}$. If $z_2 < 0$, then $|1 - e^{z_2}| = 1 - e^{z_2} < \frac{6}{\lambda^n} < \frac{1}{4}$. From this, we get $e^{|z_2|} < \frac{4}{3}$, and therefore

$$0 < |z_2| < e^{|z_2|} - 1 = e^{|z_2|} \left|1 - e^{z_2}\right| < \frac{8}{\lambda^n}.$$

In both cases, the inequality

$$0 < |z_2| < \frac{8}{\lambda^n}$$

is valid. That is,

$$0 < \left| k \log \alpha - n \log \lambda + \log \left(\frac{4\sqrt{2}}{B_m} \right) \right| < \frac{8}{\lambda^n}.$$

Dividing both sides of the above inequality by $\log \lambda$, we get

$$0 < \left| k \left(\frac{\log \alpha}{\log \lambda} \right) - n + \frac{\log \left(\frac{4\sqrt{2}}{B_m} \right)}{\log \lambda} \right| < 4.54 \cdot \lambda^{-n}. \tag{16}$$

We now put

$$\gamma := \frac{\log \alpha}{\log \lambda}, \ A := 4.54, \ B := \lambda, \ and \ w := n.$$

If we take $M := 5.81 \cdot 10^{17}$, we get $q_{39} > 6M$, where q_{39} is the denominator of the 39th convergent of γ. Hence, taking

$$\mu := \frac{\log \left(\frac{4\sqrt{2}}{B_m} \right)}{\log \lambda}$$

with $m \leq 22$, a quick computation using *Mathematica* gives us the inequality

$$\varepsilon = ||\mu q_{39}|| - M||\gamma q_{39}|| \geq 0.448824.$$

Then, let us again apply Lemma 1. With the help of *Mathematica*, we can say that the inequality (16) has no solution for $n \geq 25.6475$. In that case, $n \leq 25$. This contradicts our assumption that $n > 30$. Thus, the proof is completed.

Acknowledgements The author would like to thank anonymous referee for the valuable comments to improve the manuscript.

References

1. Baker A., Davenport H.: *The equations* $3x^2 - 2 = y^2$ *and* $8x^2 - 7 = z^2$, Quart. J. Math. Oxford Ser. (2), **20**(1) (1969),129–137.
2. Behera A., Panda G. K.: *On the square roots of triangular numbers*, Fibonacci Quarterly, **37(2)** (1999), 98–105.

3. Bugeaud Y., Mignotte M., Siksek S.: *Classical and modular approaches to exponential Diophantine equations I. Fibonacci and Lucas perfect powers*, Ann. of Math. **163**(3), (2006), 969–1018.
4. Ddamulira M., Luca F., Rakotomalala M.: *Fibonacci Numbers which are products of two Pell numbers, Fibonacci Quart.,* **54 (1)** (2016), 11–18.
5. Dujella A., Pethő A.: *A generalization of a theorem of Baker and Davenport*, Quart. J. Math. Oxford Ser. (2), **49** (3) (1998), 291–306.
6. Farrokhi D. G. M.: *Some remarks on the equation $F_n = kF_m$ in Fibonacci numbers*, Journal of Integer Sequences, **10**, 1–9 (2007)
7. Irmak N.: *Balancing with Balancing powers,* Miskolc Mathematical Notes, Vol. **14** (2013), No. 3, pp. 951–957.
8. Keskin R., Karaatlı O.: *Some New Properties of Balancing Numbers and Square Triangular Numbers,* Journal of Integer Sequences, Vol. 15 (2012), Article 12.1.4.
9. Keskin R., Demirtürk Bitim B.: *Fibonacci and Lucas Congruences and Their Applications,* Acta Mathematica Sinica, English Series, **27**(4) (2011), 725–736.
10. Keskin R., Şiar Z.: *On the Lucas Sequence Equations $V_n = kV_m$ and $U_n = kU_m$*, Colloquium Mathematicum, **130**(1) (2013), 27–38.
11. Koshy T.: *Fibonacci and Lucas Numbers With Applications*, John Wiley and Sons, Proc., New York-Toronto, 2001.
12. Matveev E. M.: *An Explicit lower bound for a homogeneous rational linear form in the logarithms of algebraic numbers II*, Izv. Ross. Akad. Nauk Ser. Mat., **64**(6) (2000), 125-180 (Russian). Translation in Izv. Math. **64**(6) (2000), 1217–1269.

Printed in the United States
By Bookmasters